Coastal Research Library

Volume 29

Series Editor
Charles W. Finkl
Department of Geosciences
Florida Atlantic University
Boca Raton, FL, USA

The aim of this book series is to disseminate information to the coastal research community. The Series covers all aspects of coastal research including but not limited to relevant aspects of geological sciences, biology (incl. ecology and coastal marine ecosystems), geomorphology (physical geography), climate, littoral oceanography, coastal hydraulics, environmental (resource) management, engineering, and remote sensing. Policy, coastal law, and relevant issues such as conflict resolution and risk management would also be covered by the Series. The scope of the Series is broad and with a unique cross-disciplinary nature. The Series would tend to focus on topics that are of current interest and which carry some import as opposed to traditional titles that are esoteric and non-controversial. Monographs as well as contributed volumes are welcomed.

More information about this series at http://www.springer.com/series/8795

Christopher Makowski • Charles W. Finkl
Editors

Impacts of Invasive Species on Coastal Environments

Coasts in Crisis

Springer

Editors
Christopher Makowski
Coastal Education and Research Foundation (CERF)
Coconut Creek, FL, USA

Charles W. Finkl
Coastal Education and Research Foundation (CERF)
Asheville, NC, USA

Department of Geosciences
Florida Atlantic University
Boca Raton, FL, USA

ISSN 2211-0577 ISSN 2211-0585 (electronic)
Coastal Research Library
ISBN 978-3-030-08239-0 ISBN 978-3-319-91382-7 (eBook)
https://doi.org/10.1007/978-3-319-91382-7

Printed on acid-free paper

This Springer imprint is published by the registered company Springer International Publishing AG part of Springer Nature.
The registered company address is: Gewerbestrasse 11, 6330 Cham, Switzerland

Preface

When examining coastal environments throughout the world, there is usually a delicate balance formed among native vegetative and animal species and the environment itself. This equilibrium helps to sustain the ecosystem as a whole and ensures that the biodiversity of a particular coastal region is preserved. However, an unfortunate imbalance is observed in this modern era where bioinvasions of alien species have infiltrated multiple coastal landscapes. This volume in the *Coastal Research Library* (CRL) focuses on the regional and localized impacts that incur to various coastal environments from nonnative, invasive species. The book has been divided into two main parts: Part I – Regional Impacts from Multiple Coastal Invasive Species; and Part II – Localized Effects of Individual Coastal Invasives. These general subject-area parts are then subdivided into chapters that describe, through either generalized overviews or specific case studies, how invasive flora and fauna create destructive cascades within coastal systems that ultimately end with substantial deleterious impacts on environmental quality. While the following collection of topics provides insight into the common threat that is coastal invasive species, it also pushes to the forefront the undeniable influence of human action, whether through urbanization, industrialization, and commercialization, to enable such detrimental bioinvasions. With so many coastal environments already compromised, it is imperative that protection against invasive species is mandated in order to rehabilitate, preserve, and sustain these delicate littoral zones.

Part I contains seven chapters highlighting regional impacts around the world from multiple coastal invasive species. Chapter 1 (Invasive Species Within South Florida Coastal Ecosystems: An Example of a Marginalized Environmental Resource Base), by Christopher Makowski and Charles W. Finkl, discusses how numerous invasive species of vegetation and wildlife have wreaked havoc over the southern Florida peninsula. Descriptions of specific invasive species are given, as well as various countermeasures used in an attempt to neutralize the alien bioinvaders. The authors also explore the notion of humans as the main invasive species in coastal environments. Chapter 2 (Invasive Species in the Sundarbans Coastal Zone (Bangladesh) in Times of Climate Change: Chances and Threats), by

Shafi Noor Islam, Sandra Reinstädtler, and Albrecht Gnauck, presents the impacts and threats of multiple invasive species to the Sundarbans deltaic region. These biological invasions are linked to vulnerabilities in mangrove forests and wetlands throughout the Sundarbans Natural World Heritage Site in Bangladesh. Chapter 3 (Threats to Sandy Shore Habitats in Sri Lanka from Invasive Vegetation), by Wasantha Rathnayake, quantifies how native plant diversity is decreasing while invasive weeds are more abundant along the sandy shorelines of Sri Lanka. Chapter 4 (Alien Species and the Impact on Sand Dunes Along the NE Adriatic Coast), by Urban Šilc, Danijela Stešević, Andrej Rozman, Danka Caković, and Filip Küzmič, continues in a similar vein by examining the results of a multifaceted approach to observe how sand dune plant communities in Montenegro have been affected by invasion of five alien species. Chapter 5 (Manila Bay Ecology and Associated Invasive Species), by Benjamin M. Vallejo Jr., Alexander B. Aloy, Melody Ocampo, Jennifer Conejar-Espedido, and Leanna M. Manubag, takes a look at how the high marine biodiversity of the Philippines' Manila Bay becomes compromised through the biological invasions of fouling organisms. Chapter 6 (Bioinvasion and Environmental Perturbation: Synergistic Impact on Coastal–Mangrove Ecosystems of West Bengal, India), by Susanta Kumar Chakraborty, reports on the prospective consequences of several bioinvasions within the coastal–estuarine network of West Bengal, India, which includes more than 100 deltas in this region. Chapter 7 (Specialized Grooming as a Mechanical Method to Prevent Marine Invasive Species Recruitment and Transport on Ship Hulls), by Kelli Z. Hunsucker, Emily Ralston, Harrison Gardner, and Geoffrey Swain, assesses the ubiquitous impact of biofouling on ship hulls and proposes an innovative countermeasure to thwart invasive species recruitment and transport.

Part II contains seven chapters and focuses on the localized effects generated by an individual invasive species, in particular. Chapter 8 (Feeding Habits of *Pterois volitans*: A Real Threat to Caribbean Coral Reef Biodiversity), by Arturo Acero P., Diana Bustos-Montes, Paula Pabón Quintero, Carlos Julio Polo-Silva, and Adolfo Sanjuan Muñoz, delves into the commercial and ecological threats caused by one invasive marine species, the lionfish, which may single-handedly be responsible for altering the biodiversity of the Caribbean Sea. Chapter 9 (Environmental Impact of Invasion by an African Grass (*Echinochloa pyramidalis*) on Tropical Wetlands: Using Functional Differences as a Control Strategy), by Hugo López Rosas, Eduardo Cejudo, Patricia Moreno-Casasola, Luis Alberto Peralta Peláez, María Elizabeth Hernández, Adolfo Campos Cascaredo, and Gustavo Aguirre León, discusses how one invasive grass species is altering the wetland and dune ecosystems in Mexico by reducing plant biodiversity, changing system hydrology, reducing faunal habitat, and causing vertical accretion of physicochemicals within the soil profiles. The authors also highlight an ongoing control strategy project to curb the bioinvader. Chapter 10 (Environmental Impacts of an Alien Kelp Species (*Undaria pinnatifida*, Laminariales) Along the Patagonian Coasts), by M. Paula Bunicontro, Silvia C. Marcomini, and Graciela N. Casas, focuses on the effects of an invasive kelp species along the Argentinean coast. This submerged aquatic bioinvader not only impacts indigenous populations but may also be responsible for collapsing commercially important benthic community structures and increasing beach erosion. Chapter 11 (Only the Strictest Rules Apply: Investigating Regulation Compliance of Beaches to Minimize Invasive Dog Impacts on Threatened Shorebird Populations), by Grainne S. Maguire, Kelly K. Miller, and Michael A. Weston, explores an unlikely coastal invasive species in domesticated dogs and how to minimize their impact on threatened populations of shorebirds in southern Australia. Chapter 12 (Evaluating How the Group Size of Domestic, Invasive Dogs Affect Coastal Wildlife Responses: The Case of Flight-Initiation Distance (FID) of Birds on Southern Australian Beaches), by S. Guinness, W.F. Van Dongen, P.-J. Guay, R.W. Robinson, and M.A. Weston, is a follow-up to the previous chapter where the flight-initiation distance (FID), a measure of wariness in shorebirds, was correlated to the group size of invasive dog packs on Australian beaches. Chapter 13 (Impact of Invasive Nypa Palm (*Nypa fruticans*) on Mangroves in Coastal Areas of the Niger Delta Region, Nigeria), by Aroloye O. Numbere, investigates one of the major bioinvading threats to mangrove and coastal systems in the Niger Delta area. This alien palm has the potential to adversely change the pedology, hydrology, and overall landscape of the deltaic environment. Chapter 14 (*Acacia* spp.: Invasive Trees Along the Brunei Coast, Borneo), by Shafi Noor Islam, Siti Mazidah Bin Haji Mohamad, and Abul Kalam Azad, probes another invasive flora, this time a non-indigenous genera of tree, that has impacted the forest ecology along the coast of Brunei Darussalam in Borneo.

This volume offers wide-ranging examples of how invasive species impact many diverse coastal environments. Chapters selected for this book selectively show that native populations of plants and animals are under constant threat of bioinvasions along the coasts of the following regions: North and South America, Australia, Southeast Asia, Bangladesh, West Africa, India, Philippines, Sri Lanka, and the Caribbean Sea. The underlining theme of this publication is to create awareness of the global impacts caused by coastal invasive species and to instill a responsibility among people that humans may in fact be the quintessential bioinvader on planet Earth. Only then can people begin to repair the damage they have unleashed in the form of exotic, alien species along the coasts. Through the dissemination of this book, researchers, managers, and the public alike can begin to collectively work together to identify the root of the problem when it comes to invasive species and to no longer put our coasts in crisis.

Coconut Creek, FL, USA
Asheville, NC, USA
Boca Raton, FL, USA

Christopher Makowski
Charles W. Finkl

Contents

Part I Regional Impacts from Multiple Coastal Invasive Species

1 **Invasive Species Within South Florida Coastal Ecosystems: An Example of a Marginalized Environmental Resource Base** 3
Christopher Makowski and Charles W. Finkl

2 **Invasive Species in the Sundarbans Coastal Zone (Bangladesh) in Times of Climate Change: Chances and Threats** 63
Shafi Noor Islam, Sandra Reinstädtler, and Albrecht Gnauck

3 **Threats to Sandy Shore Habitats in Sri Lanka from Invasive Vegetation** 87
Rathnayake Mudiyanselage Wasantha Rathnayake

4 **Alien Species and the Impact on Sand Dunes Along the NE Adriatic Coast** 113
Urban Šilc, Danijela Stešević, Andrej Rozman, Danka Caković, and Filip Küzmič

5 **Manila Bay Ecology and Associated Invasive Species** 145
Benjamin M. Vallejo Jr., Alexander B. Aloy, Melody Ocampo, Jennifer Conejar-Espedido, and Leanna M. Manubag

6 **Bioinvasion and Environmental Perturbation: Synergistic Impact on Coastal–Mangrove Ecosystems of West Bengal, India** 171
Susanta Kumar Chakraborty

7 **Specialized Grooming as a Mechanical Method to Prevent Marine Invasive Species Recruitment and Transport on Ship Hulls** 247
Kelli Z. Hunsucker, Emily Ralston, Harrison Gardner, and Geoffrey Swain

Part II Localized Effects of Individual Coastal Invasives

8 Feeding Habits of *Pterois volitans*: A Real Threat to Caribbean Coral Reef Biodiversity 269
Arturo Acero P., Diana Bustos-Montes, P. Pabón Quintero, Carlos Julio Polo-Silva, and A. Sanjuan Muñoz

9 Environmental Impact of Invasion by an African Grass (*Echinochloa pyramidalis*) on Tropical Wetlands: Using Functional Differences as a Control Strategy 315
Hugo López Rosas, Eduardo Cejudo, Patricia Moreno-Casasola, Luis Alberto Peralta Peláez, María Elizabeth Hernández, Adolfo Campos C., and Gustavo Aguirre León

10 Environmental Impacts of an Alien Kelp Species (*Undaria pinnatifida*, Laminariales) Along the Patagonian Coasts 373
María Paula Bunicontro, Silvia C. Marcomini, and Graciela N. Casas

11 Only the Strictest Rules Apply: Investigating Regulation Compliance of Beaches to Minimize Invasive Dog Impacts on Threatened Shorebird Populations 397
Grainne S. Maguire, Kelly K. Miller, and Michael A. Weston

12 Evaluating How the Group Size of Domestic, Invasive Dogs Affect Coastal Wildlife Responses: The Case of Flight-Initiation Distance (FID) of Birds on Southern Australian Beaches 413
S. Guinness, W. F. Van Dongen, P.-J. Guay, R. W. Robinson, and M. A. Weston

13 Impact of Invasive Nypa Palm (*Nypa Fruticans*) on Mangroves in Coastal Areas of the Niger Delta Region, Nigeria 425
Aroloye O. Numbere

14 Acacia spp.: Invasive Trees Along the Brunei Coast, Borneo 455
Shafi Noor Islam, Siti Mazidah Bin Haji Mohamad, and Abul Kalam Azad

Index 477

Part I
Regional Impacts from Multiple Coastal Invasive Species

Chapter 1
Invasive Species Within South Florida Coastal Ecosystems: An Example of a Marginalized Environmental Resource Base

Christopher Makowski and Charles W. Finkl

Abstract Bioinvasions from exotic flora and fauna are a constant threat to the ecological balance that allows coastal ecosystems to maintain homeostasis. Throughout the world, invasive species are responsible for a multitude of impacts upon the coastal zone, some of which include outcompetetion and displacement of native species, biochemical degradation of water resources, destabilization of the soil, overexertion of carrying capacity limits, and the overall collapse of indigenous flora-fauna boundaries. South Florida is a prime example where the successful establishment and dispersal of numerous invasive species has occurred through human disruption and interference of the natural coastal ecosystems. This chapter focuses on five species of invasive vegetation (i.e., Australian pine [*Casuarina equisetifolia*], Brazilian pepper [*Schinus terebinthifolius*], broadleaf paperbark tree [*Melaleuca quinquenervia*], water hyacinth [*Eichhornia crassipes*], hydrilla [*Hydrilla verticillata*]) and five species of invasive wildlife (i.e., red lionfish [*Pterois volitans*], marine cane toad [*Bufo marinus*], red imported fire ant [*Solenopsis invicta*], Nile monitor [*Varanus niloticus*], Burmese python [*Python molurus bivittatus*]) that have contributed to the profound ecological breakdown of a vulnerable coastal region. By reviewing how different invasive species marginalize the environmental resource base of South Florida, a spotlight is then shone on how invasions can destroy coastal biodiversity worldwide, as well as expose the role of humans, not only as the main introducing factor of alien species, but perhaps as the most invasive of all species on planet Earth.

C. Makowski (✉)
Coastal Education and Research Foundation (CERF), Coconut Creek, FL, USA
e-mail: cmakowski@cerf-jcr.com; http://www.cerf-jcr.org

C. W. Finkl
Coastal Education and Research Foundation (CERF), Asheville, NC, USA

Department of Geosciences, Florida Atlantic University, Boca Raton, FL, USA

C. Makowski, C. W. Finkl (eds.), *Impacts of Invasive Species on Coastal Environments*, Coastal Research Library 29,
https://doi.org/10.1007/978-3-319-91382-7_1

Keywords Bioinvasion · Alien fauna · Biodiversity · Beach vegetation · Indigenous flora · Biogeography · Florida Everglades · Environmental conservation · Ecological change

1.1 Introduction

Ecosystems throughout the world maintain a certain order in the specific types of native flora and fauna species found within them as a means to sustain a harmonistic balance in nature. This delicate balance, in terms of geologic time, may span entire eras before a shift or transition occurs. Studies have shown that it is usually an outside factor that gives rise to a change in the species composition of ecosystems. One famous example is the postulated K-T Mass Extinction Event, where approximately 65 million years ago, more than three-fourths of all the plant and animal species living on Earth became extinct. Named for the boundary between the Cretaceous (K) and Tertiary (T) time periods, Alvarez et al. (1980) hypothesized that an extraterrestrial meteorite impact was the main cause for such a shift in the types of organisms that were then found on the planet. Extraordinary amounts of the metal iridium in the rocks that were laid down at the time of the K-T boundary (Alvarez et al. 1982) lend credence to the theory that a punctuated outside force could be responsible for such a disruption in the composition of native plants and animals.

While the K-T Event proved to cause an ancient global shift of plant and animal species, much more common localized disruptions to native flora and fauna can be seen today with the onset of human interventions. Within the field of biogeography, a particular species is referred to as native, endemic, or indigenous to a specific ecosystem if their presence there is only the result of natural processes. Non-native, exotic, or alien species, however, are those organisms that have been introduced to a new ecosystem through direct anthropogenic influences. And finally, invasive species, by definition, are those non-natives whose presence will most likely cause economic or environmental damage to the ecosystem, with the potential to inflict harm to human health.

There has been a struggle to combat invasive species within many of the world's coastal ecosystems. With the advent of human transportation technological advancements, it has become apparent that invasive species could be easily introduced to foreign ecosystems simply by 'hitching a ride' in the cargo hold of an airplane or the ballast of a ship. What proved to be even more unforgiving was the hubris of humans to 'play God' by systematically introducing exotic species into a particular coastal ecosystem in order to disrupt the natural order of things for their own gain. Sometimes one alien species is introduced to offset a previously introduced non-native species that has turned invasive. Unfortunately, in a lot of those circumstances, both introduced exotic species ultimately turn out to be invasive within a coastal region. As a consequence, the introduction of invasive species, along with both pollution and habitat loss, are now considered the top three environmental threats in the modern era (Perrings 2005).

The coastal plain of South Florida (Fig. 1.1) is a prime example of how human interventions have completely transformed an ecosystem of natural harmony into one that is constantly under duress from invasive flora and fauna species (Tables 1.1, 1.2, and 1.3). The delicate balance of any coastal ecosystem allows for many species of native plants and animals to flourish, however, due to the wide-spread release of invasive species in South Florida, that balance has been critically disrupted. Invasive plants, such as the Australian pine (*Casuarina equisetifolia*), Brazilian pepper (*Schinus terebinthifolia*), and broadleaf paperbark tree (*Melaleuca quinquenervia*) (Austin 1978; D'Antonio and Meyerson 2002; Doren et al. 2009a, b), as well as invasive animals, such as lionfish (*Pterois volitans*), Nile monitors (*Varanus*

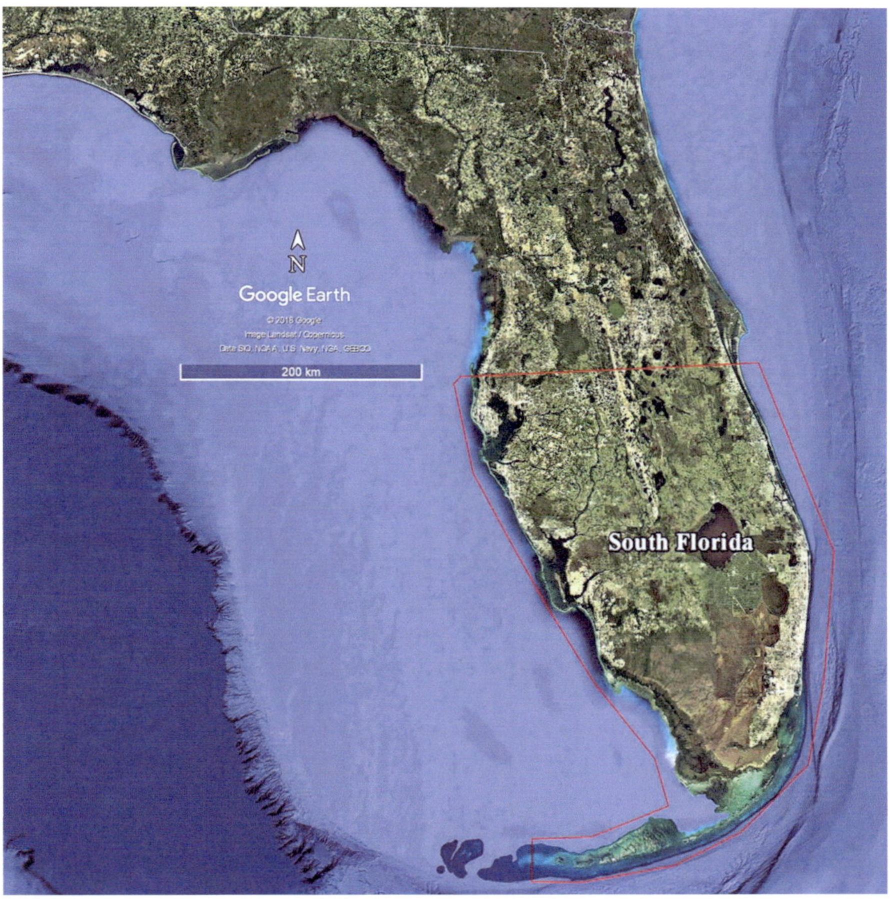

Fig. 1.1 Satellite imagery, combined with ocean floor composite renderings, showing the geographical location of South Florida in relation to the rest of the state. The red outline demarcates the southern limits of Florida from the northern limits and includes such features as the Florida Everglades, Lake Okeechobee, the Florida Keys, Biscayne Bay, Florida Bay, the Florida Reef Tract, and the Miami-metropolitan conurbation. (Credit: Google Earth)

niloticus), and Burmese pythons (*Python molurus bivittatus*), displace or eradicate native species and threaten to disrupt the entire ecosystem balance (LeSchiava et al. 2013). The purpose of this chapter is to review some of the main invasive species of plants and animals that continue to disturb and out-compete the indigenous species populations of South Florida. Through such an evaluation, one can begin to see a different type of extinction event occurring, albeit on a much smaller scale, where human outside forces cause a major shift in the species composition of a particular coastal area.

1.2 Invasive Flora

*1.2.1 Australian Pine Tree (***Casuarina equisetifolia***)*

One of the main invasive alien species plaguing South Florida is the Australian pine tree, *Casuarina equisetifolia*, which in fact, is not a pine tree at all. It is actually classified as a deciduous dicot angiosperm tree that mistakenly resembles the appearance of a typical conifer tree. *C. equisetifolia* can grow on average between 20 and 46 m in height, at a rate of 1.5–3.0 m per year, and has a maximum lifespan ranging from 40–50 years (Elfers 1988a, b; Swearingen 1997) (Fig. 1.2). Distinct features of the Australian pine include a single straight, rough-barked trunk with an open, irregular crown of branches, cone-like fruits that are small and round, and wispy needle-like branchlets (Fig. 1.3), that may or may not contain small non-descript brown flowers (Snyder 1992; Swearingen 1997; Langeland and Craddock Burks 1998).

While Australian pines are known to reproduce sexually via seed dispersal, this invasive species also has the ability to replicate vegetatively through the sprouting of new clonal trunks from existing rootstock. Usually, small, inconspicuous flowers are wind pollinated throughout a coastal area, with each of the oval cone-like fruits (i.e., nutlets) containing approximately 12 rows of seeds when they mature. Australian pines are capable of flowering for extended periods of time, even year-round occasionally, and individual trees can produce thousands of seeds a year, making them a very difficult species to control (Morton 1980; Elfers 1988a, b) (Fig. 1.4).

Within South Florida, *C. equisetifolia* are known as a Category I exotic species. This designation indicates that Australian pines have become so abundant within a particular region that entire native plant community structures are being altered and ecological functions of specific ecosystems are being negatively affected (Schmid et al. 2008; Wheeler et al. 2011). The native range of *C. equisetifolia* includes southern Asia, Malaysia, Australia, and Oceania (i.e., the islands of the Pacific between Asia and the Americas). However, the current worldwide-introduced range includes the Caribbean Territories (which include Puerto Rico and the Bahamas), Hawaii, and coastal Florida. Specifically in Florida, this invasive species ranges from north-central regions of the state southward through the Florida Keys (Wheeler et al. 2011; FLEPPC 2017).

Fig. 1.2 Australian pine trees (*Casuarina equisetifolia*) growing as part of the foredune canopy adjacent to Boca Raton beaches in South Florida. As a non-native, invasive species, *C. equisetifolia* outcompetes the indigenous vegetation below it by shielding a lot of the essential sunlight resources. (Credit: Chris Makowski)

Fig. 1.3 Distinct wispy needle-like branchlets growing from the Australian pine tree. This unique feature often has *C. equisetifolia* mislabeled as a conifer pine tree, when in fact they are deciduous dicot angiosperm trees. (Credit: Chris Makowski)

Fig. 1.4 Two juvenile Australian pine trees establishing themselves in the foreground as part of an upper dune ecosystem in South Florida. This invasive species can reproduce vegetatively through the sprouting of new clonal trunks from existing rootstock. Once exotic trees become established among the native species, management against this type of bioinvasion proves to be very difficult. For example, if these young *C. equisetifolia* individuals continue to grow, they have the potential of producing thousands of germinating seeds per year. (Credit: Chris Makowski)

The history of the Australian pine in South Florida began in 1898, when the species was intentionally introduced as both an ornamental tree and as a windbreak buffer to border agricultural groves (Morton 1980). The trees were also utilized as support in a lot of ditch and canal stabilization projects, as engineers channelized much of the Everglades in the early twentieth Century (Snyder 1992; Swearingen 1997) (Fig. 1.5). However, the species proved to be unsuitable for these purposes, as their shallow and wide-spreading root systems disrupted residential lawns and pavement areas, and ultimately made the tree susceptible to being overblown in strong wind storms (e.g., tropical storms and hurricanes). *C. equisetifolia* also grows too tall for the roots to support its own weight, especially in the sandy-rich soils along South Florida's coasts. Furthermore, not only was this species ill-suited for commercial purposes, but it quickly became evident that Australian pines were a highly-invasive species capable of high fecundity in disturbed and nutrient-poor coastal areas. *C. equisetifolia* grows at a faster rate than most indigenous species and typically form monospecific stands that can produce a dense canopy that shades out competing flora. These invaders also form a thick layer of dropped branchlets (i.e.,

Fig. 1.5 Drainage canal within the southeastern section of the Florida Everglades, where *C. equisetifolia*, an exotic invasive species, has colonized along the waterway's margins. Once thought to be a good stable tree to reinforce these canals, Australian pine trees proved to be unsuitable for this purpose due to their shallow, wide-spreading root system. The trees commonly grow too tall for its own roots to support it, thereby making this invasive foliage more of a nuisance than a benefit. (Credit: Charles W. Finkl)

needles) and fruits that blanket the ground below, which eliminates available regolith for native plants to germinate and grow. Additionally, the roots of *C. equisetifolia* harbor nitrogen-fixing microbial assemblages that allow the host tree to colonize and thrive in low nutrient soil conditions that many other native species cannot tolerate (Swearingen 1997).

Despite an ongoing statewide ban on cultivation, Australian pines are now widely established throughout South Florida and they continue to thrive as an invasive species in a variety of open coastal habitats, such as coastal strands, sand and shell beaches, and dune fields (Snyder 1992). They are usually the dominant species when in direct competition with native Florida vegetation and can prompt permanent ecological alteration of an ecosystem through the rapid displacement of indigenous flora. This loss of native vegetation (i.e., food and shelter resources) to an ecosystem has a cascade effect that can reduce the species diversity of mammals, birds, and other native coastal animals. In addition, competitive displacement of coastal mangrove stands by *C. equisetifolia* can eliminate a lot of the nursery habitat needed for recreational and commercial fishing, as well as the loss of critical nesting and roosting habitat for many waterbird species. To make matters worse, the fact that Australian pines are highly unstable during storms introduces the high potential of obstruction hazards in the form of fallen trees, which can then encumber keystone and endangered species, such as gopher tortoises (*Gopherus polyphemus*) and sea turtles, respectively (Elfers 1988a, b). These fallen trees can also increase erosion rates along beach and dune systems and further compromise the South Florida coastal region.

*1.2.2 Brazilian Pepper Tree (*Schinus terebinthifolius*)*

Often referred to as an evergreen shrub or small tree, the Brazilian pepper tree (*Schinus terebinthifolius*) is an aggressive, rapidly colonizing invader of natural communities and disturbed habits in South Florida (Ewe and Sternberg 2003; Ewe 2004). This non-native species grows on average 3–7 m tall and forms odd-pinnately compound leaves that are alternately arranged on branches (Fig. 1.6). When crushed, these leaves emit an odor that has been distinctly described as peppery or turpentine-like (Tomlinson 1980; Ferriter 1997). Specifically on female trees, flowering is followed by the production of bright red, fleshy, spherical fruits, often referred to as berries or drupes, each approximately 5–6 mm in diameter and containing a single seed (Ferriter 1997) (Fig. 1.7). Fruit production typically occurs from November to February, at which time the branches of female trees are heavily laden with the red drupes, while male trees remain bare. The survivorship of the naturally established seedlings is very high, ranging from 66–100%, and the ripe fruits can be retained on a single tree for up to 8 months. The tenacity of *S. terebinthifolius* makes it an especially difficult species to compete with, as its seedlings seem to survive for a

Fig. 1.6 Newly formed leaves growing from an invasive Brazilian pepper tree (*Schinus terebinthifolius*). The odd-pinnately compound leaves, along with intermitted spherical fruits, are indicative of the non-native flora. When crushed, the leaves of *S. terebinthifolius* give off a distinct aroma similar to turpentine. (Credit: Chris Makowski)

Fig. 1.7 The bright red berries (also known as drupes) growing from the Brazilian pepper tree are characteristic of female trees during the Northern Hemisphere's winter months (November–February). The survivorship of the newly formed fruits can be very high, making *S. terebinthifolius* a very hardy invasive species that is difficult to control. (Credit: Chris Makowski)

very long time in the dense shade of older canopy growth, where they typically develop (Ewel et al. 1982; Elfers 1988b; Ferriter 1997).

Brazilian pepper uses a variety of strategies to invade and displace native vegetation along the coast (Habeck et al. 1994; Randall 2000; Hight et al. 2002, 2003; Cuda et al. 2006). For example, *S. terebinthifolius* is believed to have allelopathic properties which aid in altering the Chl α concentration of indigenous flora, thus hindering their growth (Morgan and Overholt 2005; Hargraves 2008). Additionally, these invaders form dense monospecific stands that ultimately shade out and reduce the biological diversity of native plants and animals within the affected areas (Ewe and Sternberg 2003; Cuda et al. 2006; Donnelly and Walters 2008). Specifically in Florida, it has become one of the most widespread and problematic invasive plants, infesting approximately 280,000 ha within various ecosystems (Ewe 2004; Cuda et al. 2006; Ewe and Sternberg 2003) (Fig. 1.8). Aqueous extracts have confirmed that Brazilian pepper negatively affects the growth of two South Florida native plants, *Bromus alba* and *Rivina humilis* (Morgan and Overholt 2005), and threatens numerous mangrove swamp community species found in the Everglades, such as *Jacquemontia reclinata* and *Remirea maritime* (Doren and Jones 1997; Cuda et al. 2006). Furthermore, *S. terebinthifolius* has been found to reduce the density and species diversity of native bird populations when compared to uninvaded native pinelands and forest-edge habitats, and can even alter natural forest fire regimes because of its resultant increased shade production (Curnutt 1989; Cuda et al. 2006). Lastly, it has been postulated that Brazilian pepper can trigger a negative cascade effect on primary production, biodiversity, and the overall ecological community structure from species-specific impacts on microalgae, which usually occur at the land-sea ecotonal interface (Hight et al. 2003).

Adverse impacts to humans have also been the result of *S. terebinthifolius* exposure. Because it is a relative of poison ivy (*Toxicodendron radicans*), Brazilian pepper induces allergic skin reactions on contact (Lampe and Fagerstrom 1968; Tomlinson 1980). In fact, the concentration of volatile, aromatic monoterpenes and alkyl phenols is

Fig. 1.8 The Brazilian pepper tree commonly grows among other vegetation and begins to outcompete indigenous species for resources. One strategy that *S. terebinthifolius* uses is to alter the Chl α concentration of indigenous flora through allelopathic properties, thus inhibiting the growth of native species. (Credit: Chris Makowski)

at such a high level, individuals sitting beneath *S. terebinthifolius* trees have exhibited respiratory problems, such as sneezing, sinus congestion, chest pains, and acute headaches (Morton 1969, 1978; Ferriter 1997). Cuda et al. (2006) also showed the tripterpenes found in the fruits of Brazilian pepper can result in irritation of the throat, gastroenteritis, diarrhea, and vomiting in humans.

In order to control the spread of this invasive species, a variety of biological control agents have been investigated or released. Among the most effective include the Brazilian pepper thrip (*Pseudophilothrips ichini*), the Brazilian pepper leafroller (*Episimus utilis*), the Brazilian pepper sawfly (*Heteroperreyia hubrichi*), the torymid wasp (*Megastigmus transvaalensis*), and a variety of different fungal pathogens (Wheeler et al. 2001; Cuda et al. 2006; Cleary 2007). In the case of *M. transvaalensis*, the wasp attacks the seeds of *S. terebinthifolius* and damages them to prevent germination. Wheeler et al. (2001) found that in Florida, *M. transvaalensis* damaged up to 31% of Brazilian pepper drupes in the major winter fruiting period and 76% in the minor spring fruiting phase. Additionally, an array of fungal agents, such as *Sphaeropsis tumefaciens*, *Rhizoctonia solani*, and

Chrondostereum purpureum, are all known to infect *S. terebinthifolius* in different capacities and may also prove to be useful biological controls (Cuda et al. 2006).

*1.2.3 Broadleaf Paperbark Tree (*Melaleuca quinquenervia*)*

In South Florida, the broadleaf paperbark tree (*Melaleuca quinquenervia*), commonly known as melaleuca, is a rapidly-growing, hardy invasive tree, whose native range includes Australia, New Guinea, and the Solomon Islands (Langeland and Craddock Burks 1998). With a distinctive peeling paper-like white bark (Fig. 1.9), melaleuca can grow up to 33 m, at a rate of approximately 1–2 m per year (FLEPPC 2017). Branches occur at irregular intervals off the main trunk and support long (10–15 cm) evergreen leaves that are known to release a distinct aromatic smell when crushed (Langeland and Craddock Burks 1998) (Fig. 1.10). The flowers are small and white, and arranged with multiple stamens, while the fruits are small, round woody capsules containing approximately 200–300 seeds each (Austin 1978; FLEPPC 2017).

Fig. 1.9 *Melaleuca quinquenervia*, also known as the broadleaf paperbark tree, gets its name from the distinctive paper-like white bark that peels from the trunk. The lighter color of the bark made this invasive tree appealing as an ornamental species, prompting the transport of specimens from Australia, where they are native, to South Florida. (Credit: Ianaré Sévi)

Fig. 1.10 Branches of the broadleaf paperbark tree occur at irregular intervals and support evergreen leaves that emit an aromatic scent when crushed. This invasive species is capable of producing woody capsules that contain up to 300 seeds each. When the branches are disturbed (i.e., broken or cut), a rapid release of the seeds occurs ensuring a wide dispersal of the non-native tree. (Credit: Homer Edward Price)

Melaleuca quinquenervia primarily propagates by sexual seed production and is capable of flowering within 2–3 years of germination (Meskimen 1962; Laroche 1994). In South Florida, the species can propagate as many as five times a year, with blooms primarily occurring during the months of November through January. Flowering is known to be asynchronous among both the trees and the flowers of a single specimen (FLEPPC 2017). Large *M. quinquenervia* specimens can have a very high reproductive potential and up to 20 million seeds per year are known to be stored in the seed capsules of a single tree (Laroche 1994). Before the seeds are dispersed into the environment, seed capsules must first be dried out; however, seeds can remain viable within the seed capsules for up to 10 years (Meskimen 1962; Langeland and Craddock Burks 1998). Physical damage to the tree (e.g., broken or cut branches, whole tree falls, exposure to a hot-burning wildfire) will trigger the rapid release of seeds from capsules, which culminates in the shedding of all seeds within a few days (Woodall 1982; Flowers 1991; Stocker and Hupp 2008).

In Florida, *M. quinquenervia* is restricted to the southern half of the state, with counties at the very southern end of the state being most vulnerable. Melaleuca is considered one of the most prominent non-native plant species presently invading the natural areas of South Florida (Center and Dray 1986; Hofstetter 1991). The species had invaded more than 200,000 hectares in South Florida by 1994, including significant areas within Everglades National Park, a World Heritage Site and International Biosphere Reserve (Mazzotti et al. 1981; Langeland and Craddock Burks 1998; FLEPPC 2017).

The history of melaleuca invasion in South Florida includes multiple introduction events throughout the early twentieth century. Among the first sites recorded were Broward and Lee Counties, where melaleuca seeds were transported from Australia and planted as a landscape ornamental tree and a source of wood. The popularity of melaleuca as an ornamental species, especially as a windbreak for many properties and along fencerows, further facilitated the spread of this invasive species. As recently as 1970, *M. quinquenervia* continued to be recommended as “one of Florida’s best landscape trees” (Langeland and Craddock Burks 1998). Additionally, melaleuca was planted as soil stabilizers along canal levees bordering the southern end of Lake Okeechobee and throughout the Big Cypress National Preserve. Seeds were also intentionally scattered from airplanes over the Everglades throughout the 1930s to facilitate the rapid establishment of melaleuca forests (Austin 1978). This led to the environmentally-mediated spread of melaleuca deep into the interior of the Florida Everglades, which was facilitated by propagule transport via wind and water (FLEPPC 2017). A problem arises at the explosive speed by which melaleuca spreads and comes to dominate new areas (Hofstetter 1991). As little as 25 years is required for a 2.5 km^2 area to progress from 5% to 95% infestation of melaleuca (Laroche and Ferriter 1992). This poses a serious threat to the ongoing Everglades restoration and preservation efforts, and continues to threaten South Florida’s other natural areas (FLEPPC 2017). Almost a century later from those first introductions, the general distribution of melaleuca in South Florida still remains uncontained (Mazzotti et al. 2001).

Though melaleuca appears to have some positive economic benefits in Florida as a plant utilized by commercially-managed honeybees, the negative impacts of this coastal invasive species are of far greater consequence. For example, a cost-benefit analysis for South Florida determined that melaleuca could contribute an estimated annual benefit of approximately USD $15 million for the beekeeping and pollination service industries, however, an estimated loss of around USD $168.6 million/year would be suffered by the eco-tourism industry in the event of complete infestation by melaleuca throughout the Everglades and other South Florida wetland areas (Diamond et al. 1991).

When competing with indigenous plants, melaleuca is principally an invader of disturbed sites, where it proves to be opportunistic in those Florida habitats exhibiting a high degree of clearing or development (Ewel et al. 1976) (Fig. 1.11). Canal banks, managed pineland margins, pine savannas, sawgrass prairie marshes, and cypress marshes are among the South Florida ecosystems susceptible to *M. quinquenervia* (Richardson 1977; DiStefano and Fisher 1983; Myers 1983; Duever et al. 1986; Laroche and Ferriter 1992). Once it becomes established, melaleuca can form dense monotypic stands capable of displacing large amounts of native plants (Richardson 1977). In fact, the World Conservation Union’s Invasive Species Specialist Group (ISSG) lists melaleuca as among “100 of the world’s worst invasive alien species” and recognizes them as major drivers of ecosystem disruption.

Fig. 1.11 A common stand of melaleuca trees growing along the disturbed fringes of Interstate 75, also known as Alligator Alley, in South Florida. These invasive trees have proven very opportunistic within human-induced areas of disturbance or development. Melaleuca's ability to become rapidly established prevents many measures of control to be effective. (Credit: Forest & Kim Starr)

*1.2.4 Water Hyacinth (*Eichhornia crassipes*)*

The water hyacinth, *Eichhornia crassipes*, is an invasive non-native plant commonly found as floating dense mats in South Florida freshwater habitats. Originally native to the Brazilian Amazon Basin, the water hyacinth produces distinctive lavender-colored flowers (Fig. 1.12) and a thin walled, capsule-like fruit that can contain up to 400 seeds (Gopal 1987; Langeland and Craddock Burks 1998). This invasive is able to remain buoyant in the water through the use of bulbous, or inflated, petiole stalks and has long feathery roots that hang suspended in the water column.

Water hyacinth flourishes in freshwater ecosystems and is even capable of growing in low-salinity coastal lagoon habitats; for example, along the coastal margins of the Everglades. In fact, over 55 tropical and subtropical countries, including the southern portion of the United States, have reported *E. crassipes* as a noxious weed (Holm et al. 1977; Langeland and Craddock Burks 1998; Ramey 2001). However, increased salinity is a limiting factor in the distribution of the invasive plant. Experimental studies by de Casabianca and Laugier (1995) showed there was an inverse relationship between increased salinity and water hyacinth plant yield. Their results showed that at salinities above 6 ppt, either no plant production occurred or cankerous plants developed. Furthermore, at salinities above 8 ppt, irreversible physiological damage to the vegetation occurred (de Casabianca and Laugier 1995).

Vegetative reproduction of the water hyacinth usually occurs *via* the breaking off of clonal individuals. The stolons (i.e., the horizontal shoots capable of forming new

Fig. 1.12 Distinctive lavender flowers with an orange-yellow flame pattern on the top petal is a trademark feature of water hyacinth (*Eichhornia crassipes*). This invasive plant remains buoyant through the use of inflated, petiole stalks, keeping the pollenating parts of the flower above water. (Credit: USDA)

shoots) are easily broken by wind or wave action and dispersed, whereas, the floating clonal mats of *E. crassipes* are readily transported intact through wind or water movement (Barrett 1980; Langeland and Craddock Burks 1998). Germination typically occurs when water levels are down and the seedlings can grow in saturated soils.

The invasion history of *E. crassipes* first began in 1884, when the Brazilian native was first introduced to the United States as an ornamental aquatic plant at a New Orleans, Louisiana Exposition. Water hyacinth was first recorded in Florida by 1890, and over the next 60 years, dense mats of this highly invasive plant had altered more than 50,000 ha of the state's freshwater habitats (Gopal and Sharma 1981; Schmitz et al. 1993). Water hyacinth mats are capable of creating incredibly high plant density and biomass, with a single hectare containing more than 360 metric tons of plant material. The capacity of water hyacinth to invade and overtake aquatic habitats is remarkable, with growth rates that can double the vegetative population in as little as 1–3 weeks (Mitchell 1976; Wolverton and McDonald 1979; Langeland and Craddock Burks 1998). Because of this, water hyacinth is a considered a Category 1 invasive exotic species in Florida, capable of altering native plant communities by displacing indigenous species and changing community structures or ecological functions permanently (FLEPPC 2017). Some researchers have even gone on to describe *E. crassipes* as one of the worst weeds in the world (Holm et al. 1977).

The negative economic impacts of water hyacinth invasion include the clogging of irrigation channels, the choking off of navigational routes, smothering of native vegetation, loss of fishing areas, and the increase in breeding habitat available to

Fig. 1.13 Example of water hyacinth outcompeting natural flora to bioinvade a freshwater pond ecosystem. The incredibly high plant density and biomass of *E. crassipes* often leads to the infestation and clogging of irrigation canals, navigational channels, and other numerous waterways. (Credit: Cayambe)

disease-transmitting mosquitoes (Room and Fernando 1992) (Fig. 1.13). In the Florida Everglades, large, dense mats of *E. crassipes* can degrade water quality and obstruct essential waterways. Plant respiration and extensive biomass decay can often result in oxygen depletion, leading to hypoxic conditions and fish kills (Langeland and Craddock Burks 1998). Waterways are kept clear of dense infestations only through extraordinary management efforts involving field crews engaged in full-time mechanical removal and biocidal control. Even though the costs associated with the removal and maintenance control of water hyacinth are significant, exhaustive management efforts in the Everglades and other ecosystems over the last few decades have considerably reduced the amount of this invasive plant (Langeland 2008). Even so, complete eradication of *E. crassipes* from South Florida is nearly impossible.

*1.2.5 Hydrilla (*Hydrilla verticillata*)*

Hydrilla (*Hydrilla verticillata*) is well known in South Florida as an invasive aquatic weed that is not easily controlled or managed. A typical submerged, herbaceous perennial that exhibits seasonal winter dieback, hydrilla has long, sinewy branching stems that often reach the surface and form dense mats (Godfrey and Wooten 1979; Carter et al. 1994). Characteristic small, white flowers can be seen growing above the water on stalks, while the stems, which can reach lengths over 7.5 m, are usually covered in pointed, often serrate, leaves arranged in tiny whorls (Cook and Luond 1982; Langeland 1996). Reproductive strategies of hydrilla include several vegetative means, such as regrowth from stem fragments, clonal rhizome reproduction, and utilization of specialized axillary buds, also known as turions (Pieterse 1981; Hurley 1990; Spencer et al. 1994). It is noted that *H. verticillata* can proliferate very rapidly using these methods and remain reproductively viable for extended periods of time (Van and Steward 1990; Sutton et al. 1992).

Hydrilla verticillata has been referred as the most abundant aquatic plant in Florida's public waters, with over 70% of the state's freshwater drainage basins infested with the invasive vegetation (Schardt 1994, 1997). Early introduction into South Florida occurred in the early 1950s, when live samples of hydrilla were shipped from Sri Lanka and India for the aquarium trade and subsequently released into canals near Tampa Bay (Madeira et al. 2004). Soon after, other samples were introduced to the waterways of Miami and the establishment of hydrilla in Florida had been cemented (McCann et al. 1996). Being an aggressive vegetative invader capable of altering ecological community structures and displacing native indigenous plants, hydrilla is currently listed as a Category I invasive exotic plant in Florida and recognized as one of the most invasive weeds throughout the world (Haller and Sutton 1975; Bowes et al. 1977).

The control and management of hydrilla has proved to be both difficult and expensive. With the vast loss of recreational lake area due to *H. verticillata* infestation, the state of Florida has spent many millions of U.S. dollars in an attempt to curb their numbers (Langeland and Stocker 2001). Dense beds of hydrilla not only make recreational lakes unusable to the public, but oxygen depletion is a serious consequence from the decomposition of the plant's large biomass (Canfield et al. 1983). This can then lead to a negative ecological cascade where the water chemistry is altered, zooplankton populations drastically decline, fish populations are permanently lowered, and higher trophic animals, such as amphibians, reptiles, and mammals, are critically affected (Colle and Shireman 1980; Schmitz and Osborne 1984; Schmitz et al. 1991) (Fig. 1.14).

1.3 Invasive Fauna

*1.3.1 Red Lionfish (*Pterois volitans*)*

The red lionfish, *Pterois volitans*, is a highly invasive marine fish that has swarmed the east coast of the United States, including coastal Florida, since the turn of the twenty-first century. *P. volitans* has a very distinct appearance with red and white striped bands, elaborate fan-like pectoral fins, and long separated dorsal spines (Fig. 1.15). Fleshy tabs surrounding the mouth and above the eyes are another characteristic feature of this invasive species (Myers 1991; Whitfield et al. 2002). Lionfish have 18 spines that are used defensively against predators and to assist in

Fig. 1.14 A lagoon frog (*Lithobates grylio*), also referred to as a southern bullfrog or pig frog, forages within a hydrilla-infested canal in the Florida Everglades. The large amount of decomposition from the hydrilla biomass often leads to a permanent change in the water's chemistry, which then negatively affects higher trophic groups in search of essential food resources. (Credit: USGS)

Fig. 1.15 The red lionfish, *Pterois volitans*, is a voracious reef predator that has distinctive red and white banded stripes over its body for camouflage. Fleshy tabs can be seen around the mouth and above the eyes, and the long pectoral fins and dorsal spines contain a toxic venomous poison. (Credit: Chris Makowski)

prey capture. The long dorsal and pectoral spines of *P. volitans* are known to be venomous, as the poison is produced by glands located in grooves along the spine-covered integument (Halstead et al. 1955; Ruiz-Carus et al. 2006).

Typically growing to a size of 15–30 cm, larger lionfish specimens have been measured over 40 cm in length (Baker et al. 2004). Sexual reproduction (i.e., the external fertilization of eggs) usually occurs early in the year and involves a series of complex courtship and mating behaviors between the male and female (Ruiz-Carus et al. 2006). Overall, this invasive species is generally solitary outside of the reproductive season, but during courtship, males will aggregate with multiple females to form schools of up to ten fish. Competing males will even use their spines and fins to visually display aggression towards other suitors (Fishelson 1975). In the end, females release a pair of mucus-encapsulated clusters, each containing between 2000–15,000 eggs, to the pelagic environment where they are fertilized by the males (Ruiz-Carus et al. 2006). The fact that so many eggs are fertilized at once makes population control of this invasive species very difficult.

Lionfish are widely considered to be the first marine (non-estuarine) invasive fish in South Florida (Meister et al. 2005). As one of the most popular marine ornamental species in residential aquariums, their recent introduction to nearshore reefs was most likely the result of intentional release from unwanting owners (Whitfield et al. 2002). The first recorded lionfish in Florida was reported off Dania Beach in 1985; however, the first documented release of *P. volitans* in South Florida was in fact an accidental release of six individual specimens. This occurred in the wake of destruction from Hurricane Andrew (1992), when a large private aquarium was washed away into Biscayne Bay (Courtenay 1995). Those fish were then observed alive in the adjacent marine habitat several days later.

Pterois volitans is well established and reproducing in South Florida waters, with local populations of lionfish rapidly expanding (Whitfield et al. 2002; Ruiz-Carus et al. 2006) (Fig. 1.16). This invasive fish has high fidelity to a particular location, which means once breeding adults find a suitable habitat, they tend to remain and can reach densities of more than 500 adults per hectare. These numbers are staggering, especially since the alien species in question is known to have such a voracious appetite. Lionfish are stalking predators that often corral, or herd, prey into a corner by spreading their pectoral fins (Allen and Eschmeyer 1973). In fact, they are the only fish species known to blow water at potential prey items in an effort to get the prey to turn toward the lionfish before being eaten (Sano et al. 1984). In a single rapid motion, they can consume prey that are more than half of their own length and are known to devour more than 70 marine fish and invertebrate species, including yellowtail snapper, Nassau grouper, parrotfish, banded coral shrimp, and other cleaner species. Lionfish also compete for food with native predatory fish, such as grouper and snapper, and usually negatively impact the overall coral reef ecosystem by eliminating organisms that serve important ecological roles (e.g., herbivorous fish that limit algae growth upon the reef substrates) (Whitfield et al. 2002; Ruiz-Carus et al. 2006).

Fig. 1.16 South Florida coral reefs have been bioinvaded by swarms of red lionfish. In this image, a rogue individual combs the nearshore reef in search of prey. *Pterois volitans* is responsible for the consumption of over 70 reef fish and invertebrates, which can negatively impact the overall coral reef ecosystem. (Credit: Chris Makowski)

*1.3.2 Marine Cane Toad (***Bufo marinus***)*

The marine cane toad, *Bufo marinus*, which is the largest of Florida's frog and toad species, is considered as one of "100 Worst" global invasive organisms and among the "most introduced amphibians in the world" (Behler 1979; Carmichael and Williams 1991). Marine cane toads in Florida are relatively large, reaching sizes of 24 cm in length and weighing up to 2.5 kg (Behler 1979; Somma 2004) (Fig. 1.17). With rough and warty skin that ranges in color from red to brown to green, this coastal invasive species has a stout body and short legs (the forelimbs lack webbing, while the hind limbs are webbed). One of the most distinguishing features of *B. marinus* is the presence of two large toxic parotid glands located at the shoulder behind each eye. These glands are capable of producing a venomous milky secretion when squeezed, while smaller venom glands (i.e., warts) can be found all over the surface of the skin (Ashton and Ashton 1988; Conant and Collins 1991) (Fig. 1.18).

In South Florida, *B. marinus* typically breeds in man-made habitats, such as drainage canals and ditches, fishponds, temporary pools, and other shallow water bodies (Somma 2004). Breeding in Florida typically occurs during the wet season (i.e., June–October), and is most common during or just after rain events (Conant and Collins 1991). During breeding season, the vocalization of marine cane toads is a distinct slow, low-pitched trill, with larger males usually having deeper calls (Somma 2004). *B. marinus* exhibits high fecundity with large

Fig. 1.17 Marine cane toads are the largest frog and toad species found in South Florida and are among the most impactful invasive species. Their large size and high breeding yield make *Bufo marinus* very difficult to control and eradicate. (Credit: Chris Makowski)

Fig. 1.18 The skin of these invasive toads are covered in toxic wart glands, shown by the irregular raised bumps throughout the body. Venomous secretions given off by the glands are capable of killing domestic animals and seriously injuring humans who come in contact with it. (Credit: Chris Makowski)

females capable of producing 20,000 eggs or more (Conant and Collins 1991; Somma 2004). The high breeding yield makes it extremely difficult to curb population numbers for this alien species.

Marine cane toads have been repeatedly introduced throughout the world as a potential biological control agent for crop-damaging insects, primarily those that damage sugarcane (Krakauer 1968, 1970; McKeown 1996; Lever 2001). In South Florida, the first attempts at intentional introduction of *B. marinus* occurred in 1936, when specimens from Puerto Rico were introduced into Palm Beach County (Krakauer 1968). However, it wasn't until 1955 that marine cane toads became a permanent bioinvader in Florida. This was the result of an accidental release by an importer at the Miami Airport and subsequent intentional releases that prompted an explosive local population growth in the species (Ashton and Ashton 1988). *Bufo marinus* was officially designated as a nuisance species that required control measures as early as 1965 (Krakauer 1968, 1970).

Impacts from marine cane toads include competition with native species for food, living space, and breeding sites. This invasive species can also eat and eradicate other indigenous amphibians and reptiles, thus, completely altering the ecological paradigm of a particular ecosystem (Krakauer 1968). Defensive secretions produced by the poison glands of *B. marinus* are highly toxic and capable of killing dogs, cats, and other domestic animals that attempt to bite or sniff them (Ashton and Ashton 1988; Conant and Collins 1991). These venomous secretions are also capable of making humans seriously ill and are the known cause of serious skin and eye irritation disorders (Carmichael and Williams 1991; Conant and Collins 1991).

*1.3.3 Red Imported Fire Ant (***Solenopsis invicta***)*

Solenopsis invicta, infamously known as the red imported fire ant, is a small insect that has become a main coastal invasive species in South Florida. Physically identifiable from a distinct reddish-brown color, red fire ants also possess a two-segmented body pedicel (i.e., waist), four-toothed mandible, a pronounced stinger at the tip of the terminal abdominal segment, and a pair of 10-segmented antennae ending in two-segmented clubs (Hedges 1997) (Fig. 1.19).

Soil mounds of *S. invicta* tend to be approximately 46 cm or less in diameter and can usually be found upwards along the foredune wrack line (Cohen 1992). A mature red fire ant colony can have nearly one-quarter million workers that average between 3–6 mm in length (Fig. 1.20). It is also estimated that nearly 100,000 new queens per year can be produced for every half hectare of infested coastal land (Vinson and Sorensen 1986).

Red fire ants were accidentally introduced into the United States from South America, likely in the late 1920s, as the result of live potted plants being transported across borders (Vinson and Sorensen 1986; ARS 2003). Unfortunately, in coastal

Fig. 1.19 Exotic fire ants are physically identified by their distinct reddish-brown color, two-segmented waist, 10-segmented antennae, and pronounced stinger at the terminal end. Regardless of its small size, fire ants cause many impacts as a coastal invasive species in South Florida. (Credit: Chris Makowski)

Fig. 1.20 The entrance to an invasive fire ant colony found in the upper dune ecosystem of South Florida. Over a quarter of a million workers can be unleashed from the colony on nesting shorebirds and sea turtles, as well as, on any unsuspecting beach goers. (Credit: Chris Makowski)

areas such as South Florida, the lack of natural competitors, predators, and parasites have led to *S. invicta* becoming five times as abundant compared to South America populations, with an expanded range that nearly includes all of the state (Callcott and Collins 1996).

This invader is responsible for extreme reductions in the number of native ant populations (Porter and Savignano 1990) and causes adverse impacts on a variety of ground-nesting indigenous bird and mammal species (Allen et al. 1995; ARS 2003). *Solenopsis invicta* is particularly attracted to mucus, making native species that leave mucous trails, such as newly pipped birds, vulnerable to predation. For example, Allen et al. (1995) report several negative impacts of red fire ants on populations of the northern bobwhite (*Colinus virginianus*), a subtropical bird species. Macroscopic identification of ant mounds can typically be made when humans, or other individuals, accidentally disturb the nests. This action is quickly followed by a programmed response where a large swarm of stinging fire ants emerge from the colony mound to attack the source of the disturbance. Red fire ant stings usually result in the formation of a white pustule within 24 h and a burning sensation upon the victim, as if the skin were on *fire* (Rhoades et al. 1989).

*1.3.4 Nile Monitor (***Varanus niloticus***)*

The Nile monitor (*Varanus niloticus*), which is native to the Nile River Delta and Sub-Saharan Africa, is a large, voracious invasive predator in South Florida, capable of reaching a maximum total length of 243 cm and weighing up to 10 kg (Faust 2001) (Fig. 1.21). Nile monitors have elongated bodies and long, muscular tails that taper in the shape of a rudder to assist in swimming. Coloration of adult Nile monitors is grey-brown to olive-brown, with light yellow rings or V-shaped marks that band around the head, body, and tail. The underside of the Nile monitor is pale without yellow markings. Overall, Nile monitors are smooth and slender in appearance, almost snakelike, with a pointed snout and forked tongue (Meshaka 2006) (Fig. 1.22).

Nile monitors are now frequently observed in South Florida, which is of particular concern because these invasive reptiles are capable of achieving high densities in a short amount of time (Engeman et al. 2009). They are a threat to outcompeting Florida's native wildlife, including some endangered and threatened species like burrowing owls, sea turtles, and crocodiles, by devouring their eggs while having larger clutches (i.e., up to 60 eggs) themselves (de Buffrénil and Rimblot-Baly 1999; de Buffrénil and Hemery 2002). Nile monitors are opportunistic predators that thrive in a wide variety of habitats near water, especially in and around disturbed areas (e.g., canals and urban sprawl), where they dwell in burrows near the water's edge (Faust 2001; Campbell 2005).

Introduction of *V. niloticus* into South Florida was most likely the result of either escape or the intentional release by human owners. These *intentional* releases usually occur when either the novelty of owning an exotic pet wanes, the monitor outgrows their cage, it becomes too expensive to feed, or their temperament and size make them very difficult to handle. Furthermore, illegal releases by reptile dealers may occur when individuals are too damaged or sick to sell, or when dealers are trying to establish a local breeding population for future capture and resale.

Fig. 1.21 Nile monitors are large, non-native lizards that have bioinvaded the Everglades of South Florida. They are responsible for the decline of many mammals, birds, reptiles, amphibians, and fish. Because these invasive reptiles have very few predators, their population size and range cannot be naturally regulated. (Credit: FWC)

Fig. 1.22 With a pointed snout, forked tongue, and distinct markings on the head and body, the Nile monitor gives a smooth and slender snakelike appearance. Even though their natural range is found in the Nile River Delta and Sub-Saharan Africa, this invasive lizard has rapidly established itself throughout the Florida Everglades. (Credit: SFWMD)

Currently, *V. niloticus* has a well-established breeding population in southwest Florida (Enge et al. 2004) and individual numbers are continuing to increase along the east coast, particularly in Miami-Dade County (Dalrymple 1994). Within residential coastal areas, they are opportunistic hunters and can pose a danger to people's pets and small children.

*1.3.5 Burmese Python (*Python molurus bivittatus*)*

One of the more recent, and potentially most destructive, coastal invasive species to establish themselves in South Florida is also one of the largest snakes in the world: the Burmese python (*Python molurus bivittatus*). Capable of growing past lengths of 6 m and weighing up to 90 kg, these large, nonvenomous constrictors are typically tan in color with dark puzzle piece-like blotches on the back and sides (Ernst and

Fig. 1.23 Dark puzzle piece-like blotches on the back and sides are a distinguishing feature of exotic Burmese pythons (*Python molurus bivittatus*). This large, nonvenomous constrictor is one of the most destructive coastal invasive species in South Florida, with lengths exceeding 6 m and weights up to 90 kg. (Credit: SFWMD)

Zug 1996; Rochford et al. 2010) (Fig. 1.23). The head is pyramid-shaped and terminates with an arrowhead wedge at the snout. Burmese pythons are semi-aquatic and are often found near or in water, such as the Florida Everglades (Meshaka 2006). Being apex predators that develop quickly to sexual maturity and possess high yield reproductive clutches of up to 100 eggs, *P. molurus bivittatus* has become an invasive species whose population numbers in South Florida are uncontrollable (Snow et al. 2007a; Rochford et al. 2009).

Burmese pythons have been reported from the saline glades and mangroves at the southern end of Everglades National Park since the 1980s. While the actual mechanism of introduction is not currently known, it has been postulated that escape from a breeding facility during Hurricane Andrew (1992) is the most plausible explanation (Willson et al. 2011). It is also likely that python owners have been releasing these snakes in and around the Everglades for decades (Snow 2006). As a result, *P. molurus bivittatus* are currently listed as a conditional species in Florida and can no longer be acquired as pets in the state. They are also federally listed by the U.S. Fish and Wildlife Service as an Injurious Species under the Lacey Act, which prevents the importation of Burmese pythons into the United States (Snow et al. 2007b).

Because of its large size, adult Burmese pythons have few predators, with alligators and humans being the exceptions. They prey upon native indigenous species, including a variety of mammals and birds, drastically reducing their populations locally (Meshaka 2006). These invasive reptiless can also consume threatened or endangered mammalian species, such as endangered Key Largo woodrats (*Neotoma floridana smalli*) and Key deer (*Odocoileus virginianus clavium*) (Holbrook and Chesnes 2011; Dorcas et al. 2012), and avian species of concern, such as limpkins (*Aramus guarauna*) and white ibis (*Eudocimus albus*) (Snow et al. 2007a; Dove et al. 2011). Furthermore, Burmese pythons can pose a threat to human safety. They have the potential to prey upon domesticated pets such as cats and dogs, or could even attack small children and young or elderly adults (Meshaka 2006) (Fig. 1.24).

Fig. 1.24 A glimpse at what it is like to stare down an apex invasive predator, the Burmese python. Because of their large size and abundant egg clutch yields, Burmese pythons are decimating populations of indigenous species. Soon food resources in the Florida Everglades will be used up, prompting the invasive reptiles to move into more metropolitan areas. (Credit: SFWMD)

1.4 Discussion

Throughout the world, coastal areas have been plagued by non-native invasions of flora and fauna that are responsible for outcompeting indigenous species and permanently altering the ecological framework of the entire biome (Burkitt and Wootton 2011; Damgaard et al. 2011; Wheeler et al. 2011; Choi et al. 2013; Finkl and Makowski 2013a, b, 2017; Ren et al. 2014; Low and Anderson 2017; Efremov et al. 2018). As one navigates through the geologic time record, there are countless event markers that signify when an *introducing disruptor* takes place; for example, in the form of a bolide impact (as with the postulated K-T Event; Alvarez et al. 1980, 1982), great flood (China's Xia Dynasty Flood; (Montgomery 2016), volcanic eruption (Mount Vesuvius), tropical storm (Odisha Cyclone; Francis et al. 2001), or glacial period (Quaternary glaciation). These events serve to disrupt the current ecosystem, usually with a punctuated effect, and later allow for the introduction of invasive species that can thrive in the disturbed ecosystem. The event itself can also be a means of dispersal for alien species to reach new environments; e.g., carried by great wind and wave forces. However, all of these introducing disruptors occur without malice or intent, as they constitute either a geomorphological or exogenic force. It is only later during the most recent part of the Holocene that one biological species has taken on the role of the introducing disruptor: *Homo sapiens*.

Within the last century, humans have been responsible for the development of many of the world's coastlines. As a result, many coastal ecosystems that had maintained an ecological balance amongst the native vegetation and animal life are now completely disturbed. New anthropogenic dispersal mechanisms, such as ship ballast and airplane cargo holds, allow many more invasive species to be introduced to these altered regions. And in some cases, humans intentionally introduce alien species to a coastal area for the reasons of commerce, because they are aesthetically pleasing, or as a control measure to inhibit another species (the ultimate irony is when one invasive species has to be introduced to combat another invasive species). It can even be hypothesized that perhaps the most invasive of all species on planet Earth are in fact humans.

South Florida is an exemplary example of how an anthropogenically disturbed coastal area can become inundated with a wide variety of invasive species (Tables 1.1, 1.2, and 1.3). The following section discusses some of the coastal plant and animal invasions found in South Florida as it compares to other coastal regions worldwide and offers an introspective view into the current state of humans as the planet's main invasive species.

1.4.1 Bioinvasions from Exotic Vegetation

One does not have to venture far in South Florida to witness invasive vegetation growing amongst native species. Australian pines (*Casuarina equisetifolia*) commonly grow along the upper reaches of beach dunes (Wheeler et al. 2011), melaleuca (*Melaleuca quinquenervia*) can be seen throughout the Florida Everglades (Mazzotti et al. 1981, 2001), and water hyacinths (*Eichhornia crassipes*) are known to choke a multitude of irrigational waterways and canals (Schmitz et al. 1991). In the case of hydrilla (*Hydrilla verticillata*), while huge mats of these invasive plants can lead to hypoxic conditions in many of Florida's recreational lakes, other far reaches of the world, such as Northern Eurasia, also suffer the impacts from this alien species (Efremov et al. 2018). There have been several studies that show hydrilla to be a major aquatic weed of concern throughout many secondary ranges in Asia, including Siberia and the Russian Far East (Cook and Luond 1982; Efremov et al. 2018). It is unusual to see an invasive plant of this nature tolerate such a wide latitudinal range, but in many of these coastal areas throughout Northern Eurasia, as also seen in South Florida, hydrilla invasion leads to the outcompeting of indigenous vegetative species (Probatova and Buch 1981; Invasive Species Compendium 2016). Similarly, *Schinus terebinthifolius*, known as Brazilian pepper, has quite a worldwide range as an invasive plant. In the continental United States, Brazilian pepper not only is a nuisance in South Florida, but can also be found in Arizona, California, Louisiana, and Texas (Ewe and Sternberg 2007). It is also considered a major invasive species in coastal regions that experience moderate to high rainfall, such as Australia, southern China, and South Africa. In fact, within the KwaZulu-Natal province of South Africa, *S. terebinthifolius* has been classified as a Category 1 invasive species (Nel et al. 2004). This designation means that all Brazilian pepper plants must be removed from that particular coastal area and destroyed; otherwise, there is a risk that native species could be displaced and the structure of the entire ecological community could be permanently altered (Richardson and van Wilgen 2004).

Unfortunately, there are many other examples of where the proliferation of invasive vegetation in disturbed areas can change the composition, function, structure, and dynamics of a coastal ecosystem. Throughout Shenzhen and Leizhou Bays, both located in South China, a total of 34 invasive plant species were recorded, with most of the alien vegetation originating from tropical America, Africa, and Southeast Asia (Li and Xie 2002; Ren et al. 2014). The amount of invasive plant biomass was so great in this area that the biochemistry of the soil was altered with a net increase in organic C and total N content (Ren et al. 2010). This, in turn, led to the degradation of the entire coastal ecosystem, allowing more alien species to be established in the surrounding terrestrial areas, across ecotonal successions, and into fringing mangrove habitats (Ren et al. 2007, 2014). Another instance of a coastal region under attack from invasive vegetation can be seen along the expansive dune systems of northwestern Europe. Specifically in Denmark, where approximately 20% of the total dune area in northern Europe can be found (Doody 1994), impacts from invasive trees and shrubs, such as *Pinus mugo* and *Rosa rugose*, include a reduction of indigenous biodiversity, increased eutrophication, and the initiation of a positive feedback loop leading to the destabilization of the overall dune system (Thiele et al. 2009; Damgaard et al. 2011). Furthermore, in the mid-latitudes of the United States, along the beaches in the state of New Jersey, there is an ongoing management effort to combat the adverse effects from many invasive plant species. One invader in particular, the Asiatic sand sedge (*Carex kobomugi* Ohwi), which was first introduced to the United States through discarded ship ballast (Small 1954), has significantly reduced native species abundance and diversity by outcompeting the indigenous flora (Burkitt and Wootton 2011).

These few examples of alien species bioinvasion reveal that coastal areas throughout the world incur various impacts from non-native vegetation. From high-latitude regions to the tropics, invasive flora remains a constant threat to disrupt the ecological balance between indigenous species and their respective ecosystems.

1.4.2 Impacts from Invasive Fauna

Invasive fauna includes countless varied forms of non-native, alien animal life (e.g., fish, insects, amphibians, reptiles), many of which have created numerous impacts throughout the world's coastlines. In South Florida, there is a multitude of introduced invasive animals that not only wreak havoc in the subtropical U.S. state, but can also be seen establishing themselves in other coastal areas around the planet. For example, lionfish (*Pterois volitans*), which are originally native to the waters of the Indo-Pacific, can now be found along the eastern seaboard of the United States, throughout the Caribbean, and within the Mesoamerican Reef System (Whitfield et al. 2002; Ruiz-Carus et al. 2006). Being that the lionfish are an alien species with no natural predators in these introduced waters, their voracious appetites and uncontrolled population numbers allow them to eat a majority of the reef fish. As the number of indigenous reef fish declines, so does the natural mechanisms for limiting algae growth upon the reef itself. Without enough fish to crop the algae, the settling substrate becomes choked with a high proliferation of marine vegetation, which can either shade coral from receiving essential sunlight resources or can eliminate available bare reef patches necessary for the growth of new coral colonies (Ruiz-Carus et al. 2006). In the case of marine cane toads, *Bufo marinus*, these invasive amphibians are mostly the product of human introduction as a biological control agent against sugarcane insect parasites (Lever 2001). Unfortunately, once these non-native toads become established, which has occurred throughout Australia, the Philippines, and South Florida, they are prone to consume a majority of indigenous vertebrates (e.g., rodents, frogs, lizards, snakes) (Hinkley 1962) and pose a serious health risk to humans and domestic animals (e.g., dogs, cats) who come in contact with the toads' toxic gland secretions (Lever 2001). A less obvious coastal invasive species can be seen with the red imported fire ant (*Solenopsis invicta*). While at first it may be perceived as small and innocuous, *S. invicta* is responsible for the ecological disruption of many localized coastal environments. Red fire ants, having spread and established themselves in China, Australia, and the southern United States, particularly thrive in anthropogenically disturbed coastal areas. Tschinkel (2006) actually describes the invasive insects to have the same qualities as an intrusive weed, where the ants flourish in those landscapes disturbed by humans. Likewise, two non-native, invasive reptiles have established themselves in South Florida and are the direct result of human interference. Due to the increased popularity of the exotic reptile pet trade, Nile monitors (*Varanus niloticus*) and Burmese pythons (*Python molurus bivittatus*) have become introduced to a very hospitable environment in the Florida Everglades through the intentional and/or accidental release by pet owners and distributors (Hoover 1998; Kraus 2008; Fujisaki et al. 2010). By allowing these invasive reptilian introductions, humans have once again compromised a coastal ecosystem that may never be able to correct itself. For example, Nile monitors are very large carnivorous predatory lizards that disperse throughout ecologically sensitive areas and directly threaten native species, such as burrowing owls (*Athene cunicularia*), sea turtles, waterbirds, and a variety of other indigenous wildlife (Enge et al. 2004; Campbell 2005). Even more dangerous than exotic monitors are Burmese pythons, some of which can reach lengths past 6 m, as they become apex invasive predators within disturbed ecosystems (i.e., Everglades National Park, Loxahatchee National Refuge, Big Cypress National Preserve, and Everglades Water Conservation Areas) preying on native species, outcompeting indigenous populations for food and other resources, and disturbing the physical nature of the environment (Reed et al. 2012). As human interference continues today, the Everglades are less than half of its original extent (Finkl and Makowski 2017); making it ever less likely this reptilian bioinvasion will be contained to the park's watery borders. The favorable subtropical climate of South Florida and the fact that the Everglades are surrounded by metropolitan areas to the east (Miami, Fort Lauderdale, West Palm Beach) and to the west (Naples), sets the stage for expanded infestations of these invasive species into cities and adjacent suburban communities, which then puts human life at risk when considering exotic reptiles of this size.

1.4.3 Bioinvasion Countermeasures

Once a coastal ecosystem has been compromised with biological invasions from exotic vegetation and/or non-native animal wildlife, a wide variety of management countermeasures can be put into use. In addition to educational signage that passively informs the public (Fig. 1.25), more assertive control measures include: Legislative, Chemical, Biological, and Mechanical. In most instances, a comprehensive risk assessment and management strategy protocol is drafted to include as many of these preventative control measures as possible.

Legislative control measures are when the local or federal government passes specific regulations in response to invasive species impacts over a particular area. For example, in response to the invasion of Brazilian pepper (*S. terebinthifolius*) throughout South Florida, the Florida legislature passed a state law prohibiting the sale, cultivation, or transport of the exotic plant (Elfers 1988b; Cuda et al. 2006). Likewise, there are several Florida legislative statutes, most of which fall under title of Public Health and chapter of Environmental Control, that grant permission to eradicate all Brazilian pepper, Australian pines (*C. equisetifolia*), and melaleuca (*M. quinquenervia*) within wetland areas (The Florida Legislature 2017). Invasive fauna are also the target of legislative control measures. In Florida, regulatory rules were put into law that strictly prohibits the release of non-native animal species, such as Nile monitors (*V. niloticus*) and Burmese pythons (*P. molurus bivittatus*) (Fujisaki et al. 2010). Furthermore, a state-issued permit is required to even possess these invasive reptiles of concern and the international trade of such species requires specific regulatory procedures (Levell 1995; Branch 1998).

Chemical countermeasures against invasive plants are mostly used in the form of herbicides and aquacides. This method usually proves to be effective and cost-

Fig. 1.25 An example of educational signage describing native vegetation species in South Florida waterways. The goal is to inform the public so discernment between native versus exotic flora can be achieved. This countermeasure is a passive approach to bioinvasion occurrences. (Credit: Chris Makowski)

prohibitive. For instance, glyphosate herbicides have been applied to stop the proliferation of water hyacinth (*E. crassipes*) (Godfrey 2000), triclopyr herbicides effectively control Brazilian pepper populations (Langeland and Stocker 2001), hexazinone and tebuthiuron chemicals prohibit the growth of melaleuca (FLEPPC 2017), and endothall aquacides are used to combat bioinvasions of hydrilla (*H. verticillata*) (Netherland et al. 1991). Similarly, lethal oral toxicants are usually administered to invasive animal species. Acetaminophen is most commonly used to kill invasive reptiles, such as the Nile monitor and the Burmese python (Mauldin and Savarie 2010).

Preventative methods using biological control agents are very widely used against coastal invasive species worldwide. To control the global spread of Brazilian pepper trees, biological agents such as the Brazilian pepper thrip (*Pseudophilothrips ichini*), the Brazilian pepper leafroller (*Episimus utilis*), the Brazilian pepper sawfly (*Heteroperreyia hubrichi*), and the torymid wasp (*Megastigmus transvaalensis*) have been intentionally released (Wheeler et al. 2001; Cuda et al. 2006; Cleary 2007). For invasions of Australian pine trees, the most effective bio-control proxies are the seed-feeding wasp (*Bootanelleus orientalis*) and the defoliator moth (*Zauclophora pelodes*) in Australia (Woodward and Quinn 2011), the bark-eating caterpillar (*Arbela tetraonis*) in India (Elfers 1988a, b), the lymantriid moth (*Lymantria xylina*) in China (Li et al. 1981), and the fungus *Clitocybe tabescens* in Florida (Rhoads 1952). Likewise, herbivore-mediated reductions of melaleuca in Australia have come in the form of the melaleuca snout beetle (*Oxyops vitiosa*) and the melaleuca psyllid (*Boreioglycaspis melaleucae*), both of which have been approved for release in the United States (Gioeli and Neal 2004; Tipping et al. 2009). Other biological agents, such as parasites, have been used to control the spread of exotic faunal invasions in coastal environments. When studying invasive marine cane toad (*B. marinus*) populations in Australia, Kelehear et al. (2009) found that parasitic lungworm nematodes (*Rhabdias pseudosphaerocephala*) were an effective countermeasure against the amphibians by reducing their growth and survival rates, impairing motor function, and lessening native prey consumption. Overall, the diminished abilities in the toads to feed, evade predators, and disperse showed that the lungworm-induced infections were a viable means of invasive biological control (Goater 1992; Goater and Ward 1992; Kelehear et al. 2009).

Mechanical means of controlling invasive species have mostly been applied by either the physical removal of the exotic species from an area or the implementation of a preventative physical barrier to deny the invader access. Physical removal techniques can include the specific excavation of select invasive species or prescribed burnings in fire-tolerant plant communities. For example, controlled burns in *C. equisetifola* infested areas has shown to be an effective method in limiting dense stands of the Australian pine trees (Snyder 1992). Similarly, controlled fires at repeated intervals of 3–7 years have shown to slow the proliferation of Brazilian pepper, as their seeds fail to germinate post-burn (Doren et al. 1991; Cuda et al. 2006). For aquatic invasive species, such as water hyacinth and hydrilla, physical removal usually occurs when the surrounding water body is drained and specialized harvesting machines physically collect the invasive plants for permanent removal

(Smith and Brown 1994; Hofstra and Clayton 2014). Invasive animal species, on the other hand, are usually controlled via a physical barrier or through organized hunting parties. For instance, to effectively exclude marine cane toads from invading local ponds, canals, and dams, a barrier of at least 50 cm high is constructed using a thick hedge or wire mesh (Brandt and Mazzotti 1990). Selective hunting events have also proven to be successful in limiting the amount of non-native fish and reptiles in South Florida. In particular, lionfish and Burmese pythons have been targeted species in these organized hunting *rodeos* or *roundups*, where local government officials have encouraged the public to aid them in the crusade to eliminate these invasive threats.

In most cases, an integrated management approach that combines several of these different countermeasures proves to be the most effective way to limit bioinvasions. In fact, sometimes a whole government program or department is devoted to perfecting the ways to control invasive species. For example, Florida's Aquatic Plant Management Program is one of the oldest invasive species removal programs active today. The program's beginnings dating back to the late 1800s, when exotic water hyacinths were first introduced into South Florida. Later, this regulatory section was tasked with the removal of hydrilla in the 1950s, as well as melaleuca and Brazilian pepper in the 1960s. With over 1.5 million acres of land in Florida impacted by non-native species, this section oversees the largest invasive plant management program of its kind in the United States (Fig. 1.26).

Fig. 1.26 Preventative control measures against invasive species allow for the native vegetation of South Florida to once again flourish. Some indigenous plants that are commonly found include arrow arum (*Peltandra virginica*), arrowhead (*Sagittaria lancifolia*), fireflag (*Thalia geniculata*), pickerelweed (*Pontederia cordata*), and spatterdock (*Nuphar luteum*). (Credit: Chris Makowski)

1.4.4 Humans as an Invasive Species

After reviewing the role humans have played in the dissemination and establishment of invasive species worldwide, one must begin to consider that the most invasive of all species on planet Earth are in fact human beings. Although human population numbers have been increasing through time, the agricultural revolution and advances in disease control initiated a big demographic splurge. Both of these occurrences have allowed humans to move away from hunter-gatherer dominated societies to more agriculturally-based communities, and eventually to congregate in large population numbers within cities along the coast. The impact of the human invasions into the countryside *via* urban sprawl led to the widespread destruction of natural systems to the point where Nature was more or less eliminated by cultural citified landscapes. The human species dates back at least 3 million years and for most of human history these distant ancestors lived a precarious existence as hunters and gatherers; a way of life that kept total numbers small (probably less than 10 million). However, as agriculture was introduced, communities evolved that could support more and more people.

The rate of global population growth accelerated after World War II when the population of less developed countries began to increase dramatically. After millions of years of very slow growth, populations exploded by doubling over and over; for example, a billion people were added between 1960 and 1975, with another billion added between 1975 and 1987. In the twentieth century, each additional billion was achieved in shorter periods of time, so that in the beginning of the twentieth century there were approximately 1.6 billion people in the world and at the end there were about 7.5 billion (Population Matters 2017). Most of the population influx occurred in cities that in turn spilled over into the countryside areas with the advent of urban sprawl.

Today, urban sprawl covers huge areas of the Earth with some of the larger cities and conurbations occupying vast tracts, mostly along coasts and rivers, that account for 3% of national territories on a global basis (e.g., Tokyo-Yokohama, 37.8 million; Jakarta, Indonesia, 31.3 million; Delhi, India, 25.7 million; Mexico City, 23.9 million; Lagos, Nigeria, 24.2 million; Cairo, Egypt, 24.5 million; Karachi, Pakistan, 24.8 million; Dhaka, Pakistan, 27.4 million; Beijing, China, 27.7 million; Mumbai, India, 27.8 million, Shanghai, China, 30.8 million) (Worldwatch Institute 2000). Tokyo, Japan, for example, is the largest urban area on Earth; it consists of more than 500 connected urban settlements over 30,000 km^2 (Demographia 2017). Coastal environments have much higher concentrations of urban land area (about 10%) and urban populations (about 65%) than any other ecosystem. It is now estimated that about 70% of urban dwellers live in the world's largest coastal megacities (Shirber 2005). Aside from the megacities, according to GRUMP (Global Rural Urban Mapping Project) data, there are 75,000 distinct urban settlements worldwide. And, according to Demographia (2017), the world has recently become more than one-half urban for the first time.

Expansion of human systems into natural environments not only impacts living and working spaces, but also the surrounding areas that are manipulated to support the populous. In the 1st to 3rd centuries AD, Rome was the first city known to reach 1 million inhabitants and was at its zenith in 100 C.E. (Girardet 2000), closely followed by Baghdad, Alexandria, Constantinople, and Hangzhou. Today, there are more than 336 cities with populations greater than 1 million individuals (HarperCollins 2009). Surrounding areas are impacted by waste disposal, dams on waterways, mining, electrical power generation (including wind farms), roads and super highways, railways, airports, and so on. Although currently taking up just 3% of the world's land surface, cities draw and deplete resources from all over the biosphere. Overall, it can be determined that urban centers with their resource demands have come to dominate all life on Earth for the benefit of just one species: *Homo sapiens*.

An end result from the infrastructure development demands that support these massive human colonies is that Nature is squeezed back onto itself, as the natural ranges of flora and fauna are increasingly limited and compacted to the point where natural systems are outcompeting themselves for survival. With natural habitats shrinking because of urban sprawl, there are more and more conflicts with wildlife, as they are perceived to be invading urban space, when in fact it is quite the opposite. Nature was always there and the invading species are humans. The multiplicity of large urban centers, based on an industrializing humanity, is rapidly changing the way in which the *web of life* itself functions (Girardet 1999). Crucial to the expansion of urban centers and the influx of resources that they draw upon from all over the world is the inadvertent introduction of invasive species. It might be argued that humans no longer live in a *civilization* but in a *mobilization* of people, resources, and products because most of the world's transport routes begin and end in cities.

As an example of the ecological impact of cities, Girardet (2000) estimated the ecological footprint of London, whose population is approximately 7 million. The basis for determining the ecological impact involved the areas required to feed a city, to supply its timber and paper needs, and the surface area that would be needed to reabsorb the CO_2 output by areas growing vegetation. Putting these three variables together, London has an ecological footprint the size of the entire United Kingdom. Even more shocking was the fact that London has only 12% of the U.K.'s population and in effect London's footprint is scattered all over the world. Multiplying the effect of London on the world's resources, it becomes immediately apparent that the combined effect of global urbanization is massively deleterious to the planet, with part of the adverse impacts being related to invasive species that modify extent ecosystems around conurbations. Reese and Wackernagel (1992) estimate that if other countries adopted present consumption patterns in urbanized Europe and North America, three planets would be required to support all those urban systems. The mismatch between human demand patterns and the carrying capacity of the planet to supply such provisions will require major changes in the technical use of resources and administration of cities upon the Earth.

Perceptions of the so-called *rights* of humans to expand exponentially and without restriction are inextricably intertwined with religious beliefs, philosophical purviews, socioeconomic and political desires, and egotistical or self-righteous attitudes of both the rich and poor alike. Unfortunately, contemporary societal eidolons do not consider voluntary curtailment when it pertains to population growth. There are misguided opinions that the Earth can support an expanding unlimited human population because *science* will find ways to accommodate food production, supply urban centers, and provide housing for an ever-denser concentration of humans. However, there are many problems with these viewpoints, not the least of which is damage to the environment. The size limits of cities are presently unknown, but there must be constraints to expansion and development that will determine the maximum sizes of conurbations. Water and food supplies, waste disposal, and electrical power seem to be typical factors that regulate human expansion. A high level of entropy where the web of life hangs together in a chain of mutual benefit thus characterizes cities; however, where natural ecosystems have essentially a circular metabolism, urbanized cities tend to possess a linear metabolism. To become sustainable, cities have to adopt circular metabolisms where efficient use and re-use of resources is established and a minimum footprint of waste discharge into the natural environment occurs.

The invasion, or swarm, of the human species in search of increased living space into various environments has not gone unnoticed by researchers. Issues of population growth have been carefully studied by the United Nations and plans have been posited, as in Agenda 21, to eventually congregate people in specified large urban areas and exclude them from newly designated natural areas. Humans would be confined in these specified areas and generally not granted access to the natural environment outside of the conglomeration. As draconian as this proposal sounds, it must be taken serious among other proposals in order to protect the Earth from despoliation by uncontrolled human population growth.

Turning to the example of the southeast Florida conurbation in the modern age, it is seen that the coastal area between the cities of Miami and West Palm Beach began a population surge in the early 1900s, when dry uplands along the Atlantic Coastal Ridge were deemed suitable for habitation. Widespread advertising of cheap land in this subtropical realm by real estate developers enticed people to move from colder northern states to the sun and warmth of Florida. The lure of cheap land and a life in the sun fostered growth rates of about 240% in the 1920s and 1930s, when there were fewer than 150,000 people in the region (USCB 2017). The trend continues today with an 8.7% growth rate on a base population of 2.7 million people in Miami-Dade County. The Miami Metropolitan Area (Miami-Fort Lauderdale-West Palm Beach, 15,980 km^2) is the 73rd largest metropolitan area in the world and the 8th largest in the United States. This coastal strip with about 7 million inhabitants is the most populous in Florida and the 2nd largest in the southeastern United States.

Because the population of South Florida is largely confined to a strip of land between the Atlantic Ocean and the Everglades, the Miami urbanized area (i.e., the area of contiguous urban development) is about 160 km long (north to south), but never more than 32 km wide, and in some areas only 8 km wide (east to west).

The Miami metropolitan statistical area is longer than any other urbanized area in the United States, with the exception of the New York metropolitan area (Demographia 2017).

It is important to realize these numbers not only represent a large number of people crammed into a narrow strip of coastal land between the Atlantic Ocean and the Florida Everglades (which is designated as a Wetland of Importance, International Biosphere Reserve, and a World Heritage Site in Danger) (see Finkl and Charlier 2003a, b, Finkl and Makowski 2017), but these statistics also describe immense densities of humans as an invasive species. The average population density in Miami-Dade County is about 520 persons per km^2. The population density in Sunny Isles, a barrier island community northeast of the Miami city center, is approximately 8139 persons per km^2, a density of occupancy that compares with those found in Singapore and Hong Kong. Furthermore, there are other small, but densely settled, pockets of human invasion; for example, father north of Miami in Broward County, the towns of Franklin Park and Hallandale Beach have 5658 people per km^2 and 2810 people per km^2, respectively.

This human invasion of South Florida has obliterated the natural landscape along the coast *per se* (Fig. 1.27), exceeded the natural carrying capacity for a sustainable population, and taken up half of the Florida Everglades through a process called *dredge and drain* that facilitated westward migration (e.g., Finkl and Makowski

Fig. 1.27 Garbage washed ashore in the form of an offshore piling bumper is a constant reminder that natural coastal ecosystems are being adversely affected by human influences. Many coastal systems worldwide, like this beach in South Florida, have felt the brunt of human bioinvasions, which can ultimately alter the natural landscape, as well as indigenous flora and fauna populations, forever. (Credit: Chris Makowski)

2017). The influx of humans into the Everglades' *River of Grass* had disastrous consequences to the ecological processes of this environment (Douglas 1997; Finkl and Charlier 2003a, b; Finkl and Makowski 2017). Thus, the problem is a double-edged sword as humans themselves not only invade new habitat, but they bring other invasive species with them either deliberately or inadvertently. As indicated previously in this chapter, some non-native species (e.g., melaleuca, water hyacinths, hydrilla) were deliberately imported to the region in support of drainage efforts. Others, such as Nile monitors and Burmese pythons, were originally pets, but when they became unwanted or unmanageable, people discarded them into the Florida Everglades. These invasive species then found very suitable habitat, but at the same time became threats to native indigenous species (Finkl and Makowski 2017).

It is also important to know that the population of the Miami Metropolitan Area is about the same as that in the greater London area. Recalling that Girardet (2000) worked out the land area required to support a metropolis of 7 million people is about the size of the United Kingdom, the ecological footprint of the Miami Metropolitan Area theoretically would be about same size, which is approximately 242,500 km^2. The point here is that the conurbation of southeast Florida cannot support itself without massive imports by cargo ships from distant lands. Along with these daily shipments from all over the world also come exotic species *hitchhikers* that arrive unnoticed in the ballast of these huge vessels. This problem is not trivial by any means, yet it goes largely ignored. The extent of the environmental consequences of an expanding population and introduction of alien species by human invaders has been highlighted by Finkl and Makowski (2017). The situation in South Florida is emblematic of coastal urban centers throughout the world and as population densities continue to increase, the introduction of invasive species into coastal ecosystems becomes more apparent.

The extent of human invasion on this planet has prompted some researchers to laud this event in Earth's history by memorializing it in a new phase of evolution, commonly referred to as the Anthropocene, as a proposed epoch dating from the initiation of significant human impact on the Earth's geology and ecosystems starting with the Industrial Revolution (e.g., Dukes 2011). The reported justification to modify Holocene terminology is based on the impact of humans on the global environment over the last 11,700 years or so, since the end of the last glacial cycle marking the end of the Pleistocene. Mid-twentieth century researchers such as Thomas' (1956) landmark edition of *Man's Role in Changing the Face of the Earth* foreshadowed deleterious impacts of a burgeoning global population without realizing the extent of change that was coming. Recognition of the term Anthropocene perhaps unwittingly emphasizes the environmental ravages of an exploding global population that exploits Earth resources, increasingly appropriates living space by invading natural ecosystems, and allows exotic flora and fauna species to run rampant.

This last insult to the natural environment can be seen today in South Florida with the widespread introduction of invasive plants and animals (Tables 1.1, 1.2, and 1.3). Some alien species were deliberately introduced, while others were accidental; for example, in the case of Hurricane Andrew (1992), when some reptiles escaped from captivity. Whatever the mode of introduction to South Florida ecosystems, these non-native species have thrived in their new environments by finding it rather salubrious and causing a major detriment to the native species. Burmese pythons, for example, are now challenging alligators as the top predator population in the Everglades. Common sense suggests that as the snakes' food supply dwindles, they will search for new prey sources and eventually work their way into adjacent urban areas. This problem is not only a threat to indigenous Everglades' wildlife populations, but also to residents and their domesticated pets in urban and suburban areas.

This vignette makes the point that the Florida Everglades ecosystem is not only under threat from invasive flora and fauna that change the entire ecosystem, but that the wetlands are changed forever. Multiple attempts to control exotic populations have failed. That is say even with best management practices, adaptive management, and so on there is no way to correct, ameliorate, remediate, or rectify the damage caused by exotics that are mobile and impossible to track down or eliminate. Even with various removal programs, there has been no reduction in the numbers of invasive reptiles that are now proliferating unchecked in their new environment (e.g., Finkl and Makowski 2013a, b). The Everglades can never be put back together or restored to pre-settlement conditions and there is thus great uncertainty in the results of management scenarios (e.g., Fuller et al. 2008; Finkl and Makowski 2017). Some conditions can be improved, such as providing clean water to the Everglades (Perry 2004, 2008), but conditions involving exotic invasives are escalating and becoming direr with time. The unfortunate fact is that eradication of invasive plants and animals in South Florida is practically impossible and the ecosystem is now irreversibly changed from what it originally was due to human disruption (Finkl and Makowski 2017).

1.5 Conclusions

South Florida's environmental resource base has been marginalized, first through the disturbance of human settlement and development, and then through the establishment of invasive flora (e.g., Australian pine, Brazilian pepper, broadleaf paperbark tree, water hyacinth, hydrilla) and fauna (e.g., red lionfish, marine cane toad, red imported fire ant, Nile monitor, Burmese python). Bioinvasions of exotic vegetation and wildlife, which also occur on a global scale, negatively impact native species and disrupt the entire ecological balance of the coastal system. Once established and actively proliferating (flora) or breeding (fauna), management and control methods of eradication prove to be difficult and costly. Being that invasive species, by definition, are a product of human interference, disruption, and introduction upon natural coastal ecosystems, one can possibly conclude that human beings are in fact the planet's main invasive species.

Acknowledgements The authors would like to graciously thank Dr. K.W. Rusenko (Gumbo Limbo Nature Center, Boca Raton, Florida, USA; https://www.gumbolimbo.org/) and Dr. A.L. Gardner (South Florida Wildlife Center, Fort Lauderdale, Florida, USA; http://www.southfloridawildlifecenter.org/) for their expertise and resources in compiling this chapter.

Appendix

Table 1.1 List of exotic flora that has been designated as Category I invasive species in the state of Florida, U.S.A. The following alien species are responsible for altering native plant communities by displacing indigenous plants, changing community structures or ecological functions, and/or hybridizing with natives. The scientific name of the invasive species is given in alphabetical order along with the common name

Scientific name	Common name
Abrus precatorius	Rosary pea
Acacia auriculiformis	Earleaf acacia
Albizia julibrissin	Mimosa, silk tree
Albizia lebbeck	Woman's tongue
Ardisia crenata	Coral ardisia
Ardisia elliptica	Shoebutton ardisia
Asparagus aethiopicus	Asparagus-fern
Bauhinia variegata	Orchid tree
Bischofia javanica	Bishopwood
Calophyllum antillanum	Santa Maria, mast wood
Casuarina equisetifolia	Australian-pine
Casuarina glauca	Suckering Australian-pine
Cinnamomum camphora	Camphor tree
Colocasia esculenta	Wild taro
Colubrina asiatica	Lather leaf
Cupaniopsis anacardioides	Carrotwood
Deparia petersenii	Japanese false spleenwood
Dioscorea alata	Winged yam
Dioscorea bulbifera	Air-potato
Eichhornia crassipes	Water-hyacinth
Eugenia uniflora	Surinam cherry
Ficus microcarpa	Laurel fig
Hydrilla verticillata	Hydrilla
Hygrophila polysperma	Green hygro
Hymenachne amplexicaulis	West Indian marsh grass
Imperata cylindrica	Cogon grass
Ipomoea aquatica	Water-spinach
Jasminum dichotomum	Gold coast jasmine
Jasminum fluminense	Brazilian jasmine
Lantana camara	Lantana, shrub verbena

(continued)

Table 1.1 (continued)

Scientific name	Common name
Ligustrum lucidum	Glossy privet
Ligustrum sinense	Chinese privet
Lonicera japonica	Japanese honeysuckle
Ludwigia hexapetala	Uruguay waterprimrose
Ludwigia peruviana	Peruvian primrosewillow
Lumnitzera racemosa	Black mangrove
Luziola subintegra	Tropical American watergrass
Lygodium japonicum	Japanese climbing fern
Lygodium microphyllum	Old world climbing fern
Macfadyena unguis-cati	Catclawvine
Manilkara zapota	Spodilla
Melaleuca quinquenervia	Melaleuca, paper bark
Melinis repens	Natal grass
Microstegium vimineum	Japanes stiltgrass
Mimosa pigra	Catclaw mimosa
Nandina domestica	Nandina, heavenly bamboo
Nephrolepis brownii	Asian sword fern
Nephrolepis cordifolia	Sword fern
Neyraudia reynaudiana	Burma reed
Nymphoides cristata	Crested floating heart
Paederia cruddasiana	Sewer vine
Paederia foetida	Skunk vine
Panicum repens	Torpedo grass
Pennisetum purpureum	Napier grass, elephant grass
Phymatosorus scolopendria	Serpent fern, wart fern
Pistia stratiotes	Water-lettuce
Psidium cattleianum	Strawberry guava
Psidium guajava	Guava
Pueraria montana	Kudzu
Rhodomyrtus tomentosa	Downy rose-myrtle
Ruellia simplex	Mexican-petunia
Salvinia minima	Water spangles
Sapium sebiferum	Popcorn tree, Chinese tallow tree
Scaevola taccada	Half-flower, beach naupaka
Schefflera actinophylla	Schefflera, Queensland umbrella tree
Schinus terebinthifolius	Brazilian-pepper
Scleria lacustris	Wright's nutrush
Senna pendula	Christmas cassia
Solanum tampicense	Wetland nightshade
Solanum viarum	Tropical soda apple
Sporobolus jacquemontii	West Indian dropseed
Syngonium podophyllum	Arrowhead vine

(continued)

Table 1.1 (continued)

Scientific name	Common name
Syzygium cumini	Java-plum
Tectaria incisa	Incised halberd fern
Thelypteris opulenta	Jeweled maiden fern
Thespesia populnea	Seaside mahoe
Tradescantia fluminensis	Small-leaf spiderwort
Urena lobata	Caesar's weed
Urochloa mutica	Para grass
Vitex rotundifolia	Beach vitex

Table 1.2 List of exotic flora that has been designated as Category II invasive species in the state of Florida, U.S.A. The following alien species have increased in abundance and/or frequency, but are not yet responsible for adversely altering native plant communities. The scientific name of the invasive species is given in alphabetical order along with the common name

Scientific name	Common name
Adenanthera pavonina	Red sandalwood
Agave sisalana	Sisal hemp
Aleurites fordii	Tung-oil tree
Alstonia macrophylla	Devil tree
Alternanthera philoxeroides	Alligator-weed
Antigonon leptopus	Coral vine
Ardisia japonica	Japanese ardisia
Aristolochia littoralis	Elegant Dutchman's pipe
Asystasia gangetica	Ganges primrose
Begonia cucullata	Wax begonia
Broussonetia papyrifera	Paper mulberry
Bruguiera gymnorhiza	Large-leaved mangrove
Callistemon viminalis	Bottlebrush
Callisia fragrans	Inch plant, spironema
Casuarina cunninghamiana	Australian-pine
Cecropia palmata	Trumpet tree
Cestrum diurnum	Day jessamine
Chamaedorea seifrizii	Bamboo palm
Clematis terniflora	Japanese clematis
Cocos nucifera	Coconut palm
Crassocephalum crepidioides	Redflower ragleaf, Okinawa spinach
Cryptostegia madagascariensis	Rubber vine
Cyperus involucratus	Umbrella plant
Cyperus prolifer	Dwarf papyrus
Dactyloctenium aegyptium	Durban crowfoot grass
Dalbergia sissoo	Indian rosewood, sissoo

(continued)

Table 1.2 (continued)

Scientific name	Common name
Elaeagnus pungens	Silverthorn, thorny olive
Elaeagnus umbellata	Silverberry, autumn olive
Epipremnum pinnatum	Pothos
Eulophia graminea	Chinese crown orchid
Ficus altissima	False banyan, council tree
Flacourtia indica	Governor's plum
Hemarthria altissima	Limpo grass
Heteropterys brachiata	Red wing, Beechey's with
Hyparrhenia rufa	Jaragua
Ipomoea carnea	Shrub morning-glory
Kalanchoe x houghtonii	Mother-of-millions
Kalanchoe pinnata	Life plant
Koelreuteria elegans	Flamegold tree
Landoltia punctata	Spotted duckweed
Leucaena leucocephala	Lead tree
Limnophila sessiliflora	Asian marshweed
Livistona chinensis	Chinese fan palm
Macroptilium lathyroides	Phasey bean
Melia azedarach	Chinaberry
Melinis minutiflora	Molasses grass
Merremia tuberosa	Wood-rose
Mikania micrantha	Mile-a-minute vine
Momordica charantia	Balsam apple
Murraya paniculata	Orange-jessamine
Myriophyllum spicatum	Eurasian water-milfoil
Panicum maximum	Guinea grass
Passiflora biflora	Two-flowered passion vine
Pennisetum setaceum	Green fountain grass
Pennisetum polystachion	Mission grass, West Indian Pennisetum
Phoenix reclinata	Senegal date palm
Phyllostachys aurea	Golden bamboo
Pittosporum pentandrum	Taiwanese cheesewood
Platycerium bifurcatum	Common staghorn fern
Praxelis clematidea	Praxelis
Pteris vittata	Chinese brake fern
Ptychosperma elegans	Solitaire palm
Richardia grandiflora	Large flower Mexican clover
Ricinus communis	Castor bean
Rotala rotundifolia	Roundleaf toothcup, dwarf Rotala, redweed
Ruellia blechum	Green shrimp plant, Browne's blechum
Sansevieria hyacinthoides	Bowstring hemp
Sesbania punicea	Rattlebox

(continued)

Table 1.2 (continued)

Scientific name	Common name
Sida planicaulis	Mata-pasto
Solanum diphyllum	Two-leaf nightshade
Solanum torvum	Turkeyberry
Spermacoce verticillata	Shrubby false buttonweed
Sphagneticola trilobata	Wedelia, creeping oxeye
Stachytarpheta cayennensis	Nettle-leaf porterweed
Syagrus romanzoffiana	Queen palm
Syzygium jambos	Malabar plum, rose-apple
Talipariti tiliaceum	Mahoe, sea hibiscus
Terminalia catappa	Tropical-almond
Terminalia muelleri	Australian-almond
Tradescantia spathacea	oyster plant
Tribulus cistoides	Puncture vine, burr-nut
Vitex trifolia	Simple-leaf chaste tree
Washingtonia robusta	Washington fan palm
Wisteria sinensis	Chinese wisteria
Xanthosoma sagittifolium	Malanga, elephant ear

Table 1.3 List of exotic fauna that has been designated as invasive species in the state of Florida, U.S.A. The following non-native animal species include mammals, amphibians, freshwater and marine fish, and reptiles. The scientific names have been collectively organized alphabetically, along with their common names, and contain members of statewide established populations, regionally established populations, locally established populations, observed species (i.e., not established), reproducing species (i.e., not established), and potentially extirpated species (i.e., those species that have been actively eliminated or have naturally died out)

Scientific name	Common name
Aceros undulatus	Wreathed Hornbill
Acridotheres cristatellus	Crested Myna
Acridotheres fuscus	Jungle Myna
Acridotheres tristis	Common Myna
Acrochordus javanicus	Javan Filesnake
Agama agama	African Redhead Agama
Agapornis fischeri	Fischer's Lovebird
Agapornis personata	Masked Lovebird
Agapornis roseicollis	Peach-faced Lovebird
Aix galericulata	Mandarin Duck
Alectoris chukar	Chukar
Alopochen aegyptiaca	Egyptian Goose
Amandava amandava	Red Avadavat
Amandava subflava	Zebra Finch
Amazona aestiva	Turquoise-fronted Parrot

(continued)

Table 1.3 (continued)

Scientific name	Common name
Amazona albifrons	White-fronted Parrot
Amazona amazonica	Orange-winged Parrot
Amazona auropalliata	Yellow-naped Parrot
Amazona autumnalis	Red-lored Parrot
Amazona barbadensis	Yellow-shouldered Parrot
Amazona farinosa	Mealy Parrot
Amazona festiva	Festive Parrot
Amazona finschi	Lilac-crowned Parrot
Amazona ochrocephala	Yellow-crowned Parrot
Amazona oratrix	Yellow-headed Parrot
Amazona pretrei	Red-spectacled Parrot
Amazona ventralis	Hispaniolan Parrot
Amazona viridigenalis	Red-crowned Parrot
Ameiva ameiva	Giant Ameiva
Anabas testudineus	Climbing perch
Anas bahamensis	White-cheeked Pintail
Anas luzonica	Philippine Duck
Anas poecilorhyncha	Spot-billed Duck
Anas punctata	Hottentot Teal
Ancistrus sp.	Bristlenosed catfish
Anodorhynchus hyacinthinus	Hyacinth Macaw
Anolis chlorocyanus	Hispaniolan Green Anole
Anolis cristatellus cristatellus	Puerto Rican Crested Anole
Anolis cybotes	Largehead Anole
Anolis distichus	Bark Anole
Anolis equestris equestris	Knight Anole
Anolis extremus	Barbados Anole
Anolis ferreus	Marie Gallant Sail-tailed Anole
Anolis garmani	Jamaican Giant Anole
Anolis porcatus	Cuban Green Anole
Anolis sagrei	Brown Anole
Anser anser	Greylag Goose
Anser cygnoides	Swan Goose
Anser fabalis	Bean Goose
Anser indicus	Bar-headed Goose
Ara ararauna	Blue-and-yellow Macaw
Ara auricollis	Yellow-collared Macaw
Ara macao	Scarlet Macaw
Ara militaris	Military Macaw
Ara nobilis	Red-shouldered Macaw
Ara severa	Chestnut-fronted Macaw
Aramides cajanea	Gray-necked Wood-Rail

(continued)

Table 1.3 (continued)

Scientific name	Common name
Aratinga acuticaudata	Blue-crowned Parakeet
Aratinga aurea	Peach-fronted Parakeet
Aratinga canicularis	Orange-fronted Parakeet
Aratinga chloroptera	Hispaniolan Parakeet
Aratinga erthogenys	Red-masked Parakeet
Aratinga finschi	Crimson-fronted Parakeet
Aratinga holochlora	Green Parakeet
Aratinga leucopthalmus	White-eyed Parakeet
Aratinga mitrata	Mitred Parakeet
Aratinga pertinax	Brown-throated Parakeet
Aratinga wagleri	Scarlet-fronted Parakeet
Aratinga weddellii	Dusky-headed Parakeet
Artibeus jamaicensis	Jamaican Fruit-eating Bat
Astronotus ocellatus	Oscar
Balearica pavonia	Black Crowned Crane
Balearica regulorum	Gray Crowned Crane
Basiliscus plumifrons	Green Basilisk
Basiliscus vittatus	Brown Basilisk
Belonesox belizanus	Pike killifish
Betta splendens	Siamese fightingfish
Boa constrictor	Common Boa
Bonasa umbellus	Ruffed Grouse
Brotogeris chiriri	Yellow-chevroned Parakeet
Brotogeris jugularis	Orange-chinned Parakeet
Brotogeris sanctithomae	Tui Parakeet
Brotogeris versicolurus	White-winged Parakeet
Bucorvus abyssinicus	Abyssinian Ground-Hornbill
Buteogallus anthracinus	Common Black-Hawk
Buteogallus urubitinga	Great Black-Hawk
C. citrinellum x C. urophthalmus	Cichlasoma hybrid
Cacatua galerita	Greater Sulphur-crested Cockatoo
Cacatua moluccensis	Salmon-crested Cockatoo
Cacutua alba	White Cockatoo
Cacutua goffini	Tanimbar Cockatoo
Caiman crocodilus	Spectacled Caiman
Cairina moschata	Muscovy Duck
Callipepla squamata	Scaled Quail
Callonetta leucophrys	Ringed Teal
Calotes mystaceus	Indochinese Tree Agama
Calotes versicolor	Oriental Garden Lizard
Carduelis carduelis	European Goldfinch
Carduelis chloris	European Greenfinch

(continued)

Table 1.3 (continued)

Scientific name	Common name
Carpodacus mexicanus	House Finch
Ceratogymna brevis	Silvery-cheeked Hornbill
Cervus elaphus	Elk
Cervus unicolor	Sambar Deer
Ceryle torquata	Ringed Kingfisher
Chamaeleo calyptratus	Veiled Chameleon
Channa argus	Northern snakehead
Channa marulius	Bullseye snakehead
Chitala ornata	Clown knifefish
Chlorocebus aethiops	Vervet Monkey
Chondrohierax uncinatus	Hook-billed Kite
Chrysolophus pictus	Golden Pheasant
Cichla ocellaris	Butterfly peacock
Cichlasoma bimaculatum	Black acara
Cichlasoma citrinellum	Midas cichlid
Cichlasoma cyanoguttatum	Rio Grande cichlid
Cichlasoma managuense	Jaguar guapote
Cichlasoma meeki	Firemouth cichlid
Cichlasoma nigrofasciatum	Convict cichlid
Cichlasoma octofasciatum	Jack dempsey
Cichlasoma salvini	Yellowbelly cichlid
Cichlasoma trimaculatum	Threespot cichlid
Cichlasoma urophthalmus	Mayan cichlid
Ciconia abdimii	Abdim's Stork
Ciconia ciconia	White Stork
Ciconia episcopus	Wooly-necked Stork
Clarias batrachus	Walking catfish
Cnemidophorus lemniscatus	Rainbow Lizard
Cnemidophorus motaguae	Giant Whiptail
Colossoma macropomum	Black pacu
Columba livia	Rock Dove
Columbina inca	Inca Dove
Corvus corax	Common Raven
Coscoroba coscoroba	Coscoroba Swan
Cosymbotus platyurus	Asian Flattail House Gecko
Cricetomys gambianus	Gambian Pouch Rat
Ctenopharyngodon idella	Grass carp
Ctenopoma nigropannosum	Twospot ctenopoma
Ctenosaura pectinata	Mexican Spinytail Iguana
Ctenosaura similis	Black Spinytail Iguana
Cyanocorax caeruleus	Azure Jay
Cyanocorax yncas	Green Jay

(continued)

Table 1.3 (continued)

Scientific name	Common name
Cyanoliseus patagonus	Burrowing Parrot
Cygnus atratus	Black Swan
Cygnus cygnus	Whooper Swan
Cygnus olor	Mute Swan
Cynomys ludovicianus	Prairie Dog
Cyprinus carpio	Common carp
Dendrocygna arborea	West Indian Whistling-Duck
Dendrocygna viduata	White-faced Whistling-Duck
Eclectus roratus	Eclectus Parrot
Eleutherodactylus coqui	Coqui
Eleutherodactylus planirostris	Greenhouse Frog
Eolophus roseicapillus	Galah
Eos bornea	Red Lory
Ephippiorhynchus asiaticus	Black-necked Stork
Estrilda melpoda	Orange-cheeked Waxbill
Eudocimus ruber	Scarlet Ibis
Euplectes afer	Yellow-crowned Bishop
Euplectes ardens	Red-collared Widowbird
Euplectes franciscanus	Orange Bishop
Euplectes orix	Red Bishop
Falco tinnunculus	Eurasian Kestrel
Felis catus	Feral cats
Francolinus francolinus	Black Francolin
Gallus gallus	Red Junglefowl
Garrulax pectoralis	Greater Necklaced Laughing-thrush
Gekko gecko	Tokay Gecko
Geopelia cuneata	Diamond Dove
Geophagus sp.	Eartheater
Geranospiza caerulescens	Crane Hawk
Gonatodes albogularis fuscus	Yellowhead Gecko
Gracula religiosa	Hill Myna
Grus antigone	Sarus Crane
Gyps sp.	Griffon-type Old World Vulture
Haplochromis callipterus	Eastern happy
Hemichromis letourneuxi	African jewelfish
Hemidactylus frenatus	Common House Gecko
Hemidactylus garnotii	Indo-Pacific Gecko
Hemidactylus mabouia	Tropical House Gecko
Hemidactylus turcicus	Mediterranean Gecko
Heros severus	Banded cichlid
Hoplias malabaricus	Trahira
Hoplosternum littorale	Brown hoplo

(continued)

Table 1.3 (continued)

Scientific name	Common name
Hydrochaeris hydrochaeris	Capybara
Hypostomus sp.	Suckermouth catfish
Hypsipetes madagascariensis	Black Bulbul
Icterus icterus	Troupial
Icterus pectoralis	Spot-breasted Oriole
Iguana iguana	Green Iguana
Irena puella	Asian Fairy-bluebird
Leiocephalus carinatus armouri	Northern Curlytail Lizard
Leiocephalus personatus scalaris	Green-legged Curlytail Lizard
Leiocephalus schreibersii schreibersii	Red-sided Curlytail Lizard
Leiolepis belliana belliana	Butterfly Lizard
Leiothrix lutea	Red-billed Leiothrix
Lonchura atricapilla	Chestnut Munia
Lonchura cantans	African Silverbill
Lonchura maja	White-headed Mannikin
Lonchura malacca	Chestnut Mannikin
Lonchura nana	Madagascar Mannikin
Lonchura punctulata	Nutmeg Mannikin
Lorius garrulus	Chattering Lory
Loxigilla violacea	Greater Antillean Bullfinch
Luscinia megarhynchos	Common Nightingale
Mabuya multifasciata	Many-lined Grass Skink
Macaca fascicularis	Crab-eating Macaque
Macaca mulatta	Rhesus Macaque
Macrognathus siamensis	Spotfin spiny eel
Melopsittacus undulatus	Budgerigar
Melopyrrha nigra	Cuban Bullfinch
Metynnis sp.	Silver dollar
Misgurnus anguillicaudatus	Oriental weatherfish
Molossus molossus tropidorhynchus	Pallas's Mastiff Bat
Monopterus albus	Asian swamp eel
Mus musculus	House Mouse
Musophaga violacea	Violet Touraco
Myiopsitta monachus	Monk Parakeet
Myocastor coypus	Nutria
Nandayus nenday	Black-hooded Parakeet
Neochen jubatus	Orinoco Goose
Netta peposaca	Rosy-billed Pochard
Nothura maculosa	Spotted Nothura
Numida meleagris	Helmeted Guineafowl
Nymphicus hollandicus	Cockatiel
Oreochromis aureus	Blue tilapia

(continued)

Table 1.3 (continued)

Scientific name	Common name
Oreochromis mossambicus	Mozambique tilapia
Oreochromis niloticus	Nile tilapia
Osteopilus septentrionalis	Cuban Treefrog
Pachydactylus bibroni	Bibron's Gecko
Padda oryzivora	Java Sparrow
Parabuteo unicinctus	Harris' Hawk
Paroaria coronata	Red-crested Cardinal
Paroaria gularis	Red-capped Cardinal
Passer domesticus	House Sparrow
Passer luteus	Sudan Golden Sparrow
Pavo cristatus	Common Peafowl
Phasianus colchicus	Ring-necked Pheasant
Phelsuma madagascariensis grandis	Giant Day Gecko
Phoenicopterus chilensis	Chilean Flamingo
Phoenicopterus ruber	Greater Flamingo
Phrynosoma cornutum	Texas Horned Lizard
Piaractus brachypomus	Redbellied pacu
Pica hudsonia	American Magpie
Pionites melanocephala	Black-headed Parrot
Pionus senilis	White-crowned Parrot
Pitangus sulphuratus	Great Kiskadee
Pitta guajana	Banded Pitta
Platalea leucorodia	White Spoonbill
Platytropius siamensis	False Siamese shark
Ploceus cucullatus	Village Weaver
Ploceus velatus	Vitelline Masked Weaver
Poecilia reticulata	Guppy
Poicephalus rueppellii	Rueppell's Parrot
Poicephalus senegalus	Senegal Parrot
Polypterus delhezi	Barred bichir
Porphyrio porphyrio	Purple Swamphen
Psarocolinus montezuma	Montezuma Oropendola
Psephotus haemantonotus	Red-rumped Parrot
Pseudeos fuscata	Dusky Lory
Psittacula alexandri	Moustached Parakeet
Psittacula columboides	Malabar Parakeet
Psittacula cyanocephala	Plum-headed Parakeet
Psittacula eupatria	Alexandrine Parakeet
Psittacula krameri	Rose-ringed Parakeet
Psittacula roseata	Blossom-headed Parakeet
Psittacus erithacus	Gray Parrot
Pterois volitans	Lionfish

(continued)

Table 1.3 (continued)

Scientific name	Common name
Pterygoplichthys disjunctivus	Vermiculated sailfin catfish
Pterygoplichthys multiradiatus	Orinoco sailfin catfish
Pulsatrix perspicillata	Spectacled Owl
Pycnonotus jocosus	Red-whiskered Bulbul
Pyrrhura frontalis	Maroon-bellied Parakeet
Pyrrhura molinae	Green-cheeked Parakeet
Python sebae	Northern African Python
Python molurus bivittatus	Burmese Python
Ramphastos citreolaemus	Citron-throated Toucan
Ramphastos sulfuratus	Keel-billed Toucan
Ramphastos toco	Toco Toucan
Ramphotyphlops braminus	Brahminy Blind Snake
Rattus norvegicus	Norway Rat
Rattus rattus	Black Rat
Rhinella marina	Cane Toad
Rhynchopsitta terrisi	Maroon-fronted Parrot
Saimiri sciureus	Squirrel Monkey
Sarcoramphus papa	King Vulture
Sarotherodon melanotheron	Blackchin tilapia
Sciurus aureogaster	Mexican Red-bellied Squirrel
Serinus mozambicus	Yellow-fronted Canary
Serrasalmus humeralis	Pirambeba
Sicalis flaveola	Saffron Finch
Sphaerodactylus argus argus	Ocellated Gecko
Sphaerodactylus elegans elegans	Ashy Gecko
Streptopelia chinensis	Spotted Dove
Streptopelia decaocto	Eurasian Collared-Dove
Streptopelia risoria	Ringed Turtle-Dove
Sturnus vulgaris	European Starling
Sus scrofa	Hog
Tadorna ferruginea	Ruddy Shelduck
Tadorna tadorna	Common Shelduck
Tarentola annularis	White-spotted Wall Gecko
Tarentola mauritanica	Moorish Wall Gecko
Theraps melanurus x T. zonatus	Theraps hybrid
Thraupis episcopus	Blue-gray Tanager
Threskironis aethiopicus	Sacred Ibis
Tiaris canora	Cuban Grassquit
Tilapia buttikoferi	Hornet tilapia
Tilapia mariae	Spotted tilapia
Tilapia zillii	Redbelly tilapia
Tockus nasutus	African Gray Hornbill

(continued)

Table 1.3 (continued)

Scientific name	Common name
Trachemys scripta elegans	Red-eared Slider
Trichoglossus chlorolepidotus	Scaly-breasted Lorikeet
Trichoglossus haematod	Rainbow Lorikeet
Trichoglossus ornatus	Ornate Lorikeet
Trichopsis vittata	Croaking gourami
Tupinambis merianae	Argentine Black and White Tegu
Turaco schalowi	Schalow's Turaco
Uraeginthus bengalus	Red-cheeked Cordonbleu
Vanellus chilensis	Southern Lapwing
Varanus niloticus	Nile Monitor
Vidua macroura	Pin-tailed Whydah
Xiphophorus hellerii	Green swordtail
Xiphophorus maculatus	Southern platyfish
Xiphophorus variatus	Variable platyfish
Zenaida asiatica	White-winged Dove

References

Agricultural Research Service (ARS) (2003) Update: hot on the trail of fire ants. Agric Res Mag 51

Allen GR, Eschmeyer WN (1973) Turkeyfishes at Eniwetok. Pac Discov 26:3–11

Allen CR, Lutz RS, Demarais S (1995) Red imported fire ant impacts on northern bobwhite populations. Ecol Appl 5:632–638

Alvarez LW, Alvarez W, Asaro F, Michel HV (1980) Extraterrestrial cause for the cretaceous–tertiary boundary. Science 208:1095–1108

Alvarez W, Alvarez LW, Asaro F, Michel HV (1982) Current status of the impact theory for the terminal cretaceous extinction. In: Silver LT, Schultz PH (eds) Geological implications of impacts of large asteroids and comets on the earth. Geological Society of America Special Paper 190, pp 305–315

Ashton RE, Ashton PS (1988) Handbook of Reptiles and Amphibians of Florida. Part three, the Amphibians. Windward Publishing, Miami, 191p

Austin DF (1978) Exotic plants and their effects in Southeastern Florida. Environ Conserv 5:25–34

Baker P, Baker SM, Fajans J (2004) Nonindigenous marine species in the greater Tampa Bay ecosystem. Tampa Bay Estuary Program Technical Publication #02-04 131p

Barrett SCH (1980) Sexual reproduction in Eichhornia crassipes (water hyacinth). 1. Fertility of clones from diverse regions. J Appl Ecol 17:101–112

Behler JL (1979) The Audubon Society field guide to North American Reptiles and Amphibians. Alfred A. Knopf, New York, 743p

Bowes G, Holaday AS, Van TK, Haller WT (1977) Photosynthetic and photorespiratory carbon metabolism in aquatic plants. In: Proceedings 4th international congress of Photosynthesis. Reading, UK, pp 289–298

Branch WR (1998) A field guide to the snakes and other reptiles of Southern Africa, revised edition. Struik Publishing, Cape Town

Brandt LA, Mazzotti FJ (1990) Marine toads (Bufo marinus). University of Florida – Cooperative Extension Service, Institute of Food and Agricultural Sciences, Gainesville http://edis.ifas.ufl.edu/pdffiles/UW/UW04600.pdf

Burkitt J, Wootton L (2011) Effects of disturbance and age of invasion on the impact of the invasive sand sedge, Carex kobomugi, on native dune plant populations in New Jersey's coastal dunes. J Coast Res 27(1):182–193
Callcott AMA, Collins HL (1996) Invasion and range expansion of red imported fire ant (Hymenoptera: Formicidae) in North America from 1918–1995. Fla Entomol 79:240–251
Campbell TS (2005) Eradication of introduced carnivorous Lizards from the Cape Coral Area. Final Report to the Charlotte Harbor National Estuary Program, Ft. Myers, 30p
Canfield DE Jr, Langeland KA, Maceina MJ, Haller WT, Shireman JV, Jones JR (1983) Trophic state classification of lakes with aquatic macrophytes. Can J Fish Aquat Sci 40:1713–1718
Carmichael P, Williams W (1991) Florida's fabulous Reptiles and Amphibians. World Publications, Tampa, 120p
Carter V, Rybicki NB, Landwehr JM, Turtora M (1994) Role of weather and water quality in population dynamics of submersed macrophytes in the tidal Potomac River. Estuaries 17:417–426
Center TD, Dray FA Jr (1986) Melaleuca quinquenervia: a serious threat to South Florida wetlands. Report submitted to the U.S. Department of Agriculture. Agricultural Research Service (ARS), Aquatic Plant Laboratories, Ft. Lauderdale
Choi KH, Kim YM, Jung PM (2013) Adverse effect of planting pine on coastal dunes, Korea. In: Conley DC, Masselink G, Russell PE, O'Hare TJ (eds) Proceedings 12th international coastal symposium (Plymouth, England). Journal of Coastal Research, Special Issue No 65, pp 909–914
Cleary RL (2007) Controlling upland invasive exotic plants on public conservation land: a strategic plan. Nat Areas 27(3):218–225
Cohen PR (1992) Imported fire ant stings: clinical manifestations and treatment. Pediatr Determol 9:44–48
Colle DE, Shireman JV (1980) Coefficients of condition for largemouth bass, bluegill, and redear sunfish in hydrilla-infested lakes. Trans Am Fish Soc 109:521–531
Conant R, Collins JT (1991) Reptiles and Amphibians, Eastern/Central North America. Houghton Mifflin Company, Boston, 450p
Cook CDK Luond R (1982) A revision of the genus Hydrilla (Hydrocharitaceae). Aquat Bot 13:485–504
Courtenay WR Jr (1995) Marine fish introductions in southeastern Florida. Am Fish Soc Introd Fish Section Newsl 14:2–3
Cuda JP, Ferriter AP, Manrique V, Medal JC (eds) (2006) Interagency Brazilian Peppertree (Schinus terebinthifolius) management plan for Florida, 2nd edition: recommendations from the Brazilian peppertree task force Florida exotic Pest plant council, 81p
Curnutt JL (1989) Breeding bird use of a mature stand of Brazilian peppertree. FL Ornithol Soc 17:53–76
D'Antonio C, Meyerson LA (2002) Exotic plant species as problems and solutions in ecological restoration: a synthesis. Restor Ecol 10(4):703–713
Dalrymple GH (1994) Non-indigenous amphibians and reptiles. In: Schmitz DC, Brown TC (eds) An assessment of invasive non-indigenous species in Florida's Public Lands. Florida Department of Environmental Protection, Tallahassee, pp 67–78. Technical Report No. TSS-94-100
Damgaard C, Nygaard B, Ejrnæs R, Kollmann J (2011) State-space modeling indicates rapid invasion of an alien shrub in coastal dunes. J Coast Res 27(3):595–599
de Buffrénil V, Hemery G (2002) Variation in longevity, growth, and morphology in exploited Nile monitors (Varanus niloticus) from Sahelian Africa. J Herpetol 36(3):419–426
de Buffrénil V, Rimblot-Baly F (1999) Female reproductive output in exploited Nile Monitor Lizard (Varanus niloticus L.) populations in Sahelian Africa. Can J Zool 77:1530 1539
de Casabianca M-L, Laugier T (1995) Eichhornia crassipes production on petroliferous wastewaters: effects of salinity. Bioresour Technol 54:39–43
Demographia (2017) World urban areas report (built-up urban areas or urban agglomerations), 13th annual edition. Demographia Staff, Belleville

Diamond C, Davis D, Schmitz DC (1991) Impact statement: the addition of Melaleuca quinquenervia to the Florida prohibited aquatic plant list. In: Proceedings of the symposium on exotic pest plants, November 2–4, 1988, Miami
DiStefano JF, Fisher RF (1983) Invasion potential of Melaleuca quinquenervia in southern Florida, U.S.A. For Ecol Manag 7:133–141
Donnelly MJ, Walters LJ (2008) Water and boating activity as dispersal vectors for Schinus terebinthifolius (Brazilian pepper) seeds in freshwater and estuarine habitats. Estuar Coasts 31 (5):960–968
Doody JP (1994) The coastal dunes of Europe. In Danish Dunes: monitoring, management and research. Nordjylland, 25–27 May 1992
Dorcas ME, Willson JD, Reed RN, Snow RW, Rochford MR, Miller MA, Meshaka WE, Andreadis PT, Mazzotti FJ, Romagosa CM, Hart KM (2012) Severe mammal declines coincide with proliferation of invasive Burmese pythons in Everglades National Park. Proc Natl Acad Sci 109 (7):2418–2422
Doren RF, Jones DT (1997) Plant management in Everglades National Park. In: Simberloff D, Schmitz DC, Brown TC (eds) Strangers in paradise: impact and Management of Nonindigenous Species in Florida. Island Press, Washington, DC, pp 275–286
Doren RF, Whiteaker LD, LaRosa AM (1991) Evaluation of fire as a management tool for controlling Schinus terebinthifolius as secondary successional growth on abandoned agricultural land. Environ Manag 15:121–129
Doren RF, Volin JC, Richards JH (2009a) Invasive exotic plant indicators for ecosystem restoration: an example from the Everglades restoration program. Ecol Indic 9S(2009):S29–S36
Doren RF, Trexler JC, Gottlieb AD, Harwell MC (2009b) Ecological indicators for system-wide assessment of the greater everglades ecosystem restoration program. Ecol Indic 9S(2009):S2–S16
Douglas MS (1997) The Everglades: river of grass, 50th anniversary edition. Pineapple Press, Sarasota, 392p
Dove CJ, Snow RW, Rochford MR, Mazzotti FJ (2011) Birds consumed by the invasive Burmese python (Python molurus bivittatus) in Everglades National Park, Florida, USA. Wilson J Ornithol 123(1):126–131
Duever MJ, Carlson J, Meeder J, Duever L, Gunderson L, Riopell L, Alexander T, Myers R, Spangler D (1986) The big cypress national preserve. The National Audubon Society Research Center Report No. 8, New York
Dukes P (2011) Minutes to midnight: history and the Anthropocene Era from 1763. Anthem Press, New York, 168p
Efremov A, Bolotova Ya, Mesterhazy A, Toma C (2018) Features of distribution of Hydrilla verticillata (L. fil.) Royle (Hydrocharitaceae) in North Eurasia. J Coast Res 34(3):675–686
Elfers SC (1988a) Element stewardship abstract for *Casuarina equisetifolia*. Report to The Nature Conservancy, 15p
Elfers SC (1988b) Element stewardship abstract for Schinus terebinthifolius Brazilian Pepper-tree. Report to The Nature Conservancy, 10p
Enge KM, Krysko KL, Hankins KR, Campbell TS, King FW (2004) Status of the Nile monitor Varanus niloticus in southwestern Florida. Southeast Nat 3:571–582
Engeman RM, Constantin B, Hardin S, Smith H, Meshaka W Jr (2009) "Species pollution" in Florida: a cross-section of invasive vertebrate issues and management responses USDA National Wildlife Research Center, Staff Publications No 897
Ernst CH, Zug GR (1996) Snakes in question: the Smithsonian answer book (No. 597.96 E71s). Smithsonian Institution Press, London, 182p
Ewe SML (2004) What makes Brazilian pepper such a successful invader in Florida? A study of Schinus physiology. "West of Eden-Where Research, Policy and Practice Meet", A Joint Conference of the Southeast EPPC and Florida EPPC, Pensacola Beach, FL, 28–30 April, Abstracts

Ewe SML, Sternberg LSL (2003) Seasonal gas exchange characteristics of Schinus terebinthifolius in a native and disturbed upland community in Everglades National Park, Florida. For Ecol Manag 179:27–36
Ewe SML, Sternberg LSL (2007) Water uptake patterns of an invasive exotic plant in coastal saline habitats. J Coast Res 23(1):255–264
Ewel J, Meador R, Myers R, Conde L, Sedlik B (1976) Studies of vegetation changes on South Florida. Report to U.S. Forest Service on Research Agreement 18-492, University of Florida, Gainesville
Ewel J, Ojima D, Karl D, Debusk W (1982) Schinus in successional ecosystems of Everglades National Park. South Florida Res. Cent. Rep. T-676. Everglades National Park, National Park Service, Homestead, Florida
Faust RJ (2001) Nile monitors: everything about history, care, nutrition, handling, and behavior. Barron's Educational Series, Hauppauge
Ferriter AP (ed) (1997) Brazilian Pepper management plan for Florida. Florida EPPC, W. Palm Beach, Florida, 38p
Finkl CW, Charlier RH (2003a) Sustainability of subtropical Coastal Zones in Southeastern Florida: challenges for urbanized coastal environments threatened by development, pollution, water supply, and storm hazards. J Coast Res 19(4):934–943
Finkl CW, Charlier R (2003b) Sustainability of coastal zones in southeastern Florida: urbanized coastal environments threatened by development, pollution, water supply, and storm hazards. 30th Pacem in Maribus (27–30 October 2003, Kiev). International Ocean Institute, Sevastopol, pp 95–96
Finkl CW, Makowski C (2013a) The Southeast Florida coastal zone (SFCZ): a cascade of natural, biological, and human-induced hazards. In: Finkl CW (ed) Coastal hazards. Springer, Dordrecht, pp 3–56
Finkl CW, Makowski C (2013b) Perception of coastal hazards in terms of physical, biological, and human-induced cascades: an example from the Southeast Florida Coastal Zone (SFCZ). In: Proceedings of the 12th international coastal symposium, 9–14 April 2013, Plymouth, England. Journal of Coastal Research, Special Issue No. 65, pp 708–713
Finkl CW, Makowski C (2017) The Florida Everglades: an overview of alteration and restoration. In: Finkl CW, Makowski C (eds) Coastal wetlands: alteration and remediation. Springer, Dordrecht, pp 3–45
Fishelson L (1975) Ethology and reproduction of pteroid fishes found in the Gulf of Aqaba (Red Sea), especially Dendrochirus brachypterus (Cuvier), (Pteroidae, Teleostei). PSZN 39(Suppl. 1):635–656
Florida Exotic Pest Plant Council [FLEPPC] (2017) Florida Exotic Pest Plant Council's 2017 list of invasive species. http://www.fleppc.org/list/list.htm
Flowers JD (1991) Tropical fire suppression in Melaleuca quinquenervia. In: Center TD, Doren RF, Hofstetter RL, Myers RL, Whiteaker LD (eds) United states department of interior. National Park Service, Washington, DC, pp 151–158
Fuller MM, Gross LJ, Duke-Sylvester SM, Palmer M (2008) Testing the robustness of management decisions to uncertainty: Everglades restoration scenarios. Ecol Appl 18(3):711–723
Francis S, Gunturi PVSK, Arora M (2001) Performance of built environment in the Oct 1999 Orissa super Cyclone. Conference on Disaster Management, Birla Institute of Technology & Sciences (Pilani, March 2001)
Fujisaki I, Hart KM, Mazzotti FJ, Rice KG, Snow S, Rochford M (2010) Risk assessment of potential invasiveness of exotic reptiles imported to south Florida. Biol Invasions 12:2585–2596
Gioeli TK, Neal AS (2004) Biological control of in Florida's natural and man-made landscapes. Proc Fla State Hort Soc 117:254–255
Girardet H (1999) Creating sustainable cities. Green Books, Darlington
Girardet H (2000) Cities, people, planet. Liverpool (UK) Schumacher Lectures, Urban Sustainability, Liverpool, 15p

Goater CP (1992) Experimental population dynamics of Rhabdias bufonis (Nematoda) in toads (Bufo bufo): density-dependence in the primary infection. Parasitology 104:179–187

Goater CP, Ward PI (1992) Negative effects of Rhabdias bufonis (Nematoda) on the growth and survival or toads (Bufo bufo). Oecologia 89:161–165

Godfrey K (2000) Eichhornia crassipes. In: Bossard CC, Randall JM, Hoshovsky MC (eds) Invasive plants of California Wildlands. University of California Press. http://groups.ucanr.org/ceppc/Invasive_Plants_of_California's_Wildlands

Godfrey RK, Wooten JW (1979) Aquatic and wetland plants of the southeastern United States, the Monocotyledons. University of Georgia, Athens, 933p

Gopal B (1987) Water hyacinth. Elsevier Science Publishers, Amsterdam, 471p

Gopal B, Sharma KP (1981) Water-hyacinth (Eichhornia crassipes), most troublesome weed of the world. Hindasia Publications, Delhi, 219p

Habeck DH, Bennett FD, Balciunas JK (1994) Biological control of terrestrial and wetland weeds. In: Rosen D, Bennett FD, Capinera JL (eds) Pest Management in the Subtropics: biological control- a Florida perspective. Andover, Intercept, pp 523–547

Haller WT, Sutton DL (1975) Community structure and competition between hydrilla and vallisneria. Hyacinth Control J 13:48–50

Halstead BW, Chitwood MJ, Modglin FR (1955) The anatomy of the venom apparatus of the zebrafish Pterois volitans (Linnaeus). Anat Rec 122:317–333

Hargraves PE (2008) Allelopathy at the land/sea interface: microalgae and Brazilian pepper. Mar Environ Res 66(5):553–555

HarperCollins (2009) The times concise atlas of the world. Glasgow: HarperCollins

Hedges SA (1997) Fire ants. In: Moreland D (ed) Handbook of pest control, 8th ed. Mallis Handbook and Technical Training Company, pp 531–535

Hight SD, Cuda JP, Medal JC (2002) Brazilian peppertree. In: Van Driesche R, Lyon S, Blossey B, Hoddle M, Reardon R (eds) Biological control of invasive plants in the Eastern United States, USDA Forest Service Publication FHTET-2002-04. USDA Forest Service, Morgantown, pp 311–321

Hight SD, Horiuchi I, Vitorino MD, Wikler CW, Pedrosa-Macedo JH (2003) Biology, host specificity tests, and risk assessment of the sawfly Heteroperreyia hubrichi, a potential biological control agent of Schinus tere in Hawaii. BioControl 48:461–476

Hinkley AD (1962) Diet of the giant toad, Bufo marinus (L.), in Fiji. Herpetologica 18:253–259

Hofstetter RH (1991) The current status of Melaleuca quinquenervia in southern Florida. In: Center TD, Doren RF, Hofstetter RL, Myers RL, Whiteaker LD (eds) Proceedings of the symposium on exotic pest plants. 2–4 November 1988, University of Miami, Florida. United States Department of Interior, National Park Service, Washington, DC, pp 159–176

Hofstra D, Clayton J (2014) Native flora and fauna response to removal of the weed Hydrilla verticillata (Lf) Royle in Lake Tutira. Hydrobiologia 737(1):297–308

Holbrook J, Chesnes T (2011) The effect of Burmese pythons (Python Molurus Bivittatus) on mammal populations in southern Florida. Florida Scient 74(1):17–24

Holm LG, Plucknett DL, Pancho JV, Herberger JP (1977) The world's worst weeds: distribution and biology. University Press of Hawaii, Honolulu, 609p

Hoover C (1998) The US role in the international live reptile trade: Amazon tree boas to Zululand dwarf chameleons. TRAFFIC North America, Washington, DC

Hurley LM (1990) Field guide to the submerged aquatic vegetation of Chesapeake Bay. United States Fish and Wildlife Service, Annapolis, 51p

Invasive Species Compendium (2016) Invasive species compendium. Datasheets, Maps, Images, Abstracts and Full Text on Invasive Species of the World. http://www.cabi.org/isc/datasheet/28170

Kelehear C, Webb JK, Shine R (2009) Rhabdias pseudosphaerocephala infection in Bufo marinus: lung nematodes reduce viability of metamorph cane toads. Parasitology 136:919–927

Krakauer T (1968) The ecology of the neotropical toad, Bufo marinus, in South Florida. Herpetologica 24:214–221

Krakauer T (1970) The invasion of the toads. Florida Nat:12–14

Kraus F (2008) Alien reptiles and amphibians: a scientific compendium and analysis. Springer Science & Business Media, New York, 563p

Lampe KF, Fagerstrom R (1968) Plant toxicity and dermatitis. Williams and Wilkins Co., Baltimore

Langeland KA (1996) Hydrilla verticillata (L.F.) Royle (Hydrocharitaceae), "The perfect aquatic weed.". Castanea 61:293–304

Langeland KA (2008) Identification and biology of nonnative plants in Florida's natural areas. IFAS Communication Services, University of Florida

Langeland KA, Craddock Burks K (1998) Identification and biology of non-native plants in Floridas natural areas. University of Florida, Gainesville, 165p

Langeland KA, Stocker RK (2001) Control of non-native plants in natural areas of Florida. 2nd ed. University of Florida, IFAS, Cooperative Extension Service, Gainesville, SP-242

Laroche FB (ed) (1994) Melaleuca management plan for Florida. Task Force Report. Florida Exotic Pest Plant Council, West Palm Beach, 88p

Laroche FB, Ferriter AP (1992) Estimating expansion rates of Melaleuca in South Florida. J Aquat Plant Manag 30:62–65

LeSchiava AJ, Best RB, Burns RE, Gray S, Harwell MC, Hines MC, McLean AR, Clair TS, Traxler S, Vearil JW (2013) Lessons learned from the first decade of adaptive management in comprehensive Everglades restoration. Ecol Soc 18(4):70

Levell JP (1995) A field guide to reptiles and the law. Serpent's Tale Natural History Book Distributors 240p

Lever C (2001) The cane toad. The history and ecology of a successful colonist. Westbury Publishing, Westbury, 230p

Li ZY, Xie Y (2002) Invasive alien species in China. China Forestry, Beijing, 211p

Li YG, Chen SL, Xie QM, Cai QJ, Wu J, Li YW, Zheng XQ, Zhu ZW, Zhou BT, Zheng HQ (1981) Studies on the Lymantriid moth Lymantria xylina Swinhoe. Acta Entomol Sin 24(2):174–183

Low L, Anderson CJ (2017) The threat of a nonnative, invasive apple snail to oligohaline marshes along the northern Gulf of Mexico. J Coast Res 33(6):1376–1382

Madeira PT, Van TK, Center TD (2004) An improved molecular tool for distinguishing monoecious and dioecious hydrilla. J Aquat Plant Manag 42:28–32

Mauldin RE, Savarie PJ (2010) Acetaminophen as an oral toxicant for Nile monitor lizards (Varanus niloticus) and Burmese pythons (Python molurus bivittatus). Wildl Res 37:215–222

Mazzotti FJ, Ostrenko W, Smith AT (1981) Effects of the exotic plants Melaleuca quinquenervia and Casuarina equisetifolia on small mammal populations in the eastern Florida Everglades. Florida Sci 44:65–71

Mazzotti FJ, Pearlstine LG, Brandt LA, Moller MP, Zweig C (2001) A multi-species/habitat ecological evaluation of alternative everglades restoration plans. In: Poster presented at the 28th annual national areas association conference. Cape Canaveral

McCann JA, Arkin LN, Williams JD (1996) Nonindigenous aquatic and selected terrestrial species of Florida. Washington, DC: Smithsonian Inst. Press 301p

McKeown S (1996) A field guide to Reptiles and Amphibians in the Hawaiian Islands. Diamond Head Publishing, Inc, California, 172p

Meister HS, Wyanski DM, Loefer JK, Ross SW, Quattrini AM, Sulak KJ (2005) Further evidence for the invasion and establishment of Pterois volitans (Teleostei: Scorpaenidae) along the Atlantic Coast of the United States. Southeast Nat 4:193–206

Meshaka WE (2006) An update on the list of Florida's exotic amphibians and reptile species. J Kans Herpetol 19:16–17

Meskimen GF (1962) A silvical study of the melaleuca tree in South Florida. MS thesis (Forestry). University of Florida, Gainesville, p 177

Mitchell DS (1976) The growth and management of Eichhornia crassipes and Salvinia spp. in their native environment and in alien situations. In Varshney CK, Rzoska J (eds), Aquatic weeds in southeast Asia. Hague: W. Junk Publishers 396p

Montgomery DR (2016) Emperor Yu's great flood. Science 353(6299):538–539
Morgan EC, Overholt WA (2005) Potential allelopathic effects of Brazilian pepper (Schinus terebinthifolius Raddi, Anacardiaceae) aqueous extract on germination and growth of selected Florida native plants. J Torrey Bot Soc 132:11–15
Morton JF (1969) Some ornamental plants excreting respiratory irritants. Proc Fl Hort Soc 82:415–421
Morton JF (1978) Brazilian pepper – its impact on people, animals and the environment. Econ Bot 32(4):353–359
Morton JF (1980) The Australian pine or beefwood (*Casuarina equisetifolia* L.) an invasive "weed" tree in Florida. Proc Florida State Hortic Soc 93:87–95
Myers RL (1983) Site susceptibility to invasion by the exotic tree Melaleuca quinquenervia in southern Florida. J Appl Ecol 20:645–658
Myers RF (1991) Micronesian reef fishes, 2nd edn. Coral Graphics, Barrigada, 298p
Nel JL, Richardson DM, Rouget M, Mgidi TN, Mdzeke N, Le Maitre DC, van Wilgen BW, Schonegevel L, Henderson L, Neser S (2004) A proposed classification of invasive alien plant species in South Africa: towards prioritizing species and areas for management action. S Afr J Sci 100:53–64
Netherland MD, Green WR, Getsinger KD (1991) Endothall concentration and exposure time relationships for the control of Eurasian Watermilfoil and Hydrilla. J Aquat Plant Manag 29:61–67
Perrings C (2005) Mitigation and adaptation strategies for the control of biological invasions. Ecol Econ 52(3):315–325
Perry WB (2004) Elements of South Florida's comprehensive everglades restoration plan. Ecotoxicology 13:185–193
Perry WB (2008) Everglades restoration and water quality challenges in South Florida. Ecotoxicology 17:569–578
Pieterse AH (1981) Hydrilla verticillata – a review. Abstr Trop Agric 7:9–34
Population Matters (2017) World population growth. Population Matters, London
Porter SD, Savignano DA (1990) Invasion of polygyne fire ants decimates native ants and disrupts arthropod community. Ecology 71:2095–2106
Probatova NS, Buch TG (1981) Hydrilla verticillata (Hydrocharitaceae) in the Soviet Far East. Botanichesky Zhurnal 66(2):208–214
Ramey V (2001) Non-native invasive aquatic plants in the United States: Eichhornia crassipes. Center for Aquatic and Invasive Plants, University of Florida and Sea Grant http://plants.ifas.ufl.edu/plant-directory/eichhornia-crassipes/
Randall JM (2000) Schinus terebinthifolius Raddi. In: Bossard CC, Randall JM, Hoshovsky MC (eds) Invasive plants of California's Wildlands. University of California Press, Berkeley, pp 282–287
Reed RN, Willson JD, Rodda GH, Dorcas ME (2012) Ecological correlates of invasion impact for Burmese Pythons in Florida. Integr Zool 7(3):254–270
Reese W, Wackernagel M (1992) Our ecological footprint. New Society Publishers, Gabriola Island
Ren H, Li Z, Shen W, Yu Z, Peng S, Liao C, Ding M, Wu J (2007) Changes in biodiversity and ecosystem function during the restoration of a tropical forest in South China. Sci China Ser C Life Sci 50(2):277–284
Ren H, Chen H, Li Z, Han W (2010) Biomass accumulation and carbon storage of four different aged Sonneratia apetala plantations in southern China. Plant Soil 327(2):279–291
Ren H, Guo Q, Liu H, Li J, Zhang Q, Xu H, Xu F (2014) Patterns of alien plant invasion across coastal bay areas in southern China. J Coast Res 30(3):448–455
Rhoades RB, Stafford CT, James FK Jr (1989) Survey of fatal anaphylactic reactions to imported fire ant stings. J Allergy Clin Immunol 84:159–162
Rhoads AS (1952) The destructiveness of Clitocybe root rot to plantings of Casuarinas in Florida. Lloydia 15:161–184

Richardson DR (1977) Vegetation of the Atlantic Coastal Ridge of Palm Beach County, Florida. Florida Sci 40:281–330
Richardson DM, van Wilgen BW (2004) Invasive alien plants in South Africa: how well do we understand the ecological impacts? S Afr J Sci 100:45–52
Rochford M, Brien ML, Carrigan J, Snow RW, Mazzotti FJ (2009) Python molurus bivittatus (Burmese Python) clutch size. Herpetol Rev 40:442
Rochford M, Krysko KL, Nifong J, Wilkins L, Snow RW, Cherkiss MS (2010) Python molurus bivittatus (Burmese Python). Diet Herpetol Rev 41:97
Room PM, Fernando IVS (1992) Weed invasions countered by biological control: Salvinia molesta and Eichhornia crassipes in Sri Lanka. Aquat Bot 42:99–107
Ruiz-Carus R, Matheson RE Jr, Roberts DE Jr, Whitfield PE (2006) The western Pacific red lionfish, Pterois volitans (Scorpaenidae), in Florida: evidence for reproduction and parasitism in the first exotic marine fish established in state waters. Biol Conserv 128:384–390
Sano M, Shimizu M, Nose Y (1984) Food habits of teleostean reef fishes in Okinawa Island, Southern Japan. University of Tokyo Press, Tokyo, 128p
Schardt JD (1994) Florida aquatic plant survey report. Florida Department of Environmental Protection, Bureau of Aquatic Plant Management, Tallahassee, Report nr 942-CGA, 83p
Schardt JD (1997) Florida aquatic plant survey report. Florida Department of Environmental Protection, Bureau of Aquatic Plant Management, Tallahassee, Report nr 942-CGA, 86p
Schmid JL, Addison DS, Donnelly MA, Shirley MA, Wibbels T (2008) The effect of Australian Pine (*Casuarina equisetifolia*) removal on Loggerhead sea turtle (Caretta caretta) incubation temperatures on Keewaydin Island, Florida. In: Kennish M (ed) Research & monitoring of NERRS aquatic ecosystems. J Coast Res Special Issue No. 55:214–220
Schmitz DC, Osborne JA (1984) Zooplankton densities in a Hydrilla infested lake. Hydrobiologia 111:127–132
Schmitz DC, Nelson BV, Nall LE, Schardt JD (1991) Exotic aquatic plants in Florida: a historical perspective and review of the present aquatic plant regulation program. In: Center TD (ed) Proceedings of the symposium on exotic pest plants. National Park Service, United States Department of Interior, Washington, DC, pp 303–326
Schmitz DC, Schardt JD, Leslie AJ, Dray FA, Osbourne JA, Nelson BV (1993) The ecological impact and management history of three invasive alien aquatic plant species in Florida. In: McKnight (ed) Biological pollution-the control and impact of invasive exotic species. Indiana Academy of Science, Indianapolis, 261p
Shirber M (2005) Cities cover more of earth than realized. Live Science
Small JA (1954) Carex kobomugi at Island Beach, New Jersey. Ecology 35(2):289–291
Smith D, Brown T (eds) (1994) An assessment of invasive non-indigenous species in Florida's public lands. TSS-94-100, 10–28. Florida Department of Environmental Protection, Tallahassee
Snow RW (2006) Disposable pets, unwanted giants: Pythons in Everglades National Park. In: Proceedings from Florida Exotic Plant Council, 21st Annual Symposium. Gainesville
Snow RW, Brien ML, Cherkiss MS, Wilkins L, Mazzotti FJ (2007a) Dietary habits of the Burmese python, Python molurus bivittatus, in Everglades National Park, FL. Herpetol Bull 101:5–7
Snow RW, Krysko KL, Enge KM, Oberhofer L, Walker-Bradley A, Wilkins L (2007b) Introduced populations of Boa constrictor (Boidae) and Python molurus bivittatus (Pythonidae) in southern Florida. In: Henderson RW, Powell R (eds) The biology of Boas and Pythons. Eagle Mountain Publishing, Eagle Mountain
Snyder SA (1992) Casuarina spp. In: Fire effects information system [Online]. U.S. Department of Agriculture, Forest Service, Rocky Mountain Research Station, Fire Sciences Laboratory (Producer). http://www.fs.fed.us/database/feis/
Somma LA (2004) Bufo marinus species profile. USGS Nonindigenous Aquatic Species Database, Gainesville
Spencer D, Anderson L, Ksander G, Klaine S, Bailey F (1994) Vegetative propagule production and allocation of carbon and nitrogen by monoecious Hydrilla verticillata (L.f.) Royle grown at two photoperiods. Aquat Bot 48:121–132

Stocker R, Hupp KVS (2008) Fire and nonnative invasive plants in the southeast bioregion. USDA Forest Service Gen. Tech. Rep. RMRS-GTR-42-vol 6:91–112
Sutton DL, Van TK, Portier KM (1992) Growth of dioecious and monoecious Hydrilla from single tubers. J Aquat Plant Manag 30:15–20
Swearingen JM (1997) Australian pine. National Park Service, Plant Conservation Alliance, Alien Plant Working Group, Washington, DC
The Florida Legislature (2017) The 2017 Florida statutes. Title XXIX, public health. Chapter 403, EnvironmentalControl. http://www.leg.state.fl.us/Statutes/index.cfm?App_mode=Display_Statute&URL=0400-0499/0403/0403.html
Thiele J, Kollmann J, Andersen UR (2009) Ecological and socioeconomic correlates of plant invasions in Denmark: the utility of environmental assessment data. Ambio 38(2):89–94
Thomas WL (ed) (1956) Man's role in changing the face of the earth. University of Chicago Press, Chicago
Tipping PW, Martin MR, Nimmo KR, Pierce RM, Smart MD, White E, Madeira PT, Center TD (2009) Invasion of a West Everglades wetland by Melaleuca quinquenervia countered by classical biological control. Biol Control 48(1):73–78
Tomlinson PB (1980) The biology of trees native to tropical Florida. Harvard University Printing Office, Allston
Tschinkel W (2006) The fire ants. Harvard University/Belknap Press, Cambridge
United States Census Bureau (USCB) (2017) Current lists of metropolitan and Micropolitan statistical areas and definitions. United States Census Bureau, Washington, DC
Van TK, Steward KK (1990) Longevity of monoecious hydrilla propagules. J Aquat Plant Manag 28:74–76
Vinson SB, Sorensen AA (1986) Imported fire ants: life history and impact. Texas Department of Agriculture. College Station, Texas: Department of Entomology, Texas A&M University
Wheeler GS, Massey LM, Endries M (2001) The Brazilian peppertree drupe feeder Megastigmus transvaalensis (Hymenoptera: Torymidae): Florida distribution and impact. Biol Control 22:139–148
Wheeler GS, Taylor GS, Gaskin JF, Purcell MF (2011) Ecology and management of Sheoak (Casuarina spp.), an invader of coastal Florida, USA. J Coast Res 27(3):485–492
Whitfield PE, Gardner T, Vives SP, Gilligan MR, Coutenay WR Jr, Ray GC, Hare JA (2002) Biological invasion of the Indo-Pacific lionfish Pterois volitans along the Atlantic coast of North America. Mar Ecol Prog Ser 235:289–297
Willson JD, Dorcas ME, Snow RW (2011) Identifying plausible scenarios for the establishment of invasive Burmese pythons (Python molurus) in Southern Florida. Biol Invasions 13:1493–1504
Wolverton BC, McDonald RC (1979) Water hyacinth (Eichhornia crassipes) productivity and harvesting studies. Econ Bot 33:1–10
Woodall SL (1982) Seed dispersal in Melaleuca quinquenervia. Fla Sci 45:81–93
Woodward SL, Quinn JA (2011) Encyclopedia of invasive species: from Africanized honey Bees to Zebra mussels, volume 1. ABC-CLIO, 764p
Worldwatch Institute (2000) State of the World. Washington, DC

Chapter 2
Invasive Species in the Sundarbans Coastal Zone (Bangladesh) in Times of Climate Change: Chances and Threats

Shafi Noor Islam, Sandra Reinstädtler, and Albrecht Gnauck

Abstract The Sundarbans mangrove forests, wetlands and their native as well as invasive plant species are lying within the Bangladesh coastal region, which is gifted with vast natural resources, a delta, tidal flat, mangrove forests, marches, lagoons, bars, spilt, estuaries and coastal ecological environment. These habitats, biotopes and ecosystems also serve as habitat for especially four dominant tree species of the Sundarbans, the Sundri (*Heritiera fomes*), Gewa (*Excoecaria agallocha*), (*Ceriops decandra)* and (*Sonneratia apetala)*. But the existence of these and many more native species is endangered. The Sundarbans species are threatened by various natural and anthropogenic pressures including climate change. So the native species are approximately decreasing significantly by the year 2100 due to sea level rise (88 cm) in the Sundarbans area compared to the year 2001.There 23 invasive species, which belong to 18 families and 23 genera. These species are highly invasive, six species are moderately invasive and the remaining are potentially invasive. From the 23 invasive species only four are exotic or alien. The disturbances may arrest succession at any stage and contribute to the biological invasion of invasive plants. The within these biodiversity hotspots and vast natural or coastal water resources lying potential for communities survival and 36.8 million dependent people are some of the strongly pending managing demands next to existing Natural World Heritage Sites' already existing protective management support to be discussed in this chapter. As with the coastal natural resources drastically reduction due to unplanned use by community and stakeholders, also the Sundarbans mangrove forests, wetlands and their species are vastly affected through these developments and the most important observation for vulnerability aspects and maximum possible amplitudes is coming up, which has been investigated in within this research and chapter. Also the present situation stated that an integrated natural

S. N. Islam (✉)
Department of Geographical and Environmental Studies, Faculty of Arts and Social Sciences (FASS), Universiti Brunei Darussalam, Bandar Seri Begawan, Brunei Darussalam
e-mail: shafi.islam@ubd.edu.bn

S. Reinstädtler · A. Gnauck
Brandenburg University of Technology, Cottbus-Senftenberg, Germany

C. Makowski, C. W. Finkl (eds.), *Impacts of Invasive Species on Coastal Environments*, Coastal Research Library 29,
https://doi.org/10.1007/978-3-319-91382-7_2

resource management plan is necessary for the protection of the mangrove coastal ecosystem.

The chapter is prepared based on primary and secondary data sources. The objectives of this study are to analyze the present coastal mangrove plant species natural resources management status. The study seeks the deltaic Sundarbans Region with its natural world heritage site and mangrove forests, wetlands plant species development and management strategies for ensuring less vulnerability and a sustainable development of coastal mangrove resources in Ganges-Brahmaputra Rivers deltaic coastal floodplain region in Bangladesh.

Keywords Invasive plant species · Ecosystem services · Mangrove · Coastal wetland · Climate change · Sea level rise · Degradation and management Sundarbans · Bangladesh

2.1 Introduction

Biological invasions are now considered one of the main threats to the world's biodiversity (Mooney and Hobbs 2000). Bio-invasion is a process or phenomenon; it can mean aggressive introduction of invasive species into a new place, new environment, or within the same ecosystem with a different role (Islam and Gnauck 2009a, b). This role might be a negative one and that should affect the ecosystems adversely (IUCN 2003; Biswas et al. 2007). Focusing on the Sundarbans Region and its mangrove forests and wetlands with their plant species, the major water challenges will be described in its vulnerabilities especially for plant species in exemplary form (Gopal and Chauhan 2006). The connection to existing ecosystem services will help in understanding the topical specifications of vulnerability of mangrove and wetland plants and their focal areas for Sundarbans mangroves (Biswas et al. 2007).

The Sundarbans region inhabits the Sundarbans Natural World Heritage Site. It is lying within the Bangladesh coastal region of the Ganges-Brahmaputra Rivers Delta, which is gifted with vast natural resources, a delta, tidal flat, mangrove forests, marches, lagoons, bars, spilt, estuaries and coastal ecological environment. It is the largest floodplain wetland region worldwide and is located in the South Asian Region (comp. Morgen and Mclntire 1959; Khandoker 1987; Sarker et al. 2003; Goodbred and Kuehl 2000; Goodbred and Nicholls 2004; Islam 2016).

The inhabited, above named habitats, biotopes and ecosystems also serve potentially for community's survival: 36.8 million people are living within the coastal region of Bangladesh and being dependent on coastal water resources and their native (plant) species mixtures, for which the Sundarbans Natural World Heritage Site is giving some protective management support. Nevertheless the coastal natural resources are drastically reducing due to unplanned use by the community and stakeholders, although the Ganges-Brahmaputra-Meghna Rivers are carrying 6 million m^3/s water. So the Sundarbans mangrove forests, wetlands and their plant species are vastly affected through these developments: 'the most important observation for vulnerability aspects and maximum possible amplitudes is coming up, which has been investigated in within this research and chapter.

So it can be stated that the coastal zone of Bangladesh with its Sundarbans protective mangrove area is enormously important for the development and management of natural resources and a balanced (plant) species variety. The coastal natural resources are playing an important role to protect the coastal ecosystems with its unique Sundarbans mangrove forest and wetland species as well as the socio-economy. The present situation stated that an integrated natural resource management plan is necessary for the protection of the mangrove coastal ecosystem.

The chapter is prepared based on primary and secondary data sources. The objectives of this study are to analyze the present coastal mangrove plant species and their natural resources management status. The study seeks the deltaic Sundarbans region with its implemented natural world heritage site, mangrove forests and wetlands ecosystem development or management strategies for ensuring less vulnerability and a sustainable development of coastal mangrove species resources in Ganges-Brahmaputra Rivers deltaic coastal floodplain region in Bangladesh. Further on this chapter presents the potential impacts of climate change on the plant species, the potential as well as threats through invasive species, the coupled ecosystem services of Sundarbans and forest dependent livelihoods. Both primary data on forest dependent livelihoods and secondary information on climate change impacts were used for this assessment.

This chapter presents the impact of climate change on Sundarbans plant species and combined ecosystem services in Sundarbans Region and coastal landscapes in in the Ganges-Brahmaputra-Meghna delta region in Bangladesh.

2.2 Aim and Objectives

The coastal zone of Bangladesh with its Sundarbans protective mangrove area is enormously important for the development and management of natural resources. The coastal natural resources are playing an important role to protect the coastal ecosystems with its unique Sundarbans mangrove forests, wetlands and socio-economy. The present situation stated that an integrated natural resource management plan is necessary for the protection of the mangrove coastal ecosystem. The following specific objectives are given bellow:

- To understand the coastal ecological features in the Ganges-Brahmaputra-Meghna deltaic region in Bangladesh.
- To investigate the ecological changes in the Sundarbans deltaic region
- To investigate the status of invasive species in the Sundarbans mangrove forest wetland areas in Bangladesh
- To investigate the Chances and Threats due to climate change impacts in the Sundarbans region

- Make some applied recommendations for better management and conservation of Sundarbans mangrove species

2.3 Geographical Characteristics of the Case Area

Bangladesh is a land of water, but water is also one the most critical and major problem in the country. The deltaic coastal areas of Bangladesh are especially impacted and more vulnerable. The deltaic coastal areas are about 710 km in length with extending along the Bay of Bengal from the mouth of Teknaf River in the south East to the Mouth of Raimangal River in the south west and including the greater districts of Chittagong, Noakhali, Barisal, Patuakhali and Khulna (Nishat 1988, 1992). The coastal region is composed of the land and the sea including estuaries and islands adjacent to the land water interface of south east Bangladesh and the coast can be divided into three distinct regions (Pramanik 1983).

Most parts of the Bengal deltaic coastal areas of Bangladesh are an active or nascent delta with vigorous dynamism and the islands or chars are formed, eroded and reformed (Anwar 1988; Islam 2016). The on the outskirts of the coastal area located Swatch-of-No-Ground submarine canyon (SoNG)[1] being situated on the south of Ganges-Brahmaputra delta near to Dubla Char, is also a major feature only 24 km far from the coast water and from the Sundarbans mangrove forest (Rahman 1992) being interesting for understanding the geomorphological facts within the entire region. The Sundarbans is located in the Ganges-Brahmaputra Deltaic region (Goodbred and Kuehl 2000). The Sundarbans is the world's largest continuous single block of mangrove forest is located on the Ganges-Brahmaputra-Meghna rivers deltaic coastal floodplain region (Ali 1988; Das and Siddiqi 1985; Iffekhar 1999; Rahman 1995, 2003). It lies between 89° 30′ north at the south west corner of Bangladesh extending over 6,000,386 ha of which 189,159 ha is water (Ali 1999; Rahman 1998; Runkel and Ahmed 1997). Its rich mixture of flora, fauna and complex ecosystem function makes it a unique ecosystem in the world (Biswas et al. 2007). The Sundarbans forest is intersected by a complex network of rivers, streams and water bodies (Anon 1998; Iftekher and Islam 2004).

[1]The first marine protected area is placed over here at SoNG in Bangladesh, which is also called Ganga Trough.

2.4 Present Status of the Bengal Coastal Environment and Native Plant Species in the Sundarbans Coastal Region

Coastal wetland environment and ecosystem services play a very important role in the socio-economic development of the Bengal Deltaic coastal region of Bangladesh (Islam 2016). About 25% (35 million people) of population live in the deltaic coastal areas and most families are dependent on coastal natural resources for their livelihood (Hidayati 2000). So a coastal ecosystem such as a mangrove forest or wetland also supply these socioeconomic benefitting ecosystem services for the human population with including building material, fisheries, forage, fuel, timber, and the protection of commercial, recreational, and naval vessels (Wolanski et al. 2009). Coastal wetland environment and ecosystem services play a very important role in socio-economic development of the Bengal Deltaic coastal region of Bangladesh (Islam 2016) and emphasize the tremendous importance of the established Sundarbans Natural World Heritage Site in a part of this coastal region as balancing spatial area. About 25% (35 million people) of population live in the deltaic coastal areas and most families are dependent on coastal natural resources for their livelihood (Hidayati 2000).

In this constellation the coastal mangrove forests and wetlands provide numerous important ecological services to humanity: the tropical and sub-tropical coastal mangrove wetlands provide important ecosystem services to the population living in the hinterland such as by sheltering it from storm winds, capturing salt spray and improving crop production in arid coastal areas (Wolanski et al. 2009). Also the fact of proceeding higher flora and fauna species appearance in variety and population density is giving another multiplying factor and ecosystem service for mangrove forests and wetlands: it is at the coast where the heavily sediment-laden river water with very little salinity meets the saline sea water. This mixing creates unique ecosystems with unique flora and fauna species appearance that exhibit their presence here in unique habitat forms such as these mangroves and enhance biodiversity. Amongst these mangroves the most important once are the here to be described Sundarbans with its unique tropical mangrove forest and the wild life, the fisheries dominated by economically important species such as Hilsa and Prawns (Rahman 1992). Mangrove forests and wetlands also safeguard the coast against erosion and guard against loss of capital infrastructure and human lives (Wolanski 2007; Wolanski et al. 2009).

2.4.1 Native Plant Species in the Sundarbans Deltaic Region

Mangroves are the unique ecosystem that provides a wide range of ecosystem services. The world's largest single tract mangrove forest located in Bangladesh – the Sundarbans provides a wide range of ecosystem services and contributes to socio-economic development of the neighboring communities and the country (Uddin et al. 2013). Timber, fisheries and other non-timber forest products (NTFP) are the main products of the forest (Uddin et al. 2013). Also, the Sundarbans serves as coastal defense and reduces winds and storm surges, coastal flooding and coastal erosion (Rahman 2009; SCBD 2009, as cited in Uddin et al. 2013). Over 3.5 millions of people living around the Sundarbans are directly or indirectly dependent on ecosystem services (Giri et al. 2007; Biswas et al. 2007, as cited in Uddin et al. 2013). In addition, the forest has regional and global importance for its ecological resources. UNESCO has declared Sundarbans as "The World Heritage Site" in 1997 (Islam 2003 as cited in Uddin et al. 2013).

The Sundarbans mangrove forest of Bangladesh provides ecosystem services having great importance for local livelihoods, national economy and global environment (Uddin et al. 2013). Nevertheless, the Sundarbans is threatened by various natural and anthropogenic pressures including climate change.

Recent study revealed that the suitable area of two dominant tree species of the Sundarbans – Sundri (*Heritiera fomes*) and Gewa (*Excoecaria agallocha*) may be decreased significantly by the year 2100 due to sea level rise (88 cm) in the Sundarbans compared to the year 2001, which may be reduce the timber stock of those trees (Uddin et al. 2013). This indicates the potential loss of economic value of the key provisioning services of Sundarbans. Similarly, the other ecosystem services (e.g. fisheries, tourism, biodiversity, carbon sequestration, etc.) may be affected by climate change (Uddin et al. 2013). Consequently, the forest dependent livelihoods would be affected by the degraded ecosystem services of the forest. Further studies should quantify the impacts of climate change on all the ecosystem services and explore the potential loss and opportunities in future (Uddin et al. 2013). A new paradigm of management should look forward considering climate change, ecological integrity, sustainable harvesting and ensuring continuity of the ecosystem services of the Sundarbans (Uddin et al. 2013).

2.5 Invasive Plant Species in the Sundarbans Coastal Region

The Sundarbans mangrove forests, wetlands and their native as well as invasive plant species are lying within the Bangladesh coastal region, which is gifted with vast natural resources, a delta, tidal flat, mangrove forests, marches, lagoons, bars, spilt, estuaries and coastal ecological environment (Ali 1999; Anon 1998; Islam 2008; Islam and Gnauck 2008, 2009a; Islam et al. 2017). These habitats, biotopes and ecosystems also serve as habitat for especially four dominant tree species of the Sundarbans, the Sundri (*Heritiera fomes*), Gewa (*Excoecaria agallocha*), (*Ceriops decandra)* and (*Sonneratia apetala)*. But the existence of these and many more native species is endangered (Islam and Gnauck 2009b). The Sundarbans species are threatened by various natural and anthropogenic pressures including climate change.

Table 2.1 Recorded invasive plants from Sundarbans mangrove forests

Species	Family	LF[a], Status[b]
Potentially invasive plant species PI		
Arundo donax L.	Gramineae	G PI
Clerodendrum inerme (L.)	Gaertn Verbenaceae	S PI
Cryptocoryne ciliata (Roxb) Fischer ex Waylder	Araeceae	S PI
Dendropthoe falcata (L.f.)	Etting Loranthaceae	E PI
Flagillaria indica L.	Flagillariaceae	C PI
Hibiscus tilliaceus L	Malvaceae	S PI
Hoya parasitica (Roxb.) wall. ex Wight	Asclepiadaceae	E PI
Imperata cylindrica (L.)	Raeuschel Gramineae	G PI
Ipoemea fistulosa Mart. Ex Choisy	Convolvulaceae	S PI
Pongamia pinnata (L.)	Pierre Leguminosae	T PI
Saccahrum spontaneum L.	Poaceae	G PI
Salacia prinoides DC	Celastraceae	T PI
Sarcolobus globosus wall	Asclepiadaceae	C PI
Typha angustata Borry f	Typhaceae	H PI
Invasive plant species I		
Acrosticum aureum L.	Polypodiaceae	F I
Entada rheedii Spreng	Leguminosae	C I
Excoecaria indica (Wild.) Muell.-Arg.	Euphorbiaceae	T I
Micania scandens Willd	Compositae	C I
Syzygium fruticosum (Roxb.) DC	Myrtaceae	T I
Tamarix indica L.	Tamaricaceae	T I
Highly invasive plant species HI		
Derris trifoliata Lour	Leguminosae	C HI
Eichhornia crassipes	Pontedteriaceae	H HI
Eupatorium odoratum L.	Compositae	C HI

[a]Life forms: *T* tree, *S* shrub, *H* herb, *C* climbers, *AS* aquatic shrub, *E* epiphyte, *F* fern, *G* grass
[b]Status: *HI* highly invasive, *I* invasive, *PI* potentially invasive

So the native species are approximately decreasing significantly by the year 2100 due to sea level rise (88 cm) in the Sundarbans area compared to the year 2001. Biswas found 23 invasive species, which belong to 18 families and 23 genera (Biswas et al. 2007). These species are highly invasive, six species are moderately invasive and the remaining are potentially invasive. From the 23 invasive species only four are exotic or alien (Biswas et al. 2007). The disturbances may arrest succession at any stage and contribute to the biological invasion of invasive plants. The table (Table 2.1) displaying the status of the invasive plant species in the Sundarbans region in Bangladesh.

The result of the case is showing that gradually the native plant species are getting dangerous ecological environment in the Ganges-Brahmaputra-Meghna region as well as Sundarbans mangrove forest areas. The reasons are indicated that climate change impact and anthropogenic influences are the main causes of secondary plant species development and reducing the native plant species (Table 2.1).

2.6 Climate Change Impacts

Global warming is accelerating and cumulating sea level rise more than expected – even with less warming after IPCC (2007) by the atmospheric greenhouse effect the past 15 years (with correlating to the results of IPCC-report from year of 2007). So the mainly human induced climate change (IPCC 2007, also comp. IPCC 2014) and warming is progressing continuously and ecosystems, landscapes and its land use predestinations may change or land will be lost in total as already predicted for small island states within the next two decades (Kokkonen 2006, all as cited in Reinstädtler 2013 p 48) In 2012, a first village from the Pacific Islands of Fiji, called Vunidogoloa, had to be displaced and relocated already due to SLR to higher land in order to safe human habitat, life existence, infrastructure and cultural identity of people in Fiji (Fig. 2.1).

Climate instability and uncertainties directly correlate to more unstable climate and in the long term affecting changing weather conditions and weather phenomena less correlating to the season's disposition. Continuing and gradual climate change deals with the long-term changes in climate and so with the time scale in quantity and quality of effects by global warming, global average sea level rise, risk of coastal erosion, precipitation quantity and weather phenomena on the worlds systems (Reinstädtler 2013 p 49). These risks of for instance average seal level rise or coastal erosion and their greater impacts (comp. IPCC 2012; Islam 2016; Islam et al. 2017) can be estimated as happening in tremendous shorter sequenced time frames (comp. IPCC 2012; Islam 2016; Islam et al. 2017). Further rapid or abrupt climate changes are disturbing Sundarbans ecosystems and inhabited habitats for native flora and fauna species. Rapid or abrupt climate changes are characterised as being nonlinear changes in climate, which are per term going back to the theory of crossing a

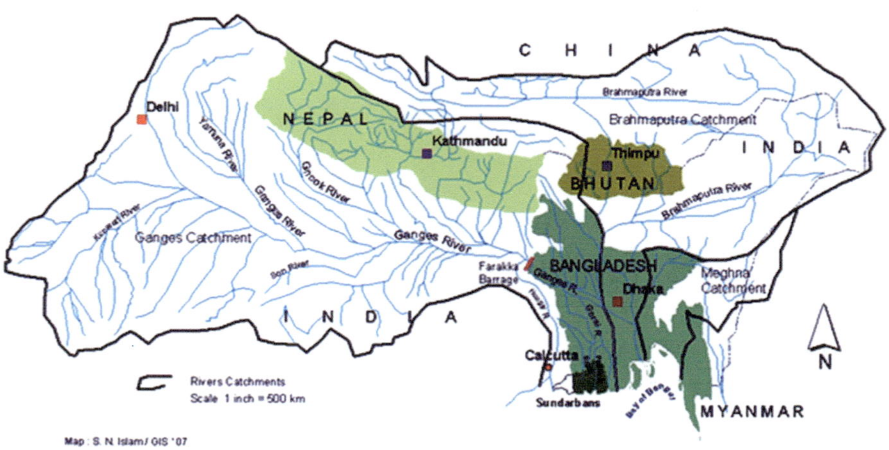

Fig. 2.1 The Geographical location of Bangladesh and Sundarbans Mangrove Forest Wetland in the Ganges-Brahmaputra-Meghna Rivers Catchment area in Bangladesh. (Map source: Islam S. N 2008)

threshold and switching to a new state on global or regional scale (IPCC 2007, 2012, 2014). Melting of glaciers (from Himalayan mountains and floating into the Ganges and GBM-delta up to the Sundarbans) and instability of monsoon rainfalls (*in Bangladesh and its regions such as Sundarbans*) or *meteorological phenomena with hurricanes, tornadoes, heat-waves, lightning, storms or fire* are globally scaled, climate related, contemporaneous phenomena (comp. Reinstädtler 2013 p 49–50; WMO 2008), but here in Sundarbans coastal zone of Bangladesh already highly disturbing facts. Also extraordinary hazards are more frequent phenomena (Reinstädtler 2013 p 49–50; comp. WMO 2008): influences of increased or decreased water yield dependent floods (e.g. flash-floods) and droughts or resulting wildfires, wind-anomalia like tornados, wind and water-anomalia like tropical cyclones (synonymous terms: hurricanes, tropical storms or depressions, typhoons) and tsunamis. Some of the most common hazards are further on: geological with volcanoes, earthquakes, mass movement (falls, slides, slumps); climate change with increased storm frequency and severity, glacial lake outburst floods (comp. IPCC 2012; Islam 2016; Islam et al. 2017).

For geographical locations as well as for the character of World Natural Heritage Sites such as Sundarbans with its remarkable and outstanding, native flora it means, to calculate also the considerably varying potential threat posed by natural disasters (see above, e.g. storms, fire, floods etc., comp. in general Reinstädtler 2013 p 50). In correlation to this worlds' largest mangrove forest of the World Heritage Natural Site of Sundarbans Region the natural disasters such as extraordinary floods or salinity intrusion as one of the impacts by flooding in this riverine and delta region have to be named (Islam and Gnauck 2008, 2009a; UNESCO 2008, all as cited in Reinstädtler 2013 p 50).

Summarising Climate Change in (and unexceptionally outside) the protected perimeter of World Heritage Sites (WHS) such as Sundarbans and within the background of "Outstanding Universal Value" (OUV), cultural and natural authenticity and/or integrity, it means to have potential for being affected most harmfully (comp Colette and UNESCO 2008; as cited in Reinstädtler 2013 p 50).

2.7 Global Warming and Future Projections

Climate change and sea level rise, induced by global warming, also compromise the ecological stability of the coastal zone; in sum, due to various natural and anthropogenic factors, the natural resource base of the zone is declining. Failing ecosystem productivity further degrades the coastal deltaic ecology, quality of life of the local communities (Dasgupta 2001). Relative Sea-Level Rise (SLR) movement has an immediate and direct effect on the coastal inter tidal ecosystems, particularly on vegetation. Arise of relative SLR decreases the influences of terrestrial processes and increases the influence of coastal marine processes (Islam 2001). The world great deltas are among the most densely populated and most vulnerable of coastal areas are threatened by sea level rise (Broadus 1993). Global warming and sea level rise and vulnerability of coastal wetland ecosystems are factors that have to be considered the long term management strategy for dealing with the coastal mangrove wetland issue. The impacts of climate change in any given region depend on the specific climatic changes that occur in that region. Local changes can differ substantially from the globally averaged climate change (Harvey 2000). The global warming and climate change a predicted sea level rise, accelerated by global warming will cause a further 'Squeezing' of the natural tidal land. In Bangladesh case it has been projected by IPCC (2007) and MoEF that 3 mm/year sea level rise which will occurs before 2030 and 2500 km^2 land (2%) will be inundated. About 20% of the net cultivable area of Bangladesh is located in the coastal and offshore island (Fig. 2.2) which is under threatened due to the above mentioned causes.

A very recent study on the Ganges deltaic coastal area in Bangladesh by IPCC reports (2007, 2012, 2014a) shows that the mean tidal level at Hiron Point is showing an increase of 4.0 mm/year which is higher than the global rate. Soils in this area are affected by different degrees of salinity (Rahman 1988). About 203,000 ha very slightly, 492,000 ha slightly, 461,000 hectare moderately and 490,200 ha strongly salt affected soils are assessed in south western part of the coastal area (Fig. 2.2). The climate change impact issue is a new threat for the coastal area of Bangladesh (Fig. 2.2). In the Sundarbans case, sea level rise would result in saline water moving further into the delta which would be the major threat for mangrove and coastal wetland ecosystems (IECO 1980). The fate of the Sundarbans with different sea level rise the potential impacts on environment if 10 cm SLR will inundate 15% of the Sundarbans and with 1.5 m SLR about 17 million (15%) of the population will be affected and will have to be displaced or homeless, whereas 22,000 km^2 (16%) land will permanently be inundated (Fig. 2.2). The figure 6 shows that with 3 m SLR would be worse scenario for Bangladesh when almost one third of land could be inundated by saline water. The reduction rate of mangrove areas will from 50% to 75% which would be more harmful for coastal ecosystems in the estuaries (IPCC 2007, 2012, 2014a). Besides, other environmental problems will arise in the coastal belt such as; water pollution and scarcity, soil degradation, deforestation, solid and hazardous wastes, loss of bio-diversity estuary landscape damage and river bank erosion which will create a lot of new challenging problems for human livelihood in the coastal region (Akter et al. 2010).

In such situation it will further create an unstable agricultural crop production, damaging fisheries and livestock and food security in the coastal riverine islands in Bangladesh. Therefore the climate change impacts on coastal region such as on the estuaries in the Sundarbans region will create new threats for estuaries ecosystem and its services (Rahman and Ahsan 2001; Sarker 2008). On the other hand the whole coastal regions of the Bengal coast is losing its cropping fertility as the whole coastal areas are concentrated by high salinity (NaCl) in water and soil. Which is making the agricultural land more unproductive and it is a huge threat for coastal agricultural crops production and as well as threat for coastal ecosystems and food security.

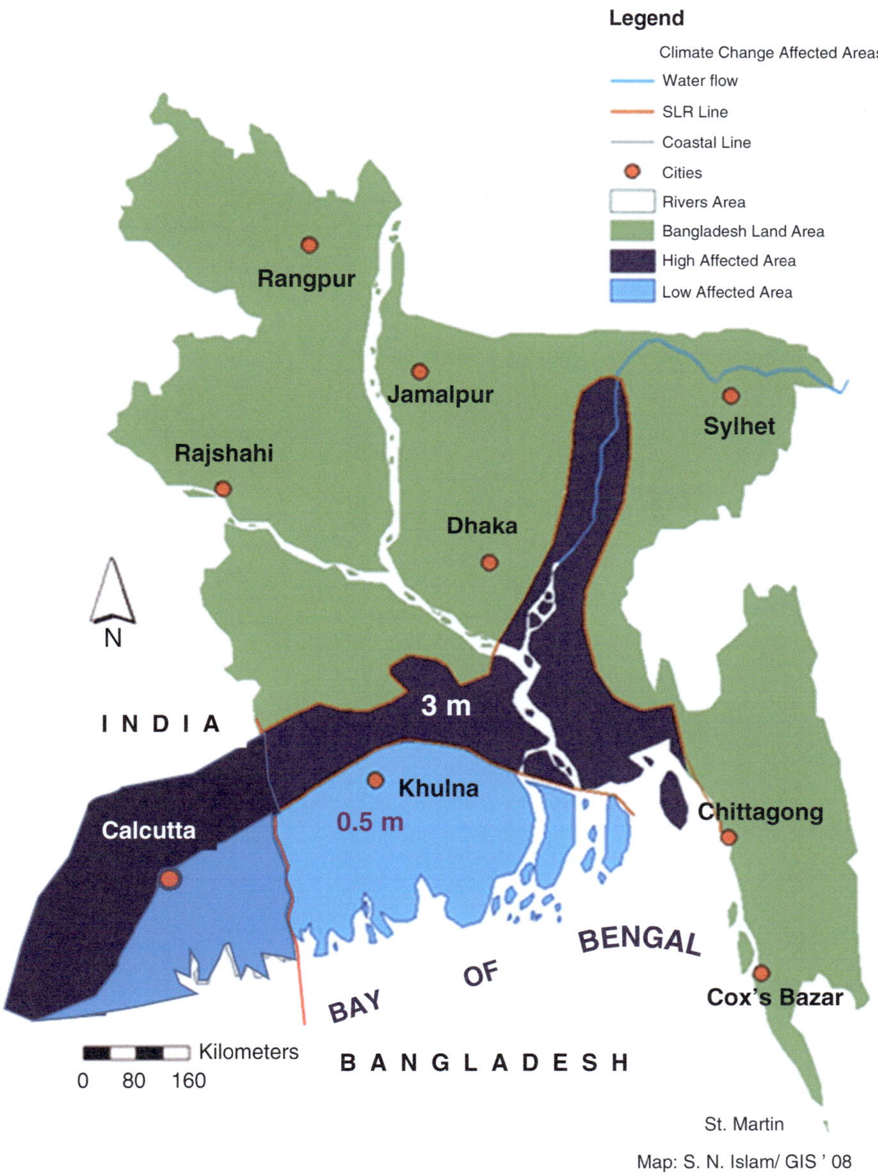

Fig. 2.2 Climate change impacts in the coastal region of Bangladesh. (Source: after Akter et al. 2010)

2.8 Threats for Native Plant Species in the Sundarbans Coastal Wetlands

Currently, Sundarbans is threatened by both anthropogenic and natural factors. The natural threats includes climate change induced sea level rise (Uddin et al. 2013). It has been anticipated that the most bio diverse areas in the Sundarbans will be reduced from 60% to 30% in the year 2100 with 88 cm sea level rise compared to the status in 2001, which would ultimately reduce the production of the forest products and dependent livelihoods (CEGIS 2006, as cited in Uddin et al. 2013).

With naming some of the water challenges combined to Ganges-Brahmaputra-Megna rivers delta (GBM), in the first row the scarcity of upstream fresh water has created serious environmental problems in the coastal region of Bangladesh (Nishat 1988). This strongly needed upstream fresh water supply to the coastal region is playing a potential role to make a balance within the complex interplay of varied existing ecosystems (Biswas et al. 2007). Also the oceanic saltwater and tidal pressures or storm winds are giving challenges to the coastal area. Due to climate change impacts the soil and coastal water quality is degrading gradually and as a result the native plant species density is reducing in different part of the Sundarbans region. The native dominating plant species *Heritera fomes* was domination 34% in the Sundarbans forest area and at present it is representing only 21%. The mangrove vegetation cover is changing and invasive plants are dominating the area, which is threat for the native plant species in site the mangrove forest areas. The coastal land area is changing due to high salinity intrusion in soil and water (Rahman and Ahsan 2001). The agricultural land areas and mangrove forest areas are using by shrimp cultivation. Therefore new fish species and agricultural crops species are introducing in the coastal areas in the Ganges-Brahmaputra –Meghnah deltaic region in Bangladesh.

2.9 Time of Climate Change – Chances and Threats

The delta front sand bodies with different shapes and sizes are now unable to hold the surge waters of advancing storm waves that can transport the bulk of sediment from seaward face to inner of the delta plains (Farley et al. 2010). The sediment lobes submerges the forests, wetlands surface and tidal creeks at a steady rate with increased frequency of storm events in the deltaic coast at present (Paul et al. 2017). Mangroves occur in the waterlogged, salty soils of sheltered tropical and subtropical shores. They are subject to the twice-daily ebb and flow of tides, fortnightly spring and neap tides, and seasonal weather fluctuations. They stretch from the intertidal zone up to the high-tide mark (GOB 2001). These forests are comprised of 12 genera comprising about 60 species of salt-tolerant trees. With their distinctive nest of stilt and prop-like roots, mangroves can thrive in areas of soft, waterlogged, and oxygen-poor soil by using aerial and even horizontal roots to gain a foothold (Islam et al. 2017). The roots also absorb oxygen from the air, while the tree's leaves can excrete excess salt. Most species typically have relatively widespread distributions; low diversity floras but overall alpha diversity very high when terrestrial and aquatic species are considered; very low beta diversity and low ecoregion endemism; some highly localized species exist; strong zonation along gradients; several distinct mangrove habitat formations. Alterations of hydrography and substrate have considerable impact, but restoration potential is high; mangroves are susceptible to pollution, particularly oil and other petroleum compounds; alteration of salinity levels can have dramatic impacts on mangroves. In Sundarbans a large number of mangrove seeds is regularly drifted into the unfavourable sandy substrate after the events of storms, tidal waves, southwest monsoon brace and HAT (Highest Astronomical Tides) phase currents along the shores of the Bay of Bengal which is creating another type of threat for mangrove biodiversity (Islam 2016; Islam et al. 2017).

2.9.1 Water and Soil Salinity in the Wetlands

The wetland areas in the coastal region of Bangladesh are affecting due to high rate of water and soil salinity intrusion. In general, the annual pattern of salinity changes in the Sundarbans region is also related with the changes of freshwater flow from upstream rivers. The adverse effects of increased salinity on the ecosystem of the Sundarbans mangrove wetlands are manifested in the drying of tops of Sundari (*Heritiera fomes*) trees retrogression of forest types, slow forest growth, and reduced productivity of forest sites (IWM 2003). The peak salinity was found to be about 56,186 dS/m in 2001 and 2002 and minimum salinity during post monsoon was found to be about 10,805 dS/m (IWM 2003). This salinity rate has crossed the salinity threshold value for the mangrove wetlands ecosystems in the coastal region. Some mangrove species have been dried and displaced due to high salinity penetration and intrusion in the coastal area in Bangladesh.

Salinity in the southern part of the Bay remains less than 10,805 dS/m during monsoon and starts to increase at a steady rate up to about 32,415 dS/m during the dry season (IWM 2003). Salinity in the western part in not reduced to low salinity range even during monsoon period; salinity increases at a steady rate during the dry periods. Almost 265 km^2 Heritiera fomes type of forest moderately and 210 km^2 areas are severely affected, which is one of the main threats for a sustainable mangrove forest management and its ecosystems (Islam 2008; Islam and Gnauck 2008, 2009a, b) (Fig. 2.3).

The highest soil salinity levels measured were ECs 41.2 dS/m at Nilkamal, ECs 40 dS/m at Mirgang and third highest rate of soil salinity is ECs 24 dS/m at Munchiganj point in the northwestern Sundarbans mangrove coastal wetlands region (Fig. 2.3).

The increasing salinity levels are major threats for both biotic and abiotic factors of mangrove wetland ecosystems in the region (Islam and Gnauck 2009a, b). The

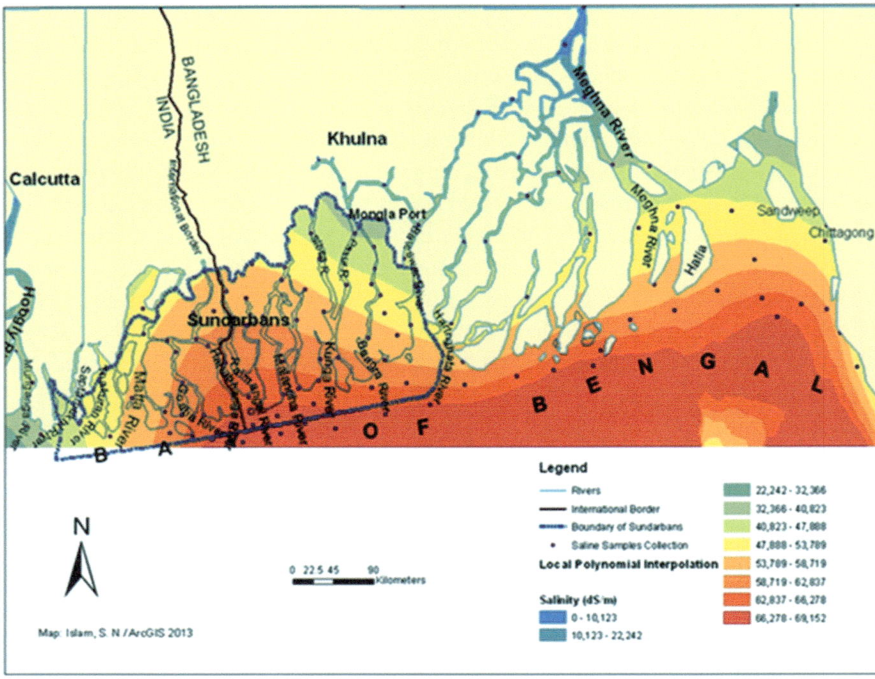

Fig. 2.3 Salinity intrusion pattern in the Sundarbans mangrove deltaic region of Bengal coast. (Source: Author 2015)

severity of salinity problem in the coastal wetlands increases with the desiccation of the soil (Fig. 2.3).

The coastal region covers an area about 29,000 km^2 or about 20% of the country. The coastal areas of Bangladesh cover more than 30% of the cultivable lands of the country. About 53% of the coastal areas are affected by salinity. Salinity causes unfavorable environment and hydrological situation that restrict the normal crop production throughout the year (Chowdhury and Haque 1990). The factors which contribute significantly to the development of saline soil are, tidal flooding during wet season, direct inundation by saline water and upward or lateral movement of saline groundwater during dry season (Rahman and Ahsan 2001). Beside salinity in ground and surface water is a major concern for many coastal urban towns (Mohiuddin 2005). A study of ESCAP (Economic and Social Council for Asia and the Pacific) in 1988 and GOB (Government of Bangladesh) has referred six sets of constraints of the development of a strategy for the coastal resource management in Bangladesh (ESCAP 1988). The constraints are as follows:

Policy making, planning for coastal resources, integrated resource management, coastal wetlands and marine resources sustainability, local environmental ecological perspective and lack of knowledge on coastal environment and its understanding (Jalal 1988; Akter et al. 2010). The National Water Policy (NWPo) of Bangladesh (1999) also gives due importance on research and development of knowledge and capacity building for sustainable management (Chowdhury 2009). In article 3 of the

Fig. 2.4 The mangrove vegetation and biodiversity in the coastal wetland areas in Sundarbans region in the Bengal Coast. (Source: Islam, et al. 2016)

NWPo, the objective is to develop a state of knowledge and capability that will enable the country to design water resources management plans by itself with economic efficiency, gender equity, social justice and environmental awareness.

The NWPo of Bangladesh (1999) is a guideline framework for the nation (Chowdhury 2009; Akter et al. 2010). As the quality of coastal water resources are dependent on the supply of upstream fresh water supply and availability in the coastal region, Sea level rise and tidal inundation factors are also potential issues, therefore some coastal issues such as coastal urban drinking water issue should be incorporated in the NWPo in Bangladesh (Akter et al. 2010).

2.9.2 Degradation of Mangrove Ecosystem and Biodiversity

The salinity investigation results show that the south west Bengal coastal regions and in these the Sundarbans Natural World Heritage Site are carrying the highest rate of water salinity which is unbalancing the coastal ecosystem and general ecology. According to salinity approximation this high rate is harmful to rural and urban biodiversity as well as for the species of humans and their urban drinking water (Islam et al. 2017). The Fig. 2.4 demonstrates the water salinity intrusion trends in the south and south west region of the Ganges deltaic region, which includes the entire Sundarbans mangroves (Islam et al. 2017).

Four major cities and 136 small towns are located in the coastal region in Bangladesh and a major portion of the inhabitants are dependent on mangrove resources in the coastal region. Most of the towns are effected through salinity intrusion and sea level rise impacts in the region as Sundarbans mangroves are also affected due to high salinity intrusion. Therefore, the investigation results of salinity modelling in the South and South west coastal deltaic regions are under threat for ecosystems and coastal rural and urban ecosystem goods and services (Islam 2014). Also several observed districts face the greater challenge of higher salinization intrusion by comparing the data of an intense of affected land in hectares from 1973 and 2000 (Islam et al. 2017).

Especially the coastal mangrove and agro-biodiversity loss is a common scenario in the Ganges-Brahmaputra-Meghna Rivers deltaic region between Bangladesh and India. The Fig. 2.3 demonstrates the scenarios of the coastal mangrove forest and wetland region in the Sundarbans and its Sundarbans Natural World Heritage Site in Bangladesh. The quality of mangrove forest and wetland water and soil are rapidly degrading due to high saline water intrusion and anthropogenic influences.

The study also found that the mangrove reduction rate is about 45% in both countries (Bangladesh – India). Deforestation is raising and land-cover is changing due to shrimp farming, salt farming, agricultural land extension, urbanization extension and settlement development (Rahman and Haque 2003). These development processes are adversely, affect coastal fish production and lead to a loss of agrobiodiversity and coastal floodplain biodiversity and of livelihood, which means to negatively influence 3.5 million people, who are dependent on natural resources in the coastal region in Bangladesh (Anon 1995). The mangrove wetland ecosystems are dependent on the water and soil salinity (Brown 1997). Almost all of the mangrove forest need freshwater supply from the upstream. In the Sundarbans Ganges deltaic coastal region the two potential rivers such as the Passur-Mongla and Chunar-Munchigannj are carrying the high rate of salinity intrusion. The Fig. 2.4 shows the high salinity intrusion trends in the coastal mangrove forest and wetland region (Giri et al. 2007). The salinity model also demonstrates that the salinity trends are much higher in the south western region of the Sundarbans mangrove wetlands regions. The salinity rate was 42,000 dS/m in 2003, whereas in 2010 the salinity rate is 53,000 dS/m in the Passur-Mongla river point. On the other hand, the south western corner is showing the highest rate of water salinity in 2010, which is over 53,000 dS/m (Fig. 2.3) (Islam and Gnauck 2009a; Islam et al. 2017).

The salinity penetration in the upstream areas of the coastal zone is one of the main obstacles to maintenance of water quality for drinking, irrigation and fisheries purposes (Islam and Gnauck 2009a) as well as for the mangrove ecosystem and biodiversity in general. And already the coastal mangrove and wetland ecosystems have been recognized as a driving force for biodiversity conservation and coastal urban socio-economic improvement (Nishat 1993; Sarkar 1993; Ahmed et al. 2008; Islam et al. 2017).

Sixty six different species of mangroves (comp. Fig 2.4) are growing in Sundarbans where 70 species have been recorded in the world, 12 species of plants and animals already vanished and Javan rhinoceros, Single horn rhinoceros, Water buffalo, Swamp deer, Mugger crocodile, Gaur and Hog deer are extinct animals.

Also in the Ganges-Brahmaputra rivers deltaic floodplain alone approximately 2.1 million ha of wetlands have been lost due to flood control, drainage, and irrigation development (Khan et al. 1994). Therefore, coastal urban wetlands biodiversity is facing serious challenges from salinity intrusion, environmental changes and anthropogenic impacts (Sarkar 1993; Sarker et al. 2003; Nair 2004; Ahmed et al. 2008) (Fig. 2.4).

The mangrove forest and wetlands in the coastal region include rivers, estuaries, mangrove swamps, marsh (haor), oxbow lakes (baor) and beels, water storage reservoirs, fish ponds, and some other lands are also facing the similar environmental problems (Khan 1993; Hughes et al. 1994; Gopal and Wetzel 1995; Islam and Gnauck 2008; Islam et al. 2017). In such problematic complicated situation the smart-use of natural resources and salinity desalination processes can solve the coastal floodplains biodiversity and ecosystem problems in the Ganges-Brahmaputra coastal surface areas in Bangladesh.

2.10 Future Management Plan

With summarizing the overall research emphasize, the question for sustainable development is a key concept in strongest driving force of the water sector (comp. Grigg 1996; Gleick 1998; Costanza et al. 1997; Islam 2007) and its specific problems and vulnerabilities in Bangladesh and especially in the Sundarbans Natural World Heritage Site. The viewpoint of peaks of vulnerability causes, pressures, states, impacts for mangrove forests and wetlands should guide into possible to be found solutions in decision-making for the specific and unique Sundarbans Natural World Heritage Site (Islam 2003).

So for creating a balance in the variety of existing plant species, their overall biodiversity and their ecosystems as being their habitats, creating a less vulnerable basis should stand within future managing plans. The present situation stated that an integrated natural resource management plan is necessary for the protection of the mangrove coastal ecosystem. Therefore a sustainable deltaic coastal wetland ecosystem management is essential for the coastal balanced species survival as well as implemented ecosystem services benefits in Bangladesh (comp. Pramanik 1983; Rahman 1992; Costanza et al. 1997; Islam 2007; Islam 2016). Especially within the multivariate existing ecosystem services being dependent on the - to be protected variety in existing, partially even endemic plant species – the Sundarbans mangrove forests, wetlands and the protected area of Sundarbans Natural World Heritage Site have to get a model region in sorts of treatments and management plans for mainly climate induced invasions of plant species.

2.11 Conclusion

Necessities in times of climate change have to be acknowledged in case of action fields for precaution and warning systems in general and multifarious impacts are combined with the global to local spatial and time scale (Reinstädtler 2013 p 50). The impact significance and the necessity for adaptive planning measures is high: as climate change affects nearly all aspects of physical, social, ecological, economic and cultural environment (Reinstädtler 2013 p 50), the strong need for adaptation as well as mitigation planning can be ascertained also in case of native flora species and balancing the invasive plant implementation. But the challenges in nature protection for species survival are that high due to insecurity reasons of exact predictability of changes of climatically stable areas, it is strongly recommended to implement into planning instances for species protection the intrusive character of invasive species. The percentages of protective plans and their recommendations for implementation of invasive species depends on natural status quo stability, the frequency of already strongly impacting climate change related hazards (and strongly needed adaptation measures in kind of ecological engineering measures) or already adjusted slow climatic and depending vegetation change syndromes (which symbolize climatic zone shifts) and the facts of existing, but strongly endangered flora as well as from each other depending fauna species.

The specific important recommendations are recommended for better development and management of the Sundarbans biodiversity and native plant species.

- To ensure the Sundarbans mangrove native and dominant plant species, and agro plant species that could ensure the food security to the coastal community in the entire Sundarbans wetlands region in Bangladesh.
- The political commitments and wills for the better management and conservation of mangrove wetland plant species and biodiversity are necessary and essential for Bangladesh context for better management in consideration of global warming and climate change impacts.
- Mangrove forest landscapes changing pattern and shrimp farming at the coastal mangrove wetlands areas should be stopped and initiatives are needed to protect the destruction of mangrove wetlands and forest biodiversity and native plant species in the south and southern region of Bangladesh.
- The destruction of sensitive mangrove ecosystem and coastal habitats causes threats to aquatic biodiversity. Significant socio-economic costs must be balanced against the direct economic benefits from the shrimp cultivation and salt beds development and anthropogenic influences on native plant species in the Sundarbans region.
- An integrated mangrove wetland ecosystems management plan and policy guideline should be developed based on the findings of this study. The findings and recommendations could be implemented for the future development and the

protection of mangrove forest biodiversity and its ecosystem services in the coastal region of Bangladesh, which could ensure the native agriculture crops and ensure the protection of native plant species in the coastal region, that could ensure livelihoods of the coastal dwellers in the Ganges-Brahmaputra mangrove deltaic region in Bangladesh.

References

Ahmed I, Deaton BJ, Sarker R, Virani T (2008) Wetland ownership and management in a common property resource setting: a case study of Hakaluki Haor in Bangladesh. Ecol Econ 68 (1–2):429–436

Akter J (2010) River system and water resources management in the Ganges Deltaic region in Bangladesh, unpublished M.Sc. thesis. Institute of Water Resources, Bangladesh University of Engineering and Technology (BUET), Dhaka

Akter J, Islam SN, Gnauck A (2010) Water resources management in the coastal region of Bangladesh. In: Modellierung und simulation von Okosystemen. Shaker-Verlag, Aachen, pp 167–185

Ali SS (1988) Sundarbans: its resources and ecosystem. In: Rahman MA, Shah MS, Murtaza GM, Matin MA (eds) Proceedings of the national seminar on integrated management of Ganges flood plains and Sundarbans ecosystem, Khulna University, Khulna, Bangladesh, 16–18 July, 1994. pp 38–49

Ali A (1999) Climate change impacts and adaptation assessment in Bangladesh. Climate Res 12:109–116

Anon (1995) Integrated resource management plan of the Sundarbans reserved forest vol.1. Draft final report of FAO/UNDP project BGD/84/056- integrated resource development of the Sundarbans resource forest, Khulna, Bangladesh, pp 1–385

Anon (1998) Integrated resource development of the Sundarbans: summary of the integrated resources management plan of the Sundarbans reserved forest. FAO/UNDP project, BGD/84/056

Anwar J (1988) Geology of Coastal area of Bangladesh and recommendation for resource development and management. In: National workshop on coastal area resource development and management, Part II. CARDMA, Dhaka, Bangladesh, pp 36–56

Biswas SR, Choudhury JK, Nishat A, Rahman M (2007) Do invasive plants threaten the Sundarbans mangrove forest of Bangladesh? For Ecol Manag 245:1–9

Broadus JM (1993) Possible impact of land adjustment to sea level rise: the cases of Bangladesh and Egypt. In: Warrick RA, Barrow EM, TML W (eds) Clomate and sea-level change; observations, projections and implication. Cambridge University Press, Cambridge, pp 263–275

Brown BE (1997) The conversion of mangroves for shrimp aquaculture development. In: By mangrove ecosystems. Ecological Economics, 29:235⍰252

CEGIS- (Centre for Environmental and Geographical Information Services) (2006) Coastal land use zoning in the southwest: impact of sea level rise on landuse suitability and adaptation options. Center for environmental and geographic information services (CEGIS), Dhaka

Chowdhury NT (2009) Water management in Bangladesh: an analytical review. Water policy uncorrected proof, IWA publishing, pp 1–20

Chowdhury ZU, Haque A (1990) Permissible water withdrawal based upon prediction of salt- water intrusion in the Meghna Delta. In: The Hydrological basin for water resources management (Proceedings of the Beijing Symposium, October 1990), IAITS Publication no. 197, pp 111⍰117

Costanza R, dArge R, deGroot R (1997) The value of the world's ecosystem services and natural capital. In: Nature, vol 387, pp 253–260

Das S, Siddiqi NA (1985) The mangrove and mangrove forests of Bangladesh. Mangrove silviculture division bulletin no. 2, BFRI and UNDP/FAO project, BGD/ 79/017, Chittagong, Bangladesh
Dasgupta P (2001) Human well-being and the natural environment. Oxford University press, Oxford
ESCAP- (Economic and Social Commission of Asia and the Pacific) (1988) Coastal environment management plan for Bangladesh, Final report, Vol. 2 ESCAPE, Bangkok
Farley J, Batker D, Torre IDL, Hudspeth T (2010) Conserving mangrove ecosystems in the Philippines: transending disciplinary and institutional borders. Environ Manag 45:39–51
Giri C, Pengra B, Zhu Z, Singh A, Tieszen LL (2007) Monitoring mangrove forest dynamics of the Sundarbans in Bangladesh and India using multi-temporal satellite data from 1973 to 2000. Estuar Coast Shelf Sci 73:91–100
Gleick P (1998) The world's water. Island Press, Washington, DC
Goodbred SL, Kuehl SA (2000) Late quaternary evolution of the Ganges –Brahmaputra River delta: significance of high sediment discharge and tectonic process on margin sequence development. Sediment Geol 133:227–248
Goodbred S L and Nicholls R (2004) Towards integrated assessment of the Ganges-Brahmaputra delta. In: Proceeding of the 5th international conference on Asian Marine Geology, and 1st Annual meeting of IGCP475 Delta and APN Mega-Delta, 13th February 2004
Gopal B, Chauhan M (2006) Biodiversity and its conservation in the Sundarban mangrove ecosystem. Aquat Sci 68:338–354
Gopal B, Wetzel G (1995) Limlogy in developing countries, vol 1. International Association of Limnology, New Delhi, pp 1–230
Grigg NS (1996) Water resources management principles, regulation and cases. McGraw Hill, Washington, DC, pp 8–14
Harvey LDD (2000) Climate and global environmental change. Pearson Education Ltd, London, pp 187–207
Hidayati D (2000) Coastal management in ASEAN countries -the struggle to achieve sustainable coastal development. UN University Press, Tokyo, pp 1–74
Hughes R, Ahmed S, Clayton BD (1994) Flood plains or flood plan? A review of approaches to water management in Bangladesh. IIED, London, pp 1–12
IECO (1980) Southwest regional plan: supplement e: saline intrusion and tidal hydraulics; supplement f: environment and ecology (part I & II), supplement g: fishery (part I & II), supplement h: forestry (part I&II), supplement i: navigation; supplement j: augmentation of the dry-season flows of the Ganges. Bangladesh Water Development Board, Dhaka
Iftekhar MS (1999) Vegetation dynamics in the Sundarbans and the contribution of salinity between 1985–1995. B.Sc. thesis, Forestry and Wood Technology Discipline, Khulna University, Khulna, Bangladesh, pp 1–54
Ifthekhar MS, Islam R (2004) Degradation of Bangladesh Sundarbans: a management issue. Int Forestry Rev 6(4):123–135
IPCC – (Intergovernmental Panel on Climate Change) (2014) Climate change 2014: synthesis report. Contribution of working groups I, II and III to the fifth assessment report of the intergovernmental panel on climate change [Core Writing Team, R.K. Pachauri and L.A. Meyer (Eds.)]. IPCC, Geneva, 151 pp
IPCC- (Intergovernmental panel on climate change) (2007) Summary for policymakers. In: Parry ML, Canziani OF, Palitikot PI et al (eds) Climate change 2007: impacts, adaptation and vulnerability, contribution of working group II to the fourth assessment report of the IPCC. Cambridge University Press, Cambridge, 1000 pp

IPCC- (Intergovernmental panel on climate change) (2012) Special report on managing the risk of extreme events and disasters to advance climate change adaptation. Cambridge university press, Cambridge, pp 1–582
IPCC- (Intergovernmental panel on climate change) (2014) Summary for policymakers. In: CB Field, VR Barros, DJ Dokken et al (eds.), Climate change 2014: Impacts,adaptation and vulnerability. Part A: Global and sectoral aspects. Contribution of working group II to the Fifth Assessment report of the Intergovernmental Panel on Climate Change, Cambridge university press, Cambridge, UK, pp 1–32
Islam MS (2001) Sea-level changes in Bangladesh: last ten thousand years, Asiatic Society of Bangladesh. Asiatic Civil Military Press, Dhaka, pp 1–185
Islam SN (2003) Sustainable eco-tourism as a practical site management policy. AHDPH, Dhaka
Islam R (2007) Pre- and Post-tsunami coastal planning and land-use policies and issues in Bangladesh. Book of proceedings of the workshop on coastal area planning and management in Asian tsunami-affected countries. workshop from 27th–29th September 2006 in Bangkok, Thailand, Session I, The regional perspective. Food and Agriculture Organization of the United Nations, regional office for Asia and the Pacific, Bangkok, Thailand
Islam S N (2008) Cultural landscape changing due to Anthropogenic influences on surface water and threats to mangrove wetland ecosystems: a case study on the Sundarbans, Bangladesh. (PhD Dissertation), Brandenburg Technical University of Cottbus, Cottbus
Islam SN (2014) An analysis of the damages of Chakoria Sundarbans mangrove and consequences on community livelihoods in South East coast of Bangladesh. Int J Environ Sustain Dev (IJESD) 13(2):153–173
Islam SN (2016) Deltaic floodplains development and wetland ecosystems management in the Ganges-Brahmaputra-Meghna Rivers Delta in Bangladesh. Sustain Water Resour Manag Springer 2(3):237–256. https://doi.org/10.1007/s40899-016=0047-6
Islam SN, Gnauck A (2008) Mangrove wetland ecosystems in Ganges-Brahmaputra delta in Bangladesh. Front Earth Sci China. Higher Education Press and Springer Verlag 2(4):439–448
Islam SN, Gnauck A (2009a) Threats to the Sundarbans mangrove wetland ecosystem from transboundary water allocation in the Ganges basin: A preliminary problem analysis. Int J Ecol Econ Stat (IJEES) 13(09):64–78
Islam SN, Gnauck A (2009b) The coastal mangrove wetland ecosystem in Ganges delta: a case study on the Sundarbans in Bangladesh. In: Proceedings of American association of petrolium geologist- AAPG Hedberg conference on variations in fluvial deltaic and reservoirs deposites deposited in Tropical Environments, April 29 to May, 2009, Jakarta Indonesia, pp 26–29
Islam SN, Reinstädtler S, Gnauck A (2016) Coastal environmental degradation and water resources management in the Ganges deltaic region. Int J Econ Ecol Stat (IJEES) 27(4):59–81
Islam SN, Reinstädtler S, Fredaush J (2017) Challenges of climate change impacts on urban water quality management and planning in coastal towns of Bangladesh. Int J Environ Sustain Dev 16 (3):228–256
IUCN-(International Union of Nature Conservation) (2003) Status of the ecological intrigrity of the Sundarbans. IUCN Bangladesh country office, Dhaka, pp 1–80
IWM – (Institute of Water Modelling) (2003) Sundarbans biodiversity conservation project surface water modelling TA No. 3158 BAN, final report, vol 1. IWM, Dhaka, Bangladesh
Jalal KF (1988) Towards a national strategy on coastal resources management in Bangladesh. In: National workshop on coastal area resource development and management, Part II. CARDMA, Dhaka, Bangladesh, pp 67–78
Kan F, Feoli E (1999) GIS technology and spatial analysis in coastal zone management. In: Belfiore S (ed) Education and training in integrated coastal area management: the Mediterranean prospect. Ocean Change Pubication, Italy
Khan AA (1993) Fresh water wetlands in Bangladesh: opportunities and options. In: Nishat A, Hussain Z, Roy MK, Karim A (eds) Freshwater wetlands in Bangladesh issues and approaches for management. IUCN, Dhaka

Khan MS, Haq E, Huq S, Rahman AA, Rashid SMA, Ahmed H (eds) (1994) Wetlands of Bangladesh. Holyday Printers, Dhaka, pp 1–88
Khandoker RA (1987) Origin of elevated Barind-Madhupur areas, Bengal basin, result of neotechtonical activities Bangladesh. J Geol 6(2):1–9
Kokkonen M (2006) Saving our seas – the world heritage marine programme. In: World heritage, vol 42. UNESCO, Paris, pp 18–22
Mohiuddin A (2005) Metropolitan government and improvement potentials of urban basic services governance in Dhaka city, Bangladesh: Rhetorie or reality? Theor Empir Res Urban Manag 5 (14):59–81
Mooney HA, Hobbs RJ (eds) (2000) Invasive species in a changing world. Island Press, Washington D.C
Morgen JP, Mclntire WG (1959) Quaternary geology of the Bengal basin, East Pakistan and India. Bull Geo Soc Am 70:319–342
Nair KS (2004) Wetlands management to meet the food and water crisis- example from a tropical region. In: Abstracts book of 7th INTECOL international conference on Wetlands, 25–30 July 2004, Utreacht, The Netherlands, p 217
Nishat A (1988) Review of the present activities and state of art of the coastal areas of Bangladesh. In: National workshop on coastal area resource development and management, Part II. CARDMA, Dhaka, Bangladesh, pp 23–35
Nishat A (1992) Review of present activities and state of art of the coastal area of Bangladesh. In: National workshop on coastal area resource development and management, Part II, Organized by CARDMA, Dhaka, Bangladesh
Nishat A (1993) Fresh water wetlands in Bangladesh: status and issues and approaches for management. IUCN, Dhaka, pp 9–22
Nishat A (2003) Management of fresh water wetlands in Bangladesh: issues and strategy overview of an integrated approach. Nat Res Forum 30:230–237
Pramanik MAH (1983) Remote sensing applications to coastal morphological investigations in Bangladesh, unpublished PhD thesis, Department of Geography and Environment, Jahangirnagar University, Savar, Dhaka, 227 p
Rahman A (1988) Bangladesh coastal environment and management. In: National workshop on coastal area resource development and management, Part II. CARDMA, Dhaka, Bangladesh, pp 1–22
Rahman A (1992) Bangladesh coastal environment and management. In: National workshop on coastal area resources development and management, Part II. Organized by CARDMA, Dhaka, Bangladesh, pp 1–22
Rahman MM (1995) Wetlands and biodiversity: a case study of common property resources in Bangladesh. Conference paper at the 5th annual common property conference held from 24–28 May 1995 at Bodo, Norway, pp 1–10
Rahman MA (1998) Diseases and wood decay of tree species with particular reference to top tying of sundari and the magnitude of its damage in the Sundarbans in Bangladesh. In: MA RMASMSMGMM (ed) Proceedings of the national seminar on integrated management of Ganges floodplains and Sundarbans ecosystem,16–18 July, 1994. Khulna University, Khulna, pp 50–76
Rahman MM (2003) Invasive plants of Sundarbans. Interim report under SBCP project, IUCN, Bangladesh, pp 1–32
Rahman AA (2009) Seal the deal in Copenhagen: the most vulnerable communities demand tear fund
Rahman M, Ahsan A (2001) Salinity constrains and agricultural productivity in coastal saline area of Bangladesh. In: Rahman M (ed) Soil resources in Bangladesh: assessment and utilization. Soil Resource Development Institute, Prokash Mudrayan, Dhaka, pp 2–14
Rahman AA, Haque N (2003) Poverty reduction and sustainable livelihoods in Bangladesh mainstreaming natural resource management into policies, institutions and process. BCAS, Dhaka, pp 1–14

Reinstädtler S (2013) Sustaining landscapes – Landscape units for climate adaptive regional planning. Conference proceeding of the international World Heritage Studies (WHS) – Alumni – Conference „World heritage and sustainable development "from 16.-19.06.2011, IAWHP (International Association for World Heritage Professionals), Cottbus, Germany, ISBN 978-3-00-041891-4, pp 45–65

Runket M, Ahmad IU (1997) Sundarbans reserved forests are distribution and status of forest types from 1985–1995, GoB/BD, pp 1–21

Sarkar SU (1993) Faunal diversity and their coservation in freshwater wetlands. In: Nishat A, Zakir H, Kanti R, Ansarul K (eds) Freshwater wetlands in Bangladesh: issues and approches for management. IUCN, Gland, pp 1–364

Sarker MH (2008) Morphological response of the Brahmaputra-Padma lower Meghna River system to the Assam Earthquake of 1950. Unpublished PhD thesis, School of Geography, University of Norttingham, pp 1–296

Sarker MH, Huque I, Alam M, Koudstaal R (2003) Rivers, chars and char dwellers of Bangladesh. J River Basin Manage 1(1):61–80

SCBD (2009) Connecting biodiversity and climate change mitigation and adaptation. Report of the second Ad Hoc technical expert group on biodiversity and climate change montreal: secretariat of the convention on biological diversity technical series no 41: 126

Uddin MS, Shah MAR, Khanom S, Nesha K (2013) Climate change impacts on the Sundarbans mangrove ecosystem services and dependent livelihoods in Bangladesh. Asian J Conserv Biol 2 (2):152–156

UNESCO- United Nations Educational Scientific and Cultural Organization (2008) Policy document on the impacts of climate change on world heritage properties. World Heritage Heritage Centre, Paris

WMO – (World Meteorological Organisation) (2008). WMO statement on the status of the global climate in 2007. WMO-No. 1031. World Meteorological Organisation, Geneva

Wolanski E (2007) Estuarine ecohydrology. Elsevier, Amsterdam, 157 p.

Wolanski E, Brinson MM, Cahon DR, Perillo ME (2009) Coastal wetlands: a synthesis. In: Coastal wetlands- an integrated ecosystem approach. Elsevier, New York, pp 1–62

Chapter 3
Threats to Sandy Shore Habitats in Sri Lanka from Invasive Vegetation

Rathnayake Mudiyanselage Wasantha Rathnayake

Abstract Sandy shore vegetation is under threat due to invasions from non-native or invasive species resulting from human interventions. As previous studies have shown, the natural zoning in Sri Lanka is so disturbed that it is possible to record only a few native species. This study was conducted at three representative study sites in the Western Province of Sri Lanka. We recorded twenty one families, 53 genera and 63 species in the sandy shore vegetation. Of these, approximately 14.28% (or 9 species) were found to be native species with *Ipomea pes-capre* and *Spinifex littoreus* as the most abundant native species at the study site. About 9.52% (or 6 species), namely, *Chromolaena odorata, Cuscuta campestris, Lantana camara, Mikania micrantha, Opuntia stricta* and *Sphagneticola trilobata* were found to be on the national list of invasive species in Sri Lanka. The study shows that a variation in plant diversity is to be found across the gradient of the shore with native species not as abundant and gradually decreasing landward while non-native weeds are more abundant and increasing landward from the sea. The Shannon Diversity Index shows that the diversity of species increases landward due to invasion. The Simpson diversity indices demonstrate that the vegetation is mainly dominated by non-native or invasive species. Thus, the study shows that the typical zoning nature and species composition found in sandy shore vegetation have been disturbed by the spread of invasive species, the main cause of which is ongoing human intervention. Under the existing policy directions, the available legislation therefore has to be enforced in order to restrict the spread of invasive species on sandy shores in Sri Lanka.

Keywords Sandy shore vegetation · Floristic composition · Zoning of vegetation · Invasive species

---

R. M. Wasantha Rathnayake (✉)
Faculty of Management Studies, Sabaragamuwa University of Sri Lanka, Belihul Oya, Sri Lanka

C. Makowski, C. W. Finkl (eds.), *Impacts of Invasive Species on Coastal Environments*, Coastal Research Library 29,
https://doi.org/10.1007/978-3-319-91382-7_3

3.1 Introduction

Sri Lanka's coastline circles 585 km of sandy beaches, extensive lagoons, estuaries, mangroves, coastal marshes and dunes. The extent of coastal habitats in Sri Lanka is around 242,024 ha (Anon. 1991). The plant diversity, structure and floristic composition of the natural vegetation types in the coastal zone are more a result of coastal habitats or geological factors than of climate. Madduma-Bandara (1989) has recorded the dominant natural vegetation types in the coastal zone of Sri Lanka as follows: sandy shore vegetation (on flat beaches or sand dunes); mangrove vegetation; salt marsh vegetation; and forest and grassland vegetation. Of these, sandy shore vegetation is the dominant vegetation type in Sri Lanka, occurring along 69% of the coastline.

The coast's substratum is very unstable because it consists of sandy soil. Moreover, due to the high percentage of sand in the soil, water retentivity is considerably low. Even where water is available, the groundwater table is more than 1 m deep. Thus the plants found in the sandy shore vegetation are mainly dependent on rainfall, dew, and airborne water fumes created by the action of wave-breaking. The vegetation thus suffers from physiological drought and shows xeromorphic characteristics.

The sandy shore vegetation extends inland from the shoreline up to 40–50 m. In general, a gradient or zoning can be observed here. Distinct zones of vegetation can usually be distinguished from the mean sea level towards inland. According to Moreno-Casasola and Espejel (1986), who have classified the community types found in the coastal sand dune systems along the Gulf coast and the Caribbean sea of Mexico, the zoning patterns are directly related to dune physiography and soil and climate differences. According to Calder and Taylor (1908), who have presented detailed species lists for a number of beach communities on the Queen Charlotte Island in Canada, the zone closest to the shore is the gravel zone, which is followed by the driftwood zone, inland species zone and forest zone.

Although Doing (1985) has identified six fore dune vegetation zones around the world, there are four easily distinguished major zones, according to Tinley (1985). These zones can be found across typical coastal dune systems in any part of the world where there is sufficiently high rainfall and the shoreline is sufficiently stable or prograding so as to allow the succession to reach conclusion.

Zone 1- Pioneers This zone, which is closest to the sea, is characterized by creeping grasses and succulent herbs with rhizomatous and stoloniferous growth. These plants fix sand above the drift line to form incipient fore dunes. The plants found on this dune type are characterized by rapid growth in order to outpace sand accumulation, succulence in order to store water, cuticular protection to ward off salt loading, and glands to exude salt. On most shores, one or two species dominate this particular zone (Doing 1985; Tinley 1985).

Zone 2- Shrubs Moderate sand movement is found in this zone. The flora may include annuals, forbs, creepers, succulents and shrubs with a wide distribution. The seeds of these forms are dispersed both by wind and by birds while the community becomes increasingly invaded by bird-dispersed species as it ages (Doing 1985; Tinley 1985).

Zone 3- Thicket The scrub-thicket zone typically has a flat canopy due to wind pruning. It is characterized by little or no sand movement, in contrast to zone 1, and only develops where rainfall exceeds (approximately) 250 mm per year. This community may consist of dwarf trees and shrubs, with a compact canopy the height of which increases with rainfall. Seed dispersal is mainly by birds while widely differing species of plants are dominant in different parts of the zone (Doing 1985; Tinley 1985).

Zone 4- Forest The development of tall thicket or forest is confined to areas of high rainfall that are located behind the shelter of large dunes. This tall thicket and/or forest includes a mixture of thicket species and true forest species as well as other minor elements. They generally only occur where rainfall exceeds 700 mm a year and where soils are mature. An increase in rain and shelter increases the canopy height of this type of vegetation. Seed dispersal is predominantly by birds (Doing 1985; Tinley 1985).

Threats to beaches arise from a range of stressors which spans a spectrum of impact scales from localized effects to those with a truly global reach (e.g. sea-level rise). These pressures act at multiple temporal and spatial scales, translating into ecological impacts that are manifested across several dimensions in time and space so that today almost every beach on every coastline is threatened by human activities (Defeo et al. 2009).

All coastal environments are part of a fragile and dynamic ecosystem. Thus, the removal of vegetation from these environments can have far reaching effects. Coastal vegetation is threatened by grazing, inappropriate firing, trampling, building construction, recreational activities and the introduction of exotic plant and animal species. Mainly due to human activities, the zoning pattern of sandy shore vegetation has come to be disturbed while plants, commonly known as weeds, as well as non-native and invasive plants have invaded all the above zones.

Sri Lanka is one of 35 global biodiversity hotspots. According to the National Red List (2012), Sri Lanka is home to about 750 vertebrates, 1500 invertebrates, 3150 flowering plants, 336 Pteridophytes (ferns) species, 240 birds, 211 reptiles, 245 butterfly species and 250 land snail species. Due to the rich coastal and marine biodiversity, there exists about 208 species of hard coral, 750 species of marine molluscs, and over 1300 species of marine fish, supported by ecosystems such as coral reefs, mangroves, sea grass beds, salt marshes, sand dunes and beaches. There is a great amount of endemism (species only found in Sri Lanka), with about 894 species of angiosperms and 43% of indigenous vertebrates (not including marine forms) being endemic.

Most plants found in Sri Lanka have originated from other parts of the world. Although some plant and animal species were introduced to Sri Lanka intentionally for various beneficial purposes, the arrival of other species in Sri Lanka has been accidental due to various other reasons. While these species were initially confined to human habitats such as home gardens, botanical gardens and farmlands, a few species have since then invaded the natural ecosystems and maintained their populations, thereby causing damage to natural and man-made habitats as pests and weeds. Furthermore, a very small number of species thus introduced have even been able to survive, colonize, and disperse beyond their areas of introduction, thus not only exceeding the original purpose of their introduction but also leading to the suppression of man-made and natural ecosystems, sometimes even replacing native plant and animal species. Such species are known as Invasive Alien Species (IAS) (Ministry of Mahaweli Development and Environment 2015).

"Invasive Alien Species (IAS) are species whose introduction and/or spread outside their natural past or present distribution threatens biological diversity." Convention of Biological Diversity.

"An alien invasive species is a species introduced into a habitat and whose establishment and spread threatens the ecosystem, habitat or species with economic or environmental harm" (McNeely 2001)

In 2015, the Ministry of Mahaweli Development and Environment announced the list of national alien species in Sri Lanka (see Table 3.1). Sri Lanka, being a small island with a rich biodiversity, the vast array of its wetland and terrestrial ecosystems are inhabited by several species of alien biota, hence causing a threat to native biodiversity. These plants would have been introduced to Sri Lanka due to different reasons such as the arrival of a timber tree, fodder, an ornamental plant, and/or through contaminated grains. Commonly, these plants were introduced as ornamental plants to the botanic gardens.

The Invasive Alien Species (IAS) have continued to affect the natural (terrestrial, aquatic and marine) and agro-ecosystems of Sri Lanka adversely impacting their biological diversity and food security. The Open Economic policies in Sri Lanka that have facilitated international trade, travel and transport as well as the natural and man-made disasters that have necessitated the free movement of international aid have seen the incidence of IAS becoming more frequent over the past several decades. Thus, the introduction of IAS to Sri Lanka has been both deliberate and accidental.

Among the alien invasive flora in Sri Lanka, some species such as *Parthenium hysterophorus*, *Eichhornia crassipes*, *Salvinia molesta*, *Chromolaena odorata*, *Lantana camara*, *Mikania micrantha, Panicum maximum*, *Mimosa pigra* and *Clidemia hirta* function as weeds in agricultural ecosystems, thereby resulting in economic losses to farmers.

Invasive species damage native biodiversity through various ways. Some invasive alien plant species directly invade and alter natural ecosystems. According to Bambaradeniya (2002), some alien invasive plants form thickets and shade out native vegetation thereby displacing them gradually. Typical examples in Sri Lanka include Mesquite (*Prosopis juliflora*) and the Prickly Pear Cactus (*Opuntia*

Table 3.1 List of national Alien invasive species

Species	Common name	Year and reason for introduction
Alstonia macrophylla	Hard milkwood	Forestry and timber
Anona glabra	Pond apple	Unknown, grafting stock
Cestrum aurantiacum	Orange cestrum	1899, ornamental
Chromalaena odorata	Siam weed	Unknown, accidental
Clidemia hirta[a]	Soap bush	1894, ornamental
Clusia rosea	Autograph tree	1886, ornamental
Cuscuta campestris	Cuscuta	Probably through contaminated grains
Dillenia suffruticosa	Shrubby dillenia	1882, ornamental
Eichchornia crassipes[a]	Water hyacinth	1905, ornamental
Lantana camara[a]	Lantana	1826, ornamental
Leucaena leucocephala[a]	White lead tree	1980, fodder and soil improvement
Mikania micrantha[a]		
Mimosa pigra[a]	Giant mimosa	1980s, strengthening river banks
Opuntia stricta[a]	Prickly pear cactus	Ornamental
Panicum maximum	Guinea grass	1801–1802, fodder
Parthenium hysterophorus	Parthenium	1980s, accidental
Prosopis juliflora[a]	Mesquite	1880, ornamental
Salvinia molesta	Salvinia	1930's, educational
Sphagneticola trilobata[a]	Creeping ox-eye	Ornamental
Typha angustifolia	Hambu pan	Unknown
Ulex europaeus[a]	Common gorse	1888, ornamental

[a]Species included in the list of 100 of the world's worst invasive alien species (IUCN ISSG 2001)

stricta) that have formed uniform stands in the arid zone scrubland and the Pond Apple (*Anona glabra*) that has invaded the coastal marshes of Sri Lanka (Bmabaradeniya 2002). They affect thereby natural and semi-natural habitats or suppress the natural succession of the ecosystems.

At the extreme level, alien invasive plants may entirely modify the structure and function of an ecosystem, which could occur in many ways. For instance, some invasive plant species can produce allelopathic substances that are toxic to other native plant species. As a result the soil is unsuitable for the native plant communities. The aquatic alien invasive plants, i.e., Water Hyacinth (*Eichhornia crassipes*) and Salvinia (*Salvinia molesta*), form dense mats that tend to accumulate greater amounts of sediment. This, coupled with high loss of water through increased transpiration, ultimately converts wetlands into terrestrial habitats. The final outcome of such uniform stands of alien invasive plants is the narrowing down of native biological diversity in a particular locality.

Scholars have found that not all alien invasive plant species are entirely bad. For instance, the thick Gorse (*Ulex europaeus*), which stands in the Horton Plains National Park in Sri Lanka, serves as a good habitat for the endemic Black-lipped Lizard (*Calotes nigrilabris*) and several amphibians, providing them with food (i.e.,

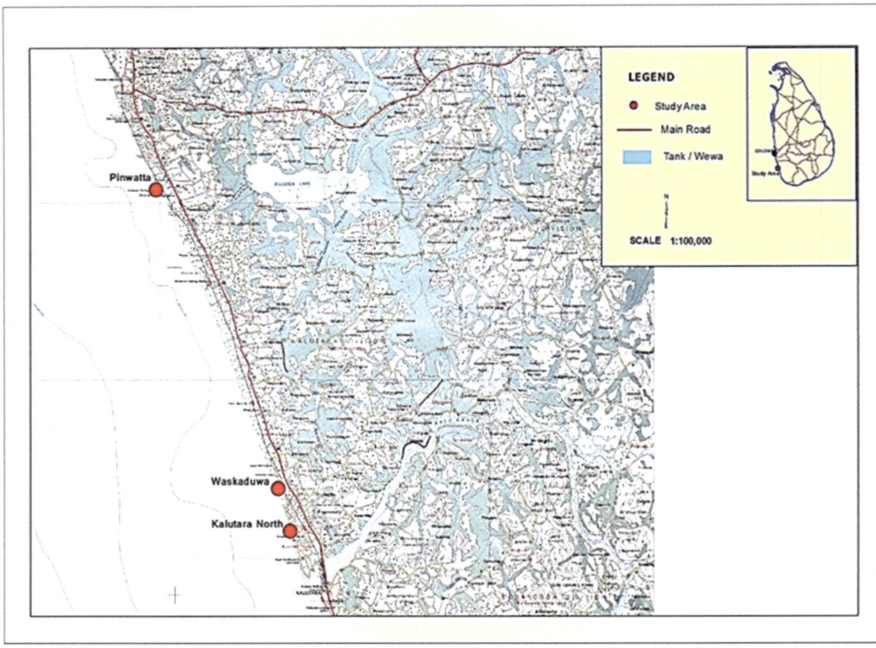

Fig. 3.1 Location map of study sites

insects attracted to flowers) and cover. It is also a nesting habitat of birds. The alien invasive shrub *Eupatorium riparium* in the montane region serves as a browsing plant for Sambar (Bambaradeniya 2002). Similarly, the berries of *Lantana camara* are fed upon by several species of birds while the fruits of the Strawberry Guava (*Psidium littorale*) and Cactus (*Opuntia stricta*) are eaten by langurs and monkeys. Therefore, when considering management options for alien invasives, land management should be given high priority that takes into consideration such beneficial attributes. However, in the case of Sri Lanka, except for *Opuntia stricta*, other beneficial alien invasive plants are not found in sandy shore vegetation. But if *Opuntia stricta* is to be found, sandy shore vegetation is not a habitat for langurs and monkeys in Sri Lanka. In the present study, the focus will be on floristic diversity and the abundance of non-native species and their diversity from the sea towards land.

3.2 Study Area

The western province of Sri Lanka is highly populated and, mainly due to coastal tourism and fishery harbor development, the sandy shore vegetation of the province is disturbed. Therefore, three coastal patches, namely, Pinwatte, Kalutara and

Waskaduwa, were selected for the present study (Fig. 3.1). In the sites selected, the sandy shore vegetation is disturbed to different levels and extents.

3.3 Methodology

3.3.1 Quantitative Survey

The herbarium specimens were prepared from the plant materials collected from each study site. The morphological characteristics were studied and identified by comparing them with those represented in the National Herbarium. The identifications were confirmed by referring to the *Flora of Ceylon* by Trimen (1974) and the revised edition of the *Flora of Ceylon* by Dassanayake and Fosberg (1980–97). The endemicity of plant species and their medicinal importance were checked by referring to the *Flora of Ceylon* by Trimen (1974) and *Medicinal Plants Used in Ceylon* by Jayaweera (1981–1982).

Ten representative sites were selected for the quantitative survey of the vegetation and, at each site, ten transects were marked with ropes. The distance between adjacent sites varied between 50 m and 250 m. Each transect ran through the sandy shore vegetation from an average boundary line of litter accumulation towards inland until it met the end of sandy shore vegetation.

In the present study, the quadrate size was 1.5 m × 1.5 m. The quadrate size was based on the species area curves drawn for each distance from the boundary line of litter accumulation. Accordingly, a quadrate (1.5 m × 1.5 m) was placed along each transect at 1.5 m intervals from the average boundary line of litter accumulation to the end of the sandy shore vegetation. The quadrate was subdivided into 100 units with nylon wires. The frequency and the relative frequency of each species at each quadrate were calculated using quantitative techniques (Greig-Smith 1957).

Dominance (cover value) is the proportion of the ground occupied by a perpendicular projection of the aerial parts of individuals of the species under consideration (Greig-Smith 1957). In the present study, the cover value was estimated by using a cover frame of which points were marked at regular intervals (15 cm). This frame was placed along each transect and the number of points at which a particular species was touched was counted. This allowed us to calculate the cover value and relative cover value of each species.

The floristic diversity landwards from the sea was studied using the Shanon Index and Simpson Index (Eqs. 3.1 and 3.2).

$$\textbf{\textit{Shanon Index}}\ (\boldsymbol{H}) = -\sum\nolimits_{i=0}^{s} p_i \ln p_i \tag{3.1}$$

$$\textbf{\textit{Simpson Index}}\ (\boldsymbol{D}) = \frac{1}{\sum_{i=0}^{s} {p_i}^2} \tag{3.2}$$

Shanon Index
The Shannon Diversity Index (H) is commonly used to characterize species diversity in a community. Shannon's Index accounts for both abundance and evenness of the species present. p is the proportion (n/N) of individual of one particular species found (n) divided by the total number of individuals found (N), ln is the natural log, $\sum$ is the sum of calculations, and **s** is the number of species.

Simpson Index
Simpson's Diversity Index is a measure of diversity. It gives more weight to common or dominant species. In ecology, it is often used to quantify the biodiversity of a habitat. It takes into account the number of species present, as well as the abundance of each species. In the Simpson Index, p is the proportion (n/N) of individuals of one particular species found (n) divided by the total number of individuals found (N), $\sum$ is still the sum of the calculations, and s is the number of species.

3.4 Results and Discussion

3.4.1 Floristic Composition

Table 3.2 provides a comprehensive list of the species composition of plants that comprise sandy shore vegetation. It also indicates their local names, life forms, endemicity and medicinal importance. A total number of 61 species belonging to 51 genera and 21 families were recorded at all three sites. Each family of Asteraceae and Poaceae represents approximately 14.28% (or 9 species) while Fabaceae represents roughly 14.28% (or 9 species) of the total species. The majority of species (i.e., 37 species or 58.73%) were herbs. The sandy shore vegetation also comprised 17.46% (or 11 species of) of shrubs, 12.70% (or 8 species) of creeping herbs and 4.76% (or 3 species) of creepers and trees. Only one species was found to be a climber. A few species, i.e., 3.27% (or 2 species) were found to be trees. About 58.73% (or 37 species) were found to be medicinally important.

Out of 63 species, only nine species, namely *Calotropis gigantea*, *Terminalia catappa*, *Ipomea pes-capre*, *Pandanus odoratissimus*, *Preman latifolia*, *Phyla nodiflora*, *Spinifex littoreus*, *Tephrosia purpurea* and *Vigna marina* were found to be native species in sandy shore vegetation while the other species could be considered as non-native species, weeds, or invasive species. The highest number of native species (or 8 species) were recorded at Pinwatte and Waskaduwa while only five native species were recorded at the Kalutara North beach. Among the non-native species, some species were found to be common weeds found in waste grounds, roadsides and farmlands, hence non-native inhabitants of sandy shore vegetation (i.e., *Bracharia* spp., *Emilia* spp. *Vernonia cinera*, *Tridax procumbens*, *Desmodium* spp. *Crotolaria* spp. and *Mikania micrantha, Euphorbia hirta, Stachytarpheta indica*). Among the non-native species found in sandy shore

Table 3.2 Floristic composition of sandy shore vegetation at study sites

Families and species	Life form	Presence of species		
		Pinwatte	Waskaduwa	Kalutara North
1. Amaranthaceae				
i. *Achyranthes aspera* L.	H+	X	X	X
i. *Aerva lanata* (L.) Juss. ex. Schult.	H+	X	X	X
ii. *Amaranthus spinosus* L.	H+	–	–	X
v. *Amaranthus viridis* L.	H+	–	–	X
2. Apocynaceae				
i. *Cerbera manghas* L.	T	X	–	–
3. Arecaceae				
i. *Cocos nucifera* L.	T+	X	X	X
4. Asclepiadaceae				
i. *Calotropis gigantea* (L.) Ait. f.	S+	X	X	X
5. Asteraceae (compositae)				
i. *Chromolaena odorata* (L.) R.M. King & H. Rob.	H#	X	X	X
i. *Emilia baldiwinii* Fosb.	*H	X	X	–
ii. *Emilia sonchifolia* (L.) DC.	H+	X	X	–
v. *Launaea sarmentosa* (Willd) Sch. Bip. ex. Kuntze	H	X		
v. *Mikania micrantha* Kunith	Cr#	X	X	X
i. *Sphagneticola trilobata* (L.) Prusk	H#	X	X	X
ii. Tridax procumbens L.	H#	X	X	X
iii. *Vernonia cinera* (L.) Less.	H+	X	X	X
6. Capparidaceae				
i. *Polanisia icosandra* (L.) Wight & Arn.	H+	X	–	–
7. Cactaceae				
Opuntia stricta (Haw.) Haw.	S#	X	X	X
8. Commelinaceae				
Commelina indehiscens	CH	X	X	x
9. Convolvulaceae				
i. *Ipomea pes-capre* (L.) R.Br.	CH+	X	X	X
i. *Ipomea pes-tigridis* L.	H+	–	X	–
10. Cuscutaceae				
i. *Cuscuta campestris* Yunck**.**	CH + #	x	x	x
11. Cyperaceae				
i. *Cyperus rotundus* L.	H+	X	X	X
i. *Mariscus pedunculatus* (R.Br.) T. Koyama	H	X	–	X
12. Euphorbiaceae				
i. *Acalypha indica* L.	H	x	X	X
i. *Agyneia bacciformis* (L.) A. Juss. ex. Wight.	H	x	–	–
ii. *Euphorbia hirta* L.	H	x	X	X

(continued)

Table 3.2 (continued)

Families and species	Life form	Presence of species		
		Pinwatte	Waskaduwa	Kalutara North
v. *Phyllanthus debelis* Klein ex. Willd.	H+	x	X	X
v. Phyllanthus gardnerianus (Wt.) Thu.	H	x	–	–
i. *Phyllanthus urinaria* L.	H+	x	X	X
13. Fabaceae (leguminosae)				
i. *Alysicarpus vaginalis* (L.) DC	H+	x	X	–
i. *Canavalia obtusifolia* DC.	Cr	x	X	X
ii. *Cassia occidentalis* L.	S+	–	–	X
v. *Crotolaria retusa* L.	S	x	X	X
v. *Crotolaria verrucosa* L.	S+	x	X	X
i. *Desmodium heterophyllum* (Willid) DC.	H+	–	X	X
ii. *Mimosa pudica* L.	H+	X	X	X
iii. *Tephrosia purpurea* (L.) Pers.	H+	X	X	X
x. *Vigna marina* (Burm f.) Merr.	Cr	X	X	X
14. Lamiaceae				
i. *Leucas zeylanica* Desf.	*H+	X	X	X
i. *Ocimum gratissimum L.*	H+	–	–	X
15. Liliaceae				
i. *Gloriosa superba* L.	C+	X	X	X
16. Malvaceae				
i. *Sida cordifolia* L.	S+	x	X	X
i. *Sida humilis* cav. Diss.	S+	x	X	X
ii. *Sida rhombifolia* L.	S+	x	–	X
17. Pandanaceae				
i. *Pandanus odoratissimus* L.f.	S+	x	X	X
18. Pedaliaceae				
i. *Pedalium murex* L.	H+	–	–	X
19. Poaceae				
i. *Brachiaria distachya* (L.) Staff.	CH	x	X	X
i. *Brachiaria paspaloides* (J. Peresl.) C.E. Hubb.	CH	x	–	–
ii. *Cynodon dactylon* (L.) Pers.	CH+	x	X	X
v. *Dactyloctennium aegyptium* (L.) Beauv.	H+	x	X	X
v. *Eleusine indica* (L.) Gaertn.	H+	x	X	X
i. *Erioshola procera* (Retz) C.E. Hubb.	H	x		
ii. *Spinifex littoreus* (Burm.f.) Merr.	CH	x	X	X
iii. *Zoysia matrella* (L.) Merr.	CH	–	X	X
20. Rubiaceae				
i. *Borreria distans* (Kunth) Cham & Schltdl.	H	–	–	X
i. *Borreria hispida* L.	H+	X	X	X
ii. *Hedyotis auricularia* L.	H+	X	X	X
v. *Richardia scabra* L.	H	x	X	X

(continued)

Table 3.2 (continued)

Families and species	Life form	Presence of species		
		Pinwatte	Waskaduwa	Kalutara North
21. Verbenaceae				
i. *Clerodendrum inerme* (L.) Gaertn.	S+	–	X	X
i. *Lantana camara* var.*aculeata* (L.) Moldenke	S#	X	X	X
ii. *Phyla nodiflora* (L.) Greene	H+	x	x	–
v. *Premna latifolia* Roxb.	T+	–	X	–
v. *Stachytarpheta indica* (L.) Vahl	H	x	X	X

Cr Creepers, *CH* Creeping herbs, *H* Herbs, *S* Shrubs, *T* Trees
* = Endemic,+ = Medicinally important, #= Exotic x- presence of species

Plate 3.1 *Chromalaena odorata*

vegetation, *Chromolaena odorata, Cuscuta campestris, Lantana camara, Mikania micrantha, Opuntia stricta* and *Sphagneticola trilobata* were listed in the national alien invasive species list in Sri Lanka (Ministry of Mahaweli Development and Environment, 2015, Plates 3.1, 3.2, 3.3, 3.4, 3.5 and 3.6). Such invasive species do not bode well for the growth of native species in sandy shore vegetation in the future as they may spread faster, thus suppressing native species. Some plant species like *Lantana camara,* which were introduced to Sri Lanka by the Botanic Gardens in 1826, were later found to be invasive (Wijesundara 1999).

Chromolaena odorata is a tropical and subtropical species of flowering shrub in the sunflower family. It is native to the Americas, from Florida and Texas in the American south, through Mexico and the Caribbean, to South America. *Chromalaena odorata* is able to quickly establish and smother plant crops, forestry, and native vegetation. It is unpleasant and noxious and may cause death if ingested by animals. *Chromalaena odorata* has been recognized as the most problematic weed in coconut plantations in Sri Lanka as early as 1944, and has become a problem

Plate 3.2 *Lantana camara*

Plate 3.3 *Mikania micrantha*

Plate 3.4 *Opuntia stricta*

Plate 3.5 *Sphagneticola trilobata*

Plate 3.6 *Cuscuta campestris*

in other cultivations since then. Further, invasion of *Chromalaena odorata* is viewed as a major agronomic problem in many countries. It expands rapidly at the onset of the rainy season and forms impenetrable tangles that may ultimately shade out indigenous vegetation. Through smothering and allelopathy, *Chromalaena odorata* reduces vegetation heterogeneity in grasslands and forests.

Lantana camara is native to the American tropics. This plant, which is often planted to embellish gardens, has spread from its native Central and South America to around 50 different countries where it has become an invasive species. It spread from the Americas to the rest of the world when it was brought back to Europe by Dutch explorers and cultivated widely, soon spreading to Asia and Oceania where it has established itself as a notorious weed. *Lantana camara* is a low, erect or subscandent vigorous shrub which can grow to 2–4 m in height. Further, it is able to climb to 15 m with the support of other vegetation. Its flower-heads can contain between 20 and 40 flowers. The color of the flower varies from white, cream or yellow to orange pink, purple and red. The fruit is a greenish blue-black color, 5–7 mm in diameter, drupaceous, and shining with two nutlets. It was introduced to Sri Lanka around 1826 through the Royal Botanic Gardens established by the

British. It was grown in home gardens as an ornamental plant for over 300 years and now has hundreds of cultivars and hybrids.

The genus ***Mikania*** is the largest genus in the tribe Eupatorieae within the family Asteraceae and contains species primarily native to tropical and temperate America. *M. micrantha* is a New World species whose full distribution in the Old World has only recently been fully realized. *Mikania micrantha* is a perennial creeping climber because of its vigorous and rampant growth. It grows well where fertility, organic matter, soil moisture and humidity are all high suppressing the growth of native species. It disturbs or kills other plants by cutting out the light and smothering them. Further, this species competes for water and nutrients but, even more importantly perhaps, it is believed that the plant releases substances that inhibit the growth of other plants. Once established, *M. micrantha* spreads at an alarming rate, readily climbing and twining around any vertical support, including crops, bushes, trees, walls and fences. Its shoots have been reported to grow up to 27 mm a day.

Opuntia stricta is a large-sized species of cactus that is endemic in the subtropical and tropical coastal areas of the Americas and the Caribbean. *Opuntia stricta* is an upright or spreading fleshy plant usually growing 50–100 cm tall. Its stems are tend to branch out a lot and consist of a relatively large, flattened and elongated (i.e., elliptic or obovate) series of stem segments which are succulent. Stem segments are hairless and sparsely covered with sharp spines (2–4 cm long) in 1–7 groups. *Opuntia stricta* was introduced to Sri Lanka during the nineteenth century as an ornamental plant but has now become a naturalized common serious weed in dry and coastal areas in Sri Lanka.

Sphagneticola trilobata has become a common weed problem in many parts of Florida. First introduced from tropical America, *Sphagneticola trilobata* is a perennial growing up to 40–60 cm high. Stems are green, rounded, rooting at nodes, 10–30 cm long, the flowering portions ascending, covered with dense hairs. This plant usually flowers throughout the year. Flowers are solitary with stalks 3–10 cm long, covering bell-shaped or hemispherical, ray florets often 8–13 per head, yellow in color and disk corolla 4–5 mm long. It is believed that *S. trilobata* was introduced to Sri Lanka as an ornamental plant from Singapore in the late 1970's. *Sphagneticola trilobata* has been used in the landscape as a groundcover. Plants like *Sphagneticola trilobata* because of its beautiful flowers and its fast growth habit.

The parasitic weed ***Cuscuta campestris*** is native to North America but has been introduced around the world and become a weed in many countries. *Cuscuta campestris* is an angiosperm parasite that adapts to the life cycle of its host. It has a very distinctive appearance. This plant has leafless, glabrous, yellow or orange stems and tendrils bearing inconspicuous scales in place of leaves. The tendrils produce haustoria–a specialized root-like sucker which penetrates another plant (a host) and absorbs water and nutrients from it. Stems are about 0.3 mm in diameter. The plant shows rapid growth and weakens the hosts so that, ultimately, the host plants die.

Some plant species such as *Tridax procumbens, Euphorbia hirta, Commelina indehiscens* and *Stachytarpheta indica* are commonly found as weeds in cultivations and natural habitats including sandy shore vegetation and forests (Plates 3.7, 3.8, 3.9 and 3.10). So far they have not been considered as invasive alien species in Sri Lanka. In the present study, too, these species are recorded at all three sites.

Plate 3.7 *Tridax procumbens*

Plate 3.8 *Commelina indehiscens*

Plate 3.9 *Euphorbia hirta*

Plate 3.10 *Stachytarpheta indica*

Further, among the recorded Invasive Alien Species, *Mikania micrantha, Opuntia stricta, Lantana camara* and *Sphagneticola trilobata* were the species included in the world's 100 worst invasive alien species lists (IUCN ISSG 2001). Therefore, the existing sandy shore vegetation is threatened by internationally recognized worst invasive alien species.

3.4.2 Abundance of Native Species against Non-native/Weed/Invasive Species

Table 3.3 shows the abundance of native species and non-native, weed or invasive species from the mean sea level towards inland, according to which no plants are to be found growing near the seaward boundary of the shore. At Pinwatte, species were recorded between 4.5 m and 37.5 m away from the boundary line of litter accumulation while at Waskaduwa species were found between 10.5 m and 43.5 m away from the boundary line of litter accumulation. In the case of Kalutara, species were confined to a small distance between 16.5 m and 40.5 m from the boundary line of litter accumulation due to ongoing tourism activities. The distance at which plants appear from the boundary line of litter accumulation also varies from one transect to another, the distance depending on shoreline erosion, stability of substratum, slope of the site, and sand and coral mining processes. The absence of a clear zoning of the vegetation, as identified by Tinley (1985), was attributable to human impacts such as construction and operation of hotels, coconut cultivation, and the establishment of fisheries harbors and beach parks.

The native creeping herbs like *Ipomea pes-capre*, *Spinifex littoreus* and *Vigna marina* were found closer to sea following the zoning pattern in sandy shore vegetation; herbs were found thereafter. Life forms like shrubs and trees, namely, *Calotropis*

Table 3.3 Abundance of species along the gradient from the boundary of litter accumulation towards land

	Distance from the boundary of litter accumulation towards inland													
	4.5 m	7.5 m	10.5 m	13.5 m	16.5 m	19.5 m	22.5 m	25.5 m	28.5 m	31.5 m	34.5 m	37.5 m	40..5 m	43.5. m
Pinwatte Beach														
Terminalia catappa	0	0	0	0	0	0	0	0	0	0	11.4	0		
Ipomea pes-capre	0	13.4	45.7	34.6	32	14.8	44.2	44.8	39.2	45.2	28	0		
Launaea sarmentosa	0	0	1.1	1.4	1.8	0.6	0.4	0	0	0	0	0		
Pandanus odoratissimus	0	0	3.8	13.8	13.6	28.4	9.8	15.8	0	0	0	0		
Phyla nodiflora														
Spinifex littoreus	0	15.8	42	46.6	35.2	27	23.2	2.4	0	0	0	0		
Tephrosia purpurea	0	0	0	0	18.2	9.8	0	2.2	2.2	7.2	0	4.4		
Vigna marina	0	0	0	0	0	0	0	2.4	0	0	0	0		
Relative frequency of native species	0.00	86.39	80.38	83.25	56.69	51.91	39.83	42.95	30.76	30.01	22.04	17.74		
Relative frequency of weeds/non- native species	0.00	13.61	19.62	16.75	43.31	48.09	60.17	57.05	69.24	69.99	77.96	82.26		
Waskaduwa Beach														
Calotropis gigantea			0	0	0	0	0	0	12.3	24.3	15.7	40.3	19.7	18.2
Ipomoea pes-capre			0	0	0	3	0	28.3	39	20.7	12.9	8.3	0	0
Pandanus odoratissimus			0	0	0	0	0	22.7	15.9	22.7	0	27.9	0	13.3
Launaea sarmentosa			0	0	0	0	0	0	1.1	1.6	.0.4	0	0	0
Premna latifolia			0	0	0	0	0	5.7	4	0	0	0	0	32
Phyla nodiflora														
Tephrosia purpurea			0	0	0	0	0	1.3	1.3	3	0	0	0	0

(continued)

Table 3.3 (continued)

	Distance from the boundary of litter accumulation towards inland													
	4.5 m	7.5 m	10.5 m	13.5 m	16.5 m	19.5 m	22.5 m	25.5 m	28.5 m	31.5 m	34.5 m	37.5 m	40..5 m	43.5. m
Vigna marina			0	0	0	0	0	2.7	0	0	0	0	0	0
Relative frequency of native species			0	0	0	6.61	0	56.20	46.58	65.73	40.45	55.39	18.67	64.27
Relative frequency of Weeds/non- native species			0	100	100	93.39	100	43.80	53.42	34.27	59.55	44.60	81.33	35.73
Kalutara North Beach														
Calotropis gigntea					0	0	0	0	0	0	0	0	9.5	
Ipomea pes-capre					0	3	21	78	59.3	60.1	52	25.5	4	
Launaea sarmentosa					0	0	0	3	1.2	0.4	0	0	0	
Pandanus odoratissimus					0	0	0	0	0	0	0	4	36	
Tephrosia purpurea					0	0	0	0	0	3.7	0	0	0	
Relative frequency of native species					0	20.69	91.30	65.32	45.83	47.21	40.94	30.41	37.36	
Relative frequency of Weeds/non -native species					0	79.31	8.69	34.68	54.17	52.79	59.05	69.59	62.64	

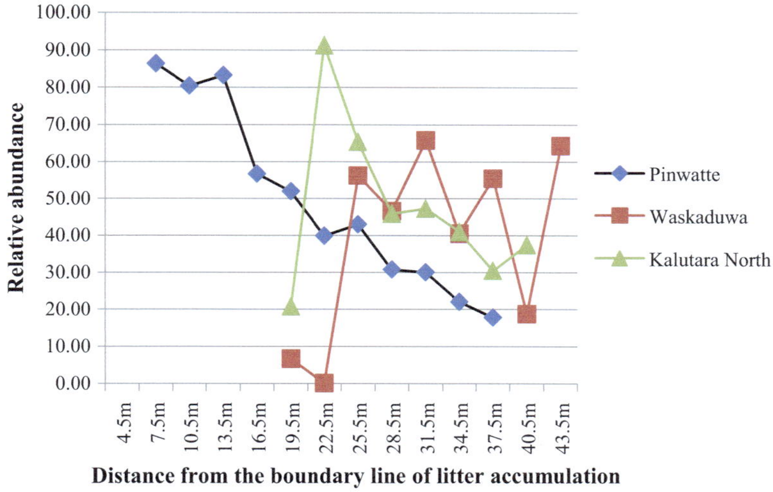

Fig. 3.2 Relative abundance of native species along the gradient from the boundary line of litter accumulation towards land

gigantea, Tephrosia purpurea, Pandanus odoratissimus, *Premna latifolia* and *Terminalia catappa* were found further away from the boundary line of litter accumulation following the zoning pattern as described by Tinley (1985). The abundance of native species gradually decreased landward while, in contrast, the abundance of weeds and non-native and invasive species increased landward from the boundary line of litter accumulation. Though zoning is found up to a certain extent with non-native species, only seven species were found in the vegetation, namely, 6 species in Pinwatte and Waskaduwa and 4 species in Kalutara North. Overall, except for the native species named *Ipomoea pes-capre,* other native species were few in number.

Figure 3.2 gives the relative abundance of native species at each site. It shows that the abundance of native species has gradually increased in Waskaduwa. In comparison, despite the fluctuation in abundance of native species in Kalutara North and Pinwatte, there was a decrease in abundance in these two sites. This type of fluctuation in abundance can be attributed to the spread of non-native, weed and invasive species influenced by human activities such as infrastructure development, trampling, and habitat modifications for recreational activities.

3.4.3 Species Diversity

The Shanon and Simpson Diversity Index values for all three sites were calculated taking into consideration the zoning and the distance from the boundary line of litter accumulation landward. Four zones were identified for each site, namely, >10.5 m zone, 10.5 m–25.5 m zone, 25.5 m–37.5 m and >37.5 m (see Table 3.4). It is found that,

Table 3.4 Shanon and Simpson Diversity Index values for all three sites

	<10.5 m zone		10.5 m–25.5 m zone		25.5 m – 37.5 m zone		> 37.5 m zone	
	Native species	Weeds/invasive species	Native species	Weeds/invasive species	Native species	Weeds/invasive species	Native species	Weeds/invasive species
Pinwatte								
Shanon index	0.831	0.310	0.988	0.354	0.862	0.458		
Simpson index	3.180	30.454	11.452	4.494	20.305	1.871		
Overall Shanon index for the zone	1.140		1.342		1.320			
Overall Simpson index for the zone	2.880		3.227		1.714			
Waskaduwa								
Shanon index			0.640	0.219	0.979	0.318	0.469	0.310
Simpson index for the zone			41.344	1.722	11.422	26.417	34.483	2.838
Overall Shanon index for the zone			0.859		1.297		0.778	
Overall Simpson index for the zone			1.722		7.974		2.621	
Kalutara North								
Shanon index for the zone			0.562	0.735	0.732	0.631	0.654	0.294
Simpson index for the zone			2.380	8.078	6.107	2.959	12.528	2.568
Overall Shanon index			1.297		1.363		0.948	
Overall Simpson index			1.839		1.993		2.131	

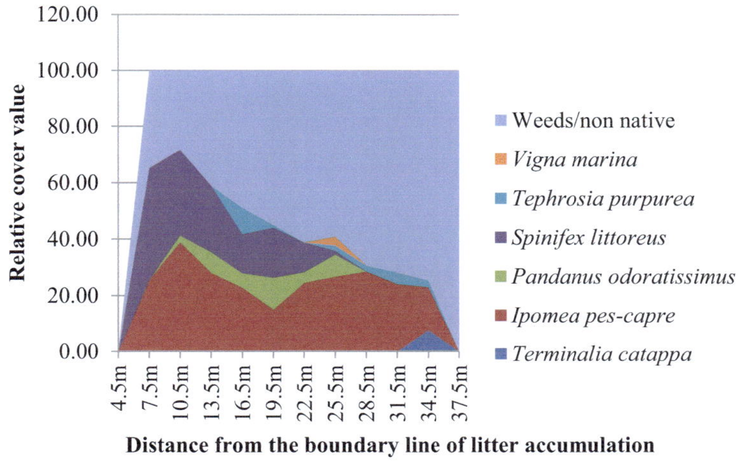

Fig. 3.3 Dominance (or cover value) of recorded native species and non-native/weeds/invasive species at Pinwatte beach

overall the Shannon Diversity Index for native species has gradually decreased from the boundary line of litter accumulation landward while it has gradually increased for non-native species, weeds or invasive species. This indicates that native species are not common at the three sites while the sites are invaded by non-native species, weeds and invasive species. This also reveals that the diversity has increased landward due to increase in the presence of different non-native species.

Similarly, the Simpson Diversity Index reveals that native species has gradually decreased from the boundary line of litter accumulation landward while non-native species has increased landward. This also proves that sandy shore vegetation has been invaded by non-native species, weeds and exotic invasive species. While a small number of native species of sandy shore vegetation is found, the vegetation is mainly dominated by non-native species, weeds and exotic species. Figures 3.3, 3.4 and 3.5 show the dominance (or cover value) of recorded native species and non-native or weedy species at each study site. The figures underscore the fact that the dominant species at each site is not native species but non-native species. Furthermore, among the native species recorded, the vegetation is dominated closer to the sea by creeping species while shrubs and trees are the dominant species landward. The most dominant tree and shrub species were *Ipomoea pes-capre* and *Pandanus ordoratissimus*. Compared to the other two sites, Waskaduwa shows zoning with more creeping plants seawards and more shrubs and trees landward.

A study by Barbour et al. (1987), which sampled the beach vegetation along the U.S. portion of the Gulf of Mexico with 63 strip transects made up of contiguous 1.5 m × 1.5 m quadrates, has shown that the environmental factors which were correlated with regional vegetation included sand texture, sand chemistry, topographic roughness, and climate. In the present study, soil physics, soil chemistry, and other environmental factors were not studied because, according to available literature, these same factors may aid the spread of non-native species at the study sites.

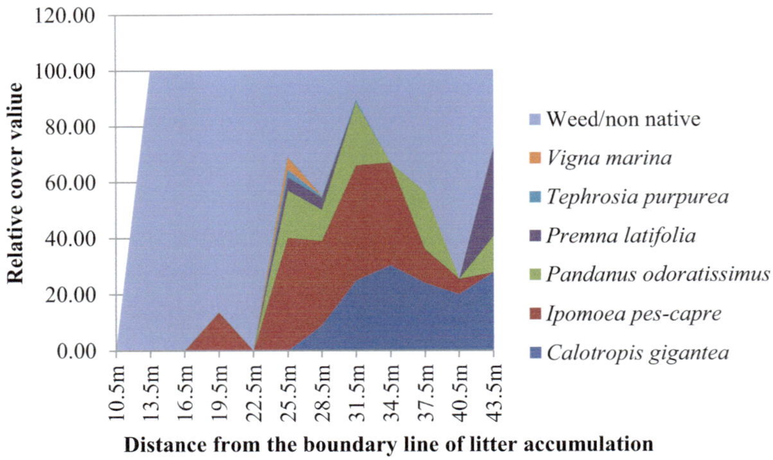

Fig. 3.4 Dominance (or cover value) of recorded native species and non-native/weeds/invasive species at Waskaduwa beach

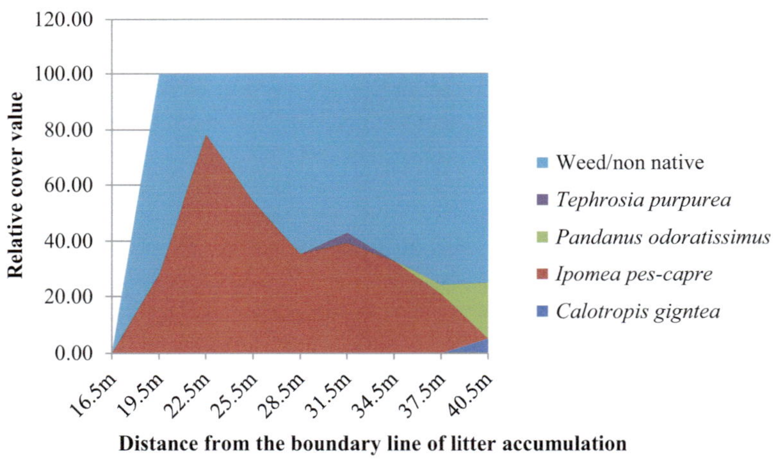

Fig. 3.5 Dominance (or cover value) of recorded native species and non-native/weeds/invasive species at Kalutara North beach

3.5 Conservation and Policy Implications

The sandy shore is the most important component of the biota on sand dunes because it is directly involved in establishing the dune forms and creating the structure of the dune habitat. Controlling and maintaining a healthy sandy shore vegetation is difficult due to ongoing human interventions like tourism activities, fishery harbor construction, recreational activities and construction. Therefore, as indicated by the findings of the present study, sandy shore vegetation will be replaced by weeds and non-native and invasive species in the near future. This will pose a threat to the presence of native species in sandy shore vegetation in Sri Lanka.

In Sri Lanka, the Coast Conservation Act No. 57 of 1981 determines the coastal set-backs. The coastal set-back is defined as an area left free of any physical modification. It is good planning practice to leave a minimum set-back of 35 meters from the mean sea level line. Such a set-back allows for the dynamics of seasonal and long term fluctuations of the coastline and to ensure both visual and actual public access to the water front. Set-back exemptions to the general rule are determined by the Coast Conservation Advisory Council taking into consideration development needs/activities of nationally important projects, fisheries-related buildings and infrastructure, and tourism-related development projects within the declared tourism zones (GOSL 1981). However, the results of the present study suggest that new policy decisions are needed to restrict even these development activities in order to protect the sandy shore vegetation which acts as a natural barrier against coastal erosion in the area under study. In fact, maintenance of the existing zoning pattern of vegetation may help with coastal conservation.

According to Gray (1997), the best way to conserve marine diversity is to conserve habitat and landscape diversity in the coastal area. Marine protected areas are only a part of the conservation strategy needed. Therefore, Gray (1997) has proposed integrated coastal area management as a framework for coastal conservation since sustainable use of coastal biodiversity is one of its primary goals. According to Defeo et al. (2009), setback and zoning strategies need to be enforced through legislation and all relevant stakeholders should be included in the design, implementation and institutionalization of these initiatives. It is evident that new perspectives for rational management of sandy beaches require paradigm shifts, which include not only basic ecosystem principles but also incentives for effective governance and sharing of management roles between government and local stakeholders. In the absence of such measures, invasion of sandy shore vegetation by non-native, weed and invasive plant species is unavoidable as at present.

In general, several factors have contributed to the spread of alien invasive plants in Sri Lanka. The biological and ecological factors include (i) absence of natural enemies (e.g. *Mikania micrantha)*, (ii) fauna that rely on alien trees for a home (e.g. *Lantana camara*), (iii) intrinsic growth and reproduction of the plant (e.g. *Chromolaena odorata)*, (iv) efficiency of seed dispersal (e.g. *Lantana camara*), (v) poor adaptation of native species (e.g. *Opuntia stricta*), (vi) presence of an empty-niche, and (vii) disturbances to the ecosystem. The social and political factors include (i) deliberate introductions to the country, (ii) exchange of seed material between countries, (iii) transport or import of organic matter (compost) and soils that are contaminated with seeds (e.g. *Commelina indehiscens*), (iv) agricultural machinery contaminated with seeds (e.g. *Parthenium hysterophorus* and *Mimosa pigra)*, (v) lack of awareness (e.g. *Lantana camara)*, and (vi) gaps in legislation to prevent entry and control. These factors have contributed in different ways and at different levels to the spread of alien invasive plants in sandy shore vegetation.

The findings of the study thus indicate that, due to various processes and activities, the growth of native species found in the sandy shore vegetation has been suppressed while diversity has been drastically changed. The results of the present study would therefore aid in the planning of agricultural and other economic development programs in the coastal belt that also take into consideration the conservation requirements of coastal vegetation. Some plant species such as *Tridax procumbens, Euphorbia hirta, Commelina indehiscens* and *Stachytarpheta indica* are commonly found as weeds in cultivations as well as natural habitats in Sri Lanka although so far they are not considered as invasive alien species in Sri Lanka. However, they will become invasive alien species in the near future.

There are enough legal provisions in Sri Lanka to control and eradicate the invasive alien species in Sri Lanka. Article 14 in Chapter VI (of the Directive Principles of State Policy and Fundamental Duties) of the Constitution of the Democratic Socialist Republic of Sri Lanka clearly states that "the State shall protect, preserve and improve the environment for the benefit of the community". This governs the activities of all state and private sectors and non-governmental organizations and individuals in protecting the environment of Sri Lanka. In keeping with the constitutional directives and international conventions that the country has been signatory to, several government institutions have developed policy statements or mechanisms to tackle the issues related to IAS in Sri Lanka. The National Environmental Policy was launched in 2003 in keeping with national needs in which Policy Statement 4 states that "environmental management systems will be encouraged to be flexible so as to adapt to changing situations (e.g. climate change, invasive species and living genetically-modified organisms) and adopt the precautionary principle as priority areas requiring action". In addition, the National Environmental Outlook 1, Caring for the Environment II 2008–2022 (the Addendum to the Biodiversity Conservation Action Plan 3 published by the Ministry of Environment) and the National Action Plan for the Haritha [Green] Lanka Programme (published by the National Council for Sustainable Development of the Presidential Secretariat 4) highlight the need to address IAS issues as priority interventions in Sri Lanka. Among other policy documents implemented by different line ministries/ departments of the government of Sri Lanka, containing relevant statements to tackle IAS issues, are the National Agriculture Policy (2007) and the National Wildlife Policy (2000). These organizations and institutions would find the results of our study useful in formulating and implementing an up-to-date policy. The results of this study would also be of use to those dealing with the succession of sandy shore vegetation as well as for comparative studies of sandy shore vegetation locally and globally.

References

Anon. (Anonymous) (1991) Natural resources of Sri Lanka, conditions and trends. Natural Resources Energy and Science Authority of Sri Lanka, Colombo

Babour MG, Rejmanek M, Johnson AF, Pavlik BM (1987) Beach vegetation and plant distribution patterns along the northern Gulf of Mexico. Phytocoenologia 15:201–233

Bambaradeniya CNB (2002) The status and implications of Alien Invasive Species in Sri Lanka. Zoos' Print J 17(11):930–935

Calder JA, Taylor RL (1908) Flora of the Queen Charlotte Island. Part I. Systematics of the vascular plants

Dassanayake MD, Fosberg FR (1980–1997) Revised hand book to the Flora of Ceylon. Parts I–VI. Amerinol Publishing Co. Pvt Ltd., New Delhi

Defeo O, McLachlan A, Scoeman DS, Schlacher TA, Dugan J, Jones A, Lastra M, Scapini F (2009) Threats to sandy beach ecosystems: a review Estuarine. Coast Shelf Sci 81:1–12

Doing H (1985) Coastal fore dune zonation and succession in various parts of the world. Vegetatio 61:65–75

Government of Sri Lanka (GOSL) (1981) Coast Conservation Act No. 57 of 1981. Government Press, Sri Lanka

Gray JS (1997) Marine biodiversity: patterns, threats and conservation needs. Biodivers Conserv 6:153–175

Greig-Smith P (1957) Quantitative plant ecology, 3rd edn. Butterworths, London

IUCN ISSG (IUCN Species Survival Commission) (2001) The status and implications of alien invasive species in Sri Lanka (PDF download available). Available from https://www.researchgate.net/publication/296229291_The_status_and_implications_of_alien_invasive_species_in_Sri_Lanka. Accessed 15 Dec 2017

Jayaweera DMA (1982) Medicinal plants (Indigenous and Exotic) used in Ceylon. Parts 1–5. National Science Council of Sri Lanka, Colombo

Madduma-Bandara CM (1989) Survey of the coastal zone of Sri Lanka. Report published by coast conservation department, Colombo

Mcneely JA (2001) The great reshuffling: human dimensions of Alien Invasive Species. IUCN, Gland

Ministry of Mahaweli Development and Environment (2015) Invasive Alien Species in Sri Lanka, training manual for managers and policy makers. Biodiversity Secretariat Ministry of Mahaweli Development and Environment, Battaramulla

Moreno-Casasola P, Espejel I (1986) Classification and ordination of coastal sand dune vegetation along the Gulf and Carribean Sea of Mexico. Vegetatio 66(3):147–182

Tinley KL (1985) Coastal dunes South Africa. National Scientific Programmes Unit, CSIR, Pretoria

Trimen MBH (1974) A hand book to the Flora of Ceylon. Parts 1-V, M/S. Bishen Singh Mahendra Pal Singh, New Connaught Place, Dehradune and M/S. Periodical experts, D-42, Vivek Vihar, Delhi-32, India

Wijesundera DSA (1999) Alien invasive plants of Sri Lanka and their history. In: Proceedings first national workshop on Alien Invasive Species of Sri Lanka (ed. B. Marambe). Ministry of forestry and environment and SLAAS-Section D

Chapter 4
Alien Species and the Impact on Sand Dunes Along the NE Adriatic Coast

Urban Šilc, Danijela Stešević, Andrej Rozman, Danka Caković, and Filip Küzmič

Abstract This chapter presents the results of a multifaceted approach to determine how sand dune plant communities have been affected by invasion of alien species. We sampled Velika plaža beach in Montenegro (E Adriatic), which is an under-researched part of the Mediterranean. Velika plaža is the largest sandy beach with both still well-developed sand dunes and moist plant communities in this part of Adriatic coast. On the other hand, this beach is also a touristic hot spot and subject to intensive land use change. We gathered maps of past and present land use by using GIS and field mapping from 1950, 1979 and 2015. The species composition of plant communities was randomly sampled and information on phylogeny and plant functional traits was gathered from several databases. Five alien plant species occurred in the sand dune vegetation, with a significant effect of the presence of aliens on native species cover. Invaded plots, when the whole beach is considered, were functionally less diverse than uninvaded ones. Plant species functional traits change along the sea-inland gradient, particularly rosette type, woodiness and nitrogen fixation. Phylogenetic diversity due to the presence of alien species was highest in sand dune slacks, but the results of phylogeny are not consistent and should be used with caution. The results of a multifaceted approach enable further management and monitoring of sand dunes and prevent the conflict between nature conservationists and landscape planners that is foreseen with further development of beaches in this part of the Mediterranean.

U. Šilc (✉) · F. Küzmič
Research Centre of the Slovenian Academy of Sciences and Arts, Institute of Biology, Ljubljana, Slovenia
e-mail: urban@zrc-sazu.si

D. Stešević · D. Caković
Faculty of Natural Sciences and Mathematics, University of Montenegro, Podgorica, Montenegro

A. Rozman
Biotechnical Faculty, University of Ljubljana, Ljubljana, Slovenia

C. Makowski, C. W. Finkl (eds.), *Impacts of Invasive Species on Coastal Environments*, Coastal Research Library 29,
https://doi.org/10.1007/978-3-319-91382-7_4

Keywords Montenegro · Phylogenetic diversity · Functional diversity · Plant communities

4.1 Introduction

4.1.1 Sand Dunes

Sand dunes provide unique ecological niches, occupied by very specialized species and plant communities, capable of surviving high salinity, drought and nutrient limitation, as well as extremely harsh disturbance determined by three interacting factors: waves, tides and sand particle size (Forey et al. 2008; McLachlan and Brown 2006). However, despite an ability to survive such stress-based ecological dynamics, dunal biota is shown to be extremely sensitive to human stressors, which are of various natures. According to Defeo et al. (2009), recreational seashore activities are recognized as the major threat to sand dune ecosystems worldwide, followed by beach cleaning, nourishment, pollution, exploitation, biological invasions, coastal development and engineering, mining and climate change. It is estimated that about 70% of dune systems of European coasts were lost during the last century as a result of increasing urbanization (Brown and McLachlan 2002). This makes coastal areas, and sandy beaches in particular, among highly endangered habitats (Defeo et al. 2009; Martínez et al. 2004). Taking into consideration the high biodiversity and key ecosystem functions of sand dunes: raw materials, coastal protection, erosion control, water catchment and purification, maintenance of wildlife, carbon sequestration, tourism, recreation, education and research (Barbier et al. 2011; van der Maarel 2003), coastal area protection has been set as one of the conservation priorities for many countries. At present, the Habitats Directive (92/43/EEC) is the most effective legal instrument for Europe in terms of biodiversity and nature conservation at a European level. The Habitats Directive describes three different beach and foredune habitat types: 'Annual vegetation of drift lines' (habitat code 1210), 'Embryonic shifting dunes' (habitat code 2110) and 'Shifting dunes along the shoreline with *Ammophila arenaria'* (habitat code 2120).

4.1.2 Impact of Alien Species, Particularly on Sand Dunes

Invasive alien species are considered to be the second major threat to biodiversity (after that caused by direct habitat loss) (Genovesi and Shine 2004), especially in coastal areas that are among highly endangered habitats (Chytrý et al. 2008b; Defeo et al. 2009). Invasive alien species cause numerous negative effects, including decline of native diversity and variation in native species abundance (Isermann 2008; Gaertner et al. 2009), reduction of fitness and growth of resident plant species (Vilà et al. 2011), and sometimes even competitive exclusion of native species (Jucker et al. 2013). Some studies have emphasized the strong ecological impact of alien species on resident ones, although the various impacts are heterogeneous and not unidirectional (Novoa et al. 2012; Vilà et al. 2011). A recent study by Marcantonio et al. (2014) suggests that alien species might carry distinct and different phylogenetic and functional characteristics, affecting the main ecological processes that drive the structuring of plant communities.

4.1.3 Which Sand Dune Communities Are Most Invaded?

Different dune habitat types show differential sensitivity to the invasion of alien species, since salinity, flooding frequency and level, trampling intensity and nitrification are considered to be the main variables that determine the abundance of many alien species (Campos et al. 2004). In fore dune, the influence of aliens is almost not observed, probably due to the extreme specialization needed to survive in such harsh conditions (Carranza et al. 2010). The highest share of aliens occurs on more stabilized dune habitats, such as transitional dunes with a perennial plant community (Carboni et al. 2010) or back and fixed dunes (Marcantonio et al. 2014). Mobile dunes sometimes have a high number of aliens, but the coverage is low (Asensi et al. 2016). Naturalised aliens found on less fluctuating sand habitats have two contrasting strategies and particular traits: annuals and perennials (Acosta et al. 2006). On Dutch sand dunes, herbaceous neophytes invade man disturbed habitats, while woody neophytes (that were mostly intentionally introduced) entered natural vegetation (Weeda 2010).

In our study, we used the multifaceted approach proposed by Marcantonio et al. (2014) to detect the impact of alien plant species on sand dune plant communities and their taxonomic, phylogenetic and functional diversity. The aims of our work were to (1) test the impact of alien plant species on sand dune plant communities, (2) test the occurrence of alien species and their impact on native species and (3) to compare the results of a multifaceted approach with classical transect survey.

4.2 Methods

4.2.1 Study Area

Our study site was Velika plaža near Ulcinj (Montenegro), which is the largest sand beach along the north-eastern Adriatic coast. The beach is 13 km long and, on average, 150 m wide (Pikelj et al. 2013).

The Adriatic Sea is a body of water (200 × 800 km) separating the Apennine and Balkan Peninsulas. The Adriatic is the northernmost part of the Mediterranean Sea, which connects with the Ionian Sea at Otranto Strait. The eastern Adriatic coast is a mainly transgressive coast of carbonate substrate, with some areas of flysch and Quaternary deposits. It follows the Dinaric range in a NW to SE direction (Pikelj et al. 2013) (Fig. 4.1).

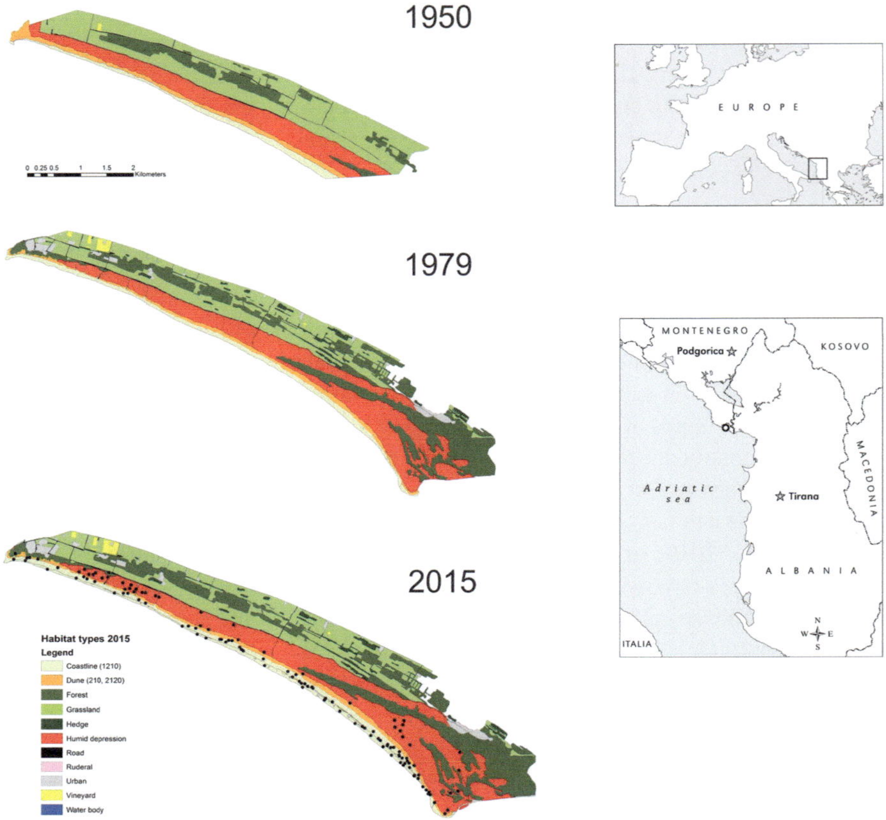

Fig. 4.1 Location of Velika plaža and generalized habitat map of sand beach and hinterland with sampling plots represented as dots (on 2015 map). The sequence of maps from 1950, 1979 and 2015 shows changes of habitat types

According to the Köppen-Geiger system, the climate of Velika plaža is classified as Csa type – Mediterranean climate with hot summers (Burić et al. 2014). The mean annual temperature in neighbouring Ulcinj is 15.5 °C, and the mean precipitation is 1258 mm. The warmest months are July and August (24.3 °C) and the coldest are January and February (6.8 °C). The prevailing winds are eastern (NE, E and ENE).

The sand dunes are low, narrow, simple in structure, with only one ridge, and the vegetation zonation follows a typical ecological sea–inland gradient (Doing 1985). Velika plaža beach is affected by erosion (Petković 2012).

The Habitats Directive is the EU legislative instrument that ensures conservation of rare plant and animal species and also some rare habitat types. The following NATURA 2000 habitats have been recorded on the beach and hinterland (numbers refer to Commission of the European Communities (2013) and an asterisk (*) indicates a priority habitat): 1210 Annual vegetation of drift lines, 1410

Mediterranean salt meadows (*Juncetalia maritimi*), 2110 Embryonic shifting dune, 2120 Shifting dunes along the shoreline with *Ammophila arenaria* (white dunes), 2190 Humid dune slack, 2220 Dunes with *Euphorbia terracina*, 2240 *Brachypodietalia* dune grasslands with annuals, 2270* Wooded dunes with *Pinus pinea* and/or *Pinus pinaster*, 3170* Mediterranean temporary ponds, and 92A0 *Salix alba* and *Populus alba* galleries (Petrović et al. 2012).

In terms of national legislation, the site was already legally protected in 1968 as a natural object (Official Gazette SRCG 1968), in 2007 as a natural monument (Official Gazette of Montenegro 2007), while in 2014 the installation of fences provided an additional conservation measure against trampling (Šilc et al. 2017).

4.2.2 Sampling

We used a stratified random sampling design. The procedure was carried out with Sampling Design Tool (Buja and Menza 2012) in ArcGis 10.4 (ArcGIS 10.4 2015). We used a detailed habitat map (1:5000) produced by the authors in spring 2015. In each EUNIS habitat type (Table 4.1), we randomly sampled 20 plots, and so 100 plots altogether were recorded. Only dune habitats from the shore to pine forest were sampled (Table 4.1). We used a standard plot size of 2 $\times$ 2 m for sampling sand dune vegetation, since this allows comparison of our results with similar vegetation studies (Carboni et al. 2009; Marcantonio et al. 2014). In the field, we located randomly selected sampling plots in GIS using GPS and we recorded all plant species in each of the plots, and visually estimated abundance as cover (according to Braun-Blanquet (1964) scale). Sampling was carried out during May 2017 when vegetation is optimally developed and the tourist season and cleaning of the beach have not yet started. Each plot was sampled only once.

The nomenclatural source for species is the Euro+Med (2006-) check list. Since transects were performed in early summer, when *Erigeron canadensis* L., *E. sumatrensis* Retz. and *Oenothera* species are only starting to develop, it was not possible to identify these species. Identification was therefore left at the genus level. The following *Oenothera* taxa were reported for the area: *O.* $\times$ *fallax* Renner, *O. glazioviana* Micheli, *O. biennis* L. and *O. suaveolens* Pers. (Stešević and Caković 2013; Stešević and Petrović 2010; Rakaj and Rostanski 2009) and own field observations).

We also estimated several environmental variables for each plot, such as distance from the sea (SeaDist), distance from a pathway or road (RoadDist) and distance from tourist facilities (SportDist). We calculated mean ecological indicator values for each plot (Pignatti 2005) and cover abundance was considered. The JUICE program (Tichý 2002) was used for calculations.

In our paper, we use the term “invaded” for plots with alien species present, regardless of the actual status or impact of these alien species locally or regionally.

Table 4.1 Sampled dune habitat types present on Velika plaža

	EU habitat types (European Habitats Directive; European Commission, 1992).
Aphytic shore	without vegetation
Strandline	1210 Annual vegetation on drift lines (*Cakiletea maritimae*)
Embryonic dune	2110 Embryonic shifting dunes
Foredune	2120 Shifting dunes along the shoreline with *Ammophila arenaria*
Dune slack	2190 Humid dune slacks
Woodland	2270* Wooded dunes with *Pinus pinea* and/or *Pinus pinaster*

For further description see (Šilc et al. 2017 and Fig. 4.2)

Fig. 4.2 Zonation of plant communities on sandy beach with embryonic and foredune, with pine forest in the background

4.2.3 Land Use

To detect changes of land use on sand dunes, we obtained aerial photos from the Real Estate Administration of Montenegro, which show Velika plaža vegetation types in 1950 and 1979. We then georeferenced these two panchromatic aerial photos using a satellite image from 2015 in ArcGis 10.4 (ArcGIS 10.4 2015). For 2015, we used a detailed habitat map (1:5000) produced by the authors in spring 2015 by field-mapping. The map was produced using EUNIS typology but, for comparison with older photos, we used the following cover categories: coastline, dune, forest, grassland, hedge, humid depression, road, urban, vineyard, water body.

4.2.4 Statistics

The species richness of five habitat types (HT) was compared using the Wilcoxon test with Bonferroni p-adjustments. Generalized linear models (GLMs) with Gaussian (total species richness) and Poisson-distributed (alien species richness) errors were used to test the relationships between explanatory variables and species richness. In total, 12 GLMs were built (total species richness and alien species richness separately in each HT and in all HTs together). We employed disturbance variables as predictors: distance to sea, roads and sport/leisure objects; stand variables: coverage of tree species, herb species and waste litter; environmental variables: ecological indicator values for acidity, continentality, humidity, light conditions, nutrient availability, salinity and temperature (Pignatti 2005); and past land use maps from 1950, 1979 and 2015. We applied a full Information-Criterion-based (IC) model selection approach, whereby in models for total species richness, the Akaike information criterion corrected for small samples (AICc) was used and in models for alien species richness, Quasi-Likelihood AIC adapted for small samples (QAICc) was used for final model selection (Calcagno and de Mazancourt 2010).

To assess vegetation patterns on dunes, DCA analysis and PCoA with the use of Bray-Curtis similarity were carried out on vegetation plots. In addition, we built multivariate GLMs with a negative binomial error distribution function to test the effect of alien species presence (AP) and HT on the abundance of native species. The univariate statistics then served to investigate which alien plant species is a significant indicator of a changed species composition within the invaded plots (Marcantonio et al. 2014; Wang et al. 2012).

4.2.5 Traits

We selected 24 basic plant functional traits available for our species (average oil content (%), average protein content (%), dicliny, duration of flowering, fruit type, hemeroby, lateral spread, leaf persistence, leaf shape, leaf size, life form, max. height, nitrogen fixation, photosynthetic pathway, pollination vector, root depth, rosette type, seed germination, seed weight, start of flowering, strategy, type of reproduction, urbanity, woodiness) from various trait databases: BIOLFLOR (Kühn et al. 2004), TRY (Kattge et al. 2011), KEW (Royal Botanic Gardens Kew 2017), CLO-PLA3 (Klimešová and De Bello 2009), CATMINAT (Julve 2017), SOFA (Aitzetmüller et al. 2003). We additionally searched for certain traits in Pignatti (1982), Tutin et al. (1964–80) and published scientific papers.

Traits have various features: qualitative (8), circular (1), ordinal (2), nominal (11) and multi-nominal (2). A trait distance matrix based on a standardized Gower coefficient was built in order to calculate four functional metrics for exploring functional diversity: functional divergence (FDiv), functional richness (FRic), functional evenness (FEve) and Rao's Quadratic Entropy (RaoQ) (Botta-Dukat 2005; Villeger et al. 2008). To render our distance matrix Euclidean, we used the transformation proposed by Lingoes (1971). All indices were calculated using the *FD* R package (Laliberté et al. 2014).

Correlations among traits were tested using principal coordinate analysis (PCoA) applied to the trait distance matrix. PCoA was applied to both global trait distance among species and for single-trait distance of the traits that contributed most to the final global distance (Marcantonio et al. 2014). In addition, we used double principal coordinate analysis (DPCoA) to test whether the trait variation influenced functional diversity (Pavoine et al. 2009). The four functional diversity indices and total species richness were plotted on the first DPCoA axis.

To ascertain the impact of the occurrence of alien species on functional diversity, two-factor ANOVA with HTs as block and AP as treatment was performed for each functional diversity metric. Furthermore, we tested the simple main effects of ANOVA models (keeping HT as a fixed factor) to study which zones underwent significant changes due to the presence of alien species (Marcantonio et al. 2014).

4.2.6 Phylogeny

A phylogenetic tree was built using a Zanne et al. (2014) tree as a reference (Qian and Jin 2016). Non-inventoried species were pruned out of the original phylogeny and missing species were bound into the phylogeny by replacing all members of the clade to which it belongs with a polytomy (Pearse et al. 2015). Phylogenetic diversity was studied using three indices: the Mean Pairwise Phylogenetic Distance (MPD), Phylogenetic Evenness-Abundance (PAE) and Imbalance of Abundances with higher cClades (IAC) (Cadotte et al. 2010).

To test whether alien species occurrence affected phylogenetic patterns, two-factor ANOVA with HTs as fixed factor and AP as treatment was performed. We also tested the simple main effects of ANOVA models (keeping HT as a fixed factor) to study in which zones significant changes had occurred due to alien species presence (Marcantonio et al. 2014). Phylogenetic analyses were carried out using *pez* (Pearse et al. 2015), *aed* (Paradis et al. 2004) and *phia* (De Rosario-Martinez 2015) R packages.

4.3 Results

4.3.1 Land Cover Changes

Comparison of aerial photos and the habitat map (Figs. 4.1 and 4.3) shows obvious changes in the cover of categories of land use. The cover of sand dunes has been decreasing, mostly because of an increase of forests (pine plantations and alluvial forests) and artificial surfaces (roads and urban).

4.3.2 Species Richness and Floristic Composition

In the 100 sampled plots, we recorded 194 vascular plants, of which 5 were neophytes. The species belonged to 58 families, and the most frequent were Poaceae (23), Asteraceae (19) and Fabaceae (17). In terms of life cycle, perennials (115) were the dominant form, followed by annuals (70); biennials (7) were only sparsely present (Tables 4.2 and 4.3).

The mean number of plant species was 9.02 (±5.19 SD) per plot. There are statistically significant differences among habitat types for all plant species and for alien species. The statistically significant highest mean number of alien species

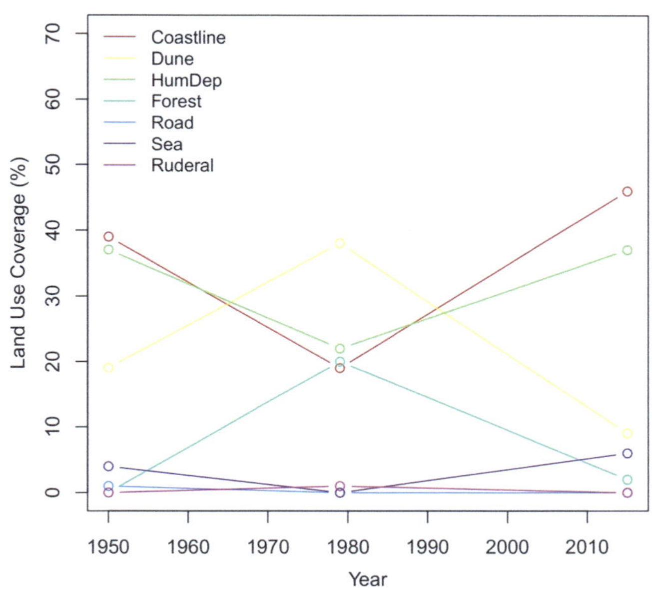

Fig. 4.3 Changes of land use cover for the studied area between 1950 and 2015. Each colour represents a particular land use category

Table 4.2 Frequency of alien species on Velika plaža and their status

Species	DAISIE status	Habitat type (nr. of plots)				
		1210	2110	2120	2190	2270
Erigeron (*canadensis*, *sumatrensis*)	Alien/ established	1			1	1
Erigeron annuus	Alien/ established		2	3	3	1
Oenothera (*biennis*, *glazoviana*, *fallax*, *suaveolens*)	Alien/ established	1	3	14	4	3
Sporobolus indicus	Alien/ unknown					1
Xanthium orientale ssp. *italicum*	Alien/ established	13	11	6		

occurred in habitat type 2120 (Shifting dunes along the shoreline with *Ammophila arenaria*), while the highest mean number of species per plot was in humid depressions (2190) (Fig. 4.4).

DCA analysis (Fig. 4.5) of vegetation data shows a clear classification of plots into five communities. Axis 1 represents distance from the sea and floristic zonation along a sea-inland gradient (Table 4.4).

The total species richness over the whole dune was affected by distance from the sea and distance from roads, while the presence of alien species was not significantly impacted by any environmental factor. Ecological indicator values show a significant impact for humidity, salinity and continentality, for all species (Fig. 4.6).

The highest native species abundance (cover) was in habitat type 2190 (Humid dune slacks) for invaded and uninvaded plots (Fig. 4.7) and lowest in 1210 (Annual vegetation on drift lines) (Fig. 4.6). The species with highest average cover were *Eleocharis palustris*, *Saccharum ravennae* and *Bolboschoenus maritimus* (all from humid dune slacks), while the most abundant aliens were *Oenothera* spp. and *Xanthium orientale* ssp. *italicum*.

The effect of the presence of alien species on native species cover is significant (Table 4.5), as is also the interaction between HT and AP. Calculations for particular habitat types (HT) revealed that alien species changed native species cover only in dune slacks ($p < 0.01$) and pine forest ($p < 0.05$, data not shown). Table 4.6 shows the results of univariate statistical tests of the most abundant native species.

4.3.3 Phylogenetic Diversity

Phylogenetic diversity because of the presence of alien species was highest in habitat 2190 and lowest in 1210, compared to other habitat types (Fig. 4.8). IAC showed the highest differentiation among habitat types and a comparable pattern to PAE metrics.

Table 4.3 Statistics of plant species richness (total and alien species) in different habitat types on Velika plaža

HT	Nr. of plots	Nr. uninvaded plots	Nr. invaded plots	Total species richness	All species			Alien species		
					Min	Mean	Max	Min	Mean	Max
1210	20	7	13	27	1.0	5.75^{a}	10.0	0.0	0.75^{abc}	2.0
2110	20	5	15	41	4.0	8.1^{ab}	19.0	0.0	0.8^{ab}	2.0
2120	20	2	18	60	6.0	10.05^{b}	17.0	0.0	1.15^{a}	2.0
2190	20	15	5	88	6.0	13.4^{bc}	27.0	0.0	0.4^{bc}	2.0
2270	20	16	4	106	3.0	11.2^{bc}	28.0	0.0	0.3^{c}	2.0
Total	100	45	55	194	1.0	9.7	28.0	0.0	0.7	2.0

Different letters next to mean values indicate significant differences between habitat types (Wilcoxon test, $p < 0.05$)

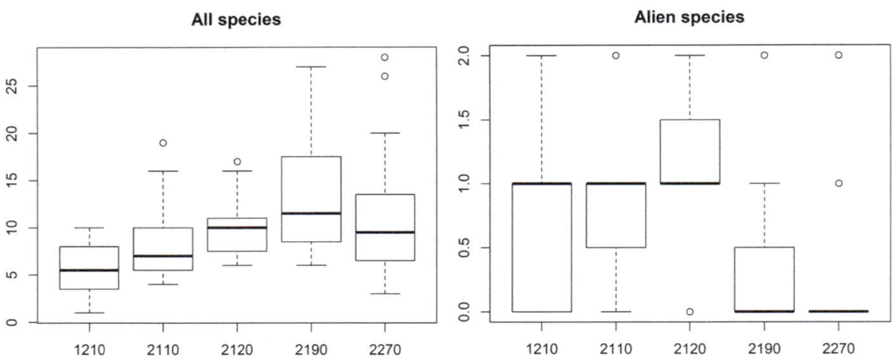

Fig. 4.4 Number of plant species per plot in different habitat types on Velika plaža

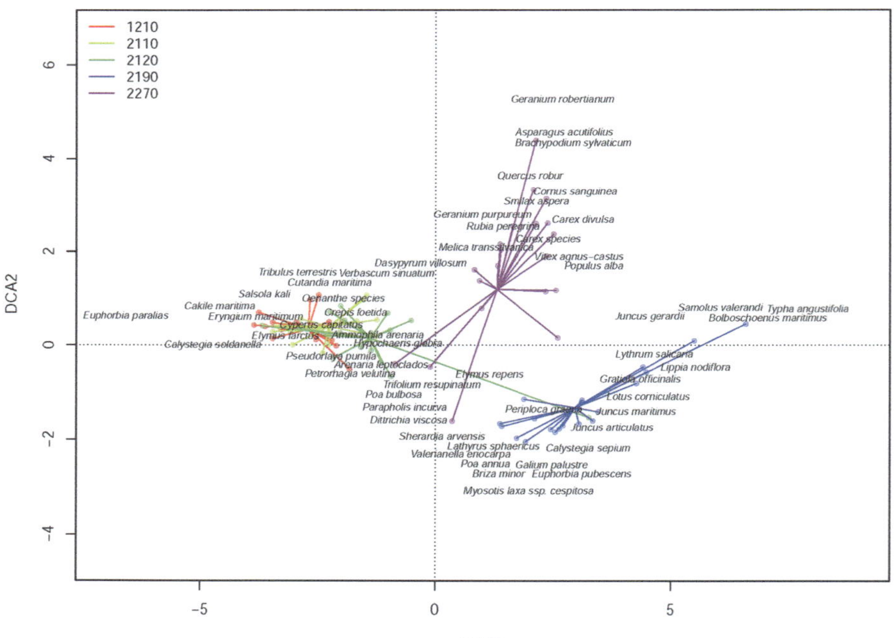

Fig. 4.5 DCA analysis of vegetation data. Points show sampling plots classified into EUNIS habitat types presented in a spider plot

The only significant increase in phylogenetic indices due to alien species presence was observed in IAC for habitat 2110 ($p = 0.08$) and the only drop in MPD for habitat 2270 ($p = 0.001$) (not shown).

Consequently, considering the whole area, phylogenetic diversity was not significantly different in uninvaded plots for any of the three phylogenetic indices (Fig. 4.8 and Table 4.7).

Table 4.4 Multivariate generalized linear model (GLM) coefficients and statistical significance of environmental factors on total and alien species richness (only statistically significant results are presented)

All species						Alien species				
Variable	Estimate	St. error	t	p		Estimate	St. error	z	p	
(intercept)	0.11	3.92	0.03	0.98	–	−6.68	2.76	−2.42	0.02	*
Sea dist	0.02	0.00	5.38	0.00	***	–	–	–	–	–
Roads dist	−0.02	0.01	−2.47	0.02	*	–	–	–	–	–
Sport dist	0.00	0.00	1.72	0.09	.	–	–	–	–	–
Continentality	3.07	0.91	3.37	0.00	**	1.05	0.36	2.95	0.00	**
Humidity	−0.78	0.29	−2.70	0.01	**	−0.22	0.11	−2.02	0.04	*
Salinity	−1.46	0.53	−2.74	0.01	**	–	–	–	–	–
Light	–	–	–	–	–	0.30	0.16	1.81	0.07	.
Adjusted R-squared	0.54	*p value* < 0.001				Adjusted D-squared	0.26			

Signif. Codes: 0 '***' 0.001 '**' 0.01 '*' 0.05 '.' 0.1 ' ' 1

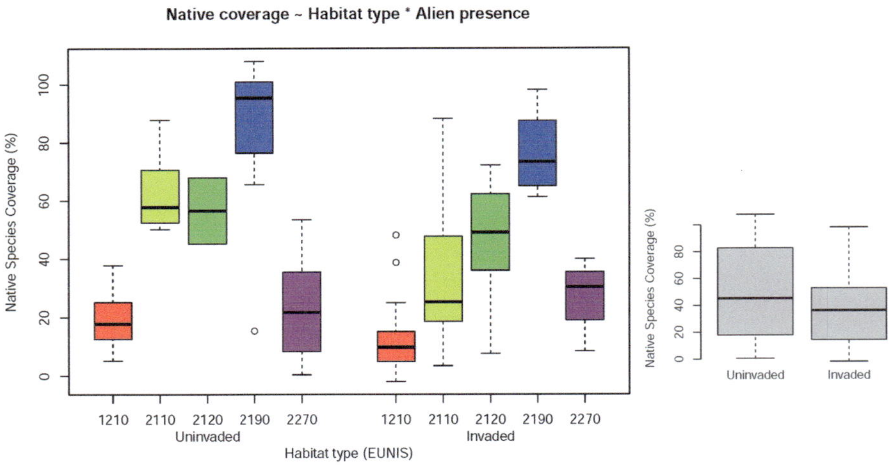

Fig. 4.6 Native species abundance in particular habitat types in plots invaded and uninvaded by alien species. In the smaller boxplot, the abundance of native species for the whole studied area divided into invaded and uninvaded plots is presented

Fig. 4.7 Humid dune slacks with dominant *Juncus maritimus* and *J. acutus*

Table 4.5 ANOVA of statistical significance for species abundances from multivariate GLM analysis

Analysis of deviance table				
Factor	Res.Df	Df.diff	Dev	Pr(>Dev)
HT	95	4	790.7	0.001***
AP	94	1	224.7	0.001***
DZ:AP	90	4	83.3	0.004**

P-value calculated using 999 resampling iterations
Signif. Codes: 0 '***' 0.001 '**' 0.01 '*' 0.05 '.'

4.3.4 Functional Diversity

The first two PCoA axes explain 3.1 and 2.6% of the variation, respectively (Fig. 4.9). Along the first axis, plant species are distinguished by rosette type, woodiness and nitrogen fixation. The second axis differentiates species according to strategy type (ruderals vs. competitors), life form, urbanity and photosynthetic pathway. PCoA analysis (Fig. 4.9) of the floristic data shows a similar pattern of plant communities zonation as DCA (Fig. 4.5).

Invaded plots, when the whole area is considered, were functionally less diverse than uninvaded plots, except for the RaoQ functional index (Figs. 4.10 and 4.11). The FRic index, which does not account for species abundance, showed an increase in functional richness in invaded plots in a gradient from the sea towards inland. Invaded dune slacks and woodland had significantly higher FRic than uninvaded and also other invaded plots. Competitive woody phanerophytes were only found in dune slacks and woodlands (Fig. 4.12). FEve and FDiv indices do not show any statistically significant differences among different habitat types nor between invaded and uninvaded subsets of a particular habitat type (the only exception is the comparison of FEve between 1210 and 2120 habitat types, where $p = 0.024$). RaoQ showed statistically significant differences among habitat types, though only small ones between 1210 and 2110, 2120 and 2270 ($p = 0.065$, $p = 0.025$ and $p = 0.089$, respectively).

The small variation in index values for uninvaded plots of habitat type 2120 was due to the small sample number ($N = 2$).

Functional richness (FRic, Fig. 4.13) showed a linear increase of invaded and uninvaded plots along the DPCoA axis. The first axis represents the gradient of temperature and humidity, indicating the sea-inland gradient. A similar pattern is observed for species richness, indicating that the most species rich plots are at the end of the sea-inland gradient. Again, woody phanerophytes prevail versus more ruderal therophytes in habitats near to the sea. Functional divergence (FDiv) and functional evenness (FEve) increased in invaded plots and decreased in uninvaded. Rao's quadratic entropy (RaoQ) did not show any trend.

Table 4.6 Univariate statistics results obtained from multivariate GLM, indicating which species contribute to differences in abundance between habitats

HT (habitat) species	Deviance	p	AP (alien presence) species	Deviance	p	HT:AP species	Deviance	p
Cyperus capitatus	68.6	0.001	*Xanthium orientale* ssp. *italicum*	48.7	0.002	*Saccharum ravennae*	34.9	0.014
Xanthium orientale ssp. *italicum*	48.7	0.001	*Oenothera* spp.	42.2	0.002	*Schoenus nigricans*	36.8	0.014
Oenothera spp.	42.2	0.001	*Juncus maritimus*	18.0	0.004			
Schoenus nigricans	36.8	0.001						
Saccharum ravennae	34.9	0.001						
Agrostis stolonifera	16.8	0.001						
Juncus maritimus	18.0	0.003						
Echinophora spinosa	51.6	0.004						
Onobrychis caput-galli	23.1	0.004						
Myrtus communis	14.6	0.011						
Vulpia fasciculata	46.8	0.023						
Rubus ulmifolius	31.9	0.025						

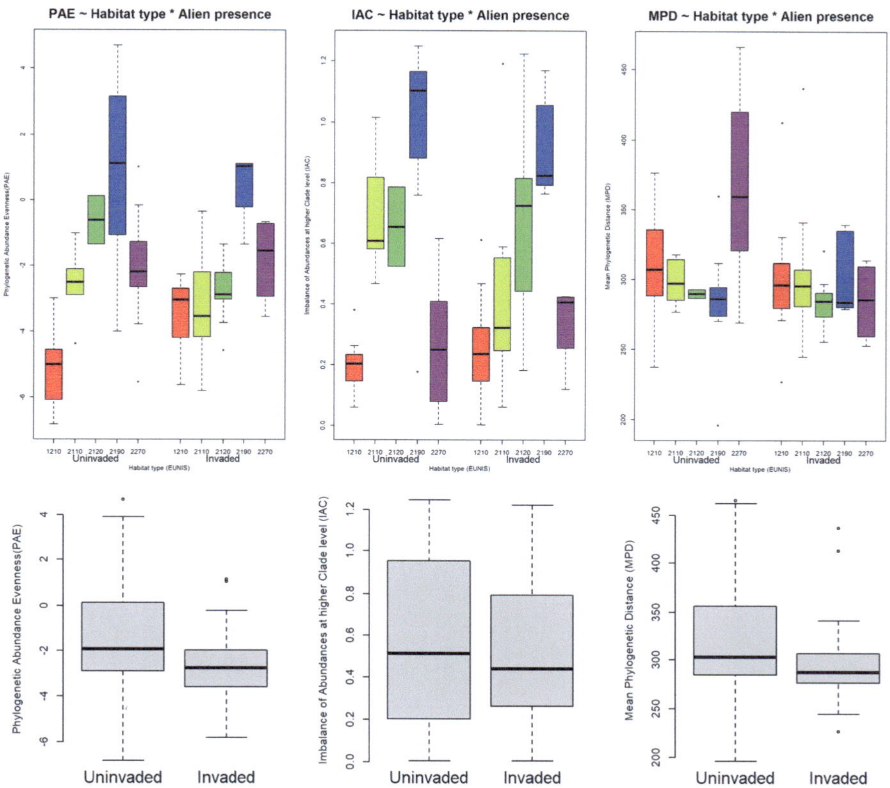

Fig. 4.8 Boxplots of phylogenetic metrics (PAE, IAC, and MPD) for five sand dune habitats (red-1210, yellow-2110, green-2120, blue-2190, and violet-2270) in uninvaded and invaded plots. In grey boxplots, the same phylogenetic metrics are presented for the whole area (all habitats) in uninvaded and invaded plots

4.4 Discussion

4.4.1 Changes of Land Use

Coastal sand dunes have undergone substantial spatio-temporal changes along all coasts in Europe in recent decades. Dune areas have been reduced in their extent and transformed from a few large, elongated patches into a mosaic of sand dune habitats (Malavasi et al. 2013). Although changes of habitat are unique to every beach, primary dune habitats are in general replaced by forests (or *Pinus* plantations), urban and agricultural use (Sciandrello et al. 2015; Bertacchi and Lombardi 2014; Malavasi et al. 2013; Marcantonio et al. 2014). Changes on Velika plaža are not influenced by agricultural expansion (in other countries mainly plastic greenhouses),

Table 4.7 Phylogenetic metrics analysed by two-factor ANOVA with habitat type (HT) as block and alien species presence (AP) as treatment

		Df	Sum	Mean	F	Pr(>F)	
PAE (phylogenetic eveness-abundance)							
	HT	4	269.125	67.281	27.7909	5.58E-15	***
	AP	1	0.013	0.013	0.0052	0.9425	
	HT:AP	4	23.68	5.92	2.4453	0.05221	.
	Residuals	89	215.468	2.241			
IAC (imbalance of abundances at higher clades)							
	HT	4	7.4218	1.85544	34.7992	<2e-16	***
	AP	1	0.0406	0.04064	0.7622	0.385	
	HT:AP	4	0.3333	0.08333	1.5629	0.1912	
	Residuals	89	4.7453	0.05332			
MPD (mean phylogenetic distance)							
	HT	4	59,350	14837.6	9.1565	3.05E-06	***
	AP	1	3477	3477.3	2.1459	0.14647	
	HT:AP	4	21,631	5407.7	3.3371	0.01355	*
	Residuals	89	144,220	1620.4			

Significance codes: 0 '***' 0.001 '**' 0.01 '*' 0.05 '.' 0.1 ' ' 1

since areas of grasslands and vineyards have remained similar through time. Most severe is the increase in tourist facilities, which in many cases are still seasonal, wooden and temporary, although there is an evident trend towards more permanent concrete buildings, like elsewhere (Bertacchi and Lombardi 2014). Humid slacks have decreased substantially, similar to that shown by a Malavasi et al. (2013) study, and this also contributes to the loss of biodiversity, although opposite trends of an increase in humid habitats have also been reported (Sciandrello et al. 2015).

In our study, alien species presence was not influenced by any environmental factor tested and this is in agreement with the study by Marcantonio et al. (2014), in which only grassland land cover type was significant. Natural disturbances and a stressful environment are most important in shaping the species composition of sand dune habitats (Forey et al. 2008), even stronger than anthropogenic disturbance (Ciccarelli 2014).

In several studies of land use, changes on sand dunes, with a significant increase in alien species, have been reported and related to urbanisation and the ornamental use of aliens in the surroundings of dunes (Bertacchi and Lombardi 2014; Malavasi et al. 2013) but this has not been statistically tested. Grasslands developed on previous agricultural land are highlighted as a potential source of alien species (Marcantonio et al. 2014). In further studies, other environmental factors or proxies for propagule pressure should be considered better to understand anthropogenic disturbance and the impact of the presence of alien species on sand dune habitats.

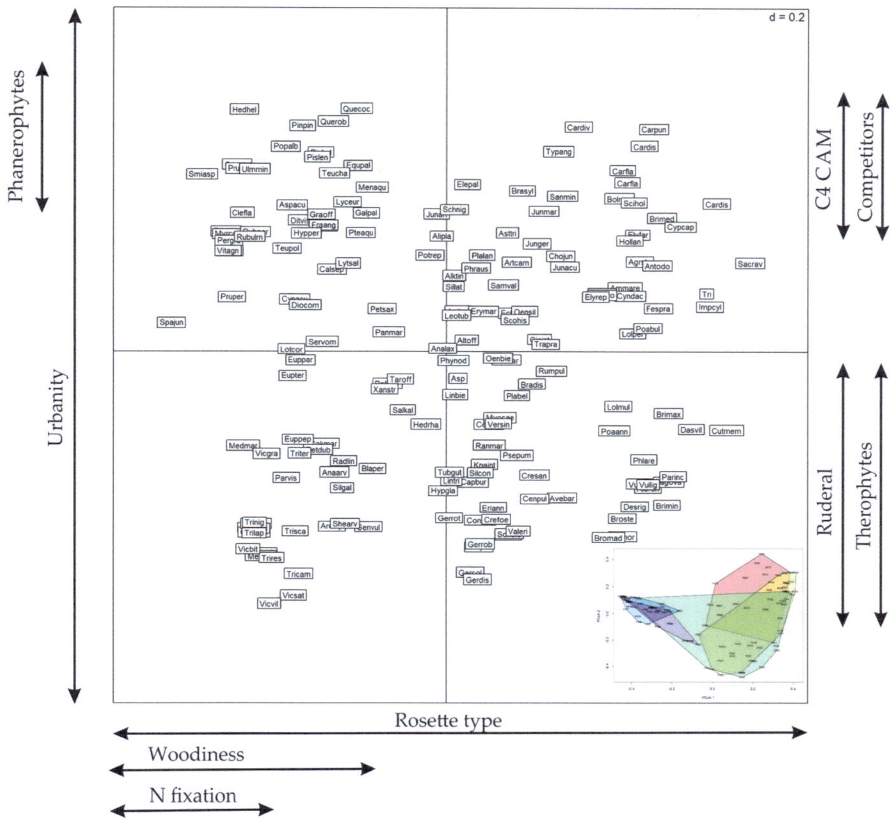

Fig. 4.9 PCoA graph of species based on traits. Outside biplot trait ranges are presented based on PCoA analysis of a particular trait. Species names are shortened to the first three letters of the genus and species name. In the right lower corner, the PCoA of samples by species matrix is shown. Different colours represent the 5 habitat types

4.4.2 Number of Alien Species

The number of alien species reported on sand dunes varies in the literature for different reasons: (i) sampling methods applied, (ii) scale, (iii) environmental characteristics of studied coast, or (iv) human impact. The alien species pool is higher when floristic/vegetation data are used with large sampling quadrats or whole areas (Acosta et al. 2008; Weeda 2010; Asensi et al. 2016; Campos et al. 2004). Surveys with random sampling based on plots obtain lower numbers, although the results may be controversial. In our study, we found 5 alien taxa (2 determined to genus level, Fig. 4.14) and this is comparable only to Carboni et al. (2010) from Central

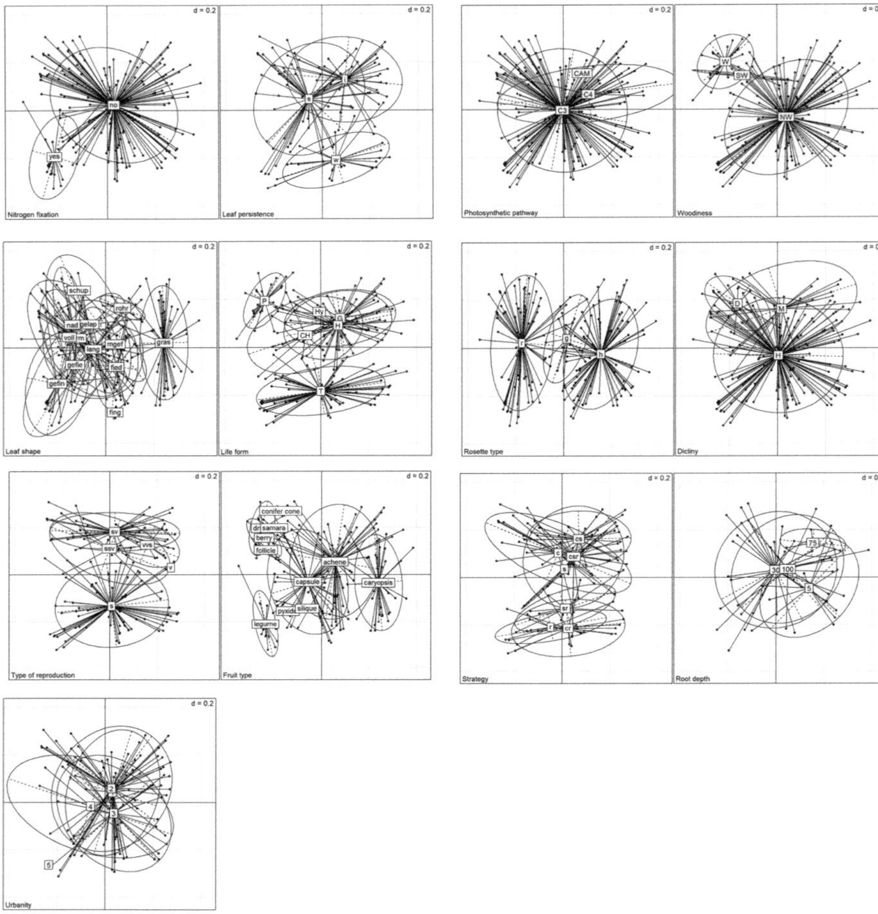

Fig. 4.10 PCoA results for 13 traits: nitrogen fixation, leaf persistence, photosynthetic pathway, woodiness, leaf shape, life form, rosette type, dicliny, type of reproduction, fruit type, strategy, root depth and urbanity. Centroids for a particular trait category are represented with abbreviations in frames

Italy, who found 8 species. Other similar studies report higher numbers (Del Vecchio et al. 2015; Marcantonio et al. 2014) and, even on the same beach, Stešević et al. (2017) using a transect method found 15 species and an additional 10 outside the sampling plots in a floristic survey. Differences in climate may be the reason for different numbers of alien species. Comparison between the warmer Tyrrhenian and colder Adriatic coast reveals a different total number of aliens and also the presence of different chorological types of alien species (Acosta et al. 2008). Human impact is very important, and coasts with high human impact host more alien species, due to more frequent and intensive disturbance and propagule pressure (Pyšek et al. 2010; Acosta et al. 2008).

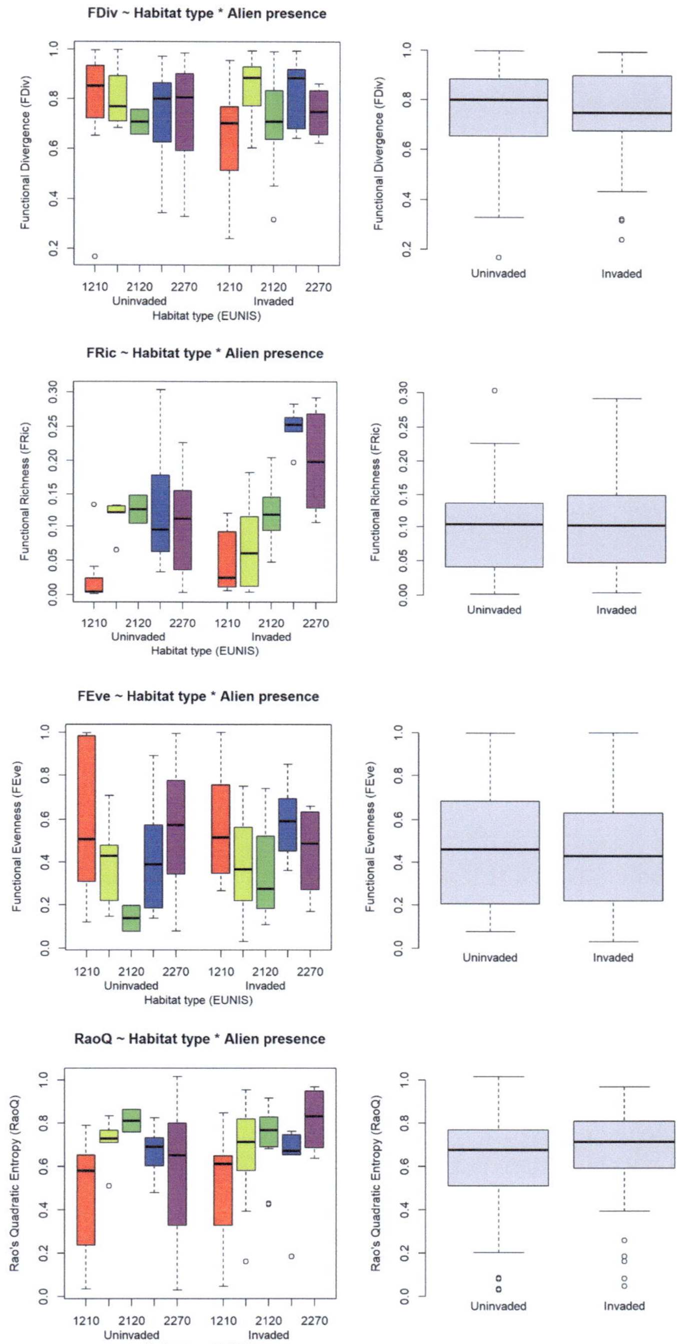

Fig. 4.11 Boxplots of functional metrics indices (FDiv, FRic, FEve and RaoQ) calculated for different dune habitats in uninvaded and invaded plots. Grey boxplots show the same indices calculated for the whole area (all habitats)

Fig. 4.12 Pine forests have a low cover of herb understory

4.4.3 Impact of Alien species on Species Composition

It is commonly accepted that alien species alter (and decrease) the abundance of native species (Santoro et al. 2012; Gaertner et al. 2009; Vilà et al. 2011). The same results were obtained on sand dunes by a multifaceted approach (Marcantonio et al. 2014) and in our study (Fig. 4.6), although we cannot be certain about the causal relationship. It is possible that some external factor influences the abundance (cover) of native species and presence/absence of alien species at the same time, thus indicating a direct correlation.

Alien species were evenly distributed in sand dune habitats (65–85% invaded plots), while their presence was lower in dune slacks and woodlands (20–25%, Table 4.2). The high alien presence on the strandline and embryonic dune is due to *Xanthium orientale* ssp. *italicum* (Fig. 4.15). The alien status of this species is doubtful (Del Vecchio et al. 2015), but we considered it to be a neophyte. Other alien species are probably not successful in invading this harsh environment. More neophytes could be found on transitional dunes, where the environment is less stressful (Carboni et al. 2010), but also disturbed. Disturbance by tourists and cattle grazing (Šilc et al. 2017; Fenu et al. 2013b; Lechuga-Lago et al. 2017) and previous agricultural use with enhanced fertile soil (Fenu et al. 2013a; Marcantonio et al. 2014) enable alien species to settle and further spread. In

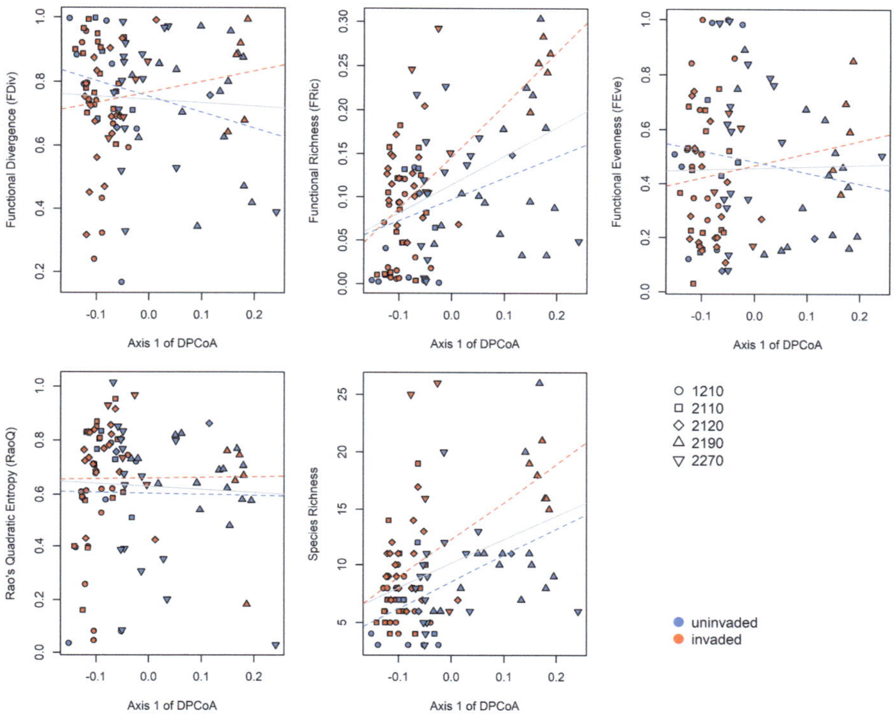

Fig. 4.13 Scatter plots of functional diversity indices and Axis 1 values of Double PCoA (DPCoA) analysis. Habitat types and invaded and uninvaded plots are shown with different symbols and colours (see legend on the right). Three lines represent least-squared regressions for all plots (grey), invaded (red) and uninvaded (blue)

our case soil conditions are also improved by nitrogen fixating species of Fabaceae family (*Onobrychis caput-galli*, *Trifolium* sp., *Medicago marina*).

4.4.4 Phylogeny

In our study, the dune zone significantly affected all three index values (unlike in Marcantonio et al. (2014)), probably because we included two additional different dune zones – dune slacks (HT 2190) have the highest IAC and PAE index values. Having the highest mean number of species per plot simultaneously influences the former, but should not affect the latter index values (Cadotte et al. 2010).

The PAE index increased along the sea-inland gradient, with the highest values in humid dune slacks and a decrease in pine forest. High values in humid slacks reflect the most stable and undisturbed plant community, with low disturbance and high species richness per plot. Lower values in pine plantations indicate man-made disturbances, due to recreational activities (Zedda et al. 2010). This pattern was

Fig. 4.14 *Oenothera* species dominate on more stable dunes

Fig. 4.15 *Xanthium orientale* ssp. *italicum* is the most abundant species on the strandline

clearer in uninvaded plots, while in invaded plots there were no differences in phylogenetic diversity among the first three sand dune habitats in the zonation.

The value of the IAC index is supposed to point to the prevailing key in forming particular communities – whether competition or environmental filtering, the so-called "competition-relatedness hypothesis" (Cadotte et al. 2010; Cahill et al. 2008). In our case, IAC index values became higher from the first to the fourth HT, which is supposed to indicate higher abundances of phylogenetically closely related species, with consequent higher relatedness in functional traits and therefore no competitive exclusion of similar species (Cadotte et al. 2010). This usually happens in habitats under higher environmental stress and disturbance, with the predominance of environmental filtering as the key in forming particular communities. For the sequence of dune HTs, the exact opposite is usually assumed – the degree of environmental stress gets smaller moving inland, with an increasing percentage of vegetation cover (confirmed but not shown for our data) and increasing competition, which should result in an ever-lower IAC index.

The MPD index represents the mean phylogenetic distances between all pairs of species in a particular community (without including abundance values). We believe that the value can be influenced by at least two factors and their interaction can yield barely interpretable results. (i) Considering the only statistically significant result of MPD, the drop in the index in invaded 2270 HT may be a consequence of the far less frequent presence of *Pinus* spp. which are the most distant relative to other species, such as gymnosperms, hence have a heavier impact (Honorio Coronado et al. 2015), thus lowering mean phylogenetic distances. (ii) Another reason may be higher species numbers in invaded plots (Kruskal-Wallis test, $p = 0.08$) that are closely related (e.g., the same families) and therefore increase the density in a few branches of the phylogenetic tree and decrease mean distances. Higher mean species numbers in invaded plots and (slightly) higher mean herb layer cover probably indicate stronger disturbance, because light cannot be the determining factor since tree cover is higher in invaded plots (Kruskal-Wallis, $p = 0.38$). The first three HTs do not differ among themselves, regardless of invasion status, which is the opposite to the findings of Marcantonio et al. (2014).

It seems that the shape of HT sequences in the graphs of functional and phylogenetic indices often follow a similar (binomial) shape (positive or negative), but an interrelated explanation and a straightforward ecological interpretation are hard to extract. For a more certain interpretation of functional and phylogenetic indices, more studies should use them (also for different habitat types) to clarify possibly underlying ecological patterns. As several studies have pointed out, phylogenetic diversity is not the best proxy for the functional diversity of plant communities in different environments (Lososová et al. 2016; Carboni et al. 2013) and should be used with caution. In addition, some theoretical works argue that mixed results from studies on the phylogenetic interpretation of ecological communities assembly show no consistent pattern due to i) other mechanisms forming plant communities, in addition to competition and environmental filtering (e.g., facilitation) and ii) alternative correlations between phylogenetic similarity and factors driving the community assembly (in some cases, phylogenetically closely related species might experience high competition and, in other cases, the opposite) (Mayfield and Levine 2010).

4.4.5 Traits

PCoA explained a lower percentage of variance compared to the study of Marcantonio et al. (2014) (17 and 14% for the first two axes). The reason may be a longer sea-inland gradient, extending to pine forest, and a larger species pool, with 25% of species occurring only once. In cases with large and noisy datasets, the explained variance can be low, but the results can still be ecologically informative (Gauch 1982).

The relatively small number of highly specialized species (for the same environment) accounts for the small functional richness (FRic) in the first dune zone and increases both in the presence of alien species (and consequently with the addition of new trait values) and in other habitat types with higher species richness (and consequently with higher trait values diversity; Table 4.3, Fig. 4.9).

In dune slacks (HT 2190) and pine forest (HT 2270), invaded plots have a higher functional richness, which occurs not only because of the inclusion of alien species trait values but also because these plots contain higher species numbers, including some less typical species for this habitat type (*Vicia sativa*, *V. grandiflora*, *Anagallis arvensis*, *Trifolium campestre*, *Lathyrus sphaericus* and *Onobrychis caput-galli, Catapodium rigidum, Lagurus ovatus, Euphorbia peplis*). The species indicate a transitional position of these plots (drier and better light conditions). This is in accordance with the theory that disturbance through resources fluctuation increases the level of invasion in different habitat types (Chytrý et al. 2008a; Davis et al. 2000).

Success of invasion along the environmental gradient depends on two processes: environmental filtering and competition. In stressful conditions, successful invaders are similar to native species and this leads to trait convergence and increases functional similarity. On the other hand, in a more productive environment, competition is important and trait divergence occurs and decreases functional similarity (Lososová et al. 2016; Gallien and Carboni 2017). In two differently disturbed (invaded) sand dune communities, different and opposing processes are acting and invaded plots on sand dunes showed clear functional and phylogenetic differentiation (Marcantonio et al. 2014).

4.5 Conclusions

Our study confirmed the importance of applying an approach that includes analysis of land-use changes, and functional and phylogenetic diversity, on the impact of alien species on sand dunes, providing a view of the problem from different perspectives. We demonstrated that the species and functional diversity of sand dune plant communities change due to alien invasion, while the results of the phylogenetic analysis should be used with caution.

Coastal regions and sand dunes are one of most endangered ecosystems worldwide and alien species are spreading and changing their natural species composition. The results of this study will therefore have important implications for the management of sandy beaches along the NE Adriatic, where human impact and propagule pressure of alien species have been increasing in recent decades, while there are still well-preserved sand dunes compared to other coasts in this part of Europe.

Acknowledgements We thank Iztok Sajko for producing the map. Martin Cregeen kindly checked our English. The research was partly financed by the Slovenian Research Agency (ARRS) through a research program (P1-0236) and bilateral project (BI-ME/16-17-018) between ARRS and the Montenegrin Ministry of Science.

The study has been supported by the TRY initiative on plant traits (http://www.try-db.org). The TRY initiative and database is hosted, developed and maintained by J. Kattge and G. Bönisch (Max Planck Institute for Biogeochemistry, Jena, Germany). TRY is currently supported by DIVERSITAS/Future Earth and the German Centre for Integrative Biodiversity Research (iDiv) Halle-Jena-Leipzig.

References

Acosta A, Izzi CF, Stanisci A (2006) Comparison of native and alien plant traits in Mediterranean coastal dunes. Community Ecol 7(1):35–41

Acosta ATR, Carranza ML, Martino LD, Frattaroli A, Izzi CF, Stanisci A (2008) Patterns of native and alien plant species occurrence on coastal dunes in Central Italy. In: Tokarska-Guzik B, Brock JH, Brundu G, Child L, Daehler CC, Pysek P (eds) Plant invasions: human perception, ecological impacts and management. Backhuys Publishers, Leiden, pp 235–248

Aitzetmüller K, Matthäus B, Friedrich H (2003) A new database for seed oil fatty acids – the database SOFA. Eur J Lipid Sci Technol 105(2):92–103. https://doi.org/10.1002/ejlt.200390022

ArcGIS 10.4 (2015) GIS ESRI. Environmental Systems Research Institute Inc, Redlands

Asensi A, Díez-Garretas B, Pereña J (2016) Alien plants of coastal dune habitats in southern Spain. Plant Biosyst Int J Deal All Asp Plant Biol 150(3):477–483

Barbier EB, Hacker SD, Kennedy C, Koch EW, Stier AC, Silliman BR (2011) The value of estuarine and coastal ecosystem services. Ecol Monogr 81(2):169–193. https://doi.org/10.1890/10-1510.1

Bertacchi A, Lombardi T (2014) Diachronic analysis (1954-2010) of transformations of the dune habitat in a stetch of the nortehern Thyrrhenian coast (Italy). Plant Biosyst Int J Deal All Asp Plant Biol 148(2):227–236

Botta-Dukat Z (2005) Rao's quadratic entropy as a measure of functional diversity based on multiple traits. J Veg Sci 16:533–540
Braun-Blanquet J (1964) Pflanzensoziologie. Grundzüge der Vegetationskunde, 3rd edn. Springer, Wien
Brown AC, McLachlan A (2002) Sandy shore ecosystems and the threats facing them: some predictions for the year 2025. Environ Conserv 29(01):62–77. https://doi.org/10.1017/S037689290200005X
Buja K, Menza C (2012) Sampling design tool for ArcGIS. NOAA's Biogeography Branch, Silver Spring
Burić D, Ducić V, Mihajlović J (2014) The climate of Montenegro: modificators and types- part two. Bull Serbian Geogr Soc 9(1):73–90
Cadotte MW, Jonathan Davies T, Regetz J, Kembel SW, Cleland E, Oakley TH (2010) Phylogenetic diversity metrics for ecological communities: integrating species richness, abundance and evolutionary history. Ecol Lett 13(1):96–105. https://doi.org/10.1111/j.1461-0248.2009.01405.x
Cahill JFJ, Kembel SW, Lamb EG, Keddy PA (2008) Does phylogenetic relatedness influence the strength of competition among vascular plants? Perspect Plant Ecol Evol Systemat 10:41–50
Calcagno V, de Mazancourt C (2010) Glmulti: an R package for easy automated model selection with (generalized) linear models. J Stat Softw 34:1–29
Campos JA, Herrera M, Biurrun I, Loidi J (2004) The role of alien plants in the natural coastal vegetation in Central-Northern Spain. Biodivers Conserv 13(12):2275–2293
Carboni M, Carranza ML, Acosta A (2009) Assessing conservation status on coastal dunes: A multiscale approach. Landsc Urban Plan 91(1):17–25. https://doi.org/10.1016/j.landurbplan.2008.11.004
Carboni M, Santoro R, Acosta ATR (2010) Are some communities of the coastal dune zonation more susceptible to alien plant invasion? J Plant Ecol 3(2):139–147. https://doi.org/10.1093/Jpe/Rtp037
Carboni M, Acosta ATR, Ricotta C (2013) Are differences in functional diversity among plant communities on Mediterranean coastal dunes driven by their phylogenetic history? J Veg Sci 24 (5):932–941. https://doi.org/10.1111/jvs.12095
Carranza ML, Carboni M, Feola S, Acosta ATR (2010) Landscape-scale patterns of alien plant species on coastal dunes: the case of iceplant in Central Italy. Appl Veg Sci 13(2):135–145. https://doi.org/10.1111/j.1654-109X.2009.01065.x
Chytrý M, Jarošík V, Pyšek P, Hájek O, Knollová I, Tichý L, Danihelka J (2008a) Separating habitat invasibility by alien species from the actual level of invasion. Ecology 89(6):1541–1553. https://doi.org/10.1890/07-0682.1
Chytrý M, Maskell LC, Pino J, Pyšek P, Vilà M, Font X, Smart SM (2008b) Habitat invasions by alien plants: a quantitative comparison among Mediterranean, subcontinental and oceanic regions of Europe. J Appl Ecol 45(2):448–458. https://doi.org/10.1111/j.1365-2664.2007.01398.x
Ciccarelli D (2014) Mediterranean coastal sand dune vegetation: influence of natural and anthropogenic factors. Environ Manag 54(2):194–204
Davis MA, Grime JP, Thompson K (2000) Fluctuating resources in plant communities: a general theory of invasibility. J Ecol 88(3):528–534. https://doi.org/10.1046/j.1365-2745.2000.00473.x
De Rosario-Martinez H (2015) Phia: post-hoc interaction analysis. R package version 0.2-1
Defeo O, McLachlan A, Schoeman DS, Schlacher TA, Dugan J, Jones A, Lastra M, Scapini F (2009) Threats to sandy beach ecosystems: a review. Estuar Coast Shelf Sci 81(1):1–12. https://doi.org/10.1016/j.ecss.2008.09.022
Del Vecchio S, Pizzo L, Buffa G (2015) The response of plant community diversity to alien invasion: evidence from a sand dune time series. Biodivers Conserv 24(2):371–392
Doing H (1985) Coastal fore-dune zonation and succession in various parts of the world. Vegetatio 61(1–3):65–75. https://doi.org/10.1007/Bf00039811

Euro+Med (2006–) Euro+Med PlantBase – the information resource for Euro-Mediterranean plant diversity. http://ww2.bgbm.org/EuroPlusMed/. Accessed Published on the Internet. Accessed 2 Apr 2016

Fenu G, Carboni M, Acosta ATR, Bacchetta G (2013a) Environmental factors influencing coastal vegetation pattern: new insights from the Mediterranean Basin. Folia Geobotanica 48 (4):493–508. https://doi.org/10.1007/s12224-012-9141-1

Fenu G, Cogoni D, Ulian T, Bacchetta G (2013b) The impact of human trampling on a threatened coastal Mediterranean plant: the case of Anchusa littorea Moris (Boraginaceae). Flora 208 (2):104–110. https://doi.org/10.1016/j.flora.2013.02.003

Flore et végétation de la France: CATMINAT (2017) http://philippe.julve.pagesperso-orange.fr/catminat.htm. Accessed Aug 2017

Forey E, Chapelet B, Vitasse Y, Tilquin M, Touzard B, Michalet R (2008) The relative importance of disturbance and environmental stress at local and regional scales in French coastal sand dunes. J Veg Sci 19:493–502

Gaertner M, Breeyen AD, Hui C, Richardson DM (2009) Impacts of alien plant invasions on species richness in Mediterranean-type ecosystems: a meta-analysis. Prog Phys Geogr 33 (3):319–338. https://doi.org/10.1177/0309133309341607

Gallien L, Carboni M (2017) The community ecology of invasive species: where are we and what's next? Ecography 40:335–352

Gauch HGJ (1982) Multivariate analysis in community ecology. Cambridge University Press, Cambridge

Genovesi P, Shine C (2004) European strategy on invasive alien species: convention on the conservation of European wildlife and habitats (Bern convention), vol 137. Nature and Environment. Council of Europe Publishing, Bern

Honorio Coronado EN, Dexter KG, Pennington RT, Chave J, Lewis SL, Alexiades MN, Alvarez E, Alves de Oliveira A, Amaral IL, Araujo-Murakami A, Arets EJMM, Aymard GA, Baraloto C, Bonal D, Brienen R, Cerón C, Cornejo Valverde F, Di Fiore A, Farfan-Rios W, Feldpausch TR, Higuchi N, Huamantupa-Chuquimaco I, Laurance SG, Laurance WF, López-Gonzalez G, Marimon BS, Marimon-Junior BH, Monteagudo Mendoza A, Neill D, Palacios Cuenca W, Peñuela Mora MC, Pitman NCA, Prieto A, Quesada CA, Ramirez Angulo H, Rudas A, Ruschel AR, Salinas Revilla N, Salomão RP, Segalin de Andrade A, Silman MR, Spironello W, ter Steege H, Terborgh J, Toledo M, Valenzuela Gamarra L, Vieira ICG, Vilanova Torre E, Vos V, Phillips OL (2015) Phylogenetic diversity of Amazonian tree communities. Divers Distrib 21 (11):1295–1307. https://doi.org/10.1111/ddi.12357

Isermann M (2008) Classification and habitat characteristics of plant communities invaded by the non-native Rosa rugosa Thunb. in NW Europe. Phytocoenologia 38(1–2):133–150

Jucker T, Carboni M, Acosta ATA (2013) Going beyond taxonomic diversity: deconstructing biodiversity patterns reveals the true cost of iceplant invasion. Divers Distrib 19(13):1566–1577

Kattge J, Díaz S, Lavorel S, Prentice IC, Leadley P, Bönisch G, Garnier E, Westoby M, Reich PB, Wright IJ, Cornelissen JHC, Violle C, Harrison SP, Van Bodegom PM, Reichstein M, Enquist BJ, Soudzilovskaia NA, Ackerly DD, Anand M, Atkin O, Bahn M, Baker TR, Baldocchi D, Bekker R, Blanco CC, Blonder B, Bond WJ, Bradstock R, Bunker DE, Casanoves F, Cavender-Bares J, Chambers JQ, Chapin FS III, Chave J, Coomes D, Cornwell WK, Craine JM, Dobrin BH, Duarte L, Durka W, Elser J, Esser G, Estiarte M, Fagan WF, Fang J, Fernández-Méndez F, Fidelis A, Finegan B, Flores O, Ford H, Frank D, Freschet GT, Fyllas NM, Gallagher RV, Green WA, Gutierrez AG, Hickler T, Higgins SI, Hodgson JG, Jalili A, Jansen S, Joly CA, Kerkhoff AJ, Kirkup D, Kitajima K, Kleyer M, Klotz S, Knops JMH, Kramer K, Kühn I, Kurokawa H, Laughlin D, Lee TD, Leishman M, Lens F, Lenz T, Lewis SL, Lloyd J, LlusiÀ J, Louault F, Ma S, Mahecha MD, Manning P, Massad T, Medlyn BE, Messier J, Moles AT, Müller SC, Nadrowski K, Naeem S, Niinemets Ü, Nöllert S, Nüske A, Ogaya R, Oleksyn J, Onipchenko VG, Onoda Y, Ordoñez J, Overbeck G, Ozinga WA, Patiño S, Paula S, Pausas JG, Peñuelas J, Phillips OL, Pillar V, Poorter H, Poorter L, Poschlod P, Prinzing A, Proulx R, Rammig A, Reinsch S, Reu B, Sack L, Salgado-Negret B, Sardans J, Shiodera S, Shipley B, Siefert A,

Sosinski E, Soussana JF, Swaine E, Swenson N, Thompson K, Thornton P, Waldram M, Weiher E, White M, White S, Wright SJ, Yguel B, Zaehle S, Zanne AE, Wirth C (2011) TRY – a global database of plant traits. Glob Chang Biol 17(9):2905–2935. https://doi.org/10.1111/j.1365-2486.2011.02451.x
Klimešová J, De Bello F (2009) CLO-PLA: the database of clonal and bud bank traits of Central European flora. J Veg Sci 20(3):511–516 https://doi.org/10.1111/j.1654-1103.2009.01050.x
Kühn I, Durka W, Klotz S (2004) BiolFlor — a new plant-trait database as a tool for plant invasion ecology. Divers Distrib 10(5–6):363–365. https://doi.org/10.1111/j.1366-9516.2004.00106.x
Laliberté E, Legendre P, Shipley B (2014) FD: measuring functional diversity from multiple traits, and other tools for functional ecology. vol R package version 1.0-12
Lechuga-Lago Y, Novoa A, Le Roux JJ, González L (2017) Understanding the influence of urbanization on invasibility: Carpobrotus edulis as an exemplar. Biol Invasions 19:3601. https://doi.org/10.1007/s10530-017-1593-z
Lingoes JC (1971) Some boundary conditions for a monotone analysis of symmetric matrices. Psychometrika 36:195–203
Lososová Z, Čeplová N, Chytrý M, Tichý L, Danihelka J, Fajmon K, Láníková D, Preislerová Z, Řehořek V (2016) Is phylogenetic diversity a good proxy for functional diversity of plant communities? A case study from urban habitats. J Veg Sci 27(5):1036–1046. https://doi.org/10.1111/jvs.12414
Malavasi M, Santoro R, Cutini M, Acosta ATR, Carranza ML (2013) What has happened to coastal dunes in the last half century? A multitemporal coastal landscape analysis in Central Italy. Landsc Urban Plan 119:54–63. https://doi.org/10.1016/j.landurbplan.2013.06.012
Marcantonio M, Rocchini D, Ottaviani G (2014) Impact of alien species on dune systems: a multifaceted approach. Biodivers Conserv 23(11):2645–2668. https://doi.org/10.1007/s10531-014-0742-2
Martínez ML, Psuty NP, Lubke RA (2004) A perspective on coastal dunes. In: Martínez ML, Psuty NP (eds) Coastal dunes. Ecology and conservation. Springer, Heidelberg, pp 3–10
Mayfield MM, Levine JM (2010) Opposing effects of competitive exclusion on the phylogenetic structure of communities. Ecol Lett 13(9):1085–1093. https://doi.org/10.1111/j.1461-0248.2010.01509.x
McLachlan A, Brown AC (2006) The ecology of sandy shores, 2nd edn. Academic, New York
Novoa A, González L, Moravcová L, Pyšek P (2012) Effects of soil characteristics, allelopathy and frugivory on establishment of the invasive plant Carpobrotus edulis and a co-occuring native, Malcolmia littorea. PLoS One 7(12):e53166. https://doi.org/10.1371/journal.pone.0053166
Official Gazette of Montenegro (2007) (30/07)
Official Gazette SRCG (1968) (30/68)
Paradis E, Claude J, Strimmer K (2004) APE: analyses of phylogenetics and evolution in R language. Bioinformatics 20:289–290
Pavoine S, Vallet J, Dufour AB, Gachet S, Hervé D (2009) On the challenge of treating various types of variables: application. Oikos 118:391–402
Pearse WD, Cadotte MW, Cavender-Bares J, Ives AR, Tucker CM, Walker SC, Helmus MR (2015) Pez: phylogenetics for the environmental sciences. Bioinformatics 31(17):2888–2890. https://doi.org/10.1093/bioinformatics/btv277
Petković S (2012) Uticaj režima reke Bojane na širu okolinu u zoni njenog ušća u Jadransko more. Paper presented at the 16. Conference SDHI and SDH, Donji Milanovac, Serbia
Petrović D, Hadžiablahović S, Vuksanović S, Mačić V, Lakušić D (2012) Katalog tipova staništa Crne Gore značajnih za Evropsku uniju. Podgorica-Beograd-Zagreb
Pignatti S (1982) Flora d'Italia, vol 1–3. Edagricole, Bologna
Pignatti S (2005) Valori di bioindicazione delle piante vascolari della flora d'Italia. Braun-Blanquetia 39:1–97
Pikelj K, Dragnic V, Malovrazic N (2013) Eastern Adriatic: Slovenia, Croatia and Montenegro. In: Pranzini E, Williams A (eds) Coastal erosion and protection in Europe. Routledge, London/New York, pp 324–344

Pyšek P, Jarošik V, Hulme PE, Kuhn I, Wild J, Arianoutsou M, Bacher S, Chiron F, Didziulis V, Essl F, Genovesi P, Gherardi F, Hejda M, Kark S, Lambdon PW, Desprez-Loustau ML, Nentwig W, Pergl J, Poboljšaj K, Rabitsch W, Roques A, Roy DB, Shirley S, Solarz W, Vila M, Winter M (2010) Disentangling the role of environmental and human pressures on biological invasions across Europe. Proc Natl Acad Sci U S A 107(27):12157–12162. https://doi.org/10.1073/pnas.1002314107

Qian H, Jin Y (2016) An updated megaphylogeny of plants, a tool for generating plant phylogenies and an analysis of phylogenetic community structure. J Plant Ecol 9:233–239

Rakaj M, Rostanski K (2009) New species of the genus Oenothera L. from coasts of the Albania and Montenegro. Natura Montenergrina 8(3):163–171

Santoro R, Jucker T, Carboni M, Acosta ATR (2012) Patterns of plant community assembly in invaded and non-invaded communities along a natural environmental gradient. J Veg Sci 23 (3):483–494. https://doi.org/10.1111/j.1654-1103.2011.01372.x

Sciandrello S, Tomaselli G, Minissale P (2015) The role of natural vegetation in the analysis of the spatio-temporal changes of coastal dune system: a case study in Sicily. J Coast Conserv 19 (2):199–212. https://doi.org/10.1007/s11852-015-0381-0

Seed Information Database (SID). Version 7.1. (2017) Available from: http://data.kew.org/sid/. Accessed Sept 2017

Šilc U, Caković D, Küzmič F, Stešević D (2017) Trampling impact on vegetation of embryonic and stabilised sand dunes in Montenegro. J Coast Conserv 21:15–21. https://doi.org/10.1007/s11852-016-0468-2

Stešević D, Caković D (2013) Contribution to the alien flora of Montenegro and Supplementum to the preliminary list of plant invaders. Biologica Nyssana 4(1–2):1–7

Stešević D, Petrović D (2010) Preliminary list of plant invaders in Montenegro. Biologica Nyssana 1(1–2):35–42

Stešević D, Luković M, Caković D, Ružić N, Bubanja N, Šilc U (2017) Distribution of alien species along sand dune plant communities zonation. Period Biol 119:239–249. https://doi.org/10.18054/pb.v119i4.4917

Tichý L (2002) JUICE, software for vegetation classification. J Veg Sci 13:451–453

Tutin TG, Heynard VH, Burges NA, Valentin DH, Walters SM, Webb DA (1964–80) Flora Europaea, 1–5. Cambridge University Press, Cambridge

van der Maarel E (2003) Some remarks on the functions of European coastal ecosystems. Phytocoenologia 33(2–3):187–202. https://doi.org/10.1127/0340-269x/2003/0033-0187

Vilà M, Espinar JL, Hejda M, Hulme PE, Jarošik V, Maron JL, Pergl J, Schaffner U, Sun Y, Pyšek P (2011) Ecological impacts of invasive alien plants: a meta-analysis of their effects on species, communities and ecosystems. Ecol Lett 14:702–708

Villeger S, Mason NWH, Mouillot D (2008) New multidimensional functional diversity indices for a multifaceted framework in functional ecology. Ecology 89:2290–2301

Wang Y, Naumann U, Wright ST, Warton DI (2012) mvabund – an R package for model-based analysis of multivariate abundance data. Methods Ecol Evol 3:471–474

Weeda EJ (2010) The role of archaeophytes and neophytes in the Dutch coastal dunes. J Coast Conserv 14(2):75–79

Zanne AE, Tank DC, Cornwell WK, Eastman JM, Smith SA, FitzJohn RG, McGlinn DJ, O'Meara BC, Moles AT, Reich PB, Royer DL, Soltis DE, Stevens PF, Westoby M, Wright IJ, Aarssen L, Bertin RI, Calaminus A, Govaerts R, Hemmings F, Leishman MR, Oleksyn J, Soltis PS, Swenson NG, Warman L, Beaulieu JM (2014) Three keys to the radiation of angiosperms into freezing environments. Nature 506(7486):89

Zedda L, Cogoni A, Flore F, Brundu G (2010) Impacts of alien plants and man-made disturbance on soil-growing bryophyte and lichen diversity in coastal areas of Sardinia (Italy). Plant Biosyst Int J Deal All Asp Plant Biol 144(3):547–562. https://doi.org/10.1080/11263501003638604

Chapter 5
Manila Bay Ecology and Associated Invasive Species

Benjamin M. Vallejo Jr., Alexander B. Aloy, Melody Ocampo, Jennifer Conejar-Espedido, and Leanna M. Manubag

Abstract As the location of the oldest and busiest international port in the Philippines, Manila Bay is a prime location to study the ecology and dynamics of marine biological invasion in a tropical high marine biodiversity environment. The bay is historically within the center of marine biodiversity in the Philippines. Rapid urbanization in the last 150 years as a result of the expansion of the City of Manila and its suburbs has changed the estuarine watershed and has resulted in environmental change. This is reflected in eutrophication, sedimentation, pollution and land reclamation that has altered fisheries, coastal oceanography and ecology. These problems are exacerbated by the governance system of the National Capital Region resulting in unplanned urban development. In this environment, most of the invertebrate marine non-indigenous species (MNIS) that are invasive in other tropical estuarine ports are not invasive in Manila Bay. It is hypothesized that monsoon driven periodic hypoxia and other pollution related events prevent invasion. While this is true for most of the MNIS, the tropical Atlantic mussel *Mytella charruana* was introduced in 2014 and is now invasive in the bay. The biological characteristics of this species are likely to displace the pollution tolerant indigenous and non-indigenous malacofauna. The Supreme Court has taken notice of the environmental condition of the bay and through a 2008 Mandamus, ordered

B. M. Vallejo Jr. (✉)
Institute of Environmental Science and Meteorology, Science and Society Program, College of Science, University of the Philippines, Diliman, Quezon City, Philippines
e-mail: bmvallejo1@up.edu.ph

A. B. Aloy
Toolijooa Environmental Restoration, Terrey Hills, NSW, Australia

M. Ocampo
Department of Biology, University of the Philippines Manila, Ermita, Manila, Philippines

J. Conejar-Espedido
Institute of Biological Sciences, University of the Philippines Los Baños, College, Los Baños, Laguna, Philippines

L. M. Manubag
Biodiversity Management Bureau, Diliman, Quezon City, Philippines

C. Makowski, C. W. Finkl (eds.), *Impacts of Invasive Species on Coastal Environments*, Coastal Research Library 29,
https://doi.org/10.1007/978-3-319-91382-7_5

the Executive Department of the Philippines Government to restore the environmental quality of the bay. The effects of the Mandamus in rehabilitating the bay still have to be seen and on this any effective management of any marine biological invasion lies.

5.1 Introduction

This chapter reviews recent studies on marine biological invasion ecology in Manila Bay, Luzon Island, the Philippines and focuses particularly on fouling organisms. Fouling species are the first to colonize port works and have larvae brought in by ship ballast water. Thus they account for many biological invasion events in the marine environment. The establishment of fouling marine non-indigenous species (MNIS) to new locations and habitats is one of the consequences of anthropogenic global environment change. For ecologists and biogeographers, the introduction of MNIS provide an unprecedented opportunity to directly study the role of environmental factors in structuring the ecological community and the consequent expansion of MNIS geographical range.

There is strong evidence to show that species interactions play an important role in the structuring of natural communities. At the initial stage of invasion, effects may range from decline of abundance in indigenous or resident species subject to direct competition by MNIS, by predation by MNIS or an increase in species abundance facilitated by the MNIS. These effects are usually observed in concert with physical modifications in the habitat or by the facilitating MNIS establishment provided by current component species in the community. In many cases of biological invasion in aquatic habitats, the invaded ecosystem is stressed and subject to chronic anthropogenic stressors. MNIS settlement in ship harbors and ports are a prime example where these events can be studied.

In Southeast Asia, Manila Bay is a focus area in studying MNIS introductions and establishment leading to biological invasion. Manila has been an important port since before the ninth century CE and one of the oldest ports in Asia. As a tropical harbor, it is an interesting area for investigating the ecology of biological invasion.

5.1.1 Manila Bay

Manila Bay is a semi-enclosed estuary that is considered to be one of the most important ports in the Philippines and in Asia. The bottom has a gentle slope with an average depth of 17 m. Its shoreline spans 190 km and includes the provinces of Bataan, Cavite and Pampanga. Varying depths, topography, direction and magnitude of circulation, salinity and nutrient composition makes the Manila Bay a complex ecosystem. Once famous for its pristine beauty, it now has notorious image of pollution and destruction brought about mainly by urbanization and increase in coastal population and various other industries around the bay. Historically, Manila Bay is part of "center of the center of marine biodiversity" which extends to the southwestern coastline of Luzon. Prior to environmental changes brought about by urbanization, the bay had extensive mudflats, sandflats, seagrass, mangroves and coral reefs at its mouth. It also has significant estuarine habitats. Some areas towards the mouth of the bay like Bataan and westernmost Cavite provincial coastlines are not as heavily developed and polluted and may still contain seagrass, mangrove and even reef ecosystems. The most common habitats in the bay however, are the mudflats and sandy beaches. Now currently littered with tons of garbage, the stretch of coasts along the City of Manila, Pasay City, Paranaque, Las Piñas to Cavite City used to be productive sandflats and mudflats.

Manila Bay is almost completely surrounded by highly urbanized and rapidly urbanizing communities. It has many environmental issues ranging from land and sea based pollution (Prudente et al. 1994, 1997; Sta Maria et al. 2009), sedimentation, harmful algal blooms (Azanza et al. 2004), overexploitation of fishery resources (Munoz 1993), reclamation, land conversion and most recently, biological invasion (Chavanich et al. 2010; Jacinto et al. 2006b). The Port of Manila is the Philippines biggest port. Manila's international ports, North and South Harbors have recorded a total of 4793 foreign ship calls in 2009 representing 73 thousand tons and an estimated 13 thousand hours of port service calls (PPA 2010). Areas near the port used for mariculture and fisheries and is the main shipping port of the country (PPA 2010). Thus the risk of MNIS biological invasion is high.

Manila Bay's major tourism attractions are no longer its beaches but commercial developments along its waterfront. Many of these are built on reclaimed land. As with any commercial development, there have been problems regarding the hazard risks of these commercial developments. Reclamation is also being blamed by environmentalist groups as responsible for the flooding problems along the coastal strip.

The Philippines Department of Environment and Natural Resources (DENR) has the mandate to monitor the environmental quality of the bay at regular intervals. Private business enterprises such as the Manila Ocean Park have monitored and recorded the water quality of the bay daily for the past 10 years as part of their operations. Their data shows evidence of a decreasing trend in PH which is possibly caused by organic acids from regular flooding of the watersheds around the bay. High levels of nitrate nitrogen have been observed near the highly urbanized regions of Metro Manila and nearby towns of Cavite. These regions also have reported the highest biological and chemical oxygen demand (BOD and COD) which are diagnostic of organic pollution. In the Manila Port area high levels of polycyclic aromatic hydrocarbons (PAH) and other hydrocarbons in the water is likely due to poor air quality in the Metro Manila airshed. High sedimentation rates have been observed at the mouth of the Pampanga River delta which contributes 49% of the freshwater input to the bay. The whole watershed of the bay which includes the provinces of Tarlac, Bulacan, Nueva Ecija, Rizal, Pampanga and Laguna is 17,000 km^2 in extent (Fig. 5.1). Hypoxia was also observed and studied in Manila Bay during the northeast monsoon (Jacinto et al. 2011) where the lowest oxygen levels were observed in the shallow areas of the Bay. It is suspected to be worse and occur more often during the southwest monsoon and intensified by increased nitrification, freshwater inflow and turnovers. The poor water quality results in an unpleasant odor that is a problem for tourism and has affected fisheries.

Fisheries remain as important economic activity in the bay. However, declining environmental quality has resulted in the decline in oyster and green mussel production as well as periodic harvesting bans due to red tide blooms in the last 30 years. The fisheries for high value shellfish like kapis *Placuna placenta* has collapsed due to pollution and the effect of reclamation of the shallow water habitats.

Reclamation and foreshore development has altered the bay's coastal circulation and may place commercial developments at risk from storm surges. Mudflats which comprise 4600 ha of the bay foreshore is most at risk from coastal reclamation. These are considered the most productive of Manila Bay ecosystems (PEMSEA 2001b).

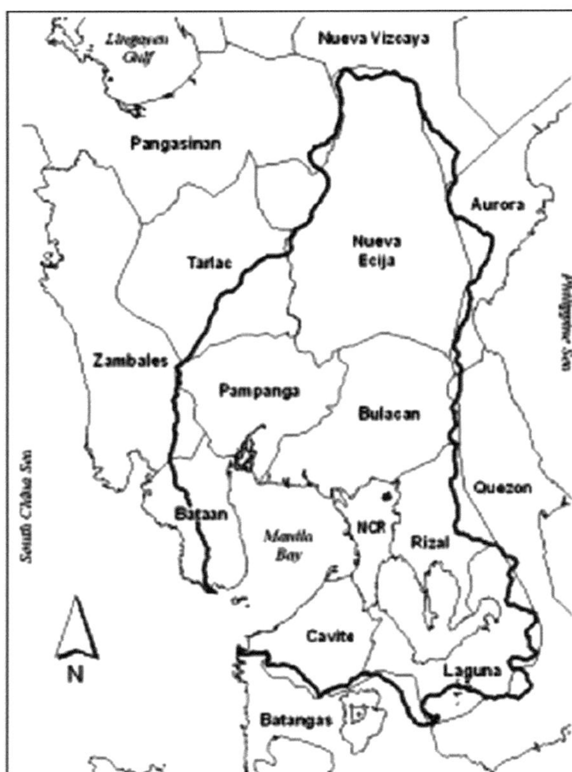

Fig. 5.1 Manila Bay watershed and coastal region. (Manila Bay Coastal Strategy 2001)

5.1.2 Political Administration and Urbanization of the Manila Bay Watershed

Urbanization of Manila Bay's watershed commenced with the city's foundation in 1571 by the Spanish conquistador Miguel Lopez de Legazpi. The city takes its name from the mangrove shrub "nilad" *Scyphiphora hydrophyllacea* which is widely distributed in the Indo-West Pacific. The original urban settlement grew from the fort of Sulayman on the southern bank of the Pasig River. Here the Spanish colonizers established the walled city of Manila, known as Intramuros or "within the walls". The Philippines as a Spanish colony was administered from this city. Growth of the city extended beyond the walls to the "arrabales" or suburbs "extramuros" or beyond the walls of Manila. This urban growth continues to the twenty-first century.

One of the early arrabales is Binondo, which is a district across Intramuros on the northern bank of the Pasig River. Binondo became the trading and business district of the expanding urban area. From Intramuros, urbanization spread eastwards and southwards and the arrabales of Quiapo, San Miguel, Paco, Pandacan, Sampaloc, Ermita and Malate were included in the city. The organization of Manila as a city corporation in 1902, under American sovereignty set the city limits and gave the city its charter. Manila's arrabales became part of the city. The city was then separated from the Spanish colonial province of Manila which later became Rizal province. The nearby towns of Pasay, Navotas, Malabon, Caloocan, Pateros, Marikina, Pasig, San Juan, Paranaque, Taguig, and Valenzuela belonged to this province. These towns are within 20 km of the city centre. In 1938, under the autonomous Commonwealth of the Philippines, Quezon City was created from the territories of Caloocan, Marikina and San Juan. Located 15 km from the city center, it was to become the capital of the Republic of the Philippines. It was named after the first Commonwealth president, Manuel Quezon. Quezon City's establishment accelerated urbanization of the region nearest Manila and to the other Rizal province towns which was arrested only by World War II. After the war, as envisioned by Quezon, most of the government departments and the Congress were transferred to the city.

By 1975, the Rizal towns within 15 km of Manila were highly urbanized and urban development departed from the plans made by the Commonwealth government resulting in urban sprawl. In that year President Ferdinand Marcos in Presidential Decree No. 824 declared that the 13 Rizal towns and 3 cities nearest Manila to be part of the Metropolitan Manila region or the National Capital Region (NCR) in order for more effective urban management of services and development. This region became one administrative unit under a governor who determined urban development policy. This remained until February 1986, when President Marcos was ousted in a civilian-military revolt. Under the government of President Corazon Aquino, no major reorganization of the NCR administrative unit was done except the position of governor was abolished. Most of the functions of the governor were devolved to the city and town mayors with certain management functions given to a regional chairman of the Metro Manila Development Authority (MMDA). This remains the NCR system of governance.

5.1.3 Geological and Climate Environment

Manila and the rest of the NCR are surrounded by two large bodies of water. Manila Bay borders the metropolis to the west and to the southeast is Laguna de Bay, a 949 km^2 freshwater lake with an average depth of 2.8 m. The lake drains through the Pasig River which bisects Manila into two. The lake drains a 45,000 km^2 watershed (Fig. 5.2).

The NCR can be subdivided into four geological zones as defined by their geomorphologies. The coastal margin occupies the zones nearest Manila Bay. The Laguna de Bay coastal lowland occupies the southeast NCR. The Manila Bay coastal margin and the Laguna de Bay coastal lowland are extremely flood prone. These areas constitute 31% of the urban region of the NCR.

The majority of the urban land area (69%) occupies the Guadalupe-Diliman plateau and the Marikina Valley. The Guadalupe-Diliman plateau rises from 20–70 m from MSL. This is drained by the Pasig-Marikina rivers to the east, the San Juan river to the northeast and the Tullahan river on the north. The San Juan is a tributary of the Pasig River. The plateau's base rock is volcanic tuff from the eruption of the prehistoric volcano that created the Laguna de Bay caldera. The alluvial Manila Bay coastal region and the Laguna de Bay coastal lowland are at risk for liquefaction during strong earthquakes.

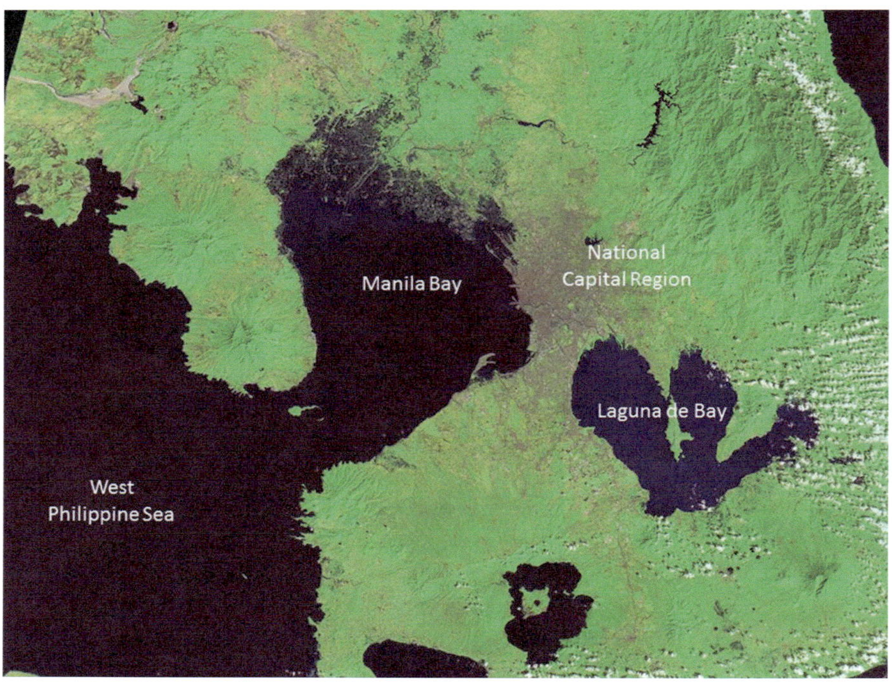

Fig. 5.2 Satellite image of the Manila Bay region. (Modified from NASA, public domain images)

The NCR has a tropical monsoonal climate bordering on a tropical savanna climate (Köppen classification Aw/Am). This is characterized by a defined dry and hot season, a cool and dry seasn and a rainy season. The cool and dry season is from November to February, the hot and dry from March to May and the rainy season from June to October.

5.2 Pollution Ecology

Manila's history as a highly urbanized international port city is the context for biological invasion. Urbanization has modified the surrounding watershed and the bay itself. The oceanographic characteristic of the bay is estuarine and is profoundly influenced by terrestrial inflows to the bay.

Manila Bay has shown decreasing dissolved oxygen trends (DO) in the last 30 years (Chang et al. 2008; Jacinto et al. 2006b; PEMSEA 2006). This is a result of rapid urbanization after World War II (PEMSEA 2001a).

The Pasig-Marikina and San Juan River systems drain the NCR. These rivers are almost completely within the highly urbanized region. The biological oxygen demand (BOD) of the Pasig River was 192,000 MT in 2003 and accounted for 58% of the total loading from domestic sewage. Forty two percent (42%) came from industrial sources. Another environmental problem is dumping of solid waste. From an estimated 34,000 m discharged per year in 1990 (Helmer et al. 1997), this has risen to 67,300 m in 2017 (Lebreton et al. 2017).

The Pasig River discharges to Manila Bay's harbor and nearby marinas. The increased organic loading to the bay is the main cause of monsoon associated anoxia and hypoxia (Chang et al. 2008; Jacinto et al. 2011) where bottom water DO levels fall below 1 mg l^{-1}. Winter (northeast) (Jacinto et al. 2006b) monsoon forcing breaks the stratification of the water column resulting in mixing that brings the low DO water to the surface. This is apparent during the transition from the summer to winter monsoons and can cause massive fish kills. Hypoxia is worse during the summer (southwest) monsoon that causes high rainfall in the Manila Bay watershed (Chang et al. 2008).

The bay's oceanographic and bathymetric characteristics also contribute to hypoxia. The central part of the bay is deep (20–25 m) (Siringan and Ringor 2007) and its bottom composed of clay sediments (Siringan and Ringor 1998). In clay sediments, sediment biological oxygen demand is greater as these bind more organic matter and pollutants (Jacinto et al. 2011). Because of high organic loading, nitrification accounts for a large fraction of BOD resulting in summer hypoxia such as has been observed in the Pearl River estuary in southern China (Zhang and Li 2010). A similar observation has been documented for Manila Bay. Monsoon forcing drives two circulation gyres, a northern one and a southern one (De las Alas 1985; Villanoy 1997). The gyres are separated at the mid-section of the bay. This is an area of convergence and may bring up anoxic or hypoxic water to the surface. Hypoxia associated nutrient levels are high for bottom water for $N0_2$, NH_4 and $N0_3$

and thus supports the hypothesis that hypoxia here is due to increased nitrification (Jacinto et al. 2006a, b, 2011). The water quality measurements do not meet ASEAN water quality standards (Cheevaporn and Menasveta 2003).

Aside from organic and solid waste pollution, Manila Bay has been well studied for heavy metal, PCB contamination (Prudente et al. 1994, 1997; Su et al. 2009) and harmful algal blooms (Farida et al. 1996; Jacinto et al. 2006b; Wang et al. 2008). Among the dominant species of harmful algae is *Noctiluca scitillans* which is characteristic of eutrophic ecosystems (Chang et al. 2008). Harmful algal blooms of *Pyrodinium bahamense* were first recorded in the mid 1970s and in from 1988 have been recurring (Siringan et al. 2008) causing major impacts on the economics of mariculture. The increase in HAB bloom frequency is attributed to greater terrestrial runoff, warmer temperatures and higher sedimentation rates in Manila Bay (Sombrito et al. 2004).

Pb and Cd concentration in the bay water column are variable but are elevated, suggesting anthropogenic inputs (Velasquez et al. 2002) and may bioaccumulate (Su et al. 2009, 2013). However the Philippine Clean Air Act which requires vehicles to use unleaded fuel has resulted in a decline in Pb concentration in sediments (Hosono et al. 2010).

5.3 The Ecology of Marine Non-indigenous Species and Biological Invasion

Historically, Manila Bay has been a very biotic rich area which supported a productive fishery (Munoz 1993). Some remaining habitats though much modified by reclamation still support high levels of coastal biodiversity (Ocampo et al. 2015).

The ecological context of Manila Bay's non-indigenous and invasive species is linked with global trade and urbanization that has significantly changed the watershed environment. This change occurred in a high biodiversity environment. This leads to scientific questions on the dynamics of species invasion in tropical high diversity systems. Are environmental conditions in Manila Bay favorable to MNIS establishment and invasion? And if it were, what are the possible effects on the existing high diversity ecological community?

Table 5.1 lists the recorded MNIS for the port of Manila. Among the previously documented mollusk MNIS from Manila Bay are *Mytilopsis sallei, M. adamsi, Brachidontes exustus, B. pharaonis* and *Crassostrea gigas* (Ocampo et al. 2014). *C. gigas* was introduced for mariculture but never established as a viable population. There is no evidence to show that the listed MNIS are invasive since they occur at low abundances. This is not the case in other ports in ASEAN (Tan and Morton 2006; Wangkulangkul and Lheknim 2008) India (Morton 1981), Hong Kong (Morton 1989a) and in Darwin, Australia where these species established themselves and became invasive (Russell and Hewitt 2000)

Table 5.1 Fouling invertebrate taxa recorded from Manila Bay from Ocampo et al. (2014), Vallejo et al. (2017)

Species	Taxon	Origin
Non-indigenous species		
Brachidontes exustus Linnaeus, 1758	Mollusca (Bivalvia)	Tropical Atlantic
*Brachidontes pharaonis (*Fischer, 1870)	Mollusca (Bivalvia)	Indian Ocean
Crassostrea bilineata (Röding 1798)	Mollusca (Bivalvia)	Indian Ocean
Crassostrea gigas Thunberg, 1793	Mollusca (Bivalvia)	North Pacific
Geukensia demissa (Dillwyn, 1814)	Mollusca (Bivalvia)	Western Atlantic
Musculista senhousia Benson, 1842	Mollusca (Bivalvia)	North Pacific
Mytilopsis adamsi Morrison, 1946	Mollusca (Bivalvia)	Eastern Pacific
Mytilopsis sallei (Recluz, 1849)	Mollusca (Bivalvia)	Tropical Atlantic
Mytella charruana d'Orbignyi, 1846	Mollusca (Bivalvia)	Tropical Atlantic
Pinctada radiata (Leach, 1814)	Mollusca (Bivalvia)	Indian Ocean
Euchelus atratus (Gmelin, 1791)	Mollusca (Gastropoda)	Indian Ocean
Styela clava Leseur, 1823	Tunicata	North Pacific
Stylela plicata Herdman, 1881	Tunicata	Tropical Atlantic
Anemonia manjano Carlgren, 1900	Cnidaria (Anthozoa)	Indian Ocean
Hydroides elegans (Haswell, 1883)	Annelida (Polychaeta)	Japan
Indigenous species	**Taxon**	**Global range**
Perna viridis Linnaeus, 1758	Mollusca (Bivalvia)	Indo-Pacific
Crassostrea iredalei Faustino, 1932	Mollusca (Bivalvia)	Indo-Pacific
Crassostrea cucullata (Born, 1778)	Mollusca (Bivalvia)	Indo-Pacific
Denostrea folium Linnaeus, 1758	Mollusca (Bivalvia)	Indo-Pacific
Placuna sella (Gmelin, 1791)	Mollusca (Bivalvia)	Indo-Pacific
Pteria sterna Gould 1851	Mollusca (Bivalvia)	Indo-Pacific
Ergalatax margariticola (Broderip, in Broderip & Sowerby, 1833)	Mollusca (Gastropoda)	Indo-Pacific
Ergalatax contracta Reeve, 1846	Mollusca (Gastropoda)	Indo-Pacific
Cronia fenestrata Blainville, 1832	Mollusca (Gastropoda)	Indo-Pacific
Chthamalus malayensis Pilsbry, 1916	Crustacea (Cirrepedia)	Indo-Pacific
Cliona sp.	Porifera (Demospongiae)	Indo-Pacific
Turbellid flatworms	Platyhelminthes (Turbellaria)	Indo-Pacific
Capitellid polychaetes	Polychaeta (Capitellidae)	Indo-Pacific
Aiptasia sp	Cnidaria (Anthozoa)	Indo-Pacific
Parasabella oculea Pillai, 1965	Polychaeta (Sabellidae)	Indo-Pacific
Globally distributed		
Bugula neritina Linnaeus. 1758	Bryozoa	Worldwide
Amphibalanus amphitrite (Darwin, 1854)	Crustacea (Cirripedia)	Pantropical

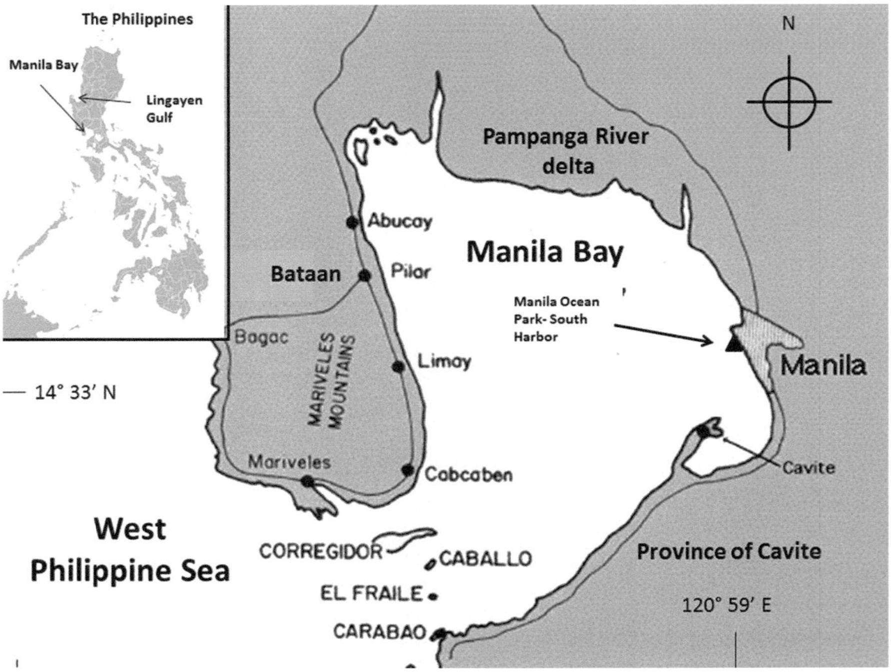

Fig. 5.3 Manila Bay fouler monitoring sites

These species were observed to persist in Manila Bay and as foulers recruit on a biogenic matrix of *Amphibalanus amphirite* barnacles and *Hyroides elegans* tubeworms. It must be noted that *H. elegans* is a common invasive in harbors worldwide (Bryan et al. 1998; Nedved and Hadfield 2009; Qui and Qian 1998). Like in many harbors, *H. elegans* in Manila Bay is associated with the bryozoan *Bugula neritina*. *H. elegans, A. aphitrite* and *B. neritina* provide a biogenic recruitment matrix for bivalves and gastropods (Fig. 5.3).

In a study conducted as part of a bay wide monitoring of foulers from February to September 2011 (Vallejo and Aloy 2012), a total of thirty two (32) fouling species were found to have recruited on plastic petri dish on fouling collector plates following the North Pacific Marine Science Organization (PICES) design (Fig. 5.4) after one year of immersion in the jetties of Manila Ocean Park- South Harbor (MOP-SH) Philippine Navy Headquarters and Manila Yacht Club basin (PN-HQ) and Heracleo Alano Naval Station-Sangley Point Cavite (HANB) Fig. 5.3. Sixteen (16) of these have been identified as non-indigenous. Polychaetes including *Hydroides elegans* (Haswell, 1883). were the most species-rich of all taxa with about three species settling in the samplers. There were at least three bivalve species with abundances at 1% of all recruits counted. These are the indigenous green mussel *Perna viridis* L., the oyster *Crassostrea bilineata (syn. iredalei)* (Röding) and the non-indigenous *Brachidontes pharoanis* (Fisher, P, 1870). The presence of *Mytilopsis sallei* (Recluz, 1849) *and M. adamsi* Morrison is confirmed in

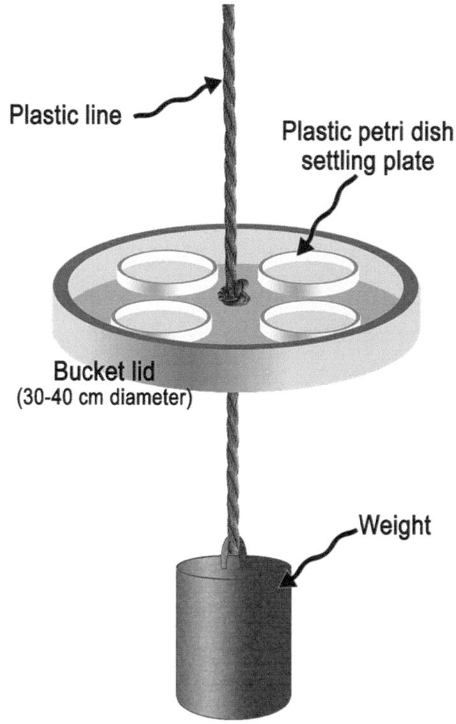

Fig. 5.4 The PICES collector

Manila Bay as this was previously reported by the Manila Ocean Park (Chavanich et al. 2010). However these species account for less than 1% of all recruits counted. The non molluscan taxa that are more than 1% in abundance were the non-indigenous tunicate *Styela*, the non-indigenous sea anemone *Anemonia majano* turbellid flatworms and, capitellid polychates.

In terms of pooled counts for all sites (N = 10,431), the cirriped *Balanus amphitrite* were the most abundant comprising 59% of recruits counted. This was followed by capitellid polychaetes at 18% followed by the non-indigenous calcareous serpulid tubeworm *Hydroides elegans* at 13% (Fig. 5.5). The polychaetes specifically colonize the surface of the plate and barnacles which were documented to be one of the pioneer species after 2 weeks of immersion. In terms of spatial dominance, *Balanus* was first since it occupied almost the entire surface of most of the plates after 2 weeks of immersion (Figs. 5.6 and 5.7).

Based on preliminary observations, the community succession of foulers is as follows. *Balanus amphitrite* are the first to recruit and establish within one month. With the barnacles forming a habitat matrix, the settlement of the bryozoan *Bugula neritina*, the *Hydroides* tubeworm recruits followed by sabellid fanworms. *Mytilopsis* recruits on the *Hydroides* matrix. Larger bivalves such as *Perna, Crassostrea, Brachidontes and, Pinctada* oysters then recruit and attach. The matrix is then colonized by filamentous green algae *Cladophora* and the red algae

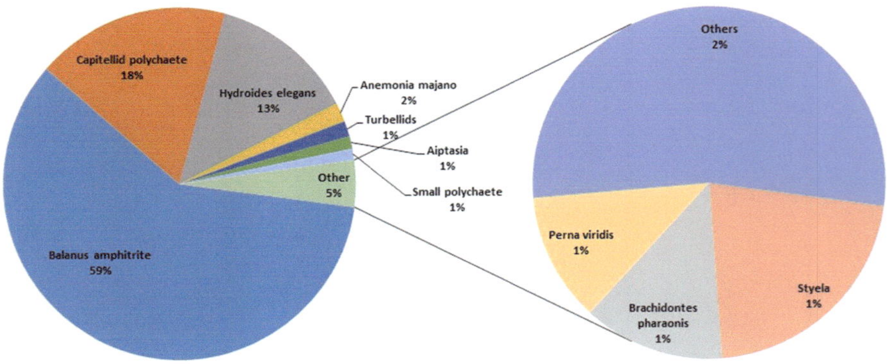

Fig. 5.5 Composition of the fouling community

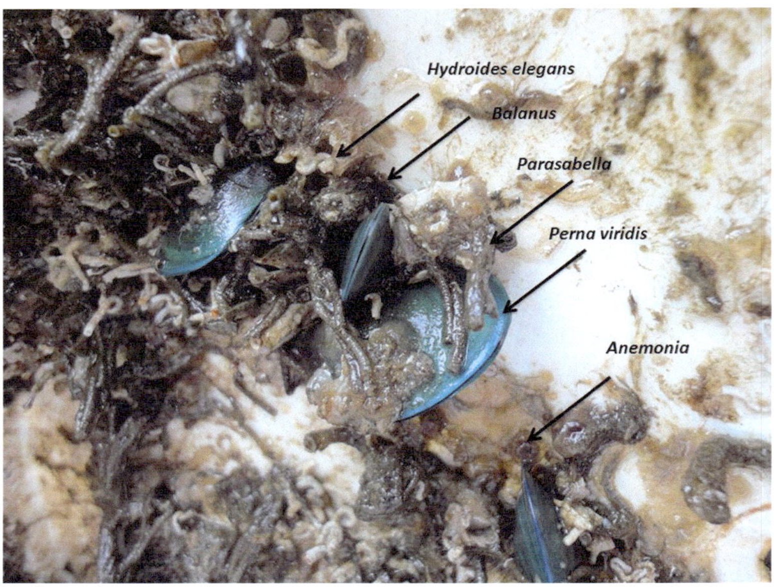

Fig. 5.6 The fouling biogenic matrix

Gracilaria, the cnidarian *Anemonia majano* (Fig. 5.5). Low salinities (<15 psu) during the rainy season may cause the collapse of the established community resulting in a recolonization by *Balanus* as what was observed at the MOP-SH plates. At this stage the plates are subject to grazing by fish namely the Manila Bay puffer *Arothron manillensis*. Several plates have been lost due to grazing.

Fig. 5.7 Community succession on the PICES collectors

5.3.1 Why Do the MNIS Fail to Invade?

The Manila Bay harbor region has a diverse fouling community, with more than half of the species considered as non-indigenous. However with the possible exception of *Hydroides* and the recently documented Tropical Atlantic mussel *Mytella charruana*, none of the fouling non-indigenous species were observed as invasive. The Caribbean and East Pacific dresseinids *Mytilopsis sallei* and *Mytilopsis adamsi* which have been recorded in Singapore, Thailand, Indonesia and Darwin, Australia as invasive and which exist there in extremely high numbers just account for 13 or 0.13% of pooled counts (10,819) of foulers in all Manila Bay sites. On the other hand, the non- indigenous tubeworm, *Hydroides elegans* accounts for 13%, the mussel *Brachidontes pharoanis* 1% and the tunicate *Styela* 1%. The other non-indigenous species are of even much lower abundances and while they have been observed have been dropped from ecological analyses since their pooled counts were less than ten.

Nonetheless even if the majority of non-indigenous species have extremely low abundances, the PICES plate collectors can record their presence. Many of the uncommon to rare foulers are characteristic of mangrove environments thereby supporting the historical observation that the Port of Manila was essentially a mangrove environment. Some of the indigenous foulers like *Placuna, Denostrea* and *Pinctada* once supported an important fishery on Manila Bay sand and mud tidal flats that existed before reclamation and port works started in the early twentieth century. Salinity measurements taken by the Manila Ocean Park curatorial department reflect a range of salinities from 10 psu in the rainy season to 33 psu in the dry season. During the flooding resulting from Typhoon Ketsana (Philippine name: Ondoy) in September 2009, the salinity in the MOP-SH dropped to 6 psu. Salinity measurements at the PN-HQ are similar in trends. The MOP-SH is however 250 m away from the Pasig River mouth and is thus more susceptible to periods of low or fluctuating salinity.

The intriguing question here however is why *Mytilopsis* does not appear to exist in high abundances unlike in Singapore where it fouled port works and harbour and storm drains thus necessitating millions of dollars in control measures (Tan and Morton 2006). In Thailand, *Mytilopsis* can thrive from 5 psu to 31 psu (Wangkulangkul and Lheknim 2008). This condition is similar to what is observed in the Port of Manila. Given the adaptations of *Mytilopsis* to reproduce, recruit and bysally colonize all surfaces from hard to muddy susbtrates, this species should have been able to easily colonize the port environment. *Mytilopsis* did not show any significant recruitment throughout the study period. In Thailand and Singapore, changes in salinity cause the gonadal maturation of *Mytilopsis* with veliger survival optimal at 25 psu. Higher salinities tend to negatively affect gonadal development. pH measurements taken by the Manila Ocean Park averaged at 7.7–8.1 with an acidifying trend over the study period.

The low abundances reported may be due to the inability of *Mytilopsis* to compete with other bivalves most notably green mussel *Perna viridis* (Fig. 5.11) and the oyster *Crassostrea bilineata* which constitute 1% to 2% each of recruits counted on the collection plates in all sites. The much larger *Perna viridis* has similar environmental requirements as *Mytilopsis* and is an important mariculture species in Manila Bay. *Mytilopsis* is also known to have colonized the Mediterranean coast of Israel via the Suez Canal (Galil and Bogi 2008) from ships coming from Southeast Asia. This may represent the establishment of a low salinity tolerant tropical fouling community in the Mediterranean. In Israel, *Mytilopsis* recruited onto *Balanus* skeletons. This was observed in Manila Bay but the *Mytilopsis* were not present in large numbers.

Brachidontes pharoanis is a documented Lessepian migrant to the Levantine coast of the Mediterranean facilitated by ship traffic (Galil 2007). In Manila Bay, it only accounts for 1% of all counts. It is found in the MOP-SH site and the HANB but is absent in the PN-HQ. On Israel's Mediterranean coast, this invasive mussel unlike *Mytilopsis*, represents a high salinity invasive community. It was reported as early as 1876 in the northern terminus of the Suez Canal (Pallary 1912). In the early 1970s this species was rare in the Levantine coast but now has spread as far as Sicily (Rilov et al. 2004). Long term changes in salinity may facilitate invasion success (Rilov et al. 2004).

The low percentage of recruits of *Perna viridis* and *Crassostrea bilineata*, both commercially valuable maricultured species in Manila Bay likely reflects on the environmental quality of the bay. It is possible that acidifying conditions mat negatively affect recruitment and spat survival which has been observed in freshwater dreisseinids (Claudi et al. 2011). This also can be a reason why *Mytilopsis* has few recruits. *Perna viridis* is a documented invasive in the Caribbean and Tropical Atlantic outside its Indo-Pacific native range. It is euryhaline and eurythermal and has displaced indigenous *Perna* in these areas where pH averages at 8.2 (Segnini et al. 1998).

Prior to the introduction of *Mytella charruana* in 2014, the calcareous serpulid tubeworm *Hydroides* was the only non-indigenous fouling species in Manila Bay that may be considered as invasive. *Hydroides* is a major tropical biofouling species worldwide and with warming oceans, threatens to invade cold water ports in Europe, Australia and North America. In places where it is considered to be invasive, *Hydroides* can form monospecific mats that are further colonized by bivalves like *Mytilopsis*. *Hydroides* is dioecious and has a generation time of approximately 3 weeks. At 25 °C, metamorphosis of larvae and recruitment can happen in 5 days (Nedved and Hadfield 2009). The arborescent bryozoa *Bugula neritina* which have been observed to have recruited in MOP-SH and HANB, is known to be a preferred recruitment substrate for *Hydroides* since the bryozoan provides chemical cues for recruitment (Bryan et al. 1998). *Hydroides* is likely killed by prolonged periods of low salinity. In Hong Kong a salinity of <15 psu causes juvenile mortality and delayed larval metamorphosis (Qui, Qian 1998). Before the onset of the 2011 rainy season when salinity averaged at 27 psu, *Hydroides* made for 59% of all counts of recruits in MOP-SH plate collectors. After the rainy season in September, *Hydroides* made up for 2% of counts. Salinity in the MOP-SH area averaged at 12 psu during this time. With its short generation time and recruitment time, *Hydroides* may be susceptible to mass die off. The bay is also subject to regular seasonally related anoxic conditions (Jacinto et al. 2006a, b, 2011) which may result in fish kills and *Perna viridis* die offs.

The apparent inability of the MNIS foulers except *Hydroides* to establish in Manila Bay may be explained by the theory of a taxon cycle in the succession of ecological communities (Wilson 1961). The cycle which starts with colonization, then establishment, geographic range expansion and ends with decline, allows the characterization of species adaptations that will make them persist and expand in the new environment. This model which was first used to explain ecological succession in tropical rain forest communities can also be used to describe MNIS establishment via human intervention. Human intervention consists of an assisted and directed selection for colonization, establishment and geographical range expansion. Presumably this is via regular ballast water exchange in ports and harbors like Manila. It is in harbor environments where MNIS can be expected to proliferate. Consequently, in a biological invasion context, a population decline in the taxon cycle must also involve human intervention to reduce the breeding population. Taxon cycle theory predicts that biological invasion is less likely in ecosystems with diverse communities or in communities subject to frequent environmental stresses (Ricklefs 2005). Environmental envelope approaches model species tolerances to these changes and predicts their range expansion or contraction (Cummings 2002). The environmental envelope may include small to fine scale factors like salinity, pH, sedimentation and nutrient levels known to affect marine invertebrate distribution. Manila Bay is an interesting case for further investigation since the bay has a relatively high level of marine biodiversity compared to similar environments and also has a high level of disturbance as a result of human activity. The interaction of these factors is likely to determine the ecological dynamics of MNIS in the bay.

It is also possible that high density independent recruitment mortality in MNIS and indigenous fouling species is due to high sedimentation rates found in the bay. In the Mediterranean, *Brachidontes pharoanis* recruitment mortality is high in areas of high sedimentation and algal fouling (Galil and Bogi 2008; Galil 2007; Rilov et al. 2004). The introduction of Lessepian migrant non-indigenous grazing fishes in the Mediterranean such as *Siganus* may have facilitated the spread of *Brachidontes* in the 1980s. While *Brachidontes* is intolerant of low salinities, *Mytilopsis* is not. Thus there is a need to further study the population dynamics of these MNIS in Manila Bay's degraded environment which are affected by high sedimentation (Siringan and Ringor 1998; Sta Maria et al. 2009), periodic hypoxia (Jacinto et al. 2011) and fluctuating salinity.

5.3.2 A Successful Biological Invasion, Mytella charruana *in Manila Bay*

While the previously observed MNIS failed to invade but just persist in Manila Bay, in 2014, an introduced mussel species was previously reported by the Philippines Bureau of Fisheries and Aquatic Resources (BFAR) as *Mytilus* spp. due to its similarity of its shape, morphometric measurements and valve color to that of *Mytilus galloprovincialis* (Rinoza 2015). BFAR noted these as "blue mussels" and that this introduced species is used for mariculture in Lingayen Gulf. However *Mytilus* is a temperate and subtropical species and, given its biological requirements, it was unlikely to reproduce successfully in tropical estuarine conditions such as found in Manila Bay (Bayne 1976; Bownes and McQuaid 2006; Branch and Nina Steffani 2004). Thus the initial diagnosis of *Mytilus* was questionable. A more likely candidate, the Charru mussel *Mytella charruana* d'Orbignyi, 1846, is indigenous to the Tropical Atlantic coastlines of Central and South America. In 2014, this species was confirmed to have established in Manila Bay. DNA barcoding confirmed the species identity as *Mytella charruana* (Fig. 5.10).

5.3.2.1 The Ecology of *Mytella* Invasion in Manila Bay

Studies on the ecology of invasion of Manila Bay suggest that in order to successfully establish and invade, the MNIS should be able to compete against established species that occupy a similar ecological niche and be able to tolerate environmental changes typical of degraded environments.

In a 2014–2016 study on the ecology of establishment of *Mytella*, the ecological niches of the invading mussel was evaluated with respect to established bivalves, indigenous *Perna viridis*, *Modiolus* including the non-indigenous *Mytilopsis, Brachidontes* and *Musculista*. The ecological niches were defined by physical and chemical water quality parameters.

Ecological ordination suggests that *Mytella*, *Perna* and *Modiolus* have separate ecological niches with *Perna, Mytella, Musculista* and *Brachidontes* able to tolerate lower pH, salinity and higher NO_3. *Modiolus* prefers a higher salinity and higher NH_4-N and NO_2 levels. It is not surprising that *Musculista* which is negatively loaded on the 1st axis and which has been observed in Hong Kong as a major fouling species, does well there in conditions of high turbidity and low salinity (Fong 2000).

Mytilopsis appears to be more positively correlated with the salinity and NH_4-N and NO_2 –N environmental variables than the other mussel species (Fig. 5.8). *Mytilopsis* has a wide range of salinity tolerance but becomes easily established in turbid estuaries with high levels of nutrients (Morton 1981, 1989b). In Hong Kong

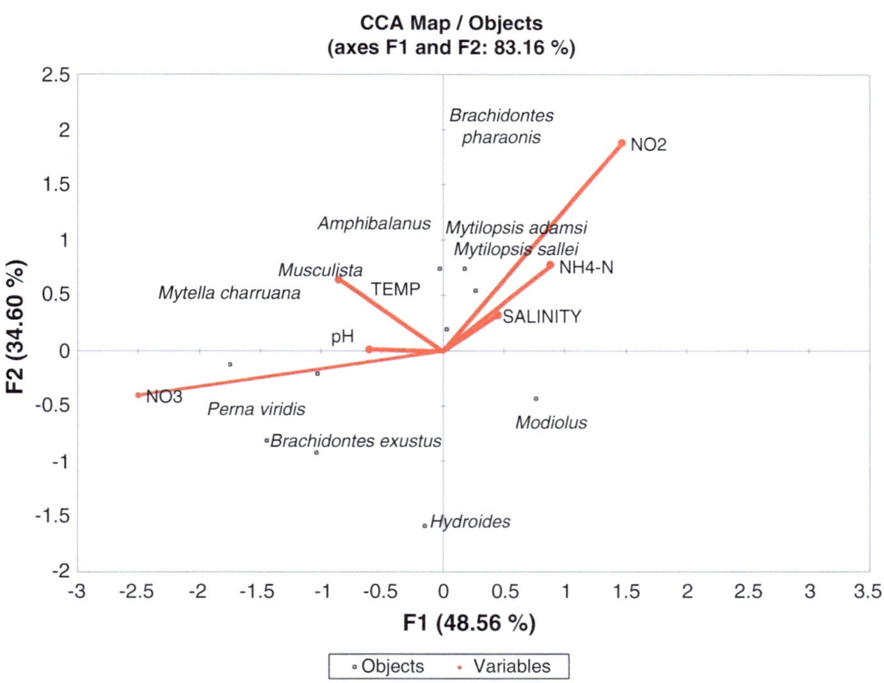

Fig. 5.8 Canonical Correlation Analysis on Manila Bay foulers

Harbour the increase in its numbers were observed during high salinity periods but in Singapore Harbour, it became established at brackishwater salinities (Tan and Morton 2006).

The major introduction vectors for *Mytella* are likely via ballast water or by translocation for mariculture (Spinuzzi et al. 2013). As in the case of the previously reported non-indigenous bivalves *Mytilopsis* and *Brachidontes* (Chavanich et al. 2010; Ocampo et al. 2014), these species have likely been introduced by ballast water or from foulers on ship hulls (Rice et al. 2016). These species have been reported only from areas closest to and at the Manila South Harbor and North Harbor. Also these species have not been recorded in high abundances in previous studies on invasive species in Manila Bay. Their persistence in the harbor marina as observed by ocean park staff and the Philippine Coast Guard is likely due to repeated introductions via ballast water. Mature gonadal individuals have not been recorded from the PICES collectors.

In contrast *Mytella* may have been introduced by translocation as part of mariculture activities. There is no report of *Mytella* mariculture in Manila Bay. But there are reports that the species has been farmed in Lingayen Gulf, 250 km north of Manila Harbor (Rice 2016; Rice et al. 2016; Rinoza 2015).

Perna and *Crassostrea* mariculture in Manila Bay is in decline due to declining environmental quality as the Manila Bay watershed rapidly urbanizes (Jacinto et al. 2006a, b). *Mytella* quickly established itself in this polluted environment. There is some evidence to show that non-indigenous species introduction has resulted in a negative effect on *Perna* settlement (Bownes and McQuaid 2006). This has been observed in *Mytilus* and *Perna perna* settlement in South African rocky reefs. This needs to be further investigated and several hypotheses can be proposed. These are all related to characterizing the ecological niche of the non-indigenous and indigenous species.

It can be presumed that the mussels in this study are all non-selective filter feeders, and are extremely tolerant of salinity changes in intertidal to estuarine environments. What may structure their communities are related to the time of spawning, and recruitment to suitable substrate (Bayne 1976). Thus this study focuses on inferring the community environmental dynamics of recruitment of these mussel species as fouling organisms.

Variations in the settlement of the mussel species are dependent on the variations of *Amphibalanus* and *Hydroides* counts. *Amphibalanus* serves as a biogenic recruitment substrate for *Mytella, Musculista, Modiolus* and *Perna*. If *Mytella* is able to colonize the plates right after the establishment of the *Amphibalanus* habitat matrix, then they can compete with *Perna* recruits. This is the likely scenario on Manila Bay since the *Perna* recruitment and settlement happens at the mid part of the rainy season from June to September. In Hong Kong Harbour, a high gonadosomatic index (GSI) was observed in *Perna viridis* before the onset of the monsoon season and by September, the GSI was lowest thus indicating post spawning condition (Cheung 1993). Similar recruitment patterns were also observed in Malaysia (Al-Barwani et al. 2007) and in India (Rajagopal et al. 1998). Thus it is possible that *Mytella* and *Perna* are in competition for recruitment space since they recruit within the same rainy season, with *Mytella* possibly recruiting at least 30 days earlier and thus having an advantage.

Mytella is likely to persist in Manila Bay as an invasive species that occupies or overlaps the niche of *Perna viridis*. The species is also likely to survive seasonal hypoxia and remain in high densities. Unlike *Mytilopsis* which exists in low abundances in Manila Bay at present, *Mytella* has demonstrated it can proliferate and spread beyond its natural range with human modification of the estuarine environment with the building of port works. The freshwater Quagga mussel *Dreisseina rostriformis bugensis* Andrusov is another example. For 150 years it remained fully within its Eastern European range until the building of port works and canals linking European Russia with Western Europe facilitated its spread starting 1940 (Karatayev et al. 2012). It is hypothesized that changes to water quality aside from increased ship traffic as a result of World War II contributed to its range expansion. There is also a need to investigate the low percentage of recruits of indigenous species like *Crassostrea bilineata* (syn. *iredalei*), *Placuna placenta* and *Pinctada margaritifera* that once formed a commercially important fishery in Manila Bay. These species could be important ecological indicators of the environmental quality of the bay.

Any improvement in the environmental quality of the bay may be unfavourable to the spread and proliferation of MNIS vis a vis the indigenous species as it may decrease its competitive advantage. In order for the MNIS to outcompete the indigenous species in the protected and less exposed environments in a cleaned up Manila Bay (Safriel and Sasson-Frostig 1988), it has to have a reproductive advantage (Branch and Nina Steffani 2004). Thus improving environmental quality may be a pro-active approach to manage biological invasion.

5.4 Cleaning Up the Bay: The Supreme Court Mandamus and the Present State of Manila Bay's Environment

Manila Bay and the Pasig-Marikina river systems have been the focus of environmental rehabilitation for the past 45 years beginning with the establishment of the Pasig River Development Council (PRDC) in 1973 whose mandate is to relocate informal settlers and improve environmental quality (Gorme et al. 2010). Lack of government funding support led to the abolition of the PRDC to be replaced by the Pasig River Rehabilitation Commission (PRCC) in 1999 under the office of the President of the Philippines. PRRC was created with a mandate to "supervise, monitor plans, programs, projects and activities, and enforce rules and regulations towards the rehabilitation of the river"(Gorme et al. 2010). The Supreme Court of the Philippines in 2008 issued a mandamus (G.R. Nos. 171947-48) to the Executive Department of the Government of the Philippines to improve the environmental quality of the bay (Velasco 2008).

The Supreme Court Mandamus is based on a 1999 suit against ten agencies of the Executive Department brought before the lower courts which the Supreme Court agreed to hear after it had determined that the 14 plaintiffs had standing. The lower courts "enjoined" rather than direct the defendants to restore Manila Bay. Unsatisfied with the decision, the plaintiffs appealed to the Supreme Court. They asked the court to order the Executive Department to clean up and rehabilitate Manila Bay and restore the water quality to Class B (fit for swimming, and other recreational activities). The Philippines Department of Environment and Natural Resources (DENR) has a water quality classification scheme from A to D based on the utility of the water resource, A being fit for drinking and D being for navigation purposes only.

The Court ordered the DENR, Bureau of Fisheries and Aquatic Resources (BFAR), Metropolitan Waterworks and Sewerage System (MWSS) and the MMDA as the leading agencies for rehabilitation, with DENR as leading agency, BFAR on ecological restoration and MMDA on urban landfill, flood control and MWSS on sewage treatment.

The 1987 Philippine Constitution has as a principle "the right [of citizens] to a balanced and healthful ecology" and on this principle the Court ordered the responsible agencies to clean up the bay. The agencies welcomed the decision since it allowed them to prosecute polluters without being hampered by lower court injunctions. The Mandamus recognized the ineffectiveness of government agencies with their overlapping functions and the lack of funding which justified inaction. The Mandamus as an act of judicial activism is a continuing one and on penalty of contempt, the agencies have to give regular reports to the court.

The Mandamus has been hailed by environmental activists worldwide as a major advance in environmental law because of its recognition of the intergenerational principle. It has been a decade since the Manila Bay decision has been handed, has the environmental quality of Manila Bay improved?

Despite the Mandamus, the water quality of the Pasig River has deteriorated since 2005 (Chang et al. 2008, Gorme et al. 2010). Consequently, Manila Bay's water quality has declined. However there is a positive development on Pb concentrations in sediment. The enforcement of the Clean Air Act has resulted in a decrease in Pb concentrations in sediment (Hosono et al. 2010). While sediment PCB levels in general are within safe limits, high levels have been observed in *Crassostrea* oysters which indicate continuing industrial pollution (Carvalho et al. 2009).

Invasion ecological theory predicts that it may be possible to manage biological invasion when the original ecosystem state is restored to the pre-invasion state (Huey et al. 2005). Here is where the Mandamus may prove useful. As the Supreme Court has decreed the government agencies to measure their efforts according to the requirements of environmental law, any improvement on environmental quality can be correlated with changes in community and ecological state of bay's invertebrate communities.

5.5 Conclusion

Given the biological invasion characteristics of the MNIS, Manila Bay is a prime site for investigating marine biological invasion in tropical estuarine environments. The environmental changes resulting from rapid urbanization of the watershed has facilitated or prevented the invasion of particular marine non-indigenous species. It is hypothesized that pollution driven events either prevent or facilitate biological invasion. With the recent Mandamus of the Philippines Supreme Court ordering the Executive Department to rehabilitate the bay, this provides the opportunity to correlate changes in environmental quality with changes in ecosystem state and the invasion potential of MNIS (Figs. 5.9, 5.10, and 5.11).

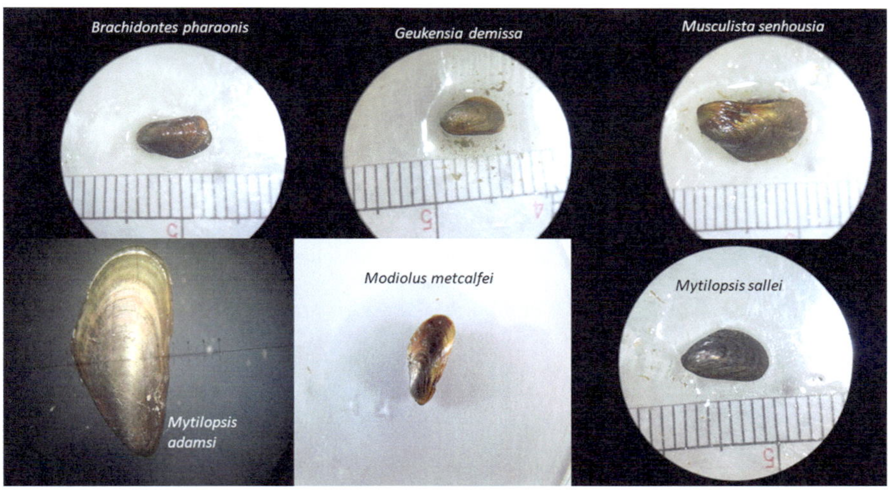

Fig. 5.9 Non-indigenous fouling bivalves recorded from Manila Bay

Fig. 5.10 *Mytella charruana*, the latest mussel to invade Manila Bay

Fig. 5.11 *Perna viridis*, the indigenous mussel commercially maricultured

References

Al-Barwani SM, Arshad A, Amin SN, Japar SB, Siraj SS, Yap CK (2007) Population dynamics of the green mussel Perna viridis from the high spat-fall coastal water of Malacca, Peninsular Malaysia. Fish Res 84(2):147–152

Azanza RV, Siringan FP, Mcglone MLSD et al (2004) Horizontal dinoflagellate cyst distribution, sediment characteristics and benthic flux in Manila Bay, Philippines. Phycol Res 52:376–386

Bayne BL (1976) Marine mussels, their ecology and physiology. Cambridge University Press, Cambridge

Bownes SJ, McQuaid CD (2006) Will the invasive mussel *Mytilus galloprovincialis* Lamarck replace the indigenous Perna perna L. on the south coast of South Africa? J Exp Mar Biol Ecol 338:140–151

Branch GM, Nina Steffani C (2004) Can we predict the effects of alien species? A case-history of the invasion of South Africa by Mytilus galloprovincialis (Lamarck). J Exp Mar Biol Ecol 300:189–215

Bryan PJ, Kreider JL, Qian P-Y (1998) Settlement of the serpulid polychaete *Hydroides elegans* (Haswell) on the arborescent bryozoan *Bugula neritina* (L.): evidence of a chemically mediated relationship. J Exp Mar Bio Ecol 220:171–190

Carvalho FP, Villeneuve J-P, Cattini C et al (2009) Organic contaminants in the marine environment of Manila Bay, Philippines. Arch Environ Contam Toxicol 57:348–358

Chang K-H, AMANO A, Miller oW et al (2008) Pollution study in Manila Bay: eutrophication and its impact on plankton community. In: Isobe YOT, Subramanian A, Suzuki S, Tanab S (eds) Interdisciplinary studies on environmental chemistry – environmental research in Asia. Terrapub, Tokyo, pp 261–267

Chavanich S, Tan LT, Vallejo B (2010) In: Chavanich S, Tan LT, Vallejo B (eds) Report on the current status of marine non-indigenous species in the Western Pacific Region. Chulalongkorn University, Bangkok, p 74

Cheevaporn V, Menasveta P (2003) Water pollution and habitat degradation in the Gulf of Thailand. Mar Pollut Bull 47:43–51

Cheung SG (1993) Population dynamics and energy budgets of green-lipped mussel *Perna viridis* (Linnaeus) in a polluted harbour. J Exp Mar Biol Ecol 168:1–24

Claudi R, Graves A, Taraborelli AC et al (2011) Impact of pH on survival and settlement of dreissenid mussels. Aquat Invasions 7:21–28

Cummings G (2002) Comparing climate and vegetation as limiting factors for species ranges of African ticks. Ecology 83:225–268

De las Alas J (1985) A model for wind driven circulation of Manila Bay. Nat Appl Sci Bull 37:159–170

Farida F, Bajarias A, Relox J (1996) Hydrological and climatological parameters associated with the Pyrodinium blooms in Manila Bay, Philippines. Harmful and Toxic Algal Blooms:49–52

Fong TC (2000) Conservation and management of Hong Kong seagrasses. Asian Mar Biol 16:109–121

Galil BS (2007) Loss or gain? Invasive aliens and biodiversity in the Mediterranean Sea. Mar Pollut Bull 55:314–322

Galil B, Bogi C (2008) *Mytilopsis sallei* (Mollusca: Bivalvia: Dreissenidae) established on the Mediterranean coast of Israel. JMBA-2 Biodivers Rec 10:1–5

Gorme JB, Maniquiz MC, Song P et al (2010) The water quality of the Pasig river in the city of Manila, Philippines: current status, management and future recovery. Environ Eng Res 15:173–179

Helmer R, Hespanhol I, Organization WH (1997) Water pollution control: a guide to the use of water quality management principles

Hosono T, Su C-C, Siringan F et al (2010) Effects of environmental regulations on heavy metal pollution decline in core sediments from Manila Bay. Mar Pollut Bull 60:780–785

Huey RB, Gilchrist GW, Hendry AP (2005) Using invasive species to study evolution. In: Sax DF, Stachowicz JJ, Gaines SD (eds) Species invasions insights into ecology, Evolution and biogeography. Sinauer Associates Inc, Sunderland, pp 139–159

Jacinto GS, Azanza RV, Velasquez IB et al (2006a) Manila Bay: environmental challenges and opportunities. In: Wolanski E (ed) The environment in Asia Pacific Harbours. Springer, Dordrecht, pp 309–328

Jacinto GS, Velazquez IB, Diego-McGlone MLS et al (2006b) Biophysical environment of Manila Bay- then and now. In: Wolanski E (ed) The environment of Asia Pacific Harbours. Springer, Dordrecht, pp 293–306

Jacinto GS, Sotto LPA, Senal MIS et al (2011) Hypoxia in Manila Bay, Philippines during the northeast monsoon. Mar Pollut Bull 63:243–248

Karatayev AY, Claudi R, Lucy FE (2012) History of Dreissena research and the ICAIS gateway to aquartic invasions science. Aquat Invasions 7:1–5

Lebreton LCM, van der Zwet J, Damsteeg J-W et al (2017) River plastic emissions to the world's oceans. 8:15611

Morton B (1981) The biology and functional morphology of *Mytilopsis sallei* (Recluz) (Bivalvia: Dreissenacea) fouling Visakhapatnam Harbour, Andhra Pradesh, India. J Molluscan Stud 47:25–42

Morton B (1989a) Life history characteristics and sexual strategy of *Mytilopsis sallei* (Bivalvia: Dreissenacea), introduced into Hong Kong. J Zool 219:469–485

Morton B (1989b) *Mytilopsis sallei* (Recluz) (Bivalvia: Dreissenacea) recorded from Hong Kong: an introduction by Vietnamese refugees? Malacol Rev 13:90–92

Munoz JC (1993) Manila Bay fisheries: status and management. Mar Pollut Bull 23:311–314

Nedved BT, Hadfield MG (2009) Hydroides elegans (Annelida: Polychaeta): a model for biofouling research. In: Flemming H-C, Murthy PS, Venkatesan R, Cooksey K (eds) Marine and industrial biofouling. Springer, Berlin/Heidelberg, pp 203–217

Ocampo M, Oliva I, Tan R et al (2014) Assessing the marine fouling community in a man-made marina at Manila Bay. Adv Environ Biol 8:597–601

Ocampo MAB, Aguila CCA, Chavez GCS (2015) Malacological survey along the intertidal zone of Las Piñas-Parañaque Critical Habitat and Ecotourism Area (LPPCHEA). Department of Biology, University of the Philippines Manila, Manila, pp 1–13

Pallary P (1912) Catalogue des mollusques du littoral méditerranéen de l'Egypte. Mem Inst Egypte 7:69–207

PEMSEA (2001a) Manila Bay initial risk assessment. In: PEMSEA (ed). PEMSEA/DENR, Quezon City, p 151

PEMSEA (2001b) Manila Bay: initial risk assessment. PEMSEA Technical Information Report No. 2001/01. In: Eng CT (ed) Global Environment Facility/United Nations Development Programme/International Maritime Organization Regional Programme on Building Partnerships in Environmental Management for the Seas of East Asia (PEMSEA) Quezon City, PH, p 112
PEMSEA (2006) Sustainable development of Manila Bay: a focus on water quality. In: PEMSEA (ed) PEMSEA, Quezon City, pp 1–5
PPA (2010) Philippine port statistics. In: Philippine Ports Authority. http://www.ppa.com.ph/ppa%20web/portstat.htm. Accessed 7 Sept 2014
Prudente MS, Ichihashi H, Tatsukawa R (1994) Heavy metal concentrations in sediments from Manila Bay, Philippines and inflowing rivers. Environ Pollut 86:83–88
Prudente M, Kim E-Y, Tanabe S et al (1997) Metal levels in some commercial fish species from Manila Bay, the Philippines. Mar Pollut Bull 34:671–674
Qui J-W, Qian P-Y (1998) Combined effects of salinity and temperature on juvenile survival, growth and maturation of the polychaete *Hydroides elegans*. Mar Ecol Prog Ser 168:127–134
Rajagopal S, Venugopalan VP, Nair KVK, Van der Velde G, Jenner HA, Den Hartog C (1998) Reproduction, growth rate and culture potential of the green mussel, Perna viridis (L.) in Edaiyur backwaters, east coast of India. Aquaculture 162(3–4):187–202
Rice MA (2016) The shellfish corner: aquaculture of exotic shellfish species. Aquac Mag 42:76–78
Rice MA, Rawson PD, Salinas AD et al (2016) Identification and salinity tolerance of the western hemisphere mussel Mytella charruana (D'Orbigny, 1842) in the Philippines. J Shellfish Res 35:865–873
Ricklefs RE (2005) Taxon cycles. In: Sax DF, Stachowicz JJ, Gaines SD (eds) Species invasions insights into ecology, evolution and biogeography. Sinauer Associates, Inc, Sunderland, pp 165–199
Rilov G, Benayahu Y, Gasith A (2004) Prolonged lag in population outbreak of an invasive mussel: a shifting-habitat model. Biol Invasions 6:347–364
Rinoza J (2015) New mussel species seen to help revolutionize PH shellfish industry. Manila Bulletin, Manila, p 14
Russell B, Hewitt C (2000) Baseline survey of the Port of Darwin for introduced marine species. A report to the Northern Territory Department of Transport and Works Marine Branch, Darwin, Northern Territory, Australia 98
Safriel UN, Sasson-Frostig Z (1988) Can colonizing mussel outcompete indigenous mussel? J Exp Mar Biol Ecol 117:211–226
Segnini MI, Chung KS, Perez JE (1998) Salinity and temperature tolerances of the green and brown mussels, *Perna viridis* and *Perna perna* (Bivalvia:Mytilidae). Rev Biol Trp 46:121–125
Siringan FP, Ringor CL (1998) Predominant nearshore sediment dispersal patterns in Manila Bay. Sci Diliman 9:29–40
Siringan F, Ringor C (2007) Changes in bathymetry and their implications to sediment dispersal and rates of sedimentation in Manila Bay. Sci Diliman 10
Siringan FP, Azanza RV, Macalalad NJH et al (2008) Temporal changes in the cyst densities of Pyrodinium bahamense var. compressum and other dinoflagellates in Manila Bay, Philippines. Harmful Algae 7:523–531
Sombrito EZ, Ad B, Sta Maria EJ et al (2004) Application of 210Pb-derived sedimentation rates and dinoflagellate cyst analyses in understanding Pyrodinium bahamense harmful algal blooms in Manila Bay and Malampaya Sound, Philippines. J Environ Radioact 76:177–194
Spinuzzi S, Schneider KR, Walters LJ et al (2013) Tracking the distribution of non-native marine invertebrates (*Mytella charruana, Perna viridis* and *Megabalanus coccopoma)* along the South-Eastern USA. Mar Biodivers Rec e55:6
Sta Maria EJ, Siringan FP, Bulos A et al (2009) Estimating sediment accumulation rates in Manila Bay, a marine pollution hot spot in the Seas of East Asia. Mar Pollut Bull 59:164–174
Su GS, Martillano KJ, Alcantara TP et al (2009) Assessing heavy metals in the waters, fish and macroinvertebrates in Manila Bay, Philippines. J Appl Sci Environ Sanit 4:187–195

Su GLS, Ramos GB, Su MLLS (2013) Bioaccumulation and histopathological alteration of total lead in selected fishes from Manila Bay, Philippines. Saudi J Biol Sci 20:353–355

Tan KS, Morton B (2006) The invasive Caribbean bivalve *Mytilopsis sallei* (Dreissenidae) introduced to Singapore and Johor Bahru, Malaysia. Raffle Bull Zool 54:429–434

Vallejo B, Aloy AB (2012) A report on the marine foulers of Manila Bay. University of the Philippines, Manila, p 20

Vallejo B Jr, Conejar-Espedido J, Manubag L, Artiaga KCC, Damatac II, Amor M, Cao EP (2017) First record of the Charru mussel Mytella charruana d'Orbignyi, 1846 (Bivalvia: Mytilidae) from Manila Bay, Luzon, Philippines. BioInvasion Records 6(1):49–55

Velasco PJ (2008) Decision on ordering the clean up of Manila Bay. In: Philippines SCot (ed) Supreme Court of the Philippines, Manila

Velasquez IB, Jacinto GS, Valera FS (2002) The speciation of dissolved copper, cadmium and zinc in Manila Bay, Philippines. Mar Pollut Bull 45:210–217

Villanoy C (1997) Modelling the circulation of Manila Bay: assessing the magnitude of wind and tide forcing. Sci Diliman 9:26–35

Wang S, Tang D, He F et al (2008) Occurrences of harmful algal blooms (HABs) associated with ocean environments in the South China Sea. Hydrobiologia 596:79–93

Wangkulangkul K, Lheknim V (2008) The occurrence of an invasive alien mussel *Mytilopsis adamsi* Morrison, 1946 (Bivalvia: Dreissenidae) in estuaries and lagoons of the lower south of the Gulf of Thailand with comments on their establishment. Aquat Invasions 3:325–330

Wilson EO (1961) The nature of the taxon cycle in the Melanesian ant fauna. Am Nat 95:169–193

Zhang H, Li S (2010) Effects of physical and biochemical processes on the dissolved oxygen budget for the Pearl River estuary during summer. J Mar Syst 79:65–88

Chapter 6
Bioinvasion and Environmental Perturbation: Synergistic Impact on Coastal–Mangrove Ecosystems of West Bengal, India

Susanta Kumar Chakraborty

Abstract Numerous ideas have emerged on the definitions, consequences of introduction, causes of damage, population dynamics and mode of propagation, prevention, and adaptibility of invasive species along with their impact on native species directly or indirectly by way of alteration of ecosystem dynamics. An Invasive species is defined as a species having been introduced outside its native range through human activities which are likely to cause economic or ecological harm. Estuaries and Coastal-Estuarine –Mangrove ecosystems being the most productive and sensitive ecosystems in the world, have appeared to be very much susceptible to introductions of non-native species because of lot of possibilities out of different ecological and people oriented activities in and around these eco-regions such as shipping and boating, ecotourism, fisheries, aquaculture etc. Alongside, several ecological perturbations such as eutrophication, global warming, biomagnification and biotransformation of persistent pollutants, etc. along with the negative impacts of introduced species on marine estuarine flora and fauna by outcompeting them for basic life support resources, human health risk associated with transmission of pathogens, and higher bioaccumulation capabilities of invasive species than native species have threatened this ecologically sensitive region considerably.

This book chapter aims at highlighting the prospective consequences of the bioinvasion on the coastal–estuarine networks of West Bengal, India which is unique because of the presence of more than hundred of deltas endowed with world's largest mangrove chunk and associated flora and fauna giving emphasis on ecosystem processes and functions. In view of the different threats as imposed by invasive species on native ones, even leading to their extinction, proper holistic eco-management strategies are being suggested giving emphasis on the self-sustaining ecosystem functioning of coastal- estuarine -mangrove ecosystem where species do not inevitably spread rapidly and extensively beyond control.

S. K. Chakraborty (✉)
Department of Zoology, Vidyasagar University, Midnapore-721102, West Bengal, India

C. Makowski, C. W. Finkl (eds.), *Impacts of Invasive Species on Coastal Environments*, Coastal Research Library 29,
https://doi.org/10.1007/978-3-319-91382-7_6

Keywords Bioinvasion · Invasive species · West Bengal coast · Mangrove ecosystem · Environmental perturbation

6.1 Introduction

Introduction of alien and invasive species is being considered as a major threat to biodiversity in the world next to habitat destruction (Coblentz 1990; Glowka et al. 1994). However, the alien environment may not be favorable for the introduced species in respect of the factors that have evolved with them in order to govern their population and propagation. On the other side, the native species may not have the ability to compete with the exotic ones for the basic life-supporting requirements such as food, shelter, reproduction etc. and may therefore be driven to extinction. Conservationists have become increasingly aware of the problem of relentless negative effects of exotic invaders on native one (Soule 1990). Proper understanding of this phenomenon is still incomplete and inconclusive, although much scientific works have been done on the ecology of invasions since the subject was first highlighted by Elton (1958). During the 1980s, the Scientific Committee on Problems of the Environment (SCOPE) engaged a large number of scientists across different scientific disciplines in an effort to document the nature of invasive species problem, the outcome of which is the publication of a book, entitled, Biological Invasion: A Global perspective (Drake et al. 1989). This has clearly highlighted that invasive species are having major impacts on ecosystem functioning leading to considerable biodiversity losses. Yet studies on the ecology of invasive species are better known in developed countries than developing one. In India, very scanty research studies have been undertaken on bioinvasion in general and aquatic invasive species in particular (Maheswari 1965; Maiti and Guhabakshi 1981; Ramkrishnan 1991; Bhakat et al. 2004; Paria et al. 2017; Bhattacharrya et al. 2017). Almost no such works are available from Indian coastal-mangrove ecosystem excepting scanty reports from Kerala coast of India (Ravinesh and Bijukumar 2016; Bijukumar and Raj 2017; Rajagopal et al. 1989) and of West Bengal coast (Bhakat et al. 2004).

6.2 Invasive Species and Bioinvasion

Invasive species are defined by the Convention of Biological Diversity (CBD 2000) as "*—those alien species which threaten ecosystems, habitats or species*". An invasive species is a nonnative species that causes damages beyond any attendant benefits (Elton 1958; Kareiva et al. 1996). The invasive species are synonyms to nonindigenous or exotic or alien species which later being evolved elsewhere, must have been introduced to new ecological set up naturally, intentionally or accidentally. An invasive species must have the attributes such as: (i) **non-native (alien or exotic) to the ecosystem under consideration** and (ii) **whose introduction causes or is likely to cause economic or environmental harm, or harm to human health. An alien species that causes (or has the potential to cause) harm to biodiversity, the environment, economics and/or human health is often referred to as Invasive Alien Species (IAS)**. Introduction and flourishing of invasive species have been increased several fold during the last few decades especially in view of globalization, lack of knowledge about them, and non-adhering to basic ecological principles towards environmental management and thereby have become the leading cause of global environmental changes e.g., biodiversity loss and ecosystem malfunction (Lodge 1993; Mack et al. 2000; Sala et al. 2000; Kolar and Lodge, 2001).

Invasion of biological organisms (plants, animals and microbes) in any ecosystem can be referred to as bioinvasion which is in other way represents the arrival, settlement, colonization and establishment of new species into an ecozone away from earlier habitats after traversing at unprecedented rates, routes and manners (Elton 1958; Williamson 1996; Vitousek et al. 1997). In biological invasions which are now considered as one of the main threats to the world's biodiversity (Mooney and Hobbs 2000), displacement of one organism or life form by another is taken place. In an increasingly globalized world, plants, animals, and microbes are introduced more and more frequently into regions that had never hosted them. These "invasive" or "exotic" or alien species can have a destabilizing influence on ecosystems that lack the natural enemies needed to check the spread of exotics.

An invasive species that is not native to a specific location, has a tendency to spread very quickly and may cause harm to the environment and also human beings. Invasion success depends upon the pre-adaptation or on characters those have evolved after the introduction and settlement in new environment, which has been described as "a grand experiment in evolution" (Ayala et al. 1989). The most significant problems, in terms of ecological damage, are usually caused by invasive species that overgrow over entire biotic communities and replace the native flora and fauna. Nonindigenous species affect native ecosystems, communities, and populations in myriad, multidimensional and cumulative ways, from plants (and a few animals) that overgrow over entire communities, and also to plants and animals that hybridize individual native species to a sort of genetic extinction. Further, nonindigenous species sometimes interact in conjunction with prevailing ecological perturbations to worsen each other's impact (Elton 1958; Crooks 2002).

Although, Biological invasions have become the second leading cause (after habitat destruction) of species endangerment and extinction in different parts of the world, most of the introduced species are not invasive as the great majority of introduced species probably do not even survive, and of those that do, only a few invade natural ecosystems (Williamson 1996).

6.3 Bioinvasion and Biodiversity

Bioinvasion being a keyword in biodiversity management and as an important contributor to biodiversity loss by virtue of notorious activities of invasive species towards jeopardising ecosystem functioning, has become one of the biggest threats to biodiversity worldwide causing unprecedented economic and environmental effects (Mooney and Hobbs 2000; Perrings et al. 2000; GISP 2004) affecting 30% of threatened birds, 11% of threatened amphibians, and 8% of threatened mammals (IUCN 2004). Also, it is responsible for a global economic loss amounting to approximately US $ 1.4 trillion annually (Pimental et al. 2001). In developing countries, especially those reliant heavily on agriculture, the negative impact of invasive species on the country's socio-economy is often intensified. Researchers have estimated India's economic loss due to bio-invasion as to be 20% of its annual GDP, as opposed to less than 1% of GDP of the USA in the same year (Pimental et al. 2001). This is why invasive species are being increasingly regarded with animosity and alarm (Paria et al. 2017; Bhattacharrya et al. 2017). The change in land use and habitat alteration is accentuating species to invade areas which ultimately threaten local and specifically adapted faunal species (Perrings et al. 2010).

6.4 What Makes a New Species Invasive?

A harmful invasive species usually displays at least one of the following characteristics in its newly settled habitat:

- Outcompetes, outnumbers and displaces local native species by competing directly for food, space or shelter.
- Substantially disrupts the ecodynamics (food chains and food webs) of the affected ecosystems.
- Enjoys prolific reproduction, recruitment, growth and survival due to 'escape' from the natural predators, grazers, parasites or pathogens that control it in its native range.
- Causes nuisance by fouling to boats, ships, fishing gears, aquaculture equipments, jetties, etc.
- Renders noxious or pathogenic effects that cause fish mortalities, disrupt aquaculture operations and/or directly threaten public health e.g. toxic 'red-tide' microalgae, waterborne diseases.

6.5 Process of Bioinvasion and Prospective Impacts

Bio-invasion being a process or phenomenon, takes into account of steady but aggressive introduction of invasive species into a new eco-zone, new or within the same ecosystem with a different role. This role might be a negative one and that should affect the ecosystems adversely (Biswas 2003; IUCN 2003). Understanding the principles of the invasion process will help predicting the mode of invasion, the pattern of establishment and intensity of of alteration and damage of ecosystems and biodiversity. The scale and impacts of this anthropogenic mixing of biotas, as well as their impacts on new environment offer an opportunity for basic biological insight, for which the invasion biology has been emerged as a rapidly developing discipline with broad ecological and conservation implications (Elton 1958; Carlton 1979). Invasion success depends on the ecological attributes of the invading organisms, the characteristics of the invasion site, and a range of stochastic short term events (Davis et al. 2000; Hobbs and Humphries 1995; Lambrions 2002; Shigesada and Kwasaki 1997). The magnitude of invasions may be apparently local, but the drivers of bio-invasion are global. According to the Convention on Biological Diversity (CBD), there is a need for "compilation and dissemination of information on alien species that threaten ecosystem, habitats, or species, which are to be used in the context of any prevention, introduction and mitigation activities (CBD 2000)". Besides imposing huge economic costs, invasive species to cause elimination and even extinction of native species, reduction of genetic diversity, and biotic homogenization through a variety of mechanisms. At a local scale, the strong impacts of introduced species are wide and different, affecting at different levels, including alteration of ecosystem structure by changing species composition and community structure of invaded areas, thereby affecting ecosystem processes. The intensity and extent of invasiveness depends on the different invasion processes which include transport, introduction, colonization (establishment) and naturalization. On establishment in an alien environment, the colonization success depends on many factors including the adversity as imposed by natural, environmental, reproductive and dispersal barriers. In order to overcome all those constrains and expand territory through population explosion, the invasive species not only get themselves adapted to newly settled habitats but exhibit foraging, reproductive, dispersal and ecological competence over native organisms.

It has been postulated that invasive plant species can only spread into natural vegetation as a result of disturbance (Biswas 2003) which in turn initiate succession as a natural process as observed in mangrove ecosystems (Das and Siddiqi 1985). A variety of biotic and abiotic processes within this very sensitive and productive coastal-estuarine ecosystem, which vary in frequency (Iftekhar 1999), magnitude, intensity, and timing, constitutes natural disturbance (Cattelino et al. 1979; Connell and Slatyer 1977; Grime 1977; Holling 1981; Levin and Pain 1974; Loucks 1970; Shugart and West 1980; Trudgill 1977; Vogl 1980; White 1979). Chronic disturbance relates more to the frequency, or return interval, of a disturbance event, which alters the existing physical environment and community of organisms at a particular site (Rejmanek 1999). Although impact of invasion in altering structure and functioning of ecosystems are well documented (Mooney and Hobbs 2000) but such linkages between invasion and ecosystem changes have so far been least documented (Charles and Dukes 2007). However, some mechanisms by which these changes occur include exploitation and competition for resources; interference competition (allelopathy) and direct predation, herbivory and parasitism (Vitousek 1990; Crooks 2002). The mechanism pertaining to the impact of invasive species operate at four levels of complexity –(1) Species, (2) Communities, (3) Ecosystems and (4) Atmosphere (Levine et al. 2003; Charles and Dukes 2007). At the species level, the negative impact because of the more aggressiveness and competitiveness the invasive species cause removal, economic decline or even loss by way of extinction process which result economic loss for food, fiber, fuel, fodder, medicine etc. and ecological damage by changing species composition, synchronized pollination, pattern of energy flow and trophic relationships (Perrings et al. 2010). At the community level alien invasion not only jeopardize the basic fabrics of ecosystem functioning but may lead to aesthetics (McNeely 2001; McNeely et al. 2001). In West Bengal coast, this is evident by the replacement of native mangroves (*Exocoeria, Avicennia* etc.) and mangrove associates (*Pandanus*, *Ipomia* etc.) by *Acasia*, *Clerodendrum*, *Eucaliptus* etc. by the NGOs, and even by the Government through afforestation program especially by means of social forestry scheme. Invasive alien species may alter the atmospheric composition by changing carbon dioxide sequestration, altering nitrogen cycle, sulpher cycle by changing the species composition of coastal-mangrove microbial species composition and thereby releasing more nitrogenous gases as green house emission from the floor of the mangroves (Lodge et al. 2006; Amanda et al. 2007).

International Union for Conservation of Nature (IUCN) considers the impacts of IAS as “immense, insidious, and usually irreversible” and the impacts are much more effervescent in the aquatic ecosystems (IUCN 2003). This is also to be noted that that not only the invasive species by virtue of their activities alter the structure and function of an ecosystem but on the other side, different environmental perturbation such as eutrophication, global warming, salinity invasion, deforestation, soil erosion etc. attract the invasive species and enable them to successfully settle down in the new ecological set up and flourish.

In such context, it can be inferred that the impact of invasive species mostly centres around ecological, environmental and economic impacts. Their mode of impacts has been broadly classified as environmental impacts on natural systems, and economic impacts on anthropogenic systems. Environmental impacts were classified into three biological levels: genetic effects, impacts on biological interactions, and ecosystem effects. For several exotic invasive species the potential to hybridize with a native species has been observed, but in no case actual hybridization in the wild has been found so far. The major environmental impact by invasive species perceived in a particular region is the reduction of the native regeneration through competition by exotic species (Perrings et al. 2010).

6.6 Bioinvasion and Role of Predictive Theories

6.6.1 *The Process of Invasion*

The movement of people around the globe and escalating trade has resulted in the translocation of animals or plants of a particular biogeographical region to other regions of the world. This newly settled biota after colonisation and acclimatization in the wild new ecological set up tend to pose threatening impact on the native biodiversity by way of competition for food and shelter which may ultimately result in the extinction of the native species, besides triggering a series of changes in the ecosystems. Nonindigenous or exotic or alien species are those that evolved elsewhere and have been purposely or accidentally relocated. An alien species that causes (or has the potential to cause) harm to biodiversity, the environment, economies and/or human health is often referred to as ***Invasive Alien Species (IAS) or Alien Invasive Species (AIS)***. In evolution, colonisation of organisms in new regions is common and always limited by natural barriers, while human-mediated introductions of species have dramatically increased the magnitude and scale of exotic species' movements across the world.

Not all introduced species become invasive. There are different phases for the invasion process, which include transport, introduction, colonization (establishment) and naturalization. When an alien species is introduced outside its range, its establishment or colonisation success depends on many factors including natural barriers. When an ecosystem is disturbed either by natural or anthropogenic factors, it provides a kind of invasion window to the alien species for establishment. Gradually, the alien species overcomes the environmental, reproductive and dispersal barriers and expands its population in order to become invasive species outnumbering and outcompetiting the native ones.

6.6.2 *Invasion Biology and Predictive Theories*

Invasion biology is the study of the reproduction, dispersal and ecological impact of organisms that occur outside of their native range. Understanding the principles of the invasion process will help to predict which species will invade, where invaders will become established, and ultimately impart negative impact on ecosystems and biodiversity. Alien Invasive Species (AIS) occurs in all taxonomic groups, including animals, plants, fungi and microorganisms, and can affect all types of ecosystems, the intensity of the impact may vary in various ecosystems.

Predictive theories are being needed to set priorities for the control of introduced invasive species (species spreading into areas where they are not native) as well as to predict the risk of future invasions as many recognized relationship between species attributes and invasiveness remain unquantified and several assumed relationships have to be tested. There lies a continuum between invasive and non-invasive species as the species which do not exhibit any invasive behavior 100 years after introduction may turn out to be somewhat invasive after 300 years. Prediction on invasiveness has to be made on time scales of even several decades as the delayed invaders may have greater impacts than already established and ready- to –spread species (Rejmanek 1999).

6.6.3 *Invasibility in Respect of Bio-Invasion*

All analyses of ecosystem invasibility based on just one-point-in-time observations are unsatisfactory (Rejmanek 1989). In most plants invasion, the quality, quantity and regime of introduction of imported propagules are not fully understood (Rejmanek 1989). It appeared to be very difficult to separate the resistance of biotic communities from resistance determined by biotic environments. Nevertheless, available evidence indicates that only a very few alien species invade successionally advanced plant communities (Rejmanek 1989, 1996). However, aquatic system is more prone to the introduction, establishment and massive propagation of notoriously bad exotic aquatic plants (Ashton and Mitchell 1989). This is in contrast to xeric environments which are not favorable for germination and seeding survival of many introduced species and undisturbed wet terrestrial habitats which do not provide open space for invaders because of the high biomass of resident species (Rejmanek 1989).

6.7 Aquatic Invasive Species: Aquatic Bioinvasion

6.7.1 *Global Perspectives*

Introduction of species outside its natural range began since antiquity and introductions of aquatic organisms to distant localities as a global phenomenon intensified in the last six decades which are for satisfying ever growing demands in globalized economies towards augmenting aquaculture, controlling vectors, supporting recreational activities, and promoting hobbies such as ornamental fish keeping and gardening. Further, there are unintentional introductions aided through various means of transportation (Ballast water of ships). The benefits provided by nonnative species to the economy and society notwithstanding, a number of species, once established, causes significant and often irreparable damage to the native ecosystems and economies of their new host countries.

Aquatic invasive species pose major ecological and economic threats to rivers, lakes and waterways worldwide through the displacement of native species and the alteration of hydrologic cycles, affecting nutrient cycles, altering food web dynamics, introducing diseases and parasites, hybridization with native species, and changes in fisheries composition and as such exotic species has been identified as one of the major reasons for biodiversity loss. The rate of invasion by exotic species has been progressing fast, which demands interventions by governments across the world, as the potential benefits of introduction of alien species are outweighed by the threat to biodiversity from new invasions (CBD 2010a, b). Although undisturbed mangrove habitats hardly provide open space for invaders to intrude and flourish because of dense canopy, high biomass of the resident species in contrast to the open coastal water bodies which are notoriously open to all kinds of exotic plants (Ashton and Mitchell 1989).

Aquatic environments may be extremely vulnerable to invasive species and eradication of such species is more difficult than in terrestrial habitats (Carlton 1999a, b, 2000). Aquatic (Alien) Invasive Species (AIS) represents non-indigenous species that threatens the diversity and/or abundance of native species, the ecological stability of infested waters, or commercial, agricultural, aquaculture or recreational activities dependent on such waters. Aquatic invasive species include non-indigenous species that may occur in inland, estuarine and marine waters, that presently or potentially threaten ecological processes and natural resources. There are three main stages in the invasion processes: (i) **Introduction**: the intentional or unintentional introduction to the wild of a species imported into a region beyond its native range; (ii) **Establishment**: the establishment of self-sustaining, reproductive populations; and (iii) **Invasion**: population growth and spreading of the species.

The Global Invasive Species Programme (GISP) is an initiative closely linked to the Convention on Biological Diversity (CBD) and is a partnership which seeks to build global cooperation. It brings together several international nongovernmental organizations (NGOs), such as the International Union for Conservation of Nature (IUCN) and the Nature Conservancy, and a network of international scientists including those from IUCN's Invasive Species Specialist Group (ISSG) (GISP 2004; CBD 2000, 2010a, b; IUCN 1989, 1997).

6.7.2 *Aquatic Invasive Species: Indian Perspective*

India being one of the 18-mega diversity countries of the world harbors around 8% of the global recorded species holding only 2.4% of the land area of the world (Ghosh 2008) but possesses all most all major types of ecosystems of the world (Chakraborty 2003). In an increasingly globalised world of today, plants, animals and microbes are being introduced more and more frequently into regions and different ecosystems that had never hosted them. The movement of people around the globe and escalating trade has resulted in the translocation of animals or plants of a particular geographical region to other regions of the globe.

Globalisation of maritime trade during last few decades has resulted in the translocation of marine species across the oceans, resulting in their introduction and invasion in many parts of the maritime ecosystems such as in estuary, corals, mangroves etc. Even though invasive species are now recognized as one of the greatest threats to marine biodiversity, pathways for species dispersal remain poorly understood in many parts, especially in countries like India. The major species recorded in the marine ecosystems as invasives in India include *Monostroma oxyspermum*, *Kappaphycus alvarezii* (algae), *Ficopomatus enigmaticus* (Annelid), *Eugymnanthea sp.* (sea anemone), *Mytilopsis sallei* (mussel), *Carcinus maenas* (crab) and many species of wood boring fauna, isopods, barnacles, amphipods bryozoans and ascidians. A survey on intertidal biodiversity along of Kerala coast of India recorded a total of 258 species of intertidal organisms, including 48 species of algae, 12 species of sponges, 7 species of cnidarians, 5 species of flatworms, 3 species of bryozoans, 109 species of molluscs, 6 species of annelids, 2 species of sipunculids, 45 species of arthropods, 14 species of echinoderms, and 7 species of ascidians. Only 1species of seaweed, *Caulerpa taxifolia,* 1 species of bryozoans, *Bugula neritina,* 1 species of mollusc, *Thecacera pennigera*, 7 species of ascidians (*Phallusia nigra, Ascidia liberata, Diplosoma listerianum, Styela plicata, Microcosmus exasperatus, Herdmania momus* and *Symplegma braken hielmi*) are thought to be invasive species from intertidal biodiversity of this part (Kerala coast) of India (Ravinesh and Bijukumar 2016; Bijukumar and Raj 2017).

Twenty five 'introduced' crustacean species have been reported to Indian waters by ballast water which includes 5 species of decapods, 6 species of isopods, 9 species of amphipods, 7 species of cirripeds and 5 species of copepods. Only one species of decapod, namely, *Litopenaeus vannamei* has been introduced to Indian water for aquaculture purposes (Fig. 6.13). The invasive status of some of these species, however, are yet to be determined (Dev Roy and Nandi 2017).

6.7.3 Need for Global Assessment of Bioinvasion on Marine Environment

Although invasive species are widely recognized as a major threat to marine biodiversity, there has been no quantitative global assessment of their impacts and routes of introduction. A recent study using over 350 databases and other sources, information pertaining to 329 marine invasive species, including their distribution, impacts on biodiversity, and introduction pathways have generated, which accounts 16% of marine eco-regions (Spalding et al. 2007) (Table 6.1) have no reported marine invasions (Molnar et al. 2008). According to Campbell and Kriesch 2003, international shipping, followed by aquaculture represent the major means of introduction of invasive species which have transformed marine habitats around the world (Table 6.2). The most harmful of these invaders displace native species, change community structure and food webs, and alter fundamental ecosystem processes, such as nutrient cycling and sedimentation. Alien marine-estuarine invasives have damaged economies by diminishing fisheries, fouling ships' hulls, and clogging intake pipes. Some can even directly impact human health by causing

Table 6.1 Total number of ecoregions along with the harmful species (%) and their pathways (%) of introduction (Molnar et al. 2008)

Ecoregions	No. of harmful species (%)	Pathways (%) of harmful species
Northern California	56 (66%)	Shipping (71%); aquaculture (7 1%)
North Sea	47 (64%)	Shipping (83%); aquaculture (57%)
Western Mediterranean	43 (66%)	Shipping (77%); aquaculture (55%)
Oregon, Washington, Vancouver	41 (65%)	Aquaculture (73%); shipping (68%)
Levantine Sea	36 (50%)	Canal (6 1); shipping (58%)
Puget Trough/Georgia Basin	35 (64%)	Aquaculture (74%); shipping (69%)
Celtic Seas	33 (66%)	Shipping (76%); aquaculture (67%)
Aegean Sea	31 (53%)	Shipping (55%); canal (52%)
Southern California Bight	31 (72%)	Shipping (81%); aquaculture (7 1%)
Hawaiian Islands	31 (42%)	Shipping (68%) aquaculture (39%)

Table 6.2 Pathways and sub-pathways for the entry and spread of invasive species (Campbell and Kriesch 2003)

Transportation-related pathways	*Commerce in living organisms pathways*
Packing materials containers – both exterior and interior	Live seafood trade
Land/terrestrial transportation	Livestock
Items used in shipping process	Aquaculture and mariculture activities
Canals that connect waterways	Enclosed facilities
Dredge spoil material	Stocking in open water
Superstructures/structures above the water line	Pet, aquarium, and water garden trade
Stowaways in holds	Bait industry
Hull/surface fouling	Biocontrol
Ballast water and sediments	Nurseries/garden/landscaping
Ballast and/or fouling	Agricultural and forestry species trade
Air transportation	Plants and plant parts as food
Freshwater/marine transportation	Other animal trade
Modes of transportation	Other plant trade
Other human-assisted pathways	
Tourism/travel/relocation	Ecosystem disturbance
Mail/internet/overnight shipping companies	Climate change
Natural spread	

disease (Ruiz et al. 1997). Although only a small fraction of the many marine species after being introduced from outside of their native range are able to thrive and invade new habitats (Mack et al. 2000), displaying drastic impacts which appear to be local, initially but with the passage of time coupled with an increasing degree of intensity of impacts tend to become global. The Convention on Biological Diversity (CBD) has identified the need for "compilation and dissemination information of alien species that threaten ecosystems, habitats, or species, which are to be used in the context of any prevention, introduction and mitigation activities" (CBD 2000). Most data pertaining to the trend of bioinvasion have been compiled at local, national, or regional scales (Ricciardi et al. 2000).

As per as coastal-marine ecosystem is concerned, once alien species become established in marine habitats, it can be nearly impossible to eliminate them (Thresher and Kuris 2004). Interception or removal of pathways is probably the only effective strategies for reducing future impacts (Carlton 1999a, b; Carlton et al. 2011).

6.8 Coastal Environment of West Bengal and its Biodiversity

Indian coast has a land frontier of 15,200 km. Coast line stretches about 5700 km on the main land and about 7500 km including the two island territories and exhibits most of the known geomorphological features of coastal zones. Presently Indian coastline is facing increasing human pressures which have resulted in substantial damage to its ecosystems (Qasim and Sengupta 1988). The coastal area of West Bengal extends over 0.28 million hectors and 220 km of coastal line shared by two districts viz. Midnapore (East) and South-24 Parganas (Fig. 6.1). The coastal plain of West Bengal reveals a long history of deposition and seaward advancement over the Bengal Basin tectonic setting by neotectonics and concomitant sea level changes in this physical domain (Paul 2002). The coastline of West Bengal is at present being eroded with the site of sedimentation shifting back to inland and the coastal plain streams have started filling their own valleys with sandy sediments perhaps as a response to a recent, slow and small rise in sea level in the region. Everywhere along the coast, sand cliffs and clipped clay banks characterize the back shore and beaches experience an alarming rate of erosion (Paul 2002).

6.8.1 The Coastal Environment of Midnapore (East) District, West Bengal, India

This coastal belt sharing 27% (60 km) of coastal tract of West Bengal, India (Fig. 6.1a) is a contiguous part of deltaic Sundarbans Mangrove Ecosystem – a

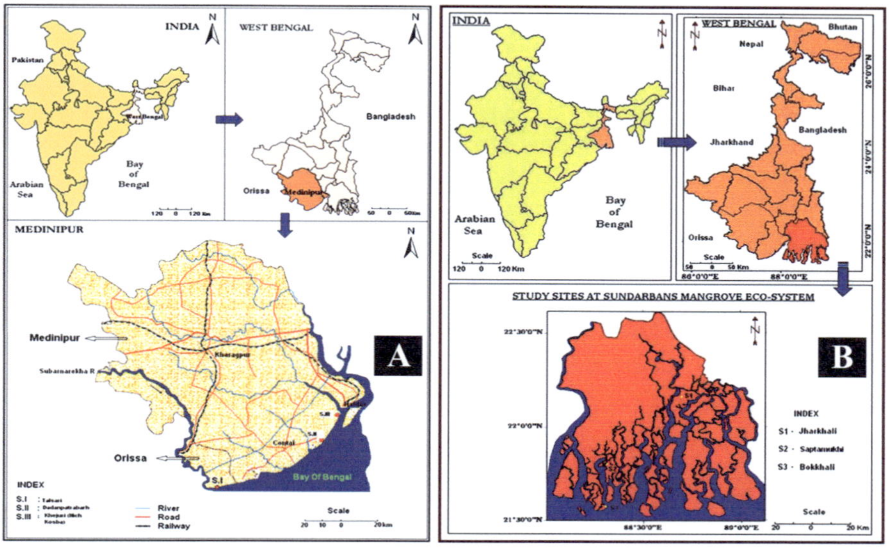

Fig. 6.1 Maps showing the coastal belt of West Bengal, India—(**a**) Midnapore (East) Coast; (**b**) South 24- Parganas Coast (Sundarbans)

world Heritage site, extending along the Hooghly estuary from New Digha (at the confluence of Subarnarekha estuary with Hooghly) to the extreme south-west point of Midnapore district and then curving around Junput, Rosulpur, Khejuri and Haldia on the east to further north east upto Tamluk or even Kolaghat on the bank of river Rupnarayan, West Bengal, India (Chakraborty 2010).

The inner boundaries of the coastal plains of Midnapore (East) littoral tract represent the corresponding shorelines of Post-Pleistocene high stand of restored sea level (Paul 2002). The coastal tract of Midnapore is characterized by sand dunes, long shore currents, high salinity, less turbidity and less vegetational coverage (Mukherjee and Chatterjee 1997; Chakraborty 2010) in comparison to its counterpart, the South 24-Pargana district, supported by Sundarbans Mangrove ecosystem (Mukherjee and Bakshi 1998). The inner boundaries of the coastal plains of Midnapore (East) littoral tract represent the corresponding shorelines of Post-Pleistocene high stand of restored sea level (Paul 2002).

This coast is dominated by a high energy macro-tidal environment with waves produced by a long fetch across the Bay of Bengal, coupled with predominant southwesterly monsoon winds blowing onshore and also easterly and southerly winds encouraged by the visiting cyclones over the Bay-head coast in the summer months. The sandy coasts of Midnapur littoral tract have been confronted with a number of coastal issues (sea level rise, coastal erosion control measures, storm hazard mitigation, environmental refugees, coastal pollution, industrial development, aquaculture, navigability of the Hoogly estuary, ecotourism etc.) which have created conflicts between various resource users and interest groups, between developers and ecologists, engineers and geoscientists and landowners and economists in West Bengal and in adjoining parts of northern Odisha (Paul 2002).

The main estuary of Midnapore (East) coast is Subarnarekha estuary which is named as "the streak of gold" is about 460 km in length. After originating from a hilly tract, 15 km southeast of Ranchi city of the Jharkhand state, it traverses through the picturesque terrain of varied landscapes of three staes of Eastern India viz. Jahrkhand, Odisha and West Bengal, before opening to Bay of Bengal (Fig. 6.1a).

6.8.2 The Coastal Environment of South-24 Parganas District, West Bengal, India (Sundarbans –Mangrove-Esuarine Complex)

The important geo-morphological settings of the Sundarbans mangrove ecosystem include mudflats, sand beaches, coastal dunes, estuarine networks, shallow creeks and mangrove swamps (Chaudhuri and Choudhury 1994; Paul 2002). Sundarbans, the only mangrove tiger-land of the globe is presently under threat of severe coastal erosion due to relative sea level rise. According to Morgan and McIntire (1959), the Bengal Basin deltaic islands of Sundarbans have been gradually tilting towards east. This has probably caused the main fresh water discharge to shift gradually eastward (towards Bangladesh) imposing severe stress on freshwater budget for Hoogly-Matla estuary. The important geo-morphological settings of the Sundarbans mangrove ecosystem include mudflats, sand beaches, coastal dunes, estuarine networks, shallow creeks and mangrove swamps (Chaudhuri and Choudhury 1994; Paul 2002). The estuarine network of Indian Sundarbans include seven major estuaries viz. Hoogly, Muriganga, Saptamukhi, Thakuran, Matla, Bidyadhari, Gosaba and Harinbhanga and they flow from west to east (Fig. 6.1b). Out of these seven estuaries, five (Muriganga, Saptamukhi, Thakuran, Mtla and Gosaba) have lost their connection with the main flow Ganga river because of siltation and their estuarine character is now maintained by the monsoonal run off alone (Cole and Vaidyaraman 1966).

6.9 Why Mangrove Is Most Productive Ecosystem of the World?

Mangrove ecosystem represents one of the most productive natural wetlands found in the intertidal zone of tropical and subtropical regions of the world (Macnae 1968; Chaudhuri and Choudhury 1994). This specialized ecosystem, dominated by intertidal salt tolerant halophytic vegetation and enjoying the influence of two high and two low tides a day, offers a dynamic ecosystem for bioresource development on one hand and maintenance of ecological balance through the protection of coastal line on the other (Chakraborty 2011). This detritus based coastal ecosystem is highly productive having an average productivity of 2500 mg cm^2 per day and plays an important role in the biogeochemical cycles of the coastal environment (Jennerjahn and Ittekkot 2002). The importance of mangrove ecosystem for its potential for fisheries and aquaculture development has received wide acceptance all over the globe mainly because of two reasons: Firstly, large quantities of energy, in the form of mangrove plant contributed detritus, is exported from the mangrove forest to open water bodies (Heald and Odum 1975) and positive correlation exists between the extent of mangroves and total fisheries yield from adjacent water (Macnae 1974; Lee 1995), Secondly, profitable regional and international markets for high quality aquaculture products are available which sustain the livelihoods of considerable number of local people (Fig. 6.2).

As mangrove ecosystem is structured by the combination of two sub systems-mangrove subsystem and its adjoining estuarine aquatic subsystem which always interact with one another by a unique physical process, the tide, the interactions always go on among different living and non living components resulting in continuous nutrient cycling (Chakraborty 2011). Fluctuating tidal water are important for transporting nutrients, controlling soil salinities, and dispersing propagules.

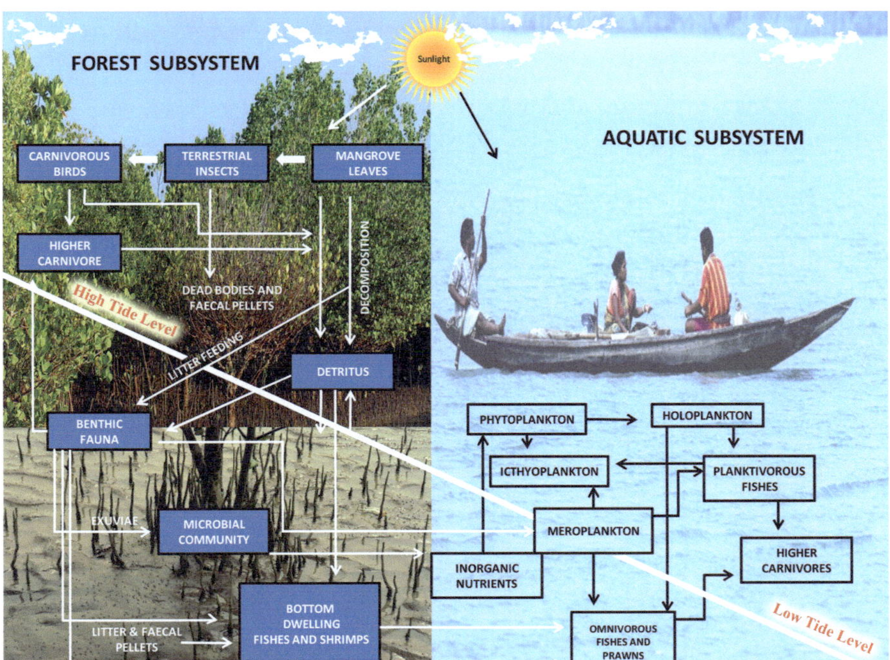

Fig. 6.2 Mangrove ecosystem's ecodynamics- interactions among different structural components of two subsystems- Forest and Aquatic

6.10 Uniqueness of Trophic Interactions in Estuarine-Mangrove Ecosystem of West Bengal, India

The trophic relationships among different biotic components of mangrove ecosystem reveal that three different types of green plants (mangrove plants, benthic algae and phytoplankton) represent the first trophic level as primary producers which are taken up as food by primary consumers representing secondary trophic level (insects, intertidal crabs, molluscs, forest inhabiting deers, wild boars, birds, monkies etc.) The fauna belonging to tertiary trophic levels are mainly carnivores and also omnivores. Such flow of energy through some other trophic levels, mostly represented by carnivores and omnivores (larger finfishes, dolphins, leopard cats, turtles, water monitors and different birds) ultimately end to highest trophic level, represented by top carnivores like Royal Bengal tiger (*Panthera tigris tigris*), estuarine crocodile (*Crocodylus porous*) etc. (Fig. 6.3).

6.11 Biodiversity of Coastal-Estuarine-Mangrove Ecosystems of West Bengal, India

6.11.1 Species Composition and Diversity in the Coastal-Estuarine-Mangrove Complex of West Bengal, India

Highly productive mangrove ecosystem supports a high abundance and diverse variety of faunal components. Three hundred twenty-two species of fin fishes both

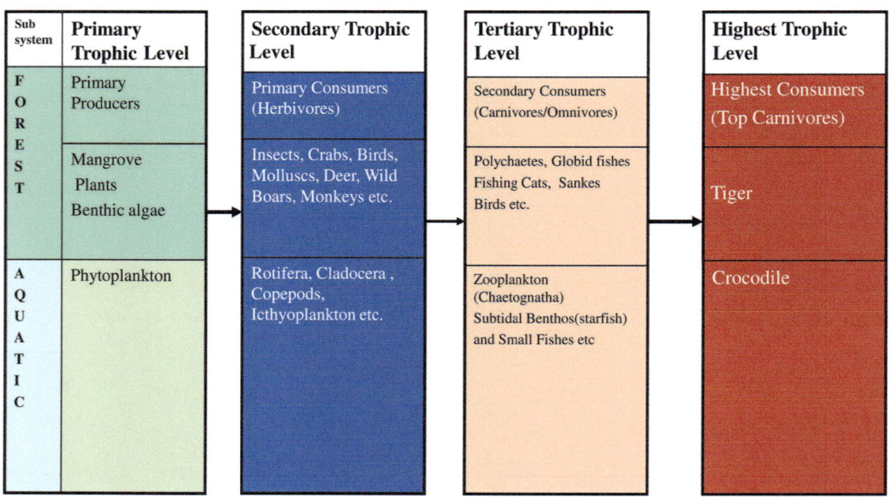

Fig. 6.3 Trophic interactions among different structural (Biotic) components of Mangrove Ecosystem

pelagic and demerseal fishery resources under 78 families and 11 orders have been reported from the Midnapore (East) coastal belt a small part (33%) of West Bengal coast (Yennawar et al. 2015). Besides, the faunal biodiversity of West Bengal coast include 24 species of shrimps, 7 species of amphibian, 59 species of reptiles (identified families include varanidae, colubridae, crocodylidae, chelonidae and emididae), more than 200 species of avifauna (cormorant, heron, egret, kingfisher, kites, stork etc.), 39 species of mammals represented by chiroptera, monkeys, fishing cats, pigs, deers, bengal fox, tiger etc., numerous species of phytoplankton (>100 species), algae (150 species), zooplankton (>100 species), ichthyoplankton (29 families), benthos (26 species of brachyuran crabs, 69 species of polychaetes, 110 species of molluscs etc), more than 300 species of soil inhibiting and mangrove plants dependent insects alongwith 44 species of micro- arthropods (Annon 2003; De et al. 1987; Sarkar et al. 1986; Chaudhuri and Choudhury 1994; Chakraborty and Choudhury 1992a, b, 1993, 1994, 1997; Chandra and Chakraborty 2008; Chakraborty et al. 2009; Dey et al. 2010; Chakraborty 2011, 2013, 2017). The commonly occurring zooplankton orders are copepoda (25 species), rotifera (4 species), chaetognatha (4 species) and cladocera (6 species). Important ichthyoplankta recorded from the estuarine networks of Sundarbans belonged to the families clupeidae, engraulidae, megalopidae etc. Seventy nine species of macrobenthos have been recorded from different study sites, among which major faunal groups were *Edwardsia jonesii* and *Paracondylactis indica*, representing phylum cnidaria; *Mastobranchas* sp., *Glycera* spp., *Perinereis* sp., *Lumbriconereis* sp. representing class polychaeta; *Scylla* sp., *Dotilla* sp., *Uca* spp., *Ocypoda* sp. etc. representing class crustacea and *Littorina, Nerita,* etc. representing phylum mollusca. Species composition, distributional pattern, population dynamics and community structure of different groups of fauna have experienced wide range of changes spatially and temporally because of the prevailing fluctuating environmental parameters like tidal exposure and inundation, salinity, temperature, pH etc. (Chakraborty 2011, 2017).

6.11.2 Diversity and Distribution Mangroves of West Bengal Coast, India

Mangroves are an 'artificial assemblage' of trees and shrubs i.e. species are not taxonomically related with one another, which demonstrate special adaptations for life in the intertidal zone (Tomlinson 1986).

These halophytic floral components create a complex and distinctive habitat with great value for biodiversity and human societies. Mangroves coexist and interact with salt marsh ecosystems in many places in the world, including the temperate regions of Australia, New Zealand, and the southern United States (West 1977; Mitsch and Gosselink 1986). In recent decades, mangroves have replaced salt marshes on the shorelines of Australia and North America. The adaptability of mangrove plants to thrive well in the stressful ecological condition is ensured by the possession of pneumetaphores and trap roots, sunken stomata, viviparous germination, thick lignin coating on leaves etc. (Davis 1938; Blasco 1975).

In 1960s, the total area of the Indian mangrove was estimated as 681,976 ha, in which nearly 45% occurs in Sundarbans and the islands of Bay of Bengal (Blasco 1975). Sixty-nine species of true mangroves belonging to 26 genera and 20 families have been recognized by Duke (1992) while Kathiresan and Bingham (2001) considered 65 species (22 genera and 16 families) as to be true mangroves. Another 80 species of plants, both herbaceous and woody showing no special adaptation for living in the intertidal environment have been recorded as mangrove associates and they bridge the gap between, mangroves and terrestrial vegetation (Duke 1995). Sundarbans mangrove ecosystem harbours 34 true mangrove species, out of the above mentioned ones, a total of 163 species of fungi (Chaudhuri and Choudhury 1994), 150 species of algae (Naskar et al. 2004), 32 species of lichens (Santra 1998) and 40 species of mangrove associates (Annon 2003).

Out of the true mangrove plant species (Tables 6.3 and 6.4), special mention may be made of *Rhizophora apiculata, Sonneratia apetala, Avicennia marina, Excoecaria agallocha, Bruguiera cylindrica, Acanthus ilicifolius* etc. The mangrove associated plants are represented by species such as *Sarcolobus carinatus, Suaeda maritima, Pandanus tectorius* etc. Important phytoplankton species include *Nitzschia* sp., *Peridinium* sp., *Ceratium* sp., *Thalassiosira* sp., *Planktoniella* sp., *Gosslorilla* sp., *Rhizosolenia* sp., *Ditylum* sp., *Triceratium* sp., *Fragilaria* sp., *Gyrosigma* sp., *Diploneis* sp., *Navicula* sp., *Dynophysis* sp., *Oedegonium* sp. etc. (Annon 2003). The algal community provide a novel source of nutrients to the whole ecosystem. The algae remain in the mangrove forest subsystem as an epiphytic assemblage living on the stems, pneumatophores of mangrove trees and on the surface of the sediment as epibenthic form. Besides, a number of algae remain in the water subsystem as phytoplankters and play a great role in the total productivity and energy flow of the system (Gopalkrishnan 1971; De et al. 1987). This coastal-mangrove ecosystem harbours 48 bacterial strains from the decomposed litter (Bhowmik et al. 1985) which play the key role in the total nutrient cycle of the

Table 6.3 The floral diversity of Indian Sundarbans

Groups of plants	No. of species	References
True mangroves	34	Chaudhuri and Choudhury (1994)
Mangrove associates	40	Annon (2003) and Chaudhuri and Choudhury (1994)
Mesophytic invasive species	10	Bhakat et al. (2004)
Algae	150	Naskar (2004)
Lichens	32	Santra (1998)
Fungies	163	Chaudhuri and Choudhury (1994)

Table 6.4 Major mangrove species of Indian Sundarbans

Sl. No.	Families	Major species	Status
1.	Rhizophoraceae	*Rhizophora mucronata* Lamk	Tree
		R. apiculata Blume	Tree
		Bruguiera gymnorhiza (L) Lamk	Tree
		B. cylindrical L.	Tree
		Ceriops tagal (Perr). Robinson	Tree
		C. decandra (Griff) Ding Hou	Tree
		Kandelia candel (L) Druce	Tree
2.	Avicenniaceae	*Avicennia officinalis* L.	Tree
		A. alba Blume	Tree
		A. marina (Forsk) Vierh	Tree
3.	Sonneratiaceae	*Sonneratia apetala* Buch Ham	Tree
		S. caseolaris (L) Engler	Tree
4.	Combretaceae	*Lumnitzera racemosa* Willd	Tree
5.	Meliaceae	*Xylocarpus granatum* Koen	Tree
		X. mekongensis Pierre P	Tree
6.	Arecaceae	*Nypa fruticans* (Thunb) Wurmb	Threatened because of less fresh water and preferred home of Tiger
		Phoenix paludosa Roxb	
7.	Sterculiaceae	*Heritiera fomes* Buch Ham	Tree; threatened
8.	Myrsinaceae	*Aegiceras corniculatum* (L) Blanco	Shrub
9.	Rubiaceae	*Scyphora hydrophyllacea* Gaertn. F	Shrub
10.	Euphorbiaceae	*Exoecaria agallocha* L.	Tree
11.	Acanthaceae	*Acanthus ilicifolius* L.	Shrub
12.	Malvaceae	*Thespesia populnea* (L) Solander	Tree
		Hibiscus tiliaceus L	Tree
13.	Tamaricaceae	*Tamarix dioica* Roxb	Tree

(continued)

Table 6.4 (continued)

Sl. No.	Families	Major species	Status
14.	Fabaceae	*Derris indica* (Lamk). Bennet	Tree/climber
15.	Convolvulaceae	*Ipomoea pes-caprae* Sw	Creepers. Sand binders pioneer species in dune formation
16.	Chenopodiaceae	*Suaeda maritima* (L) Dumort	Herb
		Salicornia brachiata Roxb	Herb
17.	Poaceae	*Porteresia coarctata* Takeoka	Grass
		Phragmites kakra Trin Ex Steud	Grass
		Cynodon dactylon (L) Pers	Grass

ecosystem by their exo-enzymatic activities leading to degradation of cellulose, hemicellulose, pectin, chitin, lignin etc. (Biswas et al. 1986). Out of 150 species of algae recorded from Indian Sundarbans, 50 species belong to Cyanophyta, 39 species belong to Chlorophyta, 2 species belong to Pheophyta, 44 species belong to Chrysophyta of which 42 species are diatoms and 15 species belong to Rhotophyta (Naskar et al. 2004). The occurrence, distribution and succession of mangrove species are determined by virtue of the adaptive ability of different species with the fluctuating ecological conditions.

Historically and geographically the Midnapore (East) coast is a contiguous part of deltaic Sundarbans of global importance, limiting the Hoogli estuary on the western front. Naturaly, this coastal tract was once covered by dense mangrove forests, the relics of which are still found which have been surviving after sustaining the century old anthropogenic pressures, especially in developing this coast as the main hub for coastal tourisms and also commercial fishing. Fifty seven species of mangroves and their associated plants (Fig. 6.3), mostly comprised of *Avicennia officinalis, A. alba, Exocoeria agallocha, Acanthus ilicifolius, Sueda maritima, Salicornia brachiata, Rizophora mucronata etc.* along with dune inhabiting plants such as Ipomea, Spinifix, Pandanus etc. which play major role to stabilise dunes, and act as buffer against erosion (Chakraborty 2010, 2013, 2017).

6.11.3 Ecology of Mangroves: Zonation and Succssesion of Mangroves of West Bengal, Coast, India

In Sundarbans, there is a distinct zonation of different mangrove species along the intertidal slope reflecting their adaptation to different prevailing ecological factors viz.

duration of tidal inundation and exposure, salinity, sediment characteristics etc. Mudflats are developed in some sheltered areas on the deltaic islands of the Sundarbans due to higher sedimentation load and lower velocity of water. The finer sediment constituting the thick mudflats are rich in organic content and ideal for penetration by plant roots. The surface of these flats are initially covered by thick algal mat. Consequently luxuriant growth of mangrove vegetation occur, showing distinct zonation pattern. Ginsberg and Lowenstam (1958) reviewed the evidence of stabilizing and salt trapping powers of filamentous algal and marine phanerograms (mangroves) in shallow sheltered waters. The species replacement of mangroves in a particular mudflats consolidates, elevates the substratum and creates shades and makes way for another. The zonation of mangrove vegetation in Sundarbans showed that the seaward edge of the forest is endowed with saline grass *Porterasia coarctata* which are followed towards landward by *Avicenia officinalis, Suaeda maritima, Bruguiera gymnorhiza, Sonneratia apetala* and *Ceriops roxburghiana*. Different floral and faunal community display different zonation pattern reflecting their adaptation to different degrees of terrestriality, duration of inundation and exposure, salinity gradients, textural and nutrients components of the soil, biotic association etc. (Chakraborty and Choudhury 1992b). *Heritiera fomes* requires more freshwater and occur together with *Phoenix pelludosa* and *Nypa fruticans*, generally above the mid tidal level. The mangrove species like *Excoecaria agallocha, Ceriops decandra, Rhizophora apiculata* and *Bruguiera gymnorhiza* are found to inhabit in relatively higher saline zones experiencing higher intensity of currents and longer exposure. Lugo and Snedaker (1974) inferred that zonation patterns in mangroves represents steady state adjustment rather than successional stages.

The succession of group of vegetations as observed in the Sundarbans, covering around 70% of forest area is as *Porteresia coarctata, Phagmites karka, Sesuvium portulacastrum, Avicennea officinalis* → *Acanthus illicifolius, Aegiceras corniculatum, Avicennia alba, A. marina, Sonneratia apetala* → *Bruguiera gymnorhiza, Salicornia brachiata, Ceriops dacandra* → *Heritiera fomes, Rhizophora apiculata, Xylocarpus mekongenesis* → *Ceriops dacandra, Excoecaria agallocha, Phoenix pelludosa, Suaeda maritime* (Chakraborty 2011). Smith (1992) reviewed factors and mechanisms that may determine local forest structure of mangroves: (1) land building processes driving autogenic succession of mangroves; (2) response of species to geomorphologic influences such as sea level rise; (3) physiological adaptations to stress gradients such as soil, salinity in the intertidal zone; (4) dispersal patterns of mangroves propagules in accordance with size and rooting time; and (5) post disposal processes determining propagule establishment, such as predation and competition.

Different mangrove species have different optimal requirements in respect of salinity, nutrient availability and soil characteristics (Hutchings and Saenger 1987), translating to correlations between soil properties and species recruitment, establishment and dominance (McKee 1995a; Lovelock et al. 2005), growth, development, availability in close proximity to parent tree, mode of tidal dispersal and retention of propagules also influence the initial community structure (McKee 1995b; Patterson et al. 1997).

A recent study in the Sundarbans of Bangladesh (Biswas et al. 2007) has revealed ecological association of mangroves with *Sonneratia apetala, Heritiera fomes, Excoecaria agallocha, Ceriops decandra,* where the invasion was found to be low. In mixed forest, the spread of invasive species was really high. Generally invasives do not prefer *Sonneratia apetala* stands but in a newly accreted char in Sundarbans, huge spread was noticed. *Eichhornia crassipes*, another floating aquatic invasive species, showed significant association with *Sonneratia* stands. The main species of Sundarbans, i.e. *Heritiera fomes* was found to be positively associated with some invasive associates such as *Derris trifoliata, Hoya parasitica and Micania scanden.* The identified ecological association of invasive species with host plants indicates that mixed forest are more susceptible to invasive species. The three most harmful invasive species in the Sundarbans ecosystem of Bangladesh are *Derris trifoliata, Eichhornia crassipes and Eupatorium odoratum.* The negative ecological impacts of invasive species were very much prominent because of their abundance and constantly spreading ability have been found to have threatened the native species by outcompeting them.

6.11.4 Ecological Uniqueness of Faunal Components of West Bengal, Coast, India

Estuarine and mangrove dwelling animals have been studied mostly for their trophic and bioturbating roles and their capacity for commercial fishing production (Connolly and Lee 2007). Ecology of two major faunal groups viz. plankton and benthos of Sundarbans Mangrove Ecosystem India, have been studied extensively during last three decades by different researchers. Sarkar et al. (1986) reported that copepods being the most dominant zooplanktonic group enjoyed bimodal type of population fluctuation with relatively higher population density during mansoon and also with a slight peak of population during last phase of winter. Some species under both chaetognatha and cladocera, the two other major zooplanktonic groups experienced their maximum population density during higher saline periods i.e. premonsoon period (Chaudhuri and Choudhury 1994).

Different macrobenthic faunal groups viz. brachyuran crabs, polychaetes, molluscs in Sundarban mangrove ecosystem displayed varying fluctuation trend of seasonal population density and community structure. The fiddlers crab's population density displayed its maximum during premonsoon followed by postmonsoon while many species of polychaetes have been found to occur in maximum density during monsoon. The species diversity indices taking into consideration of total benthic faunal groups were maximum during premonsoon and minimum during monsoon.

However, such indices with polychaetous and molluscan faunal components registered maximum values during monsoon (Chakraborty and Choudhury 1992a, b; Chakraborty et al. 2009; Chakraborty and Chaudhury 1993, 1994, 1997). Several researchers have put forward their views regarding the influence of different ecological parameters on such spatial and temporal variation of density and community of macrobenthic species (Chakraborty 2013). Barnes (1967) stressed the importance on osmoragulatory abilities of macrobenthos pertaining to their distribution in time and space. Teal (1958) and Kinne (1963) advocated the role of temperature affecting the animal distribution in the brackish water zone. Abele (1974) mentioned that substrates are important in determining the species composition of the various habitats. A species can use one substrate as a shelter, another as a feeding site and other as a source of nutrition, thus reducing competitive interactions for each one resulting in appositive co action in community existence. Thus greater number of species can inhabit in a narrow intertidal belt, resulting in the change of species diversity. Sanders (1968) postulated the role of sediments and opined that most diverse communities will always be found in the tropics and deep sea because of the constancy of their soil environments. Nutrient enrichment in the littoral zones of the mangrove ecosystem play vital role in determining the settlements, growth, population fluctuation and community interactions of fauna (Heald and Odum 1970).

6.11.5 Intertidal Macrobenthic Fauna of West Bengal Coast: Functional Relationships and Bioinvasion

Biotic community of benthic fauna (Fig. 6.4) an important and essential structural component of any marine and estuarine ecosystem (Chakraborty 2011, 2013; Chakraborty et al. 2009) are structured by arrays of ecological parameters, both living and nonliving ones which operate in an intricate manner resulting in multidimentional relationships. The biotic relationship includes trophic inter relationships and flow of energy in the form of primary production, secondary production etc. (Elliott and Taylor 1989) through inter- and intraspecific competition, feeding and predator-pray interactions, the production of biomass, and the production and delivery of recruiting stages (Gray and Elliot 2009). The benthic fauna produce millions of larvae in the form of meroplankters which not only support fish populations but also maintain the ecological equilibrium. Nearly half of the world's commercial fish catch come from the sea and estuary which supports the lives of a large number of shell fishes and demersal fishes whose main food items are derived from the benthic animals. Mangrove crabs play an essential role for leaf litter degradation in these systems (Robertson 1986; Micheli 1993a, b; Alongi 2002, 2008; Chatterjee and Chakraborty 2014; Chatterjee et al. 2014). The diversity and abundance of benthic fauna has positive relationships with demersal fisheries and aquaculture (Chakraborty 1995, 1998). Similarly, the intertidal and benthic

Fig. 6.4 Representatives of Intertidal benthic fauna—(**a**) Crawling of gastropods on mangrove floor and plant's trunk; (**b**) Ghost crab- *Ocypoda macrocera*; (**c**) Fiddler crab-*Uca acuta acuta*; (**d**) Benthic fish-*Boleopthalmus bodderti*

macrofauna are very sensitive to environmental stress hence water quality biologists use them to study the environmental changes (Washington 1984; Stark 1998).

Biotic assemblages of intertidal macrobenthic fauna in the West Bengal mangrove estuarine complex is determined because of multidimentional community interactions among themselves after being controlled by their ecological and ethological attributes such as population fluctuation synchronizing with changing ecological parameters, zonation and succession, foraging and reproductive requirements, competition for space and terretories, etc. and thereby have developed ecological resilience along with resistance stabilities which enable to maintain key ecological functions and processes in the face of ongoing environmental stresses and pressures including bioinvasion either by resisting or adapting to imposed negative changes. Resilient systems are characterized as adaptable, flexible and able to deal with change and uncertainty. Building resilience into an ecosystem means working to support the ecological health and function of associated habitats, organisms and ecosystem processes. Predations among benthic fauna represent an important determining factor for the colonization and propagation of organisms. There are three categories of predators, (1) surface predators (echinoderms, molluscs and crabs); (2) burrowing predators (nemertines, polychaetes, gastropods) and (3) digging predators (ghost crabs, thalassinoids, brachyopods etc.). The suspension feeders include cnidarians, bivalves, acron barnacles etc. while many gastropods, fiddler crabs, globiid fishes etc. are deposit/detritus feeders. All those feeding categories as found among different benthic fauna represent important determining factor for the colonization and propagation of estuarine-mangrove faunal components (Chakraborty 2011, 2017). Such community interactions and competition have been seen to trigger aggressive movement of the mobile bivalves such as *Donax*, *Anadera* etc. and gastropods like *Telescopium*, *Cirithedia* etc. on the substratum in this ecosystem as epibenthic fauna which disrupts the burrows of more sedentary infaunal bivalves, actiniarins, polychaetes etc. leading to their displacement. Burrowing activities of deposit feeders tend to disturb the substratum and make the water turbid with the suspended dispersed particles in the overlying water and that drive the filter feeders and suspension feeders out from that zone making the co-existence difficult. These deposit feeders maintain their own community and exclude suspension feeders. Such exclusion of one trophic group through modification of the environment by another group is reffered to as ***Trophic Group Ammensalism*** which is considered to be the major determining factor for limiting the settlement, distribution and colonization of these faunal components. It has been observed that different brachyuran crabs representing the key stone species in the intertidal belts of coastal West Bengal display distint zonation and succession in tune with their adaptibilities with changing b associations of biotic community as well as different degrees of terrestriality from the low tide level to high tide level (Chakraborty 1992b; Chatterjee and Chakraborty 2014; Chatterjee et al. 2014).

6.11.6 Biofouling Benthic Fauna: Possibilities for Bioinvasion

The intertidal zone, the area between high and low tides, shows greater diversity of habitats and distinct biodiversity mostly because of its changing ecological gradients vis-à-vis instability and habitat heterogenity. This eco-zone with either hard or soft bottom substrates and water rich in nutrients, provide ideal niche for a variety of taxa including sea weeds, crustaceans (brachyuran crabs, benthic shrimps), molluscs, polychaetes, brachyopods, holothuroids, bryozoans, sponges, sea anemones, polychaetes, flatworms, gastropods, sea slugs, echinoderms, protochordates and globiid fishes. The developmental activities along the coast line, especially development of ports, and coastal erosion control measures with lateritic and concrete boulders are supposed to have opened pathways for the entry and settlements of exotic species as over two-thirds of recent non-native species introductions in marine and coastal areas are likely due to ship-borne vectors. The colossal loads of ballast water may transport viruses, bacteria, protists, algae, zooplankton, benthonic invertebrates, and fishes along harbours at a faster pace (David and Gollasch 2015).

Biofouling faunal components represented by acron barnacles (*Balanus aphrodite*, belonging to the order cirripedia under the class crustacea, oysters, littorinids, predator species (Thias), grazing snails (*Nerita articulate)*, purple coloured bivalve, (*Enigmonea aenigmatica*) have been found to inhabit on hard structures like bottom of the fishing trawlers, jetties, hard lateritic boulders for embankment construction (Fig. 6.5) in West Bengal coast, India. Three species of jelly fishes *Dactylometra quinquicirrha* (L. agassiz), *Crambionella stuhlmanni* (Chun) and *Chiropsalmus buitendijki* (Horst) were recorded in such a higher densities in and around Tamilnadu

Fig. 6.5 Biofouling Faunal components—(**a**) Bivalve- *Enigmonea aenigmata*; (**b**) *Balanus aphrodite*; (**c**) Heap of Mixed shelled benthos; (**d**) *Cassostrea madranensis*; (**e**) *Saccostrea cuculita*

coast of India that such abundance tended to cause blockage of the cooling water intake screens of thermal power station (Rajagopal et al. 1989).

6.11.7 Benthic Fauna As Ecosystem Engineers: Bioturbation, Nutrient Cycling and Bioinvasion

Ecosystem engineering, a term introduced by Jones et al. (1994, 1997) is defined as the indirect or direct control of resource availability mediated by an organism's ability to cause physical state of changes in abiotic or biotic materials. Ecosystem engineering is in essence the creation, destruction or modification of habitats. Physical resources that may be affected by ecosystem engineers are varied and include living space or habitat, light, humidity, sediment, heat, water and physical materials. For example, coral reefs directly (biofowling organisms) provide living space and also modulate abiotic forces such as currents that in turn affect resource supply to other organisms.

A biological invasion being a natural process, representing the arrival of a species into a new ecozone (Carlton 1999a, b), after traversing at unprecedented rates, routes and manners (Elton 1958; Williamson 1996; Vitousek et al. 1997). In order to address the scale and impacts of this anthropogenic mixing of biotas, as well as offers an opportunity for basic biological insight, invasion biology has become a rapidly developing discipline with broad ecological and conservation implications (Elton 1958; Carlton 1999a, b).

One of the most dramatic invader effects is the alteration of ecosystems where an ecosystem is defined as the spatially explicit association of abiotic and biotic elements within which there is a flow of resources, such as nutrients, biomass or energy (Odum 1971; Carpenter and Turner 1998).

It is considered that there is three primary ways in which exotics can effect ecosystems (Vitousek 1990). First, exotics can differ from natives in their use of resources (nutrients), thus affecting resource availability for other species (Vitousek 1990) Second, exotics can alter the flow of energy or biomass by changing food web (Savidge 1987); Third exotic can affect disturbance regimes (Chapsuis et al. 1994). Another fundamental means by which invaders have ecosystem level effects is to change the physical structure of the ecosystem itself (Bertness and Callaway 1994; Crooks and Khim 1999). Among the invaders that will have the largest impacts are those that directly modify ecosystems and thus have cascading effects for resident biota. Exotics can affect ecosystems by altering system-level flows, availability, or quality of nutrients, food, and physical resources (e.g. living space, water, heat or light). When introduced into ecosystems, these exotic engineers cause physical state of changes with effects that ramify throughout the system. Although the consequences of these modifications are varied and complex, insight gained from general ecological principles offers an opportunity to predict what invaders will do upon their integration into systems.

Bioturbation, the disturbance through the stirring or mixing of sediments layers by biological activities viz. mobility, feeding, burrowing etc. of benthic animals, affects the geochemistry of sediments and their interstitial water (Woodin 1981). Brachyuran crabs, polychaetes, and globid fishes represent the major agents of bioturbation of Sundarbans and its adjoining coastal environments. Burrows and other bioturbatory structures like pseudo-pelletes, sand balls, mud balls, sand pyramids, semidome, chimney, hood etc. have been found to exihibit seasonal variations and thereby influence the rate of sediment distabilisation, intensity of soil excavation, nutrient reshuffling, pumping of water and oxygen within soil, etc. Adaptive behavioural strategies out of such bioturbatory activities have been hypothesized as their survival strategies because they dig the soil for shelter, form sediment structures in the form of pellets, domes, chimnies etc. for procuring food, maintaining territories, displaying of energy efficiencies for sexual selection etc. (Annon 2003; Chatterjee et al. 2014). One of the major bioturbatory activities is on the microbial degradation rate of sediment organic materials by microbial and benthic community. In mangrove ecosystem, macrobenthos along with several microarthropods (Dey et al. 2010) primarily acted upon mangrove litters for their ready conversion into detritus by microbial activities.

Deposit feeders are the most prominent group of bioturbators as they constantly process sediment for food, resulting in horizontal and vertical movement of particles in the sediment. In some habitats, biogenic alterations affect the erodibility of the sediments (Luckenbach 1986; Meadows and Tait 1989; Kihslinger and Woodin 2000), the distribution of grain sizes (Rhoads and Stanley 1965; Cadee 2001; Lim 2006), and the concentrations of pore water constituents (Aller 1982, 1988, 2001) which in turn determine and alter the the ability of individuals to burrow through the sediment (Brenchley 1976, 1982). Burrowing and defecation can affect infaunal recruitment patterns (Posey 1986); infaunal secretions can alter the chemical constituents of the sediment (King 1986; Woodin 1981). Biogenic modifications therefore appear to affect the distribution and abundance patterns of infauna. Defecation by large infauna has a negative effect on other infauna, causing reduced growth rates, reduced recruitment and increased emigration (Brenchley 1976). The patterns of distribution and abundance of organisms are frequently correlated with the condition of habitat heterogeneity changes which in turn lead to changes in the abundance of physical and biogenic structures (Martinez 1992, 2004; Lee 1998, 1999, 2008).

6.12 Role of Physicochemical Parameters

Different physico-chemical parameters of soil and water determine the temporal and spatial variation of biodiversity components, govern their distribution and succession, and also ensure biological productivity by maintaining the energy flow and decomposition cycle. Fluctuation of all ecological parameters, mainly because of the changes in the meteorological conditions like temperature, precipitation, wind flow etc. coupled with marine and estuarine physical processes like tides, waves and current of water, make this environment very much unstable and force the biological components to restrict their own territory in different parts of Sundarbans (Chakraborty 2011, 2017; Chakraborty et al. 2009).

6.13 Invasion in West Bengal Coast: Interactions with Driving Factors

Interactions between introduced species and other anthropogenic stressors may greatly influence the impact of introduced species. Estuaries and bays, which represent the most invaded habitats in coastal regions, are also often the most degraded coastal habitats. A common generalization is that disturbed habitats are more readily invaded than pristine ones, largely because of reduced competition or predation (Elton 1958; Lodge 1993; Simberloff and Von Holle 1999) in the West Bengal coast, both the assemblages of mesophytic plants as well as biofouling animals have been noted in the places nearer to the fishing harbors and places having ecotourism pressures. It is often the case that effects of multiple species cannot be predicted from estimates of the effect of each species alone due to complex interactions (Kareiva et al. 1996). The question of the combined effects of several introduced species has resulted the hypothesis that synergistic effects may lead to develop accelerated and intensified impact not only on native species but on ecosystems which will go on with the addition of each new species (Simberloff and Von Holle 1999; Annon 2003).

6.13.1 Do Invasive Plants Threaten the Sundarbans Mangrove? Mode of Bioinvasion

Mangroves being woody halophytic trees and shrubs that normally grow in saline intertidal zones of tropical and subtropical coastline. Salinity, therefore, appears to be one of the key environmental factors influencing the growth and survival of mangrove species. Salinity variation and duration of a particular salinity value in a year within a mangrove forest area play a vital role in the species distribution, their productivity and growth (Twilley and Chen 1998). The variations in salinity are normally controlled by climate, hydrology, rainfall, topography and the tidal flooding of an area. All these characteristics are known to undergo spatial as well as temporal variations. Accordingly, the distribution, succession, population and diversity of mangrove species along with associated flora and fauna do also vary along with the variation of salinity (Chakraborty et al. 2009).

Major ecological thrust of plant invasions is to alter detrital inputs and the structure of detritus-based food webs as detritus is being formed by the plants contributed litter decomposition. A long term research study on the detrital pathways in mangrove food webs in native and introduced *Rhizophora mangle* forests have shown that introduction of allien species not only change the biochemical composition of plant derived detritus but brought about a change on the species composition and diversity of detritivore benthic fauna (Amanda et al. 2007). Observed differences between native and invasive mangrove food webs may be due to detritivores being poorly adapted to utilizing the tannin-rich, nitrogen-poor mangrove detritus in the sector of mangrove patch having introduced alien species. In addition, differential utilization of mangrove detritus between native and introduced mangroves may be a consequence of forest age. It was postulated from this study that increasing mangrove forest age may promote diversification of bacterial food webs important in N and S cycling (Amanda et al. 2007).

Plant invasions into coastal zones have been in the process across the world, often with negative consequences (Callaway and Josselyn 1992; Posey et al. 1993; Ruiz et al. 1997). Studies of the effects of introduced plants on tidal wetlands have been limited mostly to temperate environments (Posey 1998; Ruiz et al. 1997; Rice et al. 2000; Adam 2002), and such vascular plant introductions have been regulated by water flow, sediment grain size and pore water nutrient concentrations (Neira et al. 2005).

Mangroves in their native environment provide the suitable habitat for specific, potentially coevolved faunal components with limited species overlap in the adjacent sediment-flat ecosystems (Sasekumar 1974; Shiridan 1997). It has been observed that mangroves after being introduced in a new ecozone at the coast of Hawaii without their faunal associates have led to an ecological condition where detritivores colonizing in that introduced mangroves were not able to the consumption of tannin-rich *R. mangle* detritus and thereby developed the inability to cope up up the new ecological set up resulting to the arrestation of decomposition processes (Sessegolo and Lana 1991). Tannins are toxic to many organisms (Mahadevan and Muthukumar 1980) and can interfere with detritivore digestion (Neilson et al. 1986; Alongi 1987). In addition, mangrove litter has low nutritional value compared to other marine detrital sources [e.g., phytoplankton and benthic microalgae because of high C:N ratios and lignin content (Robertson et al. 1992). The recent introduction of tannin rich, low-quality mangrove litter to detrital food webs in Hawaii may alter the food web structure of coastal communities previously unexposed to marine vascular plant detritus. While introduced mangroves may fundamentally alter Hawaiian coastal ecosystems (Demopoulos 2004), their effects on detrital food webs in Hawaii have not been evaluated. A number of food web studies have suggested a major influence of native mangrove detritus on adjacent coastal ecosystems (Heald and Odum 1975; Robertson et al. 1992; Werry and Lee 2005). However, recent research studies with stable isotopes suggests that little detritus from native mangrove stands actually enters the food webs of surrounding ecosystems (Newell et al. 1995; Bouillon et al. 2004). Infauna (animals living within the sediment column) and epifauna (animals living at the sediment-water interface or on solid substrata) have important functions in coastal habitats, e.g., they serve as food for higher trophic levels and stimulate detrital decomposition in order ensure the stability of the ecosystem by maintaining the biogeochemical cycles in the mangrove ecosystem (Parsons et al. 1985; Levin and Talley 2000; Chakraborty 2017).

In particular, following hypotheses have been put forward based on this piece of research study (Amanda et al. 2007): (1) relatively little mangrove carbon and nitrogen is assimilated by macro fauna in introduced Hawaiian mangroves compared to native stands in Puerto Rico; (2) In both introduced and native stands, deposit feeders assimilate substantially more mangrove-derived carbon and nitrogen than do suspension feeders.

6.14 Invasive Mangrove Species from Indian Sundarbans

A recent study on a comparative basis from three islands of Sundarbans Biosphere Reserve has revealed a number of potential exotic invaders covering trees, shrubs and climbers (Table 6.5 and Fig. 6.6). While the trees *Casuarina, Eucalyptus and Leucaena* have been introduced for various reasons and appeared to have been acclimatised vis-a-vis neutralised while the rests are accidental weeds which have been invaded from the adjoining inland areas through human interventions. And among these, the five most obnoxious species are *Eichornia, Eupatorium, Lantana, Micania and Parthenium* (Bhakat et al. 2004).

An interesting aspect of this study is the fact that it re-establishes the already developed hypothesis that alien invaders occupy the human altered landscapes. The major deltaic islands of sundarbans, India viz. Sagar Island, Bokkhali and Jharkhali Islands being the highly populated areas having lot of anthropogenic activities and therefore have paved the ways for the introduction and propagation of lot bioinvasive weeds whereas the Sajnekhali, Sudhanyakhali areas being reserve forests harbor virtually no exotic species.

Based on this preliminary study, it is being assumed that out of 104 deltaic islands of Sundarbans, India, almost 50% have so far been reclaimed by clearing the mangrove plants and human controlled agriculture and aquaculture activities during the last couple of centuries have paved the way for the entry and establishment of non-mangrove vegetations along with their dependable microbes and fauna which might have become invasive ones. This finding only provides a glimpse of the extent and impact of invaders in changing the floristic pattern. Since earlier baseline record on floristic study on the area infested by non mangrove species like *Acacia nilotica, A. juliflora, Azadirachita indica, Ficus bengalensis* etc. are supposed to make way to the invasion of true invasive species. Plants like *Casuarina, Eucalyptus, Thespesia populnea, Tamarix indica, Hibiscus tilliaceus, Clerodendrum viscosum, Eupatorium odoratum, Ipoemea, fistulosa, Micania scandens, Coccinia grandis, Luffa aegyptiaca, Opuntia dillenii and Leucaena* (Table 6.5 and Fig. 6.6) which are non-invasive today but might be potential invasive in the days to come (Bhakat et al. 2004). Therefore, for conservation purpose, in environmentally sensitive area

Table 6.5 List of exotic plant invaders growing in different delta of the Sundarbans Biosphere; S-sagar Island; B-bakkali; J- Jharkhali (Bhakat et al. 2004)

Species	Lifeform	Occurrence on island	Habitat invaded	Place of origin
Alternanthera philoxeroides	Herb	S-B	Aquatic	South America
Antigonon leptopus	Climber	S-B	Road side	Mexico
Casuarina equistifolia	Tree	S-B, J	Coast	Mlayasia, Australlia
Eichhornia crassipes	Herb	S-B, J	Aquatic	South America
Eucalyptus sp.	Tree	S-B, J	Road side	Australlia
Eupatorium odoratum	Shrub	S-B, J	Road side	Central Tropics
			Waste lands	
Lantana camara	Shrub	S-B, J	Road side	New World Tropics
			Waste lands	
Leucaena leucocephala	Tree	S-B, J	Road side	Central America
			Home stead	
Micania scandens	Climber	S-B, J	Road side	USA
Parthenium hysterophours	Herb	S-B, J	Road side	South America
			Waste lands	
			Home stead	

like Sundarbans, all aliens should perhaps be considered as potential threats and therefore monitored carefully (Cronk and Fuller 1995).

6.15 Invasive Mangrove Species from Bangladesh Sundarbans

A study conducted in the Sundarbans of Bangladesh has highlighted different types of plant invasive species, the rate and pattern of invasion, its intensity, association of invaders and their habitat preference. Among the 23 recorded species, 3 species are highly invasive, 6 species are moderately invasive and the remaining are potentially invasive. Climbers (6 out of 23) were the most frequently encountered invasive species followed by trees (5 out of 23) and shrubs (4 out of 23). The three highly invasive plants were *Derris trifoliata* (climber), *Eichhornia crassipes* (aquatic shrub) and *Eupatorium odoratum*. It is considered that invasive species can only spread into natural vegetation as a result of disturbance (Biswas 2003). Disturbance, which initiates succession, is a natural process in mangrove ecosystems (Das and Siddiqi 1985). A variety of biotic and abiotic processes, which vary in frequency (Iftekhar 1999), magnitude, intensity, and timing, constitutes natural disturbance (Cattelino et al. 1979; Connell and Slatyer 1977; Grime 1977; Holling 1981; Levin and Pain 1974; Loucks 1970; Shugart and West 1980; Trudgill 1977; Vogl 1980;

Fig. 6.6 Different bioinvasive plant species in West Bengal Coast, India—(**a**) Climber- *Micania scandens*; (**b**) Mangrove plam, *Phonoex pelludosa* is under threat from invasive Acacia and Eucalyptus, a major floral component for social forestry. (**c**) Tree-*Thespesia populnea*; (**d**) Casuarina and Acacia; (**e**) Mixed Mesophytic plants which include *Azadirachta indica* and *Thespesia populnea;* (**f**) Climber-*Coccinia grandia*

White 1979). Chronic disturbance relates more to the frequency, or return interval, of a disturbance event, which alters the existing physical environment and community of organisms at a particular site (Ameen 1999); in turn, this may lead to invasions of alien species (Biswas 2003; Fox and Fox 1986). However, successful invasion depends on the extent and type of disturbance (Rejmanek 1989). In addition, cryptic ecological degradation, in which introgressive mangrove associated vegetation or minor mangrove species slowly start to dominate a forest of true mangrove species without loss of spatial extent (Dahadough-Guebas et al., 2005) This is a also common feature of the Sundarbans of Bangladesh (Biswas 2003). Chronic disturbance is of special concern in Sundarbans and may alter the species composition (Ameen 1999), relative abundances of selected species (Ameen 1999; Hossain 2003), ecosystem structure (Biswas 2003; Mack et al. 2000; OTA 1993; Mooney and Hobbs 2000; Pimentel et al. 2000; Vitousek et al. 1996), function (Ameen 1999; Mack et al. 2000; Mooney and Hobbs 2000; Pimentel et al. 2000; OTA 1993; Vitousek et al. 1996), or provide a platform for some native species to become invasive (IUCN 2003).

Investigation on biological invasion in Sundarbans has revealed that mangrove development follows land formation (Lugo 1980). This theory suggests that once vegetation establishes on the new substrate, mangroves contribute to the accretion of land (Davis 1938; Bird 1971; Biswas 2003) and thereby increase the extent of mangrove coverage alongside triggering the settlement of associated faunal components. All those structural components of mangrove ecosystem function (Fig. 6.2) in order to ensure the stability of the ecosystem which can resist the penetration followed by settlement and propagation of any alien species (Annon 2003).

6.16 Enlisting of Alien Species: A Tool for Management of Invasive Species

Lists of alien and invasive alien species (IAS) have been considered as an important bioinvasion management strategies which include early warning, prevention and control measures by way of pinpointing the mode of impacts biological invasions (Ricciardi et al. 2000; Kolar and Lodge 2001) as well as in the development of appropriate and effective control methods (Wittenberg 2005). Proper identification along with studying of associated biology and distribution of alien species also form the foundation for developing effective practices for controlling those species (Rejmánek and Richardson 1996).

Furthermore, governments and management agencies commonly make alien species listing decisions as the basis for implementing IAS legislation (Shine et al. 2008; Lodge et al. 2006). Lists of alien and invasive alien invasive species have also more recently been used for reporting on biodiversity targets and indicating the status of, and trends in, biological invasions with the purpose of informing policy

(Shine et al. 2008; Simberloff 2009; Ricciardi and Rebekah, 2008; Hulme et al. 2009) in order to use of IAS information to measure progress towards the Convention on Biological Diversity (Butchart et al. 2010).

Designation of alien species as invasive using this information, a formal, systematic decision making process was followed for the inclusion of each species onto the **Invasive Alien Species** list for a particular country i.e., to designate an alien species as invasive in that country. This process was adopted to ensure that species inclusion was as standardized, transparent, and repeatable as possible (Pulin and Stewart 2006). Three evidence based criteria were used to designate established alien species as invasive in a particular country.

Criterion (1) Alien species that had a demonstrated impact on indigenous biodiversity in the country in question; for example, via hybridization, competition, prédation, change in fire regimes, or altered food web dynamics (Spear and Chown 2009; Vilà et al. 2000). This criterion was considered the most robust evidence of invasiveness at the country level.

Criterion (2) Alien species that had an extensive distribution range, are very abundant, or had a high population growth rate in the country. Widespread and abundant alien species were assumed to impact biodiversity (Pysek and Hulme 2005, Vilà et al. 2000) via. (1) the displacement of individuals of indigenous species, and (2) the alteration of ecosystem functioning (e.g., nutrient cycling, shade, and water cycling) of the system that they invade. This criterion was considered to provide some evidence of alien species invasiveness at the country level.

Criterion (3) Invasive elsewhere, i.e., the established alien species was invasive anywhere else (other than the country in question) in the introduced range of the species based on criterion one or two. "Invasiveness elsewhere" has been demonstrated to be a strong (albeit not perfect) predictor for the likely invasiveness of an alien species at a new location (Reichard and Hamilton 1997; Duncan et al. 2003), and history of invasion success has been found to be related to likelihood of establishment (Hayes and Barry 2008).

6.17 Ecological Peturbations in West Bengal Coast: Paving the Way for Bioinvasion

Anthropogenic activities tend to impart negative impact on natural ecosystem dynamics and biotic communities (Atkinson and Cameron 1993; Carlton 1989) Coastal zone representing the junction of terrestrial with marine ecosystems, harbors diversified flora, fauna, and microbes in the form of mangroves and its associates, benthos (macro, meio and micro benthos) nekton and plankton in different geo-morphological units like estuaries, creeks, intertidal and sub-tidal zones, mangroves, delta etc. All these faunal and floral inhabitants have been found to display varied patterns of succession, distribution, and eco-dynamics in tune with the changing ecological gradients and also by enjoying definite ecological niche. Biodiversity of coastal zones and their hydrologically linked coastal areas have come under tremendous environmental pressures during recent decades. The resilience of coastal ecosystem in order to cope up with ongoing environmental stresses is being threatened by pollution and over exploitation of resources which have not only aggravated the environmental stresses but also accelerated the pace of bioinvasion (Cohen and Carlton 1998) (Figs. 6.10, 6.12, 6.15, 6.16, and 6.17).

6.17.1 Aquaculture Related Environmental Perturbations

Aquaculture-the farming of fish, shellfish, and aquatic plants is among the fastest growing segments of the world food economy. Global aquaculture production has become more than doubled in volume and value during the past couple of decades and now supplies one-third of seafood consumed worldwide. Plans are under way for a fivefold increase in domestic aquaculture output by 2025 with more lenient regulatory oversight in accordance with the National Aquaculture Act (Crooks 2002; McCoy 2000). In the United States and abroad, aquaculture has led to introductions of unwanted seaweeds, fish, invertebrates, parasites, and pathogens and without proper monitoring, care and management, the rapid expansion of this sector is supposed to result in the spread of unwanted ecological consequences (Bijukumar and Raj 2017). Aquaculture has become a leading vector of aquatic invasive species worldwide (Welcome 1998; Dextrase and Coscarelli 2000). Accidental escapes and even purposeful releases create "biological pollution" with irreversible and unpredictable ecological impacts. Ecological impacts of farming oysters, clams, scallops, and other molluscs which is an important aquaculture industrial activities in U.S.A. having an annual turnover worth more than $100 million (FAO 1998) is not that much relative to other forms of aquaculture (Naylor et al. 2000). This kind of aquaculture activities relying exclusively on clean water are in need of proper ecomonitoring in order to avoid invasion of exotic species and environmental protection. Besides, dispersal of living organisms from aquaculture and fishery mediated activities have found to cause negative ecological consequences in the marine-estuarine environment.

The CBD (convened by the IUCN and ratified in 1992 by 170 countries excluding the United States) holds signatory members accountable for conducting scientific risk assessments for introductions and advocates use of native species in aquaculture. In many cases, the aquaculture industry itself has an economic stake in preventing introductions of exotic species that harm their products. Comprehensive guidelines for preventing introductions of invasive species exist through the IUCN and ICES (International Council for the Exploration of the Sea) (Shine 2003).

The coastal West Bengal has experienced a chequered history of the development and evolution of aquaculture from traditional culture (bheri) to intensive monoculture through paddy cum fish culture and polyculture (Chakraborty 1998) during last four decades with the conversion of swamps/marsh/mangroves/coastal wetlands for monoculture of *Paenius monodon* considering its international market values but ended up with total closure following the legal verdict from the highest judiciary of India because of massive ecodegradation coupled with miserable failure of crop production due to white spot diseases (Chakraborty 1995, 1998).

The scientifically managed aquaculture operations in mangrove estuarine systems with minimum capital investment and application of simpler aquaculture biotechnology, mainly modified extensive may yield substantial profit (Chakraborty 1995). However, in West Bengal, large scale aquaculture development during 1980s resulted considerable eco-degradation as such aquaculture practices required valuable but rare biological resources which had to interact with soil and water. Application of modern technologies with higher usage of water, food, fertilizers and chemicals brought about adverse ecological conditions not only in the aquaculture sites but also in the adjoining areas (Chakraborty 1998) (Fig. 6.16).

A study from the estuarine networks of Sundarbans has shown that in order to collect just 1 seed of *Paeneus. monodon*, 37 seeds of other fin and shell fishes were destroyed during early 2000s (Annon 2003) (Fig. 6.11). After spending a nonproductive lean period for about a decade, this aquaculture sector has restarted its activities with the utilisation of derelicted water bodies mostly with exotic finfishes such as african catfish (*Clarius garipinus*), pangas (*Pangasius hypothalmus*), nilotica (*Oreochomis niloticus*), grass carp (*Ctenopharyngodon idella*) etc. along with scanty culture of *Paeneus monodon* in some restricted areas. Following extensive modified farming. Extensive mass aquaculture practices with newly introduced white vannamii shrimp, *Litopenaeus (Penaeus) vannamii* have been developed during last one decade in different parts of coastal West Bengal, India (Table 6.6 and Figs. 6.13 and 6.14).

Table 6.6 Different exotic fishes used in aquaculture in the coastal belt of West Bengal, India

Sl. No.	Scientific name	Systematic position	Feeding habit
1.	*Oreochromis mossambicus*	Cichlidae	Omnivorous
2.	*Hypophthalmicthys molitrix*	Cyprinidae	Plankyonivorus
3.	*Pangasius hypothalmus*	Pangasiidae	Omnivorous
4.	*Oreochomis niloticus*	Cichlidae	Omnivorous
5.	*Barbonymus gonionotus*	Cyprinidae	Planktonivorus
6.	*Anabus testudineus*	Anabantidae	Omnivorous
7.	*Aristicthys nobilis*	Cyprinidae	Planktonivorus
8.	*Cyprinus carpio* var. *communis*	Cyprinidae	Omnivorus
9.	*Ctenopharyngodon idella*	Cyprinidae	Herbivorus
10.	*Clarius garipinus*	Claridae	Carnivorous
11.	*Cirrhinus molitorella*	Cyprinidae	Omnivorus
12.	*Mylopharyngodon piceus*	Cyprinidae	Omnivorus
13.	*Liptopenaeus (Panaeus) vannamei*	Paeniidae	Omnivotus

Aquaculture should also be regulated with a total ban on conversion of mangrove area, controlled abstraction of groundwater and appropriate treatment of effluents before being discharged into the surface water system. Similarly, regulatory measures should also be adopted to control dredging activities, discharge of burnt oil, leakage of oil due to bad maintenance of vessels, limiting setback lines for coastal construction etc.

6.17.2 Eutrophication and Its Contribution Towards Bioinvasion

Eutrophication, the process of tranformation of an oligotrophic (nutrient poor) aquatic environment to an eutrophic one (nutrient rich), triggers a number of ecological changes in respect of species composition, alteration of biogeochemical cycles, depletion of dissolved oxygen, biodiversity loss etc. This environmental perturbation mostly because of anthropogenic activities, have converted the aquatic environment challenging and non -conducive for the native species and thereby paving the ways for the invasive species not only to intrude but outnumbered the native species because of their more tolerance and resistance capabilities to withstand the adverse changing ecological consequences. Cultural eutrophication is the process that speeds up natural eutrophication because of human activities. Fertilizers cause problems with water quality when they runoff into rivers or percolate into groundwater (Cloern et al. 2007). Rich nutrient input stimulates growth of algae which change the environment when their populations increase. This is particularly the case when they undergo population explosions, referred to as "blooms." The biotic community composition of the water body changes, with fish that can tolerate low Dissolved Oxygen (D.O.), such as carp predominating. Further, some of the blooming algal species produce toxins that render the water unpalatable leading to the elimination of indigenous species and threby changes in the species composition and food web dynamics.

Changes of land use and vegetation cover have impacted fresh water flow, sediment transport, and nutrient dynamics of coastal ecosystems (Walsh 1991). In The Bay of Bengal and its surrounding coastal area, similar relationship has been experienced due to very high anthropogenic activity. Anthropogenic input of excess nutrients to many coastal watersheds has increased dramatically over the last three to four decades (Walsh et al. 1981; Nixon and Pilson 1983; Dugdale and Goering 1967; Eppley and Peterson 1979), resulting in changes in ecosystem structure and function (Cloern 1996). The rapid human settlements, intensive boating and tourist activities, deforestation, and ongoing agricultural and aquacultural practices due to high human population density and rapid economic growth of the countries surrounding the Bay of Bengal make the coastal environment vulnerable to a range of anthropogenic stress factors (Millennium ecosystem Assessment 2005; Ganguly et al. 2008). Different eutrophicate waterbodies, once used for monoculture of shrimps are being used for culturing several exotic species and are infested with a number macrophytes such as *Alternanthera philoxeroides, Pistia stratiotes, Eichornia crassipes, Wolffia microscopic, Lemna minor, Vallisneria spiralis, Ipomoea aquatic, Jussiaea repens, Nymphaea stellata, N rubra, N pubescens* and *Cyperus compressus* (Fig. 6.13b).

In general, frequent disturbance, slow recovery rate, and fragmentation of successionally advanced communities promote plants invasions (Rejmanek 1989; Schiffman 1997). Disturbance-mediated susceptibility to invasions seems particularly pronounced in fertile/eutrophic environments (Burke and Grime 1996; Fensham and Cowie 1998). This is apparently one of the reasons why many riparian habitats are so heavily infested by exotic plants (Planty-Tabacchi et al. 1996). Increasing world wide eutrophication and in particular, atmospheric nitrogen fertilization (Vitousek et al. 1997) provides a frightening prospect since aggressive and highly competitive invaders are pre-adapted to nutrient enriched environments (Odum 1971). Higher species richness is supposed to act as protector of native biotic assemblages by showing negative relationship between richness and invisibility (Prins and Gordon 2014).

6.17.3 Global Warming and Its Contribution Towards Bioinvasion: Sundarbans

Global warming is a growing threat to biodiversity all over the world (IPCC 2000). Rise in temperature affects the biology of a species at its molecular, physiological and biochemical levels and thereby altering its distribution patterns as well as community interactions (Das et al. 2004). Besides, temperature elevation, some other factors like fragmentation of habitat because of ongoing changing patterns of land uses have compounded the problem and resulted to the non adaptability of species and ecosystem to adapt. The (IPCC 2007) concludes citing evidential facts that the climate driven extinctions and other change in biodiversity result invasion of exotic species from warmer regions. The impact of global warming in the oceans include a pole ward shift of different species of fishes, plankton and algae alongside warming of ocean water (Archer and Rahmstorf 2010). Coastal zones are particularly vulnerable to the impact of climate change mainly because of warming of water, a reduced nutrient supply of the sunlit surface waters due to more stable layering of water (a thin layer of warm water tends to float on top, preventing mixing), sea- level rise, increased risk of different diseases and acidification. Moreover, coastal ecosystems are threatened by suffocation as the increased stratification and reduced mixing cause a critical loss of oxygen in the water-hypoxic event (Archer and Rahmstorf 2010). Increased air, soil, and water temperature may also increase growth and distribution of coastal salt marshes and forested wetlands. For many species, including mangroves, the limiting factor for the geographic distribution is not mean temperature, but rather low temperature or freezing events that exceed tolerance limits (McMillan and Sherrod 1986; Snedaker 1995). Climate change will likely influence the vulnerability of estuaries to eutrophication in several ways, including changes in mixing characteristics caused by alterations in freshwater runoff, and changes in temperature, sea level, and exchange with the coastal ocean (Kennedy 1990; Peterson et al. 1995; Moore et al. 1997; Najjar et al. 2000). A direct effect of changes in temperature and salinity may be seen through changes in suspension feeders such as mussels, clams, and oysters. The abundance and distribution of these consumers are affected in response to new temperature or salinity regimes which in turn significantly alter both phytoplankton abundance and water clarity (Alpine and Cloern 1992; NRC 2000).

Global climate change is hypothesized to lead to the increased invasion of communities by nonnative species (Dukes and Mooney 1999), thus compounding threats to biodiversity (Vitousek et al. 1997; Sala et al. 2000). Correlative evidence from terrestrial systems suggests that invasive species have larger latitudinal ranges than native species, which may be indicative of their ability to tolerate a broader range of environmental conditions and their potential for greater success at increased temperatures (Dukes and Mooney 1999).

Research studies have indicated that temperature increases similar to those predicted by climate change models can strongly impact marine species (Sanford 1999; Sanford et al. 2003; Phillips 2005; Harley et al. 2006; Wethey and Woodin 2008), but less is known about responses of marine invaders relative to native species (Carlton 2000; Fields et al. 2006).

Fouling communities comprise species that colonize human-made structures including ships' hulls, mariculture farms, and seawater pipelines, as well as natural hard substrata (Harris and Tyrrell 2001, Valentine et al. 2007). Fouling communities have long been models for community assembly studies (Boyd 1972; Sutherland 1974; Sutherland and Karlson 1973, 1977), and they can be dominated by nonnative species, especially in ports and marinas where human-mediated colonization is frequent (Lambert and Lambert 1998).

Space for settlement represents an important limiting resources for the juveniles of the fouling biotic communities (Stachowicz et al. 1999; Dunstan and Johnson 2004), and oscillation of temperature govern the survival and growth of the juveniles as this parameter controls and allows juveniles to initially acquire and maintain spacein such early stages. Competition becomes important in later stages and is strongly size dependent (Buss 1980, Sebens 1982); thus, there is a direct relationship between initial acquisition of bare space and adult abundance.

In the coast of West Bengal all fouling faunal components have mutually coexisted sharing their preferred space on hard structures after being governed by ecological parameters such as moisture, temperature, inundation pressures, foods etc. (Connell 1961; Connell and Slatyer 1977).

6.17.3.1 Global Warming Mediated Changes in Sundarbans

The present trend of global warming has imparted considerable impact on the mangrove-dominated Indian Sundarbans as the elevated temperature has shown considerable influence from the molecules to ecosystem especially for both pre-monsoon and monsoon periods (Maiti Dutta et al. 2013/2014; Raha et al. 2012). Temperatures have risen by 6.14% in the western sector and by 6.12% in the eastern sector of the coastal Sundarbans over the past 27 years, at a rate of approximately 0.05 °C/year (Mitra et al. 2009). This rate is, in fact, much higher than the observed and documented warming trends in the tropical Pacific Ocean (0.01–0.015 °C/year), tropical Atlantic Ocean (0.01–0.02 °C/year) and the planet itself (0.006 °C/year) (Mitra et al. 2009). Surface air temperature anomaly data over the Sundarbans and adjacent parts of the Bay of Bengal after being analyzed have shown an increasing trend in the yearly rise in temperature. This finding corroborates with the existing global warming phenomena (Hazra et al. 2002). Of the Indian Sundarbans, the western part showed a significant and continuous decrease in salinity (1.67 psu/decade) whereas the eastern sector exhibited an increase in salt (~6 psu over 30 years) because of the differential flood of freshwater (Mitra et al. 2009).

The Sundarbans Mangrove Forest is particularly critical and a highly fragile ecosystem because of its complex geo-morphological and environmental settings, enormous population density and gradual shrinking of the islands under the rising Sea level (DasGupta and Shaw 2013). Global warming mediated salinity invasion coupled with massive erosion have caused the shifting of freshwater loving mangroves such as *Heritiera fomes*, locally called Sundari and *Nypa fruticans*, locally named as Golpata from Indian part of Sundarbans towards its Bangladesh counterpart having more freshwater dominance (Fig. 6.7). Moreover, settlement of some mesophytic bioinvasive plant species has enhanced ecological instability many-fold of this sensitive eco-region forcing some other plant species such as *Sonneratia apetala*, *Avicennia alba* and *Acanthus ilicifolius* to

Fig. 6.7 Two threatened mangrove plants of Sundarbans, West Bengal, India—(**a**) *Hereteria fomes*; (**b**) *Nipa fruticans*

experience landward movement as evident from their presence in the bank of Ganges near Kolkata (Garden-Reach, Strand-Road, Aheritala etc.) All those factors together with considerable pressure from ecotourism, unplanned development of fisheries and aquaculture etc. have caused almost total loss of biodiversity of mangroves and other associated flora and fauna from the Midnapore (East) coast (Bhakat et al. 2004; Chakraborty et al. 2009; Chakraborty et al. 2014).

6.17.3.2 Time Series Analysis on Geomorphology of Sundarbans in Respect of Global Warming

A time series analysis of the change in the shape, size and geomorphic features of the island over the past 32 years (1969–2001) have depicted some important changes like degradation of mangrove swamp and mud flats, salinity invasion, development of saline banks within mangrove swamp, and overall reduction of land area inspite of feeble delta outbuilding processes (Hazra et al. 2002) Coastal erosion and accretion processes have changed in shoreline dynamics. Coastal erosion is constantly reshaping the islands of Indian Sundarbans. Rate of coastal erosion in the Indian Sundarbans have been measured to be about 5.70 km^2 year within the time span of 1989–2014 and eventually it is most dominant in the south western edges of the individual islands. Total land area of 6673 km^2 (approx) of Indian Sundarbans in the year 1989 has been found to be reduced to 5869 km^2 (approx.) in 2014. Erosional zones have been found to be more prominent among the 12 sea facing southern islands including Jambu Island, Sagar Island etc. (Hazra et al. 2002; Raha et al. 2012). Few islands, like Lohachara and Bedford (6.212 km^2), have already vanished from the map. Total erosion over the 30 years time span is estimated to be 162.879 km^2 (Hazra et al. 2002). An evaluation of Satelite imageries has revealed a trend of erosion and accretion patterns in a time span of 34 years (1989–2014) in the length and breadth of deltaic estuarine complex of Sundarbans in the tune of 116,335,800 m^2 (erosion) and 28,275,300 m^2 (accretion) (Chakraborty et al. 2014).

6.17.3.3 Mean Sea Level Variation in Sundarbans

Comparing the annual sea level variation, it is observed that the annual mean sea level has risen steadily between 1985 and 1998. This indicates a minimum 4 cm rise in relative sea level during a period of 14 years. In the Ganga-Bramhaputra delta, taking into consideration of high sedimentation load as to be 0.1 mm per year, the net rate of sea level rise would be 3.14 mm per year. This is significantly more than the present trend of average global sea level rise of 2 mm per year. Considering the present relative sea level rise @ 3.14 mm per year, it is estimated that by the year 2050, the compounded sea level elevation will become close to the 1 m (Hazra et al. 2002; Chakraborty et al. 2014).

6.17.3.4 Changing Pattern of Cyclones over the Sundarbans

The analysis of available records of cyclones over the Bay of Bengal and adjoining Sundarbans, exhibits an increasing trend in the degree of their intensity and decrease in the frequency of occurrence during last couple of decades. This has significant bearing on the extent of coastal flooding, erosion and saline water intrusion due to storm surges and such increase in the intensity also implies increase in the precipitation pattern over this part of the Bay of Bengal (Hazra et al. 2002; Chakraborty et al. 2014).

6.17.3.5 Changing Trend of Turbidity

Turbidity levels and distribution were estimated by using image processing techniques on the Landsat TM imageries. The application of remote sensing tools have shown that turbidity levels have remained low at the upper sections and extreme southern portions of the estuaries and high at the mouth of the estuaries. Suspended solids in the lower estuary zone occurred in high concentrations (in excess of 150 mg/l) whereas the middle estuary zone was characterized by lower concentrations (less than 136 mg/l); and the upper estuary zone exhibited high suspended-solid concentrations (in excess of 150 mg/l) (Ray et al. 2013; Chakraborty et al. 2014).

6.17.3.6 Changing Trend of Salinity (Chloride Concentration) Based on Remote Sensing

Distinct tidal variations of salinity were observed only in western sector whereas such results were less pronounced in other parts having negligible salinity difference (1.4–2.0 ppt) between high and low tides. In contrast, very marked seasonal variations of salinity were encountered (9.34–30.83 ppt) in the entire ecosystem. The major portion of the Sundarbans, India, used to experience almost equal level of salinity in monsoon (12.0–14.0 ppt) and summer (29.0–30.0 ppt) indicating less degree of spatial variations. Significant salinity differences between upstream and downstream were however, recorded in winter and pre-summer which exhibited relatively higher values in most of the estuarine networks, highlighting more fresh water influx in this region (Sarkar et al. 2013; Chakraborty et al. 2014).

6.17.3.7 Changing Trend of Dissolved Oxygen

The concentration of dissolved oxygen (D.O.) in the Western Sector of the Indian Sundarbans showed an increasing trend in contrast to the eastern part over the last 30 years. The observed increase in the dissolved oxygen levels is around 1 mg/l over this period. The increase of D.O. concentration in the western side is in contrast to the prevalent notion of decrease in the D.O. levels with increasing temperature (Mitra et al. 2009; Chakraborty et al. 2014).

6.17.3.8 Changing Trend of pH

Over the past three centuries, the concentration of carbon dioxide has been increasing in the Earth's atmosphere because of human influences. Owing to gradual increase of CO_2 concentration in the atmosphere, a large fraction tended to be dissolved into the ocean and thereby increased the total amount of dissolved inorganic carbon which contributed a shifting of seawater chemistry towards a lower pH condition. This indicates rising acidification of coastal waters and a decrease in the carbonate ion [CO_3^{2-}], which is believed to affect the ability of marine animals to build up shells (Mitra et al. 2009). The IPCC in its 4th Assessment Report estimates that by the end of the century, ocean pH will decline from current level of 8.1 to 7.8 due to rising concentration of CO^2 (Chandra 2013; Chakraborty et al. 2014).

6.17.4 Ecotourism and Its Impact on Coastal Environment of West Bengal

The West Bengal is prone to the impact of unplanned and unorganised ecotourism which has appeared to be a major contributor for developing eutrophication and other related ecodegradations (Figs. 6.8, 6.9, 6.12, and 6.17). Progress of tourism in the two coastal districts (Midnapore (East) and South 24 Parganas) renders valuable contribution to the economy of the state as well as help developing awareness among the local people and tourists regarding the values of nature, especially its gifted flora and fauna. The major tourist hub of coastal Midnapore (East) district is the township of Digha which has changed from a small village to a tourist resort over a period of last four decades. Initially, lack of communication and transport had kept the influx of the tourists at a low order. During the last four decades road connections have improved and a fleet of transport operations led to significant influx of tourists, which in turn necessitated development of hotels, holiday houses, private lodges, etc. Active processes of erosion and accretion have been accelerated by several manmade interventions including removal of sand dunes, clearing of mangrove associates especially *Pandanus, Ipomya* etc. and mushrooming of construction of built structures near the coastline. Apart from coastal erosion caused by wave actions and storms, removal of sand for construction of roads and hotels, exploitation of different mangroves along with *Casuarina* trees on the dune-tops for fuel wood and building materials also cause destruction of sand dunes and erosion of beach (Fig. 6.8).

Fig. 6.8 Ecotorism—Habitat destruction by illegal constructions—(**a**) Mandarmoni at Midnapore (East); (**b**) at Jharkhali in the Sundarbans

Tourism in Sundarbans of South 24-Parganas has mainly concentrated on Western sundarbans (Sagar Island, Fresergunj, Bakhkhali etc) and eastern part (Jharkhali and its adjoining deltaic islands). Better communication facilities and grid electricity are required to make this wonder of the world an attractive tourist destination for pristine, unspoilt and quiet beach resort to unwind. The diverse flora and fauna, which include some rare/endangered species in the vast pristine tract of the Sundarbans, have made it a leading ecotourism hub in West Bengal. It has significantly high ecotourism value being the only natural habitat of the Royal Bengal Tiger, spotted deer and estuarine crocodile, along with varities of birds, fishes, crabs, molluscs, insects etc. Unplanned and unorganized tourism has resulted lot of environmental problems such as construction unauthorized hotels (Fig. 6.10) violating

Fig. 6.9 Ecotourism at (**a**) Sundarbans; (**b**) Digha at Midnapore (East)

Coastal Zone Regulations as imposed by Government of India, dumping of nondegradable plastics, disposal of sewage etc. making the way for bioinvasion. All those environmental stresses not only will weaken the natural ecosystem dynamics but enable the entry of invasive species, especially plant species as a pioneer intruder followed by associated organisms (Cronk and Fuller 1995) (Fig. 6.11).

6.17.5 Chemical Pollution and Its Contribution (Heavy Metals/Pesticides)

6.17.5.1 Heavy Metals and Related Stresses

Hooghly estuary and Coastal environment of West Bengal and its estuarine networks (Fig. 6.1) in the north east coast of India (Lat. 21_530N, Long. 88_150E) are also not free from toxic metal pollution (Ghosh and Choudhury 1989; Sarkar et al. 2002; De

Fig. 6.10 Construction of Hotels Violating regulation of Government of India on Coastal regulatory zone—(**a**) At Mandarmoni of Midnapore (East) Coast destabilizing Dunes; (**b**) At Jharkhali of Sundarbans clearing mangroves

et al. 2009). Different anthropogenic activities such industrialization, urbanization, aquaculture etc. has imposed several stresses by disposing of wastes containing appreciable amount of heavy metals which find their way into the coastal water and adjacent estuaries (Ghosh and Choudhury 1989; Millennium Ecosystem Assessment 2005; Ganguly et al. 2008; Chakraborty 2011). This estuarine complex is considered possibly the most polluted estuary in the world (Mukherjee and Kashen 2007). The increasing usage of heavy metals in industry has led to serious environmental pollution through effluents and emanations during the past several decades (Sericano et al. 1995). Subsequently, these activities have increased the release of harmful

Fig. 6.11 Fishing and Prawn seed collection-causing loss of biodiversity of many fin and shell fishes

heavy metals into the aquatic environment (Agusa et al. 2005, 2007; Hajeb et al. 2009) which are well known environmental pollutants (Gulec et al. 2004). Heavy metals are a global concern, due to their potential toxic effect and ability to bioaccumulate in aquatic ecosystems (David et al. 2012; Hall 2014; Batvari et al. 2015). Pb, Ba, Cd, Hg, Cr, and As are classified as toxic heavy metals, and maximum residual levels have been prescribed for humans (FAO 1983; EC 2008; FDA 2001) and have no established role in biological system (Canli and Atli 2003), whereas metals such as Cu, Na, K, Ca, Mn, Se, Fe, and Zn are essential metals for fish metabolism but may also bioaccumulate and reach toxic levels that can potentially destroy the ecological environment (Agusa et al. 2005, 2007; Hajeb et al. 2009). The bioaccumulation of heavy metals in living organisms and biomagnifications describe the processes and pathways of pollutants from one trophic level to another. Heavy metals can enter the food web through direct consumption of water or organisms taken as food (zooplankton, phytoplankton, and faunal of the bottom) or by uptake through the gills and skin and be potentially accumulated in edible fish in aquatic ecosystem.

Being non-biodegradable, metals once get into soil or water, enjoy long residue time and spend for several years, before they are removed for other compartment of aquatic system (Walker et al. 1996). These heavy metals after being taken up by aquatic animals get bioaccumulated and subsequently biotransferred through food chains and food webs (Chandra 1999). A comprehensive research study on the bioaccumulation of heavy metals has shown that the average bioaccumulation level of the metals in different fishes along with their range of variation have revealed that the mean values of the non toxic groups of metal like Cu and Zn showed comparatively higher level of occurrence in all fishes than other metals following the order of Zn[Cu[Pb[Ni[Cd[Cr, respectively. Among the fish, the average level of Zn varied from 12.13 to 44.74, Cu from 16.22 to 47.97, Pb from 12.40 to 19.96, Ni from 1.15 to 3.69, Cd from 0.62 to 1.20 and Cr from below detection limit to 3.89 mg kg-1 DW (dry weight). The muscles of some important marine fishes collected in and around Hooghly estuarine coastal areas were analyzed for the heavy metals Cu, Zn, Ni, Cd, Cr and Pb. The concentration range of Cu (16.22–47.97 ppm), Pb (12.40–19.96 ppm) and Zn (12.13–44.74 ppm) were recorded comparatively higher and were similar to that found in contaminated areas. On the other hand the ranges of Ni (2.20–3.69 ppm), Cr (0–3.89 ppm) and Cd (0.62–1.20 ppm) were almost equal to those carried out over a wide range of geographical areas. The degree of bioaccumulations was metal-specific as well as species specific in nature (Ghosh and Choudhury 1989).

Among the various classes of pesticides, insecticides enjoy the major share in the Indian market (Adityachaudhury et al. 1997). Fifty percent of such insecticides in India belong to the chemical group Organophosphate (OP). Although the use of Organochlorine (OC) insecticides has been restricted by most of the countries for their long term persistence in the environment, this category of pesticides is being used profusely in the country like India (Tuncer et al. 1998; FAO/WHO 1978). Such pesticidal chemicals after being introduced into water systems from agricultural run off (Scott et al. 1999), sewage treatment plants (Honnen et al. 2001), industrial effluents or accidental spills (Lambert 1997 cause tremendous harm as persistent chemical to aquatic life by way of biomagnification (Murty 1986). Furthermore, the presence of pesticides in rainfall and atmosphere indicated the possibility of agricultural pesticides playing a role in forest decline (Trevisan et al. 1993). The presence and accumulation of organochlorine pesticides like DDT, HCH, etc. in various aquatic resources is well documented including river water in India (Halder et al. 1989, 1990; Kole and Bagchi 1995; Agnihotri et al. 1994; Mohapatra et al. 1995) and abroad (Eichelberger and Lichtenberg 1971; Badawy 1998; Honnen et al. 2001; Rovedatti et al. 2001; Schulz 2001).

An investigation on the monitoring and surveillance of pesticide residues with five different pesticide residues on fishes viz., HCH, DDT, Endosulfan, Dimethoate and malathion has generated substantial data which have revealed that total-HCH concentration crossed the MRL value with a much higher frequency than the remaining four pesticides. The main reasons are the sewage water and sludge may wash out heavily by rain and may be deposited in the river and other aquatic bodies. This may also contaminate the aquatic flora and fauna. Moreover, long term disposal of sewage and sludge may result the accumulation of toxic pesticides in the Hoogly estuary of Sundarbans (Aktar et al. 2009).

6.17.6 Navigation/Shipping/Trawling/Ballust Transportation (Crude Oil)

6.17.6.1 Ballast Transportation

During the last three decades human related redistributions of marine organisms in the harbours, ports andother coastal structures in tropical and temperate regions of the world have become more frequent and increasingly important in their impacts on native communities. Although increased movement of larval and adults of hull fouling organisms in cargo ship's ballast water usually attributed to be the principal cause of the proliferation of non-indigenous species as bioinvadors. Shipping has been identified as a major pathway for introducing invasive aliens to the new environment. Presently, 90%of global shipping transports are of overseas trade and is expected to see an increase throughout the world (David and Gollasch 2015). Ballast water is thus recognized as one of the prinicipal vectors of potentially invasive alien species, and is estimated to be responsible for the transfer of between 7000 and 10,000 different species of marine microbes, plants and animals globally each day (Carlton 1999a, b).

The coastal water along West Bengal over a stretch of 300 km receives voluminous wastes from a variety of industries including refineries and petrochemical complexes which are mostly located by the side of Hoogly estuary. Besides in West Bengal, 8 fishing habors such as Shankarpur, Digha, Petuaghat, Fresergaunj, Diamond Harbour, Canning, Namkhana and Kakdwip are the sites of lot of fishing

Fig. 6.12 Fishing jetties with fishing activities source of petroleum products

activities mostly by the movements of around 3000 fishing trawlers through estuarine-marine pathways (Fig. 6.12). These fishing trawlers along with another 5000 (Approximate) of mechanised boats, few hundred of regular passenger vessels and tourist launches tend to release considerable amount of crude oil and petrolium hydrocarbon (PHC) into the water system. Marine animals, plants and other organisms are being transported or translocated accidentally or intentionally in the different maritime zones by virtue of their movement, migration, transportation by different means such as drifting in the currents of the oceans, getting shelter in the ballust water, anchoring on the hard parts of vessels etc. (Panigrahy et al. 2014). Their settlements and subsequent colonization in the alien habitats far away from their own has led to a profound alteration of the diversity and structure of many shallow coastal marine and estuarine communities (Cohen and Carlton 1998). Since the 1400s, on an inter oceanic scale and much longer on an intra oceanic basis, ships have been an effective transport vector not only for humans but also for the movements of other mammals, birds, and plants. However, neretic faunal components which tend to spend their life in open water condition enjoying conducive physical, chemical and ecological set up in that environment along with their planktotrophic and teleplanic larval forms do not prefer ship transport. These organisms may undergo transitory physiological stress and/or undergo reduced metabolic functioning. During loading and unloading, dry docking in the ports/harbours, the fouling organisms get chances to renewal their colony assemblages in the coastal zone with the resumption of more active feeding by suspension feeders in the fouling community. These activities tend to trigger the denser phytoplankton blooming that characterize the shallow waters. In this manner, these animals/organisms would potentially build energy reserves for the next open-ocean leg if such should come to pass (Figs. 6.13 and 6.14).

India, the third largest importer of crude oil after the United States and China – increased its imports, while increasingly diversifying sources of supply, including Latin America and Western Africa. The failure to effective management of ballast water can jeopardize both fishing economies and marine biodiversity. Developing maritime countries such as India should view ballast water management as part of a larger obligation to protect marine bio-resources and ocean health. These countries should accede to the Ballast Water Convention (BWC) and make ballast water regulation a cornerstone of comprehensive maritime laws in their efforts to protect marine biodiversity, coastal areas, and the global oceans (Puthucherril 2011).

The BWC will enter into force on 8th September 2017. Under the Convention's terms, ships will be required to manage their ballast water to remove, render harmless, or avoid the uptake or discharge of aquatic organisms and pathogens within ballast water and sediments. The Convention will enter into force 12 months after ratification by 30 States, representing 35% of the world merchant shipping tonnage. Presently, it has been ratified by 44 states constituting 32.86% of the world's merchant shipping fleet. Enormous shipping activity with the increasing demand is challenging the potential for the introduction of more alien biota. But till date suspected ballast water invaders is small and this may be related to (a) lack of taxonomic expertise; (b) different environmental conditions between the ballast

Fig. 6.13 Water bodies for Auaculture—(**a**) For Shrimp Culture with chemical inputs; (**b**) Eutrophicated; (**c**) Abandoned shrimp ponds for exotic fish culture

Fig. 6.14 Major Exotic fin and shell fishes—(**a**) Grass carp; (**b**) Pangas; (**c**) African cat fish; (**d**) White Vannamei Shrimp

water donor with the recipient areas; (c) inadequate spatial and temporal coverage of sampling; (d) variable environmental conditions etc. It is recommended to check routine sampling of ballast holdings for biota to establish their viability under local conditions. Ancient civilization was began and shaped by watery framework. Today and in the near future the water is going to be just more than water, possibly because of non-indigenous algae, fishes, viruses, bacteria, zooplankton and numerous benthic invertebrates will find their new destination by the ship's ballast water as fast as one can imagine (Bijukumar and Raj 2017; Carlton 1979, 1989).

6.17.6.2 Status/Levels of Petroleum Hydrocarbons

Petroleum hydrocarbons after being discharged from different sources such as marine operations, land-based discharges, and atmospheric and natural inputs (IMCO 1977; GESAMP 1993; Law et al. 1997; Laws 2000). Municipal and industrial wastes, urban and river runoffs, oceanic dumping, and atmospheric fallout; transportation, dry docking, tanker accidents, deblasting etc. (GESAMP 1993) enter into marine coasta zone and affect biodiversity through bioaccumulation, biotransformation and biomagnification.

A considerable fraction of petroleum hydrocarbon (PHC) entering into the marine environment is removed by evaporation, a portion gets distributed in water, accumulated in sediment, and transferred to biota. It also receives indiscriminate release of untreated or partially treated sewage and riverine discharges. The coastal water of Digha contained increased level of PHC in offshore water which varied from 0.15 to 2.33 μg/l in Saptamukhi and 0.11 to 1.30 μg/l in Matla estuaries and in Sandheads, it ranged between 0.10 and 5.50 μg/l. Haldia port water contained much higher values of PHC (1.60–20.11 μg/l; than Diamond Harbour (1.35–1.71 μg/l), port (0.25–0.91 μg/l). It is attributed mainly to the impact of hydrocarbon from the port activities and spillages during shipping operation. PHC in Hoogly estuary water varied from 1.17–18.5 μg/l with wide fluctuation during 2002–2009. Like Mahanadi estuary, low tide water of Hoogly estuary contained higher values than the high tide. It may be due to the impact of hydrocarbon of port and fishing activities, wastes from oil refineries and petrochemicals industries, etc. The highest concentration of PHC was recorded at Haldia port and these values were fluctuating and exhibiting no significant trend. The concentration of PHC in coastal area of Sandheads is comparatively low, though these stations are in the vicinity of Haldia port and are under the influence of anthropogenic inputs. It may be due to the occurrence of maximum number of petroleum degrading bacteria in the water of Haldia port and surrounding areas (Roy et al. 2002).

6.18 Discussion

A primary focus of invasion biology is assessing the impacts of invaders (Williamson 1996). Negative interactions between exotics and natives are among the most commonly considered consequences of invasions. Exotics can have genetic effects through hybridization or altering gene flow of native species and at larger spatio-temporal scales they homogenize biotas across biogeochemical realms and alter evolutionary pathways (Williamson 1996). Invaders also can benefit natives by way of providing food sources for resident biota (Carlton 1979; Singer et al. 1993; Reusch 1998; Crooks 2002; Paria et al. 2017) or facilitate natives through indirect or commercial relationship (Posey 1998; Crooks 2002).

Many nations across the world have developed strategies for management of invasive species, considering the colossal economic, ecosystem and social damages (Figs. 6.15, 6.16, and 6.17). In a mega diversity nation like India encompassing four biodiversity hotspots and showing high levels of endemism, Invasive Alien Species (IAS) may cause irreparable damages. In marine-estuarine environment, it is very difficult to eradicate the established invasive species (Bax 1999) and base line information pertaining to biology and ecology of interacting biotic communities and non-biotic components may act as a pre-requisite for the management of the ecosystem by way of controlling the invasive species (Williamson 1996; Culver and Kuris 2000). Monitoring and eco-assessment of an aquatic ecosystem for the early detection and prevention of species invasion have appeared to be a pre requisite for maintaining the ecological integrity of uninvaded habitats (Ricciardi et al. 2000; Ricciardi and Rebekah 2008). However, there is no bench mark data on aquatic invasive species of India at population level and their specific implications on ecosystem services and other indigenous biodiversity, as well as socio-economic impacts.

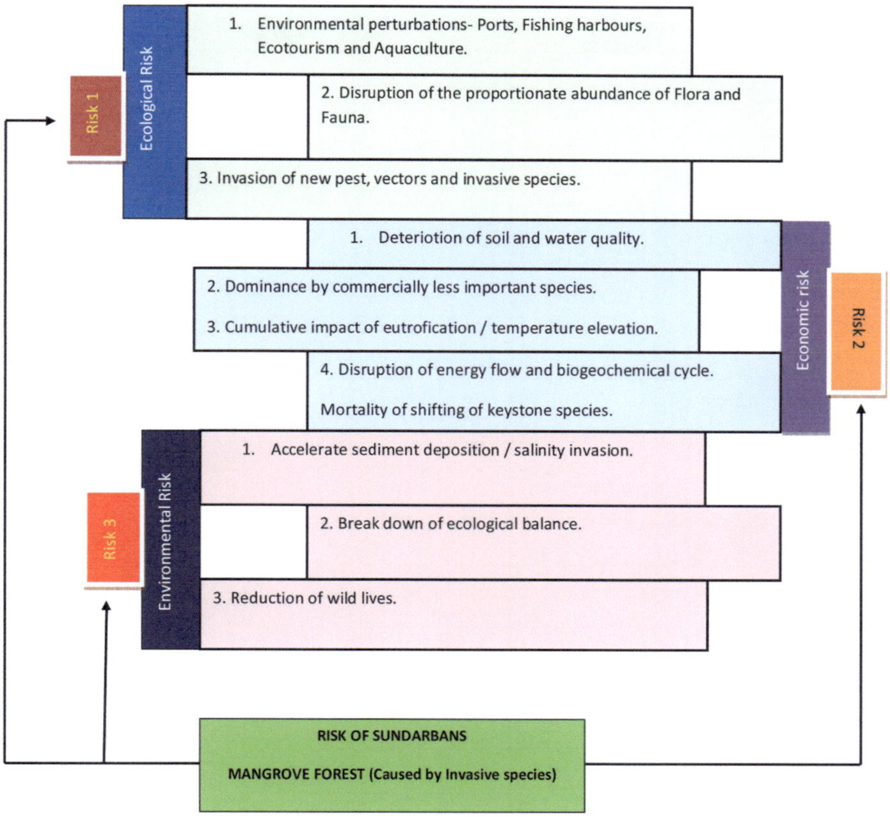

Fig. 6.15 Pathways of risk factors leading to ecodegradation

However, this task is very difficult as almost no previous data and information pertaining to population, distribution and biotic-abiotic interactions with invasive species are available in the coastal belt of West Bengal, India. A thorough comprehension of relevant environmental parameters is a prerequisite to preventing coastal degradation and environmental balance in coastal zones. Most coastal areas have a great propensity of being affected by climatic changes and other environmental pollution. Along the coast, one finds a constant cycle of erosion and deposition. The degree to which a particular process controls coastal change depends on local factors like sediment deposition, river, topography, tectonic activity, composition of land, prevailing winds and weather patterns.

India is counted as one of the major 17 megadiverse countries of the world. An 'Invasive Alien Species' can be referred to the species that is (i) non-indigenous to the ecosystem under consideration and (ii) introduction of whom causes or is likely to cause economic or environmental harm, or harm to human health (ICAE 2016). Invasive Aliens may occur in inland, estuarine and marine waters that presently or

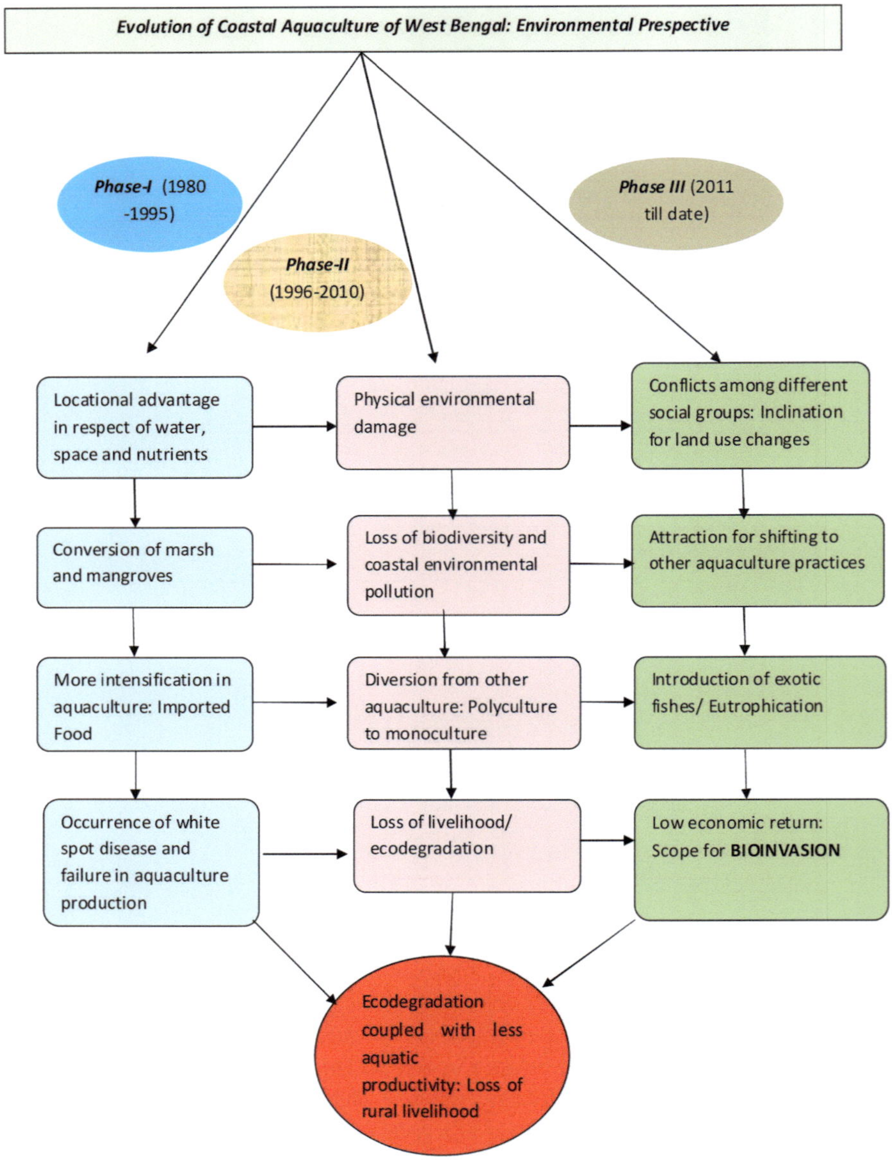

Fig. 6.16 Aquaculture and prospective ecodegradation pertaining to coastal ecosystem

potentially threaten ecological processes and natural resources. Three main stages can be recognized in the invasion processes: (i) **Introduction**: the intentional or unintentional introduction imported into a region beyond its native range; (ii) **Establishment**: the establishment of self-sustaining, reproductive populations; and (iii) **Invasion**: population growth and spreading of the species.

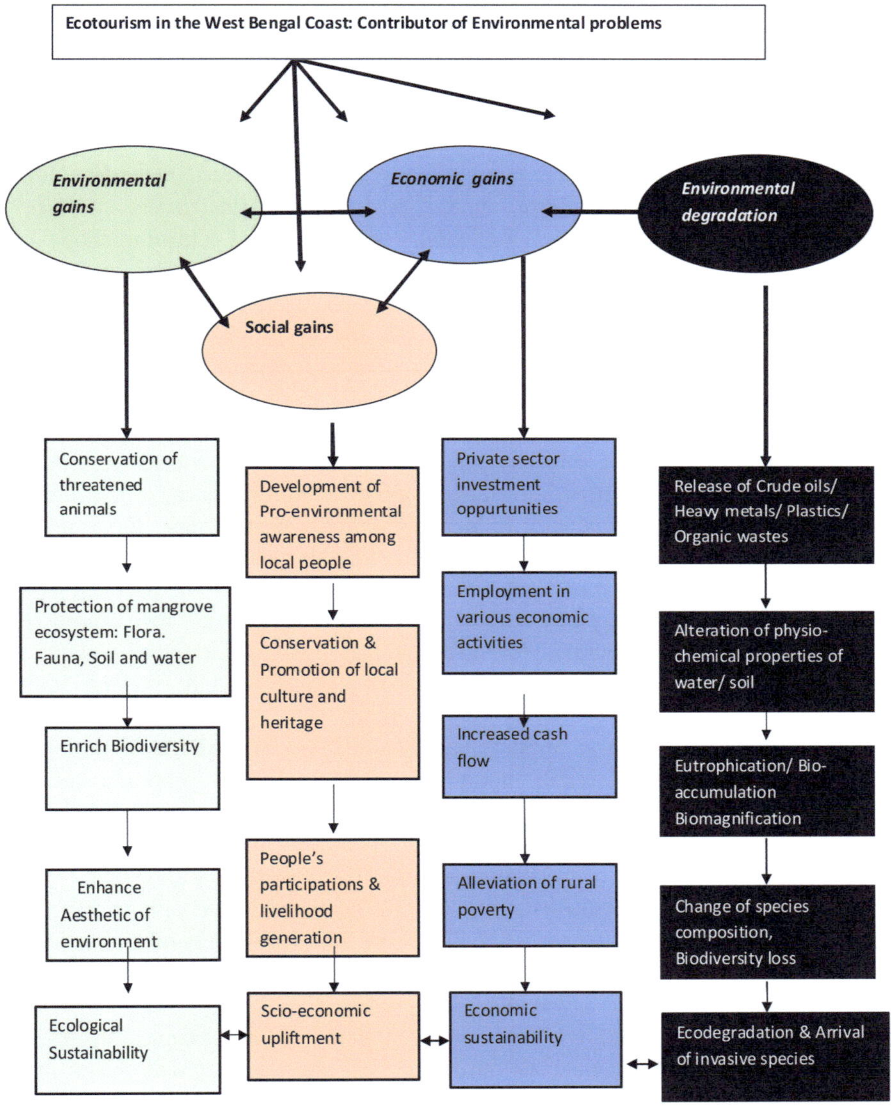

Fig. 6.17 Ecotourism and its connection with bioinvasion in respect of West Bengal coast

In a mega diversity nation like India encompassing four biodiversity hotspots and showing high levels of endemism, AIS may cause irreparable damages. Further, the documented knowledge on role of biodiversity in ecosystem functioning in India is rather fragmentary, restricting predictions on the possible impacts of AIS. Non-native species comprise 5–25% of the ichthyodiversity in the river basins of peninsular India and the Himalayas, making them priority areas for research, advocacy, policy and conservation actions related to biological invasions. For example, more than 20 species of non-native freshwater fish occur in peninsular India including in the rivers of the Western Ghats, but little information is available on their actual ecosystem-level and economic impacts (Bijukumar and Raj 2017). While the Convention on Biological Diversity (CBD 2000) Aichi Target 9 provides an excellent framework for identification, control and eradication, as well as managing pathways of invasive alien species, only moderate progress has been made in these directions. The management of invasive species include phases such as prevention, eradication (biological, chemical and physical methods), containment and control methods.

Restoration of damaged habitats also will play a key role in preventing the re-entry of invasive into aquatic ecosystems. Early detection and prevention of species invasion are vital for maintaining the ecological integrity of uninvaded habitats. However, there is no bench mark data on aquatic invasive species of India at population level and their specific implications on ecosystem services and other indigenous biodiversity, as well as socio-economic impacts. Moreover, estimating the full extent of the environmental damage caused by exotic species and the number of species extinctions they have caused is difficult because the aquatic ecosystems and their resources in India is poorly known and many thousands of species are yet to be described. The challenge for the future, besides estimating the economic and environmental costs of the impacts of invasive species, is in preventing further damage to natural and managed ecosystems caused by non-indigenous species. Sound prevention and management strategies and action plans are therefore needed to prevent further damages and to prevent future introductions.

The National Biodiversity Strategy and Action Plan (NBSAP), with a view of achieving Aichi Targets 2020, incorporating the strategies proposed by the concerned agencies in India, propose the following measures for the regulation of introduction of invasive alien species and their management.

Because of the serious consequences that can result from non-indigenous introductions, invasions on marine-estuarine biota have been ranked among the most serious potential sources of stress to the global environment in general and national/regional ecosystems in particular. India is a Party many treaties including the Convention on Biological Diversity (CBD) (1992). Article 8 (g and h) of the Convention on Biological Diversity recognizes biosafety and biological invasions. Legislative measure to regulate the introduction of planting materials, plant products, soil, living organisms, etc., only to prevent inadvertent introduction of exotic pests and establishment and further spread of the introduced pests. National legislation on Invasive alien species is the Environment (Protection) Act 1986; the Biological Diversity Act, 2002; Plant Quarantine (Regulation of Import into India) Order 2003; Livestock Importation Act, 1898 and Livestock Importation (Amendment) Ordinance 2001.

The invasiveness and bioinvasion of invasive species has established its roots and ramification into the field of biodiversity research. Out of so many types of ecosystems of the world, coastal ecosystem alongwith two major associated ecosystem – coral and mangroves represents the most sensitive, productive but vulnerable ecosystem of the world. The West Bengal coast is very unique in respect of its geological formation, diversity of geomorphological components and biodiversity in its small stretch in comparison to other coastal zones of india (Chakraborty 2010; Chakraborty 2017). However, very scanty information pertaining to bioinvasion and invasive species are available in this ecozone but based on research works, several hypothesis have been put forward by leading researchers in this field from different parts of the world, have established the hypothesis that invasion i.e. introduction, establishment, colonization and flourishing of invasive species especially in coastal estuarine mangrove ecosystem is possible if the natural ecodynamics of the ecosystem is disstabilised because of the ongoing perturbations of the environment viz. global warming mediated climate change, eutrophication, acidification, chemical pollution (crude petroleum oils, heavy metals and pesticides), disposal and accumulation of oxygen demanding wastes (sewage) mostly generated by the municipalities/industries/ecotourisms, aquaculture, fishings, port and harbor etc. All those invasive species take the advantage of such ecodegradation and outcompete the native one, proliferate and expand territories. Only holistic approach adhering to basic principles of ecology should be adopted giving due emphasis on sustainability and integration of coastal zone management.

6.19 Concluding Remarks

Studies on Bioinvasion have gained tremendous momentum during the last couple of decades as because it has been recognized as the second major cause of biodiversity loss in the global perspective. However, very scanty research works have so far been undertaken on this very emerging field of biodiversity research, especially from the tropical countries like India. Sustainable biodiversity conservation in view of such threat, does not attach much importance on just enlisting or cataloguing of invasive floral and faunal components from different ecosystems but also demands in highlighting the aggressive, disrupting and destabilizing negative roles of those alien invasive species in altering the ecodynamics of an ecosystem and thereby leading to result eco-degradation coupled with biodiversity loss.

Research orientations during the last couple of decades mostly emphasis on the under mentioned aspects:

1. **Enlisting**: Preparing list of Invasive Alien Species (IAS) is now being considered as an important tool for ecomanagement of an invaded area which include early signaling coupled with warning, prevention, eco-remediation and eco-restoration.
2. **Bioinvasion Process and Pathways**: Successful bioinvasion by an aquatic non-native species takes place through (i) inoculation, (ii) settlement (iii) competition with the natives species (iv) establishment, (v) propagation and (vi) range expansion
3. **Trend of adaptive changes of invasive species**: Time since invasion also influences the level of impact, through temporal changes in the abundance of the alien species, adaptation by the recipient community, post-invasion evolution,

alteration of biogeochemical cycles coupled with variation in the physicochemical environment in the invaded range. Besides, established nonnative species may require considerable period to become invasive one in new environments because of their small populations and dearth of resource availabilities.

4. **Synergistic Impact of Invasive species**: Explanations concerning the impact of invasive species centre around the arguments that the invasive alien species does not only impart its negative effect on native ones in an isolated manner but in conjunction with other such introduced species alongwith utilizing the advantage of instability of the invaded habitats, synergistically (Ricciardi 2005) which may lead more accelerated and intensified impacts on ecosystems with the addition of each new species.
5. **Mode/Nature of Impact**: Negative ecological effects caused by aquatic invasive species include: (1) habitat disruption; (2) Changes in species composion and community interactions (3) Modification in the food chain and food web dynamics by playing the roles as predators or pathogens (4) Biodiversity loss; (5) Hybridization leading to genetic reshuffling which may bring forth genetic pollution.
6. **Types of Impact by Aquatic Invasive Species**: This includes two broad categories- ecological and socio-economical impacts which are in turn impart impacts on each other. Interactions between introduced species and other anthropogenic stressors may cause synergistic impact having more intensity on native species Finally, other anthropogenic stressors that simultaneously alter the physical and biological environment can lead to many more interactions and obscure the effects of alien species.
7. **The prime objective**: The principal goal of examining biological invaders in respect of their biology, behavior, mode of infestation and impact is to develop ecological principles that apply generally on ecosystem and specifically on concerned species in order to reach to a better understanding on the role of bioinvadors not only on indigenous/native ones but also on the entire ecosystem in a holistic manner, taking into consideration of socio-economic-political activities of the human being so that sustainable eco-management strategies can be adopted.

Considering all the aspects of the bioinvasion as discussed above, a number of hypotheses are being proposed to explain the pattern and processes of bioinvasion in the coastal-estuarine-mangrove environment in a tropical country like India.

1. An alien species avoids to invade to an ecozone saturated with native biotic assemblages having same broad trophic level and enjoying community interactions in an stable ecosystem setup
2. The success and extent of an invasion is negatively correlated to species diversity of functional guild competitors in the invaded environment.
3. Mangrove-Estuarine ecosystem representing a very stable and productive ecosystem mainly because of an array of interactions among different structural components of its two interlinked subsystems-forest and water (Fig. 6.2) resist

the intrusion and settlement of any alien invasive species revealing the phenomenon of Eco-Immunology.

4. The invasive alien species can only find chance to settle and propagate in disturbed mangrove ecosystem which is caused mainly because of different environmental perturbations out of anthropogenic activities.
5. Removal/clearing of some mangrove plant species will create open space devoid of native ones and thereby allowing more sunlight to penetrate and attracting invasive plant species along with their associated animal and microbial associates. Social forestry with less canopied plant species provide more space under the plants so that sunlight mediated invasion takes place.
6. An invasive species will not be able to replace a native species if they occupy the same niche and invaded ecozone having abiotic conditions that are outside the physiological tolerance levels of the alien ones.
7. A species that occur at low population densities (low reproductive potential) in their native range will not be invasive because of the resistance from a native species endowed with higher density, biomass, advantageous foraging behavior and reproductive potential.

In the context of multidimensional researches undertaken during the last few decades on the **Invasive alien Species (IAS)** all over the world with regard to the danger associated with this very unique biota towards different ecosystems, especially on the very sensitive, vulnerable but productive marine-estuarine-mangrove ecosystem, it can be inferred that sustainable management of biodiversity in particular and environment in general can only be possible by adhering to the underlying ecological principles operating in the intricate interaction pathways among all structural components of the same ecosystem.

Acknowledgment The author is personally thankful to his 12 research scholars who have successfully completed their Ph.Ds on different aspects of coastal environment of West Bengal, India. He is also thankful to his research student, Shri Tridip Kumar Datta for his commited support and help in computer based preparation of the manuscript. The library and laboratory facilities provided by the Vidyasagar University, Midnapore, West Bengal, India is thankfully acknowledged.

References

Abele LG (1974) Species diversity of decapod crustaceans in marine habitats. Ecol Soc Am 55 (1):156–161

Adam P (2002) Salt marshes in a time of change. Environ Conserv 29:39–61

Adityachaudhury N, Banerjee H, Kole RK (1997) An appraisal of pesticides use in Indian agriculture with special reference to their consumption in West Bengal. Sci Cult 63:223–228

Agnihotri NP, Gajbhiye VT, Mohapatra SP, Mukesh K, Kumar M (1994) Organochlorine insecticide residues in Ganga river water near Farrukhabad, India. Environ Monit Assess 30 (2):105–112

Agusa T, Kunito T, Yasunaga G, Iwata H, Subramanian A, Ismail A, Tanabe S (2005) Concentrations of trace elements in marine fish and its risk assessment in Malaysia. Mar Pollut Bull 5:896–911

Agusa T, Kunito A, Sudaryanto T, Monirith SK, Klap A, Iwata H (2007) Exposure assessment for trace elements from consumption of marine fish in Southeast Asia. Environ Pollut 145:266–777
Aktar MW, Paramasivam M, Sengupta D, Purkait S, Ganguly M, Banerjee S (2009) Impact assessment of pesticide residues in fish of ganga river around Kolkata in West Bengal. Environ Monit Assess 157(1–4):97–104
Aller RC (1982) The effects of macrobenthos on chemical properties of marine sediment and overlying water. In: PL MC, Tevez MJS (eds) Animal–sediment relations. Plenum Press, New York
Aller RC (1988) Benthic fauna and biogeochemical processes in marine sediments: the role of burrow structures. In: Blackburn TH, Sorensen J (eds) Nitrogen cycling in coastal marine environments. Wiley, Chichester
Aller RC (2001) Transport and reactions in the bioirrigated zone. In: Boudreau BP, Jørgensen BB (eds) The benthic boundary layer: transport processes and biogeochemistry. Oxford University Press, New York
Alongi DM (1987) The influence of mangrove–derived tannins on inter–tidal meiobenthos in tropical estuaries. Oecologia 71:537–540
Alongi DM (2002) Present state and future of the world's mangrove forests. Environ Conserv 29:331–349
Alongi DM (2008) Mangrove forests: resilience, protection from tsunamis, and responses to global climate change estuarine. Coast Shelf Sci 76:1–13
Alpine AE, Cloern JE (1992) Trophic interactions and direct physical effects control phytoplankton biomass and production in an estuary. Limnol Oceanogr 27:946–955
Amanda WJ, Demopoulos BF, Smith CR (2007) Food web structure in exotic and native mangroves: a Hawaii–Puerto Rico comparison. Oecologia 153(3):675–686
Ameen M (1999) Development of guiding principles for the prevention of impacts of alien species. Paper presented at a consultative workshop in advance of the 4th meeting of SBSTTA to the CBD organized by IUCN Bangladesh at Dhaka on May 25, 1999
Annon (2003) Mangrove ecosystem: biodiversity and its influence on the natural recruitment of selected commercially important finfish and shellfish species in fisheries. National Agricultural Technology Project (NATP). Indian Council of Agriculture Research (ICAR). Principal Investigator: George JP; Co–PI: Chakraborty SK, Damroy SN: 1–514
Archer D, Rahmstorf S (2010) The climate crisis– an introductory guide to climate change. Cambridge University Press, Cambridge, pp 1–248
Ashton PJ, Mitchell DS (1989) Auatic plants, patterns and modes of invasion, attributes of invasive species and assessment of control programmes. In: Drakes JAA, Mooney HA, di Castri F, Groves RH, Kruger FJ, Rejmanek M, Williamson M (eds) Biological invasions: a global perspective. Wiley, New York, pp 111–154
Atkinson IAE, Cameron EK (1993) Human influence on the terrestrial biota and biotic communities of New Zealand. Trend Ecol Evol 8:447–451
Ayala FL, Serra L, Prevosti A (1989) A grand experiment in evolution: the Drosophila subobscura colonisation of the Americas. Genome 31:246–255
Badawy MI (1998) Use and impact of pesticides in Egypt. Int J Environ Health Res 8:223–229
Barnes RSK (1967) The osmotic behaviour of a number of grapsoid crabs with respect to their differential penetration of an estuarine system. J Exp Biol 47:535–552
Batvari B, Prabhu D, Kamalakannan S, Krishnamurthy RR (2015) Heavy metals accumulation in two fish species (Labeo rohita and Cirrhina mrigala) from Pulicat Lake, North of Chennai, Southeast Coast of India. J Chem Pharm Res 7:951–956
Bax NJ (1999) Eradicating a dreissenid from Australia. Dreissena 10:1–5
Bertness MD, Callaway R (1994) Positive interaction in communities. Trend Ecol Evol 9:191–193
Bhakat RK, Chakravarty G, Giri S, Chakraborty SK (2004) Invasive species in Sundarbans Mangrove Ecosystem, India. In: Ayyappan S, Malik DS, Dhanze R, Chauhan RS (eds) Fish diversity in protected habitats. Natcon Publication, Muzaffarnagar, pp 219–240
Bhattacharrya M, Chakraborty SK, Acharya SK (2017) Indigenous and introduced honeybees of India: brief overview of interaction. ENVIS 2:5–10

Bhowmik, SK, Roy GN, Burman S (1985) Alist of 24 strains of bacteria from the decomposed litter of mangrove areas of Sundarbans complex. In: National Symposiun on Biology, Utilization and Conservation of Mangroves, pp 389–392
Bijukumar A, Raj S (2017) Aquatic invasive alien species of India. ENVIS 23(2):13–17
Bird ECF (1971) Mangroves as land builders. Victorian Nat 88:189–197
Biswas A, Bhowmik SK, Roy GN (1986) Chitinolytic bacteria of mangrove swamps of the Sundarbans. In: National Symposium on Biology, Utilization and Conservation of Mangroves, pp 382–384
Biswas SR (2003) Invasive plants of Sundarbans. In: Interim report under SBCP project. IUCN Bangladesh, 34 pp
Biswas SR, Choudhury JK, Nishat A, Rahman MM (2007) Do invasive plants threaten the Sundarbans mangrove forest of Bangladesh? For Ecol Manag 245:1–9
Blasco F (1975) The mangroves of India. Institut Francais de Pondicherry. Pondicherry Travaux de la section, Scientific et Technique 14:1–175
Bouillon S, Koedam N, Baeyens W, Satyanarayana B, Dehairs F (2004) Selectivity of subtidal benthic invertebrate communities for local microalgal production in an estuarine mangrove ecosystem during the post–monsoon period. J Sea Res 51:133–144
Boyd MJ (1972) Fouling community structure and development in Bodega Harbor, California. Ph. D. thesis, University of California, Davis
Brenchley GA (1976) Predator detection and avoidance: ornamentation of tube–caps of *Diopatra* spp. (Polychaeta: Onuphidae). Mar Biol 38(2):179–188
Brenchley GA (1982) Predation on encapsulated larvae by adults: effects of introduced species on the gastropod *Ilyanassa obsolete*. Mar Ecol Prog Ser 9:255–262
Burke MJW, Grime JP (1996) An experimental study of plant community invisibility. Ecology 77:776–790
Buss LW (1980) Competitive intransitivity and size–frequency distributions of interacting populations. Proc Natl Acad Sci U S A 77:5355–5359
Butchart SH, Walpole M, Collen B, van Strien A, Scharlemann JP, Almond RE, Baillie JE, Bomhard B, Brown C, Bruno J, Carpenter KE, Carr GM, Chanson J, Chenery AM, Csirke J, Davidson NC, Dentener F, Foster M, Galli A, Galloway JN, Genovesi P, Gregory RD, Hockings M, Kapos V, Lamarque JF, Leverington F, Loh J, McGeoch MA, McRae L, Minasyan A, Hernández Morcillo M, Oldfield TE, Pauly D, Quader S, Revenga C, Sauer JR, Skolnik B, Spear D, Stanwell-Smith D, Stuart SN, Symes A, Tierney M, Tyrrell TD, Vié JC, Watson R (2010) Global biodiversity: indicators of recent declines. Science 328:1164–1168
Cadee GC (2001) Sedimentary dynamics by bioturbating organisms. In: Reise K (ed) Ecological comparisons of sedimentary shores. Springer, Berlin, pp 127–148
Callaway JC, Josselyn MN (1992) The introduction and spread of smooth cordgrass (Spartina alterniflora) in south San Fransisco Bay. Estuaries 15:218–216
Campbell F, Kriesch P (2003) Final report by the National Invasive Species Council's Invasive Species Pathways Team of the Prevention Working Group. wwwinvasivespeciesinfogov/council/wrkgrpsshtml. Viewed 30 Aug 2006
Canli M, Atli G (2003) The relationships between heavy metal (Cd, Cr, Cu, Fe, Pb, Zn) levels and the size of six Mediterranean fish species. Environ Pollut 121:129–136
Carlton JT (1979) History, biogeography, and ecology of the introduced marine and estuarine invertebrates of the Pacific Coast of North America. Ph.D. thesis, University of California, Davis, 904 pp
Carlton JT (1989) Man's role in changing the face of the ocean: biological invasions and implications for conservation of near–shore environments. Conserv Biol 3:265–273
Carlton JT (1999a) Molluscan invasions in marine and estuarine communities. Malacolovia 41:439–454
Carlton JT (1999b) The scale and ecological consequences of biological invasions in the world's oceans. In: Sandulund O, Schei P, Viken A (eds) Invasive species and biodiversity management. Kulwer Academic Publishers, Dordrecht, pp 195–212

Carlton JT (2000) Global change and biological invasions in the oceans. In: Mooney HA, Hobbs RJ (eds) Invasive species in a changing world. Island Press, Covelo, pp 31–53
Carlton JT, Newman WA, Pitomba FB (2011) Barnacle invasions: introduced, cryptogenic, and range expanding Cirripedia of North and South America. In: Galil BA, Clark PF, Carlton JT (eds) In the wrong place – alien marine crustaceans: distribution, biology and impacts, vol XVI. Springer Series in Invasion Ecology, Dordrecht, pp 159–214 716p
Carpenter SR, Turner MG (1998) A decade of ecosystem. Ecosystem 10(4):1–5
Cattelino PJ, Noble IR, Slatyer RO, Kessel SR (1979) Predicting the multiple pathways of plant succession. Environ Manag 3:41–50
CBD (1992) Convention on biological diversity United Nation Environment Programe, Nairobi, Kenya
CBD (Convention on Biological Diversity) (2000) Alien species that threaten ecosystems, habitats or species, as adopted by conference of the Parties 5 Decision V/8. http://cbd.int/decisions/cop–04.shtml?id=7150&lg=0&m=COP–04. Viewed 23 Feb 2007
CBD (Convention on Biological Diversity) (2010a) Global biodiversity outlook 3
CBD (Convention on Biological Diversity) (2010b) Revised and updated strategic plan: technical rationale and suggested mile stones and indicators
Chakraborty SK (1995) Aquaculture potential of mangrove ecosystem of Sundarbans, India. In: Proceedings of seminar on fisheries – a multimillion dollar industry, pp 72–83
Chakraborty SK (1998) Coastal aquaculture and environment in the context of West Bengal, India. Vidyasagar Univ J Biol Sci 4:13–20
Chakraborty SK (2003) Biodiversity and conservation. In: Mukhopadhyay AD (ed) Perspectivesand issues in environmental studies. Vidyasagar University, Midnapore (West), pp 267–286
Chakraborty SK (2010) Coastal environment of Midnapore, West Bengal: potential threats and management. J Coast Environ 1(1):28–40
Chakraborty SK (2011) Mangrove ecosystem of Sundarbans, India: biodiversity, ecology, threats and conservation. In: Metras JN (ed) Mangroves: ecology, biology and taxonomy. NOVA Publishers, New York, pp 83–112
Chakraborty SK (2013) Interactions of environmental variables determining the biodiversity of coastal mangrove ecosystem of West Bengal, India. Ecoscan Spec Issue 3:251–265
Chakraborty SK (2017) Ecological services of intertidal benthic fauna and the sustenance of Coastal Wetlands along the Midnapore (East) Coast, West Bengal, India. In: Finkl CW, Makowski C (eds) Coastal wetlands alteration and remediation. Springer, Cham, pp 777–866
Chakraborty SK, Choudhury A (1992a) Population ecology of fiddler crabs (Uca. Spp.) of the mangrove estuarine complex of Sundarbans, India. Trop Ecol 33(1):78–88
Chakraborty SK, Choudhury A (1992b) Ecological studies on the zonation of brachyuran crabs in a virgin mangrove island of Sundarbans, India. J Mar Biol Assoc India 34(1 and 2):189–194
Chakraborty SK, Choudhury A (1993) Community organization of macrobenthic molluses of Sagar Island, Sundarbans, India. Indian Biol 25(1):33–36
Chakraborty SK, Chowdhury A (1994) Community structure of macrobenthic polychaetes of intertidal region of Sagar Island, Hooghly estuary, Sundarbans, India. Trop Ecol 35(1):97–104
Chakraborty Sk Choudhury A (1997) Occurrence and abundance of benthic polychaetes in Hoohhly estuary, Sagar Island, India. J Mar Biol Assoc India 39(1&2):140–147
Chakraborty SK, Giri S, Chakravarty G, Bhattacharya N (2009) Impact of eco–restoration on the biodiversity of Sundarbans mangrove ecosystem, India. Water Air Soil Pollut Focus 9:303–320
Chakraborty SK, Maiti Dutta S, Ghosh PB, Ray R, Paul AK (2014) Impact of global warming on Sundarbans mangrove ecosystem, India: role of different assessment tools from ecosystem monitoring to molecular markers. In: Proceedings of the international conference on Green India: strategic knowledge for combating climate change: prospects and challenges, 181–200
Chandra K (1999) Status of heavy metals and pesticides in aquatic food chain in context of Indian rivers. Summer School on Ecology, Fisheries and Fish Stock Association in Indian Rivers, CIFRI, Barrackpore, pp 92–98

Chandra A, Chakraborty SK (2008) Distribution, population and community ecology of macrobenthic intertidal polychaetes in the coastal tract of Midnapore, West Bengal, India. J Mar Biol Assoc 50(1):7–16
Chandra S (2013) Climate change and coral vulnerability. Everyman Sci XLVIII(1):69–73
Chapsuis JL, Bousses P, Barnaud G (1994) Alien mammals,impact and management in the French Butantartic Island. Biodivers Conserv 67:97–104
Charles H, Dukes J (2007) Impacts of invasive species on ecosystem services. In: Nentwig W (ed) Biological invasions. Springer, Berlin, pp 217–237
Chatterjee S, Chakraborty SK (2014) Feeding behaviour and functional role of some selected species of Brachyuran crabs in nutrient cycle of coastal belt of Midnapore (East), West Bengal, India. J Biol Life Sci 5(1):106–129
Chatterjee S, Das TK, Chakraborty SK (2014) Reproductive biology and bioturbatory activities of two sympatric species of fiddler crabs *Uca lactea annulipes* and *Uca triangularis* Bengali (Decopoda: Ocypodidae) at the East Midnapore Coastal belt of West Bengal, India. J Biol Life Sci 5(2):106–126
Chaudhuri AB, Choudhury A (1994) Mangroves of the Sundarbans, India, vol 1. IUCN, Bangkok, pp 1–247
Cloern JE (1996) Phytoplankton bloom dynamics in coastal ecosystem: a review with some general lessons from sustained investigation of San Francisco Bay, California. Rev Geophys 34 (2):127–168
Cloern JE, Krantz T, Hogan CM, Duffy JE (2007) Eutrophication. In: Cleveland CJ (ed) Encyclopedia of earth. Environmental Information Calition, National Council for Science and the Environment, Washington, D.C
Cloern JE, Jassby AD, Thompson JK, Hieb KA (2007) A cold phase of the East Pacific triggers new phytoplankton blooms in San Francisco Bay. Proc Natl Acad Sci 104(47):18561–18565
Coblentz BE (1990) Exotic organisms: a dilemma for conservation biology. Conserv Biol 4 (3):261–265
Cohen AN, Carlton JT (1998) Accelerating invasion rate in a highly invaded estuary. Science 279:555–558
Cole CV, Vaidyaraman PP (1966) Salinity distribution and effect of fresh water flows in the Hoogly River. In: Proceedings of tenth conference on coastal engineering, Takyo. American Society of Civil Engineers, New York, pp 1312–1434
Connell JA (1961) The influence of interspecific competition and other factors on the distribution of the barnacle, Chthalamus stellatus. Ecology 42:710–723
Connell JH, Slatyer RO (1977) Mechanisms of succession in natural communities and their role in community stability and organization. Am Nat 111:1119–1144
Connolly RM, Lee SY (2007) Mangroves and saltmarsh. In: Connell SD, Gillanders BM (eds) Marine ecology. Oxford University Press, Oxford, pp 485–512
Cronk QCB, Fuller JL (1995) Plant invaders: the threat to natural ecosystems. Trend Ecol Evol 10 (12):508–509
Crooks JA (2002) Characterizing ecosystem–level consequences of biological invasions: the role of ecosystem engineers. Oikos 97:153–166
Crooks JA, Khim HS (1999) Architectural vs biological effects of a habitat alluring, exotic mussel, Musculista senhousia. J Exp Mar Biol Ecol 240:53–75
Culver CS, Kuris AM (2000) The apparent eradiction of a locally established introduced marine pest. Biol Invasions 2:245–253
Dahadough-Guebas F, Hettiarachchi S, Leo Seen D, Batelaan O, Sooriyarachchi S, Jayatissa LP, Koedam N (2005) Transition in ancient inland freshwater resource management in Sri Lanka affect biota and human populations in and around coastal lagoons. Curr Biol 15:579–586
Das S, Siddiqi NA (1985) The mangrove and mangrove forests of Bangladesh. Mangrove Silviculture Division, Bulletin no. 2. BFRI and UNDP/FAO project, BGD/79/017, Chittagong, Bangladesh
Das T, Pal AK, Chakraborty SK, Manush SM, Chatterjee N, Mukherjee SC (2004) Thermal tolerance and oxygen consumption of Indian major carps acclimated to four temperatures. J Therm Biol 29:157–163

DasGupta R, Shaw R (2013) Changing perspectives of mangrove management in India – an analytical overview. Ocean Coast Manag 80:107–118
David M, Gollasch S (2015) Introduction. In: Global maritime transport and ballast water management, invading nature – Springer series in invasion ecology, vol 8. Springer, Dordrecht, p 306
David IG, Matache ML, Tudorache A, Chisamera G, Rozylowicz L, Radu GL (2012) Food chain biomagnification of heavy metals in samples from the lower prut floodplain natural park. Environ Eng Manag J 11:69–73
Davis JH (1938) Mangroves makers of land. Nature 31:551–553
Davis MA, Grime JP, Thompson K (2000) Fluctuating resources in plant communities: a general theory of invasiblity. J Ecol 88:528–534
De TK, Mitra A, Choudhury A (1987) Studies on phytoplankton and productivity around Sagar Island and Prentice Island, Sundarbans, West Bengal. Rec Zool Surv India 81(3&4):81–92
De TK, De M, Das S, Chowdhury C, Ray R, Jana TK (2008) Phytoplankton abundance in relation to cultural eutrophication at the land-ocean boundary of Sunderbans, NE Coast of Bay of Bengal. India J Env Stud Sci 1:169–180
De TK, De M, Das S, Ray R, Ghosh PB (2009) Level of heavy metals in some edible marine fishes of mangrove dominated tropical estuarine areas of Hooghly River, North East Coast of Bay of Bengal, India. Bull Environ Contam Toxicol 85:385–390
Demopoulos AWJ (2004) Aliens in paradise: a comparative assessment of introduced and native mangrove benthic community composition, food–web structure, and litter–fall production. Ph. D. dissertation, University of Hawaii, Honolulu
Dev Roy MK, Nandi NC (2017) Marine invasive alien crustaceans of India. J Aquac Mar Biol 5 (20):1–3
Dextrase AJ, Coscarelli MA (2000) Intentional introductions of non–indigenous freshwater organisms in North America. In: Claudi R, Leach JH (eds) Nonindigenous freshwater organisms. Lews Publishers, Boca Raton, pp 61–98
Dey MK, Hazra AK, Chakraborty SK (2010) Functional role of microarthropods in nutrient cycling of mangrove–estuarine ecosystem of Midnapore coast of West Bengal, India. Int J Environ Technol Manag 12(1):67–84
Drake J, Mooney HA, Di Castri F, Groves R, Kruger FJ, Rejmánek M, Williamson M (1989) Biological invasions: a global perspective. Wiley, New York
Dugdale RC, Goering JJ (1967) Uptake of new and regenerated forms of nitrogen in primary productivity. Limnol Oceanogr 12:196–206
Duke NC (1992) Mangrove floristics and biogeography. In: Rebertson AI, Alongi DM (eds) Tropical mangrove ecosystems. American Geophysical Union, Washington DC, pp 63–100
Duke NC (1995) Genetic diversity, distributional barriers and rafting continents-more thoughts on the evolution of mangroves. Hydrobiologia 295:167–181
Dukes JS, Mooney HA (1999) Does global change increase the success of biological invaders? Trends Ecol Evol 14:135–139
Duncan RP, Blackman TM, Sol D (2003) The ecology of bird introductions. Annu Rev Ecol Evol Syst 34:71–98
Dunstan PK, Johnson CR (2004) Invasion rates increase with species richness in a marine epibenthic community by two mechanisms. Oecologia 138(2):285–292; Ecology 59:274–284
EC (2008) Developing an EU framework for invasive alien species discussion paper, final. EC, 29 pp. http://ec.europa.eu/environment/nature/invasivealien/docs/ias_discussion_paper.pdf. Accessed 1 December 2016
Eichelberger JW, Lichtenberg JJ (1971) Persistence of pesticides in river water. Environ Sci Technol 5:541–544
Elliott M, Taylor CJL (1989) The structure and functioning of an estuarine/marine fish community in the forth estuary, Scotland. In: Klekowski RZ et al (eds) Proceedings of the twenty-first European marine biology symposium. Polish Academy of Sciences, Institute of Oceanology, Wroclaw, pp 227–240
Elton CS (1958) The ecology of invasions by animals and plants. Metheun, London

Eppley RW, Peterson BJ (1979) Particulate organic matter flux and planktonic new production in the deep ocean. Nature 282:677–680

FAO (1983) Compilation of legal limits for hazardous substances in fish and fishery products. Fish Circ 464:5–100

FAO (1998) Food and Agriculture Organization of the United Nations, Rome, 2000, vol 86/2

FAO/WHO (1978) Acceptable daily intake, residue limits and guideline levels proposal at Joint Committee meeting: 1–23

FDA (2001) Fish and fisheries products hazards and controls guidance, 3rd edn. Center for Food Safety and Applied Nutrition, US Food and Drug Administration, Washington, DC

Fensham RJ, Cowie ID (1998) Alien plant invasions on the Tiwi Islands. Extent, implications and priorities for control. Biol Conserv 83:55–68

Fields PA, Rudomin EL, Somero GN (2006) Temperature adaptation of cytosolic malate dehydrogenases fron native and invasive success. J Exp Biol 209:656–667

Fox MD, Fox JB (1986) The susceptibility of natural communities to invasion. In: Groves RH, Burdon JJ (eds) Ecology of biological invasions. Cambridge University Press, Cambridge, pp 57–66

Ganguly D, Dey M, Mandal SK, De TK, Jana TK (2008) Energy dynamics and its implication to biosphere–atmosphere exchange of CO2, H2O and CH4 in a tropical mangrove forest canopy. Atmos Environ 42:4172–4184

GESAMP (IMO/FAO/UNESCO/WMO/WHO/IAEA/UN/UNEP (1993) Joint Group of Experts on the scientific aspects of marine pollution) (1993). Impact of oil and related chemicals and wastes on the marine environment. GESAMP Reports and Studies, vol 50, p 180

Ghosh AK (2008) Biodiversity conservation: issues and concern. Zool Res Hum Welf 2:19–22

Ghosh PB, Choudhury A (1989) Cu, Zn and Pb in the sediments of Hooghly Estuary. Environ Ecol 7:427–430

Ginsberg RN, Lowenstam HA (1958) The influence of marine bottom communities on the depositional environment of sediments. J Geol 66:310–318

GISP (2004) Tropical Asia invaded: the growing danger of invasive alien species. The Global Invasive species programme: 1–64. GISP Secretariat, Cape Town

Glowka L, Burhenne-Guilmin F, Synge H (1994) A guide to the convension on biological diversity. IUCN, Gland

Gopalkrishnan V (1971) The biology of the Hoogly-Matlah estuarine system (West Bengal, India) with special reference to its fisheries. J Mar Biol Assoc India 13:182–194

Gray JS, Elliot M (2009) Ecology of marine sediments–from science to management. Oxford University Press, London, pp 1–225

Grime GP (1977) Evidence for existence of three primary strategies in plants and its relevance to ecological and evolutionary theory. Am Nat 111:1119–1194

Gulec CW, Lol CP, Jang CS, Wang SW, Hsueh YM (2004) Bioaccumulation of arsenic compounds in aquaculture clams (Meretrix lusoria) and assessment of potential carcinogenic risks to human health by ingestion. Chemosphere 4:38

Hajeb P, Jinap S, Ismail A, Fatimah AB, Jamilah J, Rahim MA (2009) Assessment of mercury level in commonly consumed marine fishes in Malaysia. Food Control 20:79–84

Halder P, Raha P, Bhattacharyya P, Chowdhury A, Adityachaudhury N (1989) Studies on the residues of DDT and Endosulfan occurring in Ganga water. Indian J Environ Health 31:156–161

Halder P, Kole RK, Bhattacharya A, Chowdhury A, Adityachaudhury N (1990) Studies on the residues of the BHC isomers (α, β, γ & δ) occurring in Ganga waters. Pollut Res 9:51–56

Hall JE (2014) Bioconcentration, bioaccumulation, and the biomagnification in Puget sound biota: assessing the ecological risk of chemical contaminants in Puget sound. Tahoma West Lit Art Mag 8:40–51

Harley CDG, Hughes AR, Hultgren KM, Miner BG, Store CJB, Thornber CS, Rodriguez LF, Tomanek L, Williams SL (2006) The impacts of climate change in coastal marine systems. Ecol Lett 9:228–241

Harris L, Tyrrell M (2001) Changing community states in the Gulf of Maine: synergism between invaders, overfishing and climate change. Biol Invasions 3:9–21
Hayes KR, Barry SC (2008) Are their any consistent prediction of invasion success? Biol Invasions 10(4):483–506
Hazra S, Ghosh T, DasGupta R, Sen G (2002) Sea level and associated changes in the Sundarbans. Sci Cult 68:309–312
Heald EJ, Odum WE (1970) The contribution of mangrove swamps to Florida Fisheries. Proc Gulf Caribb Fish Inst 22:130–135
Heald EJ, Odum WE (1975) The contribution of mangrove swamps to Florida Fisheries. Proc Gulf Carribian Fish Inst 22:130–135
Hobbs RJ, Humphries SE (1995) An integrated approach to the ecology and management of plant invasions. Conserv Biol 9:761–770
Holling CS (1981) Forest insects, forest fires and resistance. In: Mooney H, Bonnicksen TM, Christensen NL, Lotan JE, Reiners WA (eds) Fire regimes and ecosystem properties. USDA Forest Service, General Technical Report WO–26. US Department of Agriculture, Washington, DC, pp 445–464
Honnen W, Rath K, Schlegel T, Schwinger A, Frahne D (2001) Chemical analyses of water, sediment and biota in two small streams in Southwest Germany. J Aquat Ecosyst Stress Recover 8:195–213
Hossain MK (2003) Exotic invasive plants in south-eastern hill forests of Bangladesh. In: A paper presented at the 7th international conference on the ecology and management of Alien Plant Invasions, November 3–11, at Wyndham Bonaventure Resort, Ft. Lauderdale and Miami University, Florida, USA
Hulme P, Pyšek P, Nentwig W, Vilà M (2009) Will threat of biological invasions unite the European Union? Science 324:40–41
Hutchings PA, Saenger P (1987) Ecology of mangroves. University of Queensland Press, Brisbane
ICAE (2016, March 28–30) Aquatic invasive species: management strategies and action plans. In: Raghavan R, Kiran Pramod RB, Suvarna Devi S (eds) International conference on aquatic exotics: trends, challenges and policies. Presented as draft in Conference paper. Organized by Department of Aquatic Biology and Fisheries, University of Kerala, Thiruvananthapuram
Iftekhar MS (1999) Vegetation dynamics in the Sundarbans and the contribution of salinity between 1985–1995. B.Sc. thesis. Forestry and Wood Technology Discipline, Khulna University, Khulna, Bangladesh: 1–54
IMCO/FAO/UNESCO/WMO/WHO/IAEA/UN/GESAMP (1977) Impact of oil on the marine environment. Joint Group of Experts on the scientific aspects of marine pollution (GESAMP). Report and Studies No. 6, 250 pp
IPCC (2000) In: Nakicenovic N, Swart R (eds) Special report on emissions scenarios: a special report of Working Group III of the fourth assessment report of the Intergovernmental Panel on Climate Change. Cambridge University Press, Cambridge/New York
IPCC (Intergovernmental Panel on Climate Change) (2007) Climate change 2007: the physical science basis. Contribution of Working Group I to the fourth assessment report of the IPCC. Cambridge University Press, Cambridge
IUCN (1989) Marine protected areas needs in the South Asian Seas Region 2, India
IUCN (1997) Sundarban Wildslife Sactuaries (Bangladesh): World Heritage Nomination–IUCN Technical Evaluation. http://whc.unesco.org/archive/advisory_body_evaluation/798.pdf. 12
IUCN (2003) Status of the ecological integrity of the Sundarbans. IUCN Bangladesh Country Office, 80
IUCN (2004) Red list of threatened species: a global species assessment (eds. by Baillie J, Hilton–Taylor C, Stuart SN)
Jennerjahn TC, Ittekkot V (2002) Relevance of mangroves for the production and deposition of organic matter along tropical continental margins. Naturwissenschaften 89:23–30
Jones CG, Lawton JH, Shachak M (1994) Organisms as ecosystem engineers. Oikos 69:373–386
Jones CG, Lawton JH, Shachak M (1997) Positive and negative effects of organisms as physical ecosystem engineers. Ecology 78:1946–1957

Kareiva P, Parker IM, Pascual M (1996) Can we use experiments and models in predicting the invasiveness of genetically engineered organisms. Ecology 77:1670–1675
Kathiresan K, Bingham BL (2001) Biology of mangroves and mangrove ecosystems. Adv Mar Biol 40:81–251
Kennedy VS (1990) Anticipated effects of climate change on estuarine and coastal fisheries. Fisheries 15:16–25
Kihslinger RL, Woodin SA (2000) Food patches and a surface deposit feeding spionid polychaete. Mar Ecol Prog Ser 201:233–239
King GM (1986) Inhibition of microbial activity in marine sediments by a bromophenol from a hemichordate. Nature 323:257–259
Kinne O (1963) The effect of temperature and salinity on marine and brachish water animals. Oceanogr Mar Biol Annu Rev 1:301–340
Kolar CS, Lodge DM (2001) Progress in invasion biology: predicting invaders. Trend Ecol Evol 16 (4):199–204
Kole RK, Bagchi MM (1995) Pesticide residues in the aquatic environment and their possible ecological hazards. J Inland Fish Soc India 27:79–89
Lambert MRK (1997) Environmental effects of heavy spillage from a destroyed pesticide store near Hargeisa (Somaliland) assessed during dry season, using reptiles and amphibians as bioindicators. Arch Environ Contam Toxicol 32:80–93
Lambert CC, Lambert G (1998) Non–indigenous ascidians in Southern California harbors and marinas. Mar Biol 130:675–688
Lambrions JG (2002) The variable invasive success of Cortaderia species in a complex landscape. Ecology 83:518–529
Law RJ, Kelly CA, Graham KL, Woodhead RJ, Dyrynda PEJ, Dyrynda EA (1997) Hydrocarbons and PAH in fish and shellfish from SouthwestWales following the Sea Empress oil spill in 1996. In: Proceeding, 1997, international oil spill conference. American Petroleum Institute, Washington, DC, pp 205–211
Laws EA (2000) Aquatic pollution: a introductory text. Wiley, New York, pp 432–479
Lee SY (1995) Mangrove outwelling– a review. Hydrobiologia 295:203–212
Lee SY (1998) Ecological role of grapsid crabs in mangrove ecosystems: a review work. Mar Fresh Water Res 49:335–343
Lee SY (1999) The effect of mangrove leaf litter enrichment on macrobenthic colonization of defaunated sandy substrates. Estuar Coast Shelf Sci 49(5):703–712
Lee SY (2008) Mangrove macrobenthos: assemblages, services, and linkages. J Sea Res 59:16–29
Levin SA, Paine RT (1974) Disturbance, patch formation, and community structure. Proc Natl Acad Sci 71(7):2744–2747
Levin LA, Talley TS (2000) Influences of vegetation and abiotic environmental factors on salt marsh benthos. In: Weinstein MP, Kree-ger DA (eds) Concepts and controversies in tidal marsh ecology. Kluwer, Amsterdam
Levine JM, Vila M, D'Antonio CM, Dukes JS, Grigulis K, Lavorel S (2003) Mechanisms underlying the impacts of exotic plant invasions. Proc R Soc Lond Ser B Biol Sci 270:775–781
Lim SLS (2006) Fiddler crab burrow morphology: how do burrow dimensions and bioturbative activities compare in sympatric populations of *Uca vocans* (Linnaeus, 1758) and *U. annulipes* (H. Milne Edwards, 1837)? Crustaceana 79(5):525–540
Lodge DM (1993) Biological invasions: lessons for ecology. Trend Ecol Evol 8:133–211
Lodge DM, Williams S, MacIsaac HJ, Hayes KR, Leung B, Reichard S, Mack RN, Moyle PB, Smith M, Andow DA, Carlton JT, McMichael A (2006) Biological invasions: recommendations for US policy and management. Ecol Appl 16:2035–2054
Lovelock CE, Feller IC, McKee KL, Thompson R (2005) Variation in mangrove forest structure and sediment characteristics in Bocas del Toro, Panama. Caribb J Sci 41:456–464
Loucks OL (1970) Evolution of diversity, efficiency and community stability. Am Zool 10:17–25
Luckenbach MW (1986) Sediment stability around animal tubes: the roles of hydrodynamic processes and biotic activity. Limnol Oceanogr 31:779–787

Lugo AE (1980) Mangrove ecosystems: successional or steady state. Trop Succession 12:65–72
Lugo A, Snedaker SC (1974) The ecology of mangroves. Annu Rev Ecol Syst 5:39–64
Mack RN, Simberloff D, Lonsdale WM, Evans H, Clout M, Bazzaz F (2000) Biotic invasion: causes, epidemiology, global consequences and control, Issues in ecology, no. 5. The Ecological Society of America, Washington, DC
Macnae W (1968) A general account of the fauna and flora of mangrove swamps and forests in the Indo-west-Pacific region. Adv Mar Biol 6:73–270
Macnae W (1974) Mangrove forest and fisheries Indian Ocean Programme, 34. Indian Ocean Fisheries Commission, Rome
Mahadevan A, Muthukumar G (1980) Aquatic microbiology with reference to tannin degradation. Hydrobiologia 72:73–79
Maheswari JK (1965) Alligator weed in Indian lakes. Nature 206:1270
Maiti GG, Guhabakshi DN (1981) Invasion of exotic weeds in West Bengal since 1903: dicotyledons and monocotyledons. J Econ Taxon Bot 2:158–168
Maiti Dutta S, Banerjee SB, Raha S, Chakraborty SK (2014) Assessment of thermal stress adaptation by monitoring HSP 70 and superoxide dismutase in fresh water gastropod *Bellamya bengalensis*. Environ Monit Assess 186(12):8961–8967
Martinez ND (1992) Constant connectance in community food webs. Am Nat 139(6):1391–1401
Martinez C (2004) Food and Niche overlap of the Scarlet Ibis and Yellow Crowned Night Heron in a typical mangrove swamp. Wetlands 27:1–8
McCoy HD (2000) American and international aquaculture law. Supranational Publishing Company, Peterstown
McKee KL (1995a) Seedling recruitment patterns in a Belizean mangrove forest: effects of establishment ability and physic–chemical factors. Oecologia 101:448–460
McKee KL (1995b) Mangrove species distribution patterns and propagule predation in Belize: an exception to the dominance-predation hypothesis. Biotropica 27:334–345
Mcmillan RL, Sherrod CL (1986) The chilling toler– ance of black mangrove, *Avicennia germinans*, from the Gulf of Mexico coast of Texas, Louisiana and Florida. Contrib Mar Sci 29:9–16
McNeely JA (2001) The great reshuffling: human dimensions of invasive alien species. IUCN, Gland/Cambridge
McNeely JA, Mooney HA, Neville LE, Schei P, Waage JK (2001) A global strategy on invasive alien species. IUCN, Gland
Meadows P, Tait J (1989) Modification of sediment permeability and shear strength by two burrowing invertebrates. Mar Biol 101:75–82
Micheli F (1993a) Feeding ecology of mangrove crabs in North Eastern Australia: mangrove litter consumption by Sesarma messa and Sesarma smithii. J Exp Mar Biol Ecol 171:165–186
Micheli F (1993b) Effect of mangrove litter species and availability on survival, moulting, and reproduction of the mangrove crab Sesarma messa. J Exp Mar Biol Ecol 171:149–163
Millennium Ecosystem Assessment (2005) Ecosystems and human well–being. Washington, DC, World Resources Institute
Mitra A, Gangopadhyay A, Dube A, Schmidt Andri CK, Banerjee K (2009) Observed changes in water mass properties in the Indian Sundarbans (North Western Bay of Bengal) during 1980–2007. Curr Sci 97(10):1445–1452
Mitsch WJ, Gosselink JG (1986) Wetlands. Van Nostrand Reinhold Company, New York, pp 93–95
Mohapatra SP, Gajbhiye VT, Agnihotri NP, Manju R, Raina M (1995) Insecticide pollution in Indian rivers. Environmentalist 15:41–44
Molnar JL, Rebecca LG, Carmen R, Spalding MD (2008) Assessing the global threat of invasive species to marine biodiversity. Front Ecol Environ 6(9):485–492
Mooney HA, Hobbs RJ (2000) Invasive species in a changing world. Island Press, Washington, DC
Moore MV, Pace ML, Mather JR, Murdoch PS, Howarth RW, Folt CL, Chen CY, Hemond HF, Flebbe PA, Driscoll CT (1997) Potential effects of climate change on freshwater ecosystems of the New England/mid–Atlantic region. Hydrol Process 11:925–947

Morgan JP, Mcintire WG (1959) Quaternary geology of the Bengal basin, East Pakistan and Burma. Bull Geol Soc Am 70:319–342
Mukherjee AD, Chatterjee S (1997) Coastal erosion and accretion at and around Digha in Medinipur District of West Bengal. Indian J Geogr Environ 12:1–11
Mukherjee AD, Bakshi A (1998) Integrated coastal zone management plan. In: Mukherjee AD, Datta K, Sanyal P (eds) Coastal zone problems, Proceedings of the National Workshop, 20th to 22nd March, 1998. Jadavpur University, Calcutta, pp 181–189
Mukherjee M, Kashen A (2007) Sundarban wetlands. Department of Agriculture, Aquatic Resources and Fishing Harbours, Government of West Bengal, West Bengal, p 159
Murty AS (1986) Toxicity of pesticides to fish, vol 1. CRC Press, Boca Raton
Najjar RG, Walker HA, Anderson PJ, Barro EJ, Bord RJ, Gibso JR, Kennedy VS, Knight CG, Mego-nigal JP, O'connor RE, Polsky CD, Psuty NP, Rich-ards BA, Sorenson LG, Steele EM, Swanson RS (2000) The potential impacts of climate change on the mid–Atlantic coastal region. Clim Res 14:219–233
Naskar K, Sen Sarkar N, Ghosh A, Dasgupta M, Sengupta B (2004) Status of the mangroves and mangrove ecosystem of Sundarbans in West Bengal: its impact on estuarine wet lands fisheries. Bulletine No. 134 CIFRI, Barrackpore, Kolkata: 1–40
National Research Council (NRC) (2000) Clean coastal waters: understanding and reducing the effects of nutrient pollution, vol 35. National Academy Press, Washington, DC, pp 3457–3461
Naylor RL, Goldburg RJ, Primavera JH, Kautsky N, Beveridge MCM, Clay J, Folke C, Lubchenco J, Mooney H, Troell M (2000) Effect of aquaculture on world fish supplies. Nature 405:1017–1024
Neilson MJ, Giddins RL, Richard GN (1986) Effects of tannins on the palatability of mangrove leaves to the tropical sesarminid crab Neosarmatium smithi. Mar Ecol Prog Ser 34:185–186
Neira C, Levin LA, Grosholz ED (2005) Benthic macrofaunal communities of three sites in San Francisco Bay invaded by hybrid Spartina, with comparison to uninvaded habitats. Mar Ecol Prog Ser 292:111–126
Newell RIE, Marshall N, Sasekumar A, Chong VC (1995) Relative importance of benthic microalgae, phytoplankton, and mangroves as sources of nutrition for penaeid prawns and other coastal inver- tebrates from Malaysia. Mar Biol 123:595–606
Nixon SW, Pilson MG (1983) Nitrogen in estuarine and coastal marine ecosystems. In: Carpenter EJ, Capone DG (eds) Nitrogen in the marine environment. Academic, New York, pp 565–648
Odum EP (1971) Fundamentals of ecology. W.B. Saunders and Co, Philadelphia, pp 1–297
OTA (1993) Harmful non–indigenous species in the United States. OTA–F–565, Office of Technology Assessment, US. Government Printing Office, Washington, DC
Panigrahy PK, Satapathy DR, Panda CR, Kar RN (2014) Dispersion pattern of petroleum hydrocarbon in coastal water of Bay of Bengal along Odisha and West Bengal, India using geospatial approach. Environ Monit Assess 186:8303–8315
Paria S, Bhattacharyya M, Halder Mallick P, Chakraborty SK (2017) Positive impact of bioinvasive host plant of butterfly diversity: the other side of the coin. ZSI ENVIS Newslett 23(1):5–9
Parsons TR, Sharp JC, Li WKW (1985) The cultivation of marine amphipods and their use as food for young salmonids. J Appl Ichthyol 1:77–84
Patterson S, McKee KL, Mendelssohn IA (1997) Effects of tidal inundation and predation on *Avicennia germinans* seedling establishment and survival in asub–tropical mangal/salt marsh community. Mangrove Salt Marshes 1:103–111
Paul AK (2002) Coastal geomorphology and environment. ACB Publication, Kolkata, pp 1–582
Perrings C, Williamson MH, Dalmazzone S (2000) The economics of biological invasions. Edward Elgar, Cheltenham
Perrings CM, Mooney H, Williamson M (2010) Bioinvasions and globalization–ecology, economics and policy. Oxford University Press, Oxford, pp 1–267
Peterson D, Cayan D, DiLeo J, Noble M, Dettinger M (1995) The role of climate in estuarine variability. Am Sci 83:58–67

Phillips E (2005) Growth of filter–feeding benthic invertebrates from a region with variable upwelling intensity. Mar Ecol Prog Ser 295:79–89
Pimental D, Lach L, Zuniga R, Morison D (2001) Environmental and economic costs associated with alien-invasive species in the United States. Ecol Econ 52:273–288
Planty-Tabacchi A-M, Tabacchi E, Naiman RJ, Deferrari C, Décamps H (1996) Invasibility of species–rich communities in riparian zones. Conserv Biol 10(2):598–607
Posey MH (1986) Changes in a benthic community associated with dens beds of a burrowing. Mar Ecol Prog Ser 15:22–31
Posey MH (1998) Community changes associated with the spread of an introduced seagrass, *Zostera japonica*. Ecology 69:974–983
Posey MH, Wigand C, Stevenson JC (1993) Effects of an introduced aquatic plant, Hydrilla verticillata on benthic communities in the upper Chesapeake Bay. Estuar Coast Shelf Sci 37:539–555
Prins Herbert HT, Grden LJ (2014) Testing hypothesis about biological invasions and Charles Darwin's two creators rumination, pp 1–20
Pulin AS, Stewart GB (2006) Guidelines for systematic review in conservation and environmental management. Conserv Biol 20:1647–1656
Puthucherril TG (2011) Ballast waters and aquatic invasive species: a model for India. Colo J Int Environ Law Policy 19(3):381–425
Pysek P, Hulme PE (2005) Spatio–temporal dynamics of plant invasions: linking pattern to process. Ecoscience 12:345–354
Qasim SZ, Sengupta R (1988) Present status of marine pollution in India. Indian J Fish Assoc 18:310–329
Raha A, Das S, Banerjee K, Mitra A (2012) Climate change impacts on Indian Sunderbans: a time series analysis (1924–2008). Biodivers Conserv 21:1289–1307
Rajagopal S, Nair NVK, Azariah J (1989) Some observations on the problem of jelly fish ingress in a power station cooling system at Kalpakkam, east coast of India. Mahasagar 22(4):151–158
Ramkrishnan PS (1991) Ecology of biological invasions in the tropics. National Institute of Ecology and International Science Publisher, New Delhi
Ravinesh R, Bijukumar A (2016) Invasive species along the intertidal habitats of Kerala coast, India. Abstract No–22: International Conference on Aquatic Exotics (ICAE):47–48
Ray R, Mandal S, Dhara A (2013) Environmental monitoring of estuaries: estimating and mapping various environmental indicators in Matla estuarine complex, using Landsat TM digital data. Int J Geomat Geosci 3(3):570–581
Rejmanek M (1989) Invasiblity of plant communities. In: Drake JA et al (eds) Biological invasions: a global perspective. Wiley, Chichester, pp 369–388
Rejmanek M (1996) Species richness and resistance to invasions. In: Orians GH, Dirzo R, Cushman JH (eds) Diversity and processes in tropical forest ecosystems. Springer, Berlin, pp 153–172
Rejmanek M, Richardson DM (1996) What attributes make some plant species more invasive? Ecology 77(6):1655–1661
Rejmanek M (1999) Invasive plant species and invisible ecosystems. In: Sandlund OT et al (eds) Invasive species and biodiversity management. Kluwer Academic Publishers, Dordrecht, pp 79–102
Reichard SH, Hamilton CW (1997) Predicting in-vasions of woody plants introduced into North America. Conserv Biol 11:193–203
Reusch TBH (1998) Native predators contribute to invasion resistance to the non–indigenous bibalve Musculista senhousia in southern California USA. Mar Ecol Prog Ser 170:159–168
Rhoads DC, Stnley DJ (1965) Biogenic graded bedding. J Sediment Petrol 35:956–963
Ricciardi A (2005) Facilitation and synergistic interactions among introduced aquatic species. In: Mooney HA, Mack RN, McNeely J, Neville LE, Schei PJ, Waage JK (eds) Invasive alien species: a new synthesis. Island Press, Washington, DC, pp 162–178
Ricciardi A, Rebekah K (2008) Predicting the number of ecologically harmful exotic species in an aquatic system. Diversity and Distributions 14(2):374–380 Special issue: fifty years of invasion ecology: the legacy of Charles Elton (Mar., 2008)

Ricciardi A, Steiner WWM, Mack RN, Simberloff D (2000) Toward a global information system for invasive species. Bioscience 50:239–244
Rice D, Rooth J, Stevenson JC (2000) Colonization and expansion of *Phragmites australis* in upper Chesapeake Bay tidal marshes. Wetlands 20:280–289
Robertson AI (1986) Leaf–burying crabs: their influence on energy flow and export from mixed mangrove forests (*Rhizophora* spp.) in northeastern Australia. J Exp Mar Biol Ecol 102:237–248
Robertson AI, Alongi DM, Boto KG (1992) Food chains and carbon fluxes. In: Robertson AI, Alongi DM (eds) Tropical mangrove ecosystems. American Geophysical Union, Washington, DC, pp 293–326
Rovedatti MG, Castane PM, Topalian ML, Salibian A (2001) Monitoring of organochlorine and organophosphorus pesticides in water of the Reconquista River (Buenos Aires, Argentina). Water Res Oxf 35:3457–3461
Roy S, Hens D, Biswas D, Biswas D, Kumar R (2002) Survey of petroleum–degradiny bacteria in coastal waters of Sundarban biosphere reserve. World J Microbiol Biotechnol 18:575–581
Ruiz GM, Carlton JT, Grosholz ED, Hines AH (1997) Global invasions of marine and estuarine habitats by non–indigenous species: mechanisms, extent and consequences. Am Zool 37:621–632
Sala OE, Chapin FSIII, Armesto JJ, Berlow E, Bloomfield J, Dirzo R, Huber-Sanwald E, Huenneke LF, Jackson RB, Kinzig A (2000) Global biodiversity scenarios for the year 2100. Science 287:1770–1774
Sanford E (1999) Regulation of keystone predation by small changes in ocean temperature. Science 283:2095–2097
Sanford E, Roth MS, Johns GC, Wares JP, Somero GN (2003) Local selection and latitudinal variation in a marine predator–prey interaction. Science 300:1135–1137
Santra SC (1998) Mangrove lichens. Indian Biol 30(2):76–78
Sarkar SK, Singh BN, Choudhury A (1986) The ecology of copepods from Hoogly estuary of West Bengal. India Mahasagar-Bull Nat Inst Oceanogr 19(2):103–112
Sarkar SK, Bhattacharya B, Debnath S (2002) Heavy metal concentration from biota of Sunderban mangrove ecosystem. Aquat Ecosyst Health Manage 5:467–472
Sarkar S, Ghosh PB, Das TM, Som Mazumdar S, Saha T (2013) Environmental assessment in terms of salinity distribution in the Tropical Mangrove forest of Sundarban, North East Coast of Bay of Bengal, India. Arch Appl Sci Res 5(6):109–118
Sasekumar A (1974) Distribution of macrofauna on a Malayan mangrove shore. J Anim Ecol 43:51–69
Savidge JA (1987) Extinction of an island forest avifauna by an introduced sharke. Ecology 68:660–668
Schiffman PM (1997) Animal –mediated dispersal and disturbance: driving forces behind alien plant naturalization. In: Luken JO, Thieret JW (eds) Assessment and management of plant invasions. Springer, New York, pp 87–94
Schulz R (2001) Rainfall –induced sediment and pesticide input from orchards into the Lourens river, Western Cape, South Africa: importance of a single event. Water Res 35:1869–1876
Scott GI, Fulton MH, Moore DW, Wirth EF, Chandler GT, Key PB, Daugomah JW, Strozier ED, Devane J, Clarke JR, Lewis MA, Finley DB, Ellengerg W, Karnaky KJ (1999) Assessment of risk reduction strategies for the management of agricultural nonpoint source pesticide runoff in estuarine ecosystem. Toxicol Ind Health 15:200–213
Sebens KP (1982) Competition for space: growth rate, reproductive output and escape in size. Am Nat 120:189–197
Sericano JL, Wade TL, Jackson TJ (1995) Trace organic contamination in theAmericas: an overview of the US National Status, trends and the international mussel watch programmes. Mar Pollut Bull 31:214

Sessegolo GC, Lana PC (1991) Decomposition of *Rhizophora mangle*, *Avicennia schaueriana* and *Laguncularia racemosa* leaves in a mangrove of Paranagua Bay (Southeastern Brazil). Bot Mar 34:285–289
Shiridan P (1997) Benthos of adjacent mangrove, seagrass and nonvegetated habitats in Rookery Bay, Florida, USA. Estuar Coast Shelf Sci 44(4):445–469
Shigesada N, Kwasaki K (1997) Biological invasions: theory and practice. Oxford University Press, Oxford
Shine MC (2003) Overview of international instruments addressing invasive alien species. In: Pallewatta N, Reaser JK, Gutierrez A (eds) Prevention and management of invasive alien species, Proceedings of a workshop on forging cooperation throughout South and Southeast Asia. Global Invasive Species Programme, Cape Town, pp 19–35
Shine C, Kettunen M, Genovesi P, Gollasch S, Pagad S, Starfinger U (2008) Technical support to EU strategy on invasive species (IAS) – Policy options to control the negative impacts of IAS on biodiversity in Europe and the EU. Final report for the European Commission. Institute for European Environmental Policy (IEEP), Brussels, Belgium
Shugart HHJR, West DC (1980) Forest succession models. Bio Sci 30:308–313
Simberloff D, VonHolle B (1999) Positive interactions of nonindigenous species: invasional meltdown? Biol Invasion 1:21–32
Simberloff D (2009) We can eliminate invasions or live with them. Successful management projects. Biol Invasions 11:149–157
Singer MC, Thomas CD, Parmesan C (1993) Rapid human induced evolution of insect–host associations. Nature 366:681–683
Smith TJ III (1992) Forest structure. In: Robertson AI, Alongi DM (eds) Tropical mangrove ecosystems. Am Geophys Union, Washington DC, pp 101–136
Soule ME (1990) The onslaught of alien species and other challenges in the coming decades. Cons Biol 4:233–239
Spalding MD, Fox HE, Allen GR, Davidson N, Ferdaña ZA, Finlayson M, Halpern BS, Jorge MA, Lombana A, Lourie SA, Martin KD, McManus E, Molnar J, Recchia CA, Robertson J (2007) Marine ecoregions of the world: a bioregionalization of coast and shelf areas. Bioscience 57:573–583
Spear D, Chown SL (2009) Non–indigenous ungulates as a threat to biodiversity. J Zool 279 (1):1–17
Stachowicz JJ, Whitlatch RB, Osman RW (1999) Species diversity and invasion resistance in a marine eco-system. Science 286:1577–1579
Stachowicz JJ, Terwin JR, Whitlatch RB, Osman RW (2002) Linking climate change and biological invasions: ocean warming facilitates nonindigenous species invasions. P Natl Acad Sci USA 99:15497–15500
Stark JS (1998) Heavy metal pollution and macrobenthic assemblages in soft sediments in two Sydney estuaries, Australia. Mar Freshw Res 49:533–540
Sutherland JP (1974) Multiple stable points in natural communities. Am Nat 108:859–873
Sutherland JP, Karlson RH (1973) Succession and seasonal progression in the fouling community at Beaufort, North Carolina. In: Third international congress on marine corrosion and fouling. Northwestern University Press, Evanston, pp 906–929
Sutherland P, Karlson R (1977) Development and stability of the fouling community at Beaufort, North Carolina. Ecol Monogr 47:425–446
Teal JM (1958) Distribution of fiddler crabs in Georgia Salt Marshes. Ecology 39:185–193
Thresher RE, Kuris AM (2004) Options for managing invasive marine species. Biol Invas 6:295–300
Tomlinson PB (1986) The botany of mangroves. Cambridge University Press, Cambridge, p 414
Trevisan M, Montepiani C, Ragozza L, Bartoletti C, Ioannilli E, Del Re AAM (1993) Pesticides in rainfall and air in Italy. Environ Pollut 80:31–39
Trudgill ST (1977) Soil and vegetation systems. Clarendon Press, Oxford

Tuncer G, Karakas T, Balkas Tİ, Gökçay CF, Aygnn S, Yurteri C, Tuncel G (1998) Land–based sources of pollution along the Black Sea coast of Turkey: concentrations and annual loads to the Black Sea. Mar Pollut Bull 36:409–423

Twilley RR, Chen R (1998) A water budget and hydrology model of a basin mangrove forest in Rookery Bay, Florida. Mar Freshw Res 49:309–323

Valentine PC, Carman MR, Blackwood DS, Heffron EJ (2007) Ecological observations on the colonial ascidian Didemnum sp. in a New England tide pool habitat. J Exp Mar Biol Ecol 342:109–121

Vila M, Weber E, D" Antonio C (2000) Conservation implications of invasion by plant hybridization. Biol Invasions 2:207–217

Vitousek PM (1990) Biological invasions and ecosystem processes: towards an integration of population biology and ecosystem studies. Oikos 57:7–13

Vitousek PM, D'Antonio CM, Loope LL, Westbrooks R (1996) Biological invasions as global environmental change. Am Sci 84:468–478

Vitousek PM, Mooney HA, Lubchenco J, Melillo JM (1997) Human domination of earth's ecosystems. Science 277(494):99

Vogl B (1980) The ecological factors that produce perturbation–dependent ecosystems. In: Crains J (ed) The recovery process in damaged ecosystems. Ann Arbor Science Publishers, Ann Arbor, p 63

Walker CH, Hopkin SP, Sibly RM, Peakall DB (1996) The fate of etals and radioactive isotopes in contaminated ecosystem. In: Principles of ecotoxicology. Taylor and Francies, London

Walsh JJ (1991) Importance of the continental margins in the marine biogeochemical cycling of carbon and nitrogen. Nature 350:53–55

Walsh JJ, Rowe GT, Iverson RE, Mcroy CP (1981) Biological export of shelf carbon is a sink of the global CO2 cycle. Nature 291:196–201

Washington HG (1984) Diversity, biotic and similarity indices: a review with special reference to aquatic ecosystems. Water Res 18:653–694

Weathy DS, Woodin SA (2008) Ecological hindcasting of biogeographic responses to climate change in the European intertidal zone. Hydrobiologia 606:139–151

Welcome RL (1998) International introduction of inland aquatic species. FAO Fisheries Technical Paper: 290–318

Werry J, Lee SY (2005) Grapsid crabs mediate link between mangrove litter production and estuarine planktonic food chains. Mar Ecol Prog Ser 293:165–176

West RC (1977) Tidal salt-marsh and mangal formations of middle and South America. In: Chapman VJ (ed) Ecosystems of the world. 1. Wet coastal ecosystems. Elsevier Scientific Publications Company, Amsterdam, pp 193–213

White PS (1979) Pattern, process and natural disturbance in vegetation. Bot Rev 45:229–299

Williamson MH (1996) Biologicalo invasions. Chapman and Hall, London

Wittenberg R (2005) An inventory of alien species and their threat to bioinvasion and economy in Switzerland. Federal Office for the Environment; CABI BioScience Switzerland Centre, Delémont

Woodin SA (1981) Disturbance and community structure in a shallow water sand flat. Ecology 62 (4):1052–1066

Yennawar P, Mohapatra A, Ray D, Tudu PC (2015) Ichthyofauna of Digha Coast, India. In: Marine faunal diversity in India: taxonomy, ecology and conservation. Elsevier, London, pp 235–248

Chapter 7
Specialized Grooming as a Mechanical Method to Prevent Marine Invasive Species Recruitment and Transport on Ship Hulls

Kelli Z. Hunsucker, Emily Ralston, Harrison Gardner, and Geoffrey Swain

Abstract Biofouling on ship hulls is one of the primary vectors of non-indigenous species transport. The most common method to prevent biofouling settlement is through the application of ship hull coatings. However, there is no perfect coating and the ship hull will eventually become colonized by biofouling. Hull husbandry techniques are often employed to remove the biofouling from the ship hull, which adds in restoring the ships functional abilities and prevents the transport of biofouling organism as invasive species. Two such techniques are in-water cleaning and grooming. The cleaning of a ship hull may damage hull coatings, release both biocides and fouling organisms into the local environment, and is regulated or banned in many ports around the world. A more recent mechanical approach to biofouling, is grooming, a frequent and gentle wiping of the hull, which works in synergy with ship hull coatings to prevent the growth of biofouling organisms. By incorporating grooming into the ship maintenance, invasive species recruitment and transport is prevented.

Keywords Grooming · Ship hulls · Coatings · Invasive species

7.1 Introduction

Biofouling, or the accumulation of unwanted plants and animals, is problematic in the marine environment. This is especially true for the marine shipping industry, where the accumulation of growth leads to functional, financial, and environmental setbacks. Biofouling on ship hulls can lead to increases in drag, fuel consumption, greenhouse gas emissions, operational costs, and invasive species transport (Swain et al. 2007; Schultz et al. 2011; Hewitt and Campbell 2010). The marine shipping

K. Z. Hunsucker (✉) · E. Ralston · H. Gardner · G. Swain
Center for Corrosion and Biofouling Control, Florida Insitute of Technology, Melbourne, FL, USA
e-mail: khunsucker@fit.edu

C. Makowski, C. W. Finkl (eds.), *Impacts of Invasive Species on Coastal Environments*, Coastal Research Library 29,
https://doi.org/10.1007/978-3-319-91382-7_7

industry, while transporting about 90% of the world trade (IMO 2006) also is one of the major pathways for invasive species via ballast water (Ruiz et al. 2000) and ship hull fouling (Drake and Lodge 2007). As is seen with the other chapters in this book, the spread of invasive species may cause detrimental impacts on ecosystems. While there are many vectors for invasive species transport, this chapter will focus on ship hull fouling, and review the concept of grooming as a mechanical approach which can be implemented to prevent the global spread of these non-indigenous species via marine shipping.

7.1.1 Biofouling

Biofouling is generally thought of as a several step process. A clean surface will adsorb an organic layer within seconds of submersion, called a conditioning film. This is followed by the adhesion of microbes such as bacteria and algal cells, thus developing a slimy layer (also known as a biofilm). Following biofilm formation is the settlement of larval spores and higher organisms. These stages can be successional, occur in parallel, or occur simultaneously.

Higher organisms, or macrofouling, consist of plants and animals which grow in many different forms Macrofouling can be divided in soft and hard fouling. The fouling organisms which have soft bodies are things such as seaweed, tunicates, hydroids, and arborescent bryozoans. Hard foulers consist of those organisms which have a hard outer layer, generally comprised of calcium carbonate. Examples of hard fouling organisms are barnacles, calcareous tubeworms, mussels, oysters, encrusting bryozoans, and calcareous algae. There are also organisms, referred to as fouling associates, that have the ability to move around submerged surfaces, often times hiding in cracks and crevices created by fouling. Examples include crabs, snails, and sea stars.

The type and intensity of biofouling accumulation is dependent on the geographical location, time of year, environmental conditions (e.g. salinity, temperature, nutrient levels), substrate, and if present, the biofouling prevention system.

7.1.2 Ship Hull Coatings

In the marine environment, the most common method of preventing biofouling is through the application of ship hull coatings. The most effective coatings were the tributyltin (TBT) self-polishing copolymers. These were biocidal at low levels (which prevented colonization of biofouling), self-smoothing, and had a long lifetime, thus extending a ship's time between dry docking. Unfortunately, TBT had many negative impacts on the environment, such as oyster shell thickening and imposex in dogwhelks. This lead to the ban for commercial use in 2008 (IMO 2001).

Today's commercially available ship hull coatings are typically divided into three main categories: those coatings that either use a biocide to kill the fouling organisms (antifouling systems), create a surface to which they find it difficult to attach (silicone-based fouling release systems) or form a hard-inert surface that requires cleaning (Swain 2010). Antifouling coatings are broadly separated into copper systems and copper-free systems. Many fouling organisms are copper tolerant, for example *Balanus amphitrite* has been shown to be able to settle on copper based coatings. As a result, booster biocides (e.g. zinc pyrithione) are now incorporated within the paint to enhance both the life and the performance of the copper systems. Copper and copper free antifouling coatings can be effective at preventing settlement of fouling organisms, as long as they consistently leach the biocide. However as with TBT, there is a growing movement to eliminate the use of copper based antifouling paints due to its persistence in sediments, bioaccumulation, and other environmental impacts (Chambers et al. 2006; Thomas and Brooks 2010).

The most environmentally friendly commercially available coating systems are considered to be the silicone based fouling release coatings. Fouling release systems have a low surface energy, low modulus, low micro roughness and may contain additives, which prevent a fouling organism from generating a strong bond to the surface. This weak bond allows for removal of the organism either through the weight or hydrodynamic pressure as a ship moves through water (Swain 1999; Omae 2003). These coatings are considered to be environmentally friendly, offer a smooth surface when application is correct, and reduce skin friction drag, and are easy to clean (Swain 2010). However, they are costly, not easy to apply, are not as durable as their biocidal counterparts and the environmental impacts of many of the additives are unknown (Swain 2010).

7.1.3 Type of Fouling on Ship Hulls

Specific classes of ships may have very different fouling communities in terms of extent and composition. This is primarily due to their duty cycle, maintenance schedule and region of operation. Container ships, an example of a constant service vessel with relatively fast speed and short port duration, tend to have low to no cover of macrofouling on the exposed hull surface. These ships are commonly covered with biofilm and algal communities, and if macrofouling organisms that are present, they are found in the protected niche areas (Davidson et al. 2009). For in-service ships, the variables that determine the extent of fouling include: paint type and age, geographic area, port duration, vessel speed, voyage duration and route (Swain 2010). The niche areas of ships are nearly universally fouled (Coutts and Taylor 2004; Davidson et al. 2009).

Semi-submersible rigs, drill ships and platforms represent vessels at the other end of the fouling spectrum. These vessels spend long periods of time static and are often heavily fouled with macrofouling organisms when they are transported from one location to another. These macrofouling communities may include barnacles, bryozoans, tunicates, mussels, corals, hydroids, anemones and often include mobile associates. Several of these groups contain notorious invaders such as the coral *Tubastrea coccinea*, the mussel *Perna perna*, the circumtropical barnacle invader *Amphibalanus amphitrite* and *Corophid* amphipods (Ferreira et al. 2004; Sammarco et al. 2004; Hopkins and Forrest 2010; Wanless et al. 2010). In one study, a rig was being towed from Brazil to Singapore when it broke free and ran aground on an island in a remote oceanic archipelago with a depauperate native faunal community. Sixty two alien species were recorded on the rig, most of which were alive and potentially able to reproduce (Wanless et al. 2010).

Because prevention is the best solution for non-indigenous species (NIS), the ideal situation would be to have ships traveling with clean hulls and niche areas (Wanless et al. 2010; Campbell and Hewitt 2011). However, in reality this is a more difficult goal than it may seem at first glance. When the USS Missouri was decommissioned and sent to Hawaii as a museum ship, it was moved to low salinity brackish water for 9 days in an effort to kill or incapacitate fouling before transit. Upon arrival in Hawaii, an inspection showed that 90% of the hull was free of fouling (Brock et al. 1999). However, within hours of its arrival, spawning of a non-native mussel was observed with subsequent recruitment to the ballast tank of a submarine in port nearby (Apte et al. 2000).

7.1.4 Shipping as a Mechanism for Invasive Species Transport

Shipping accounts for approximately 90% of global trade and that trade has more than doubled over the past 30 years (Fig. 7.1; Bax et al. 2003; Dafforn et al. 2011). Shipping is one of the primary vectors for the transport of non-indigenous species

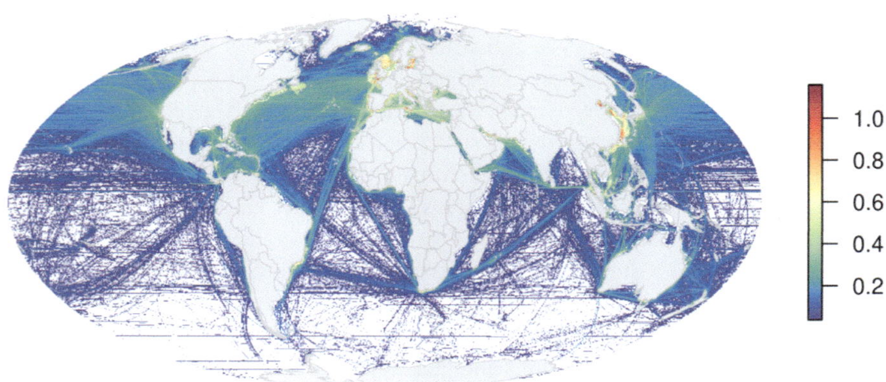

Fig. 7.1 Shipping routes and human impact. Colors indicate the increased human impact including invasive species transport. (Halpern et al. 2013)

(Fig. 7.2; Ruiz et al. 1997; Molnar et al. 2008; Piola et al. 2009; Hewitt and Campbell 2010). In the United States alone, 200 of 450 reported NIS were transported through shipping, either in ballast water or as hull fouling. The transport of 30% of those 200 species was attributed solely to biofouling (Ruiz et al. 2015). Ports and harbors are known to be hot spots for the occurence and high abundance of NIS (Fig. 7.3; Ruiz et al. 1997; Glasby et al. 2007; Piola et al. 2009). This is due to their susceptibility to invasion because most ports and harbors have vacant or underutilized ecological niches, depauperate native communities, high levels of disturbance, low habitat complexity, are subject to many transport vectors and therefore repeated inoculation and have a high proportion of anthropogenic substrates which favor NIS (Ruiz et al. 1997; Cohen and Carlton 1998; Mack et al. 2000; Hutchings et al. 2002; Floerl and Inglis 2005; Wasson et al. 2005; Glasby et al. 2007; Tyrrell and Byers 2007; Westphal et al. 2008; Dafforn et al. 2009a). In many regions as much as 55–85% of recorded NIS are attributed to transport on vessel hulls and floating structures (Piola et al. 2009). The rate of introductions is increasing (Bax et al. 2003). A conservative estimate of the wetted surface area of the global commercial fleet was calculated as 325×10^6 m^2 excluding niche areas (Moser et al. 2016). Niche areas are often disproportionately heavily fouled when compared to the hull surface because they are protected from the strongest hydrodynamic stresses and may have damaged or missing antifouling protection (Coutts and Taylor 2004; Davidson et al. 2009; Ruiz et al. 2015; Moser et al. 2016).

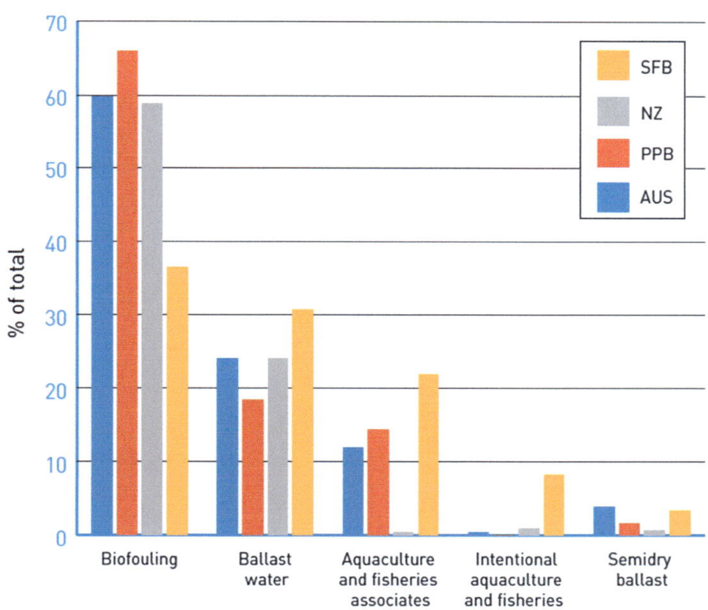

Fig. 7.2 Evaluation of historic marine invasions by vector split into regions: *AUS* Australia, *PPB* Port Philip Bay, *NZ* New Zealand, *SFB* San Francisco Bay. (Hewitt and Campbell 2010)

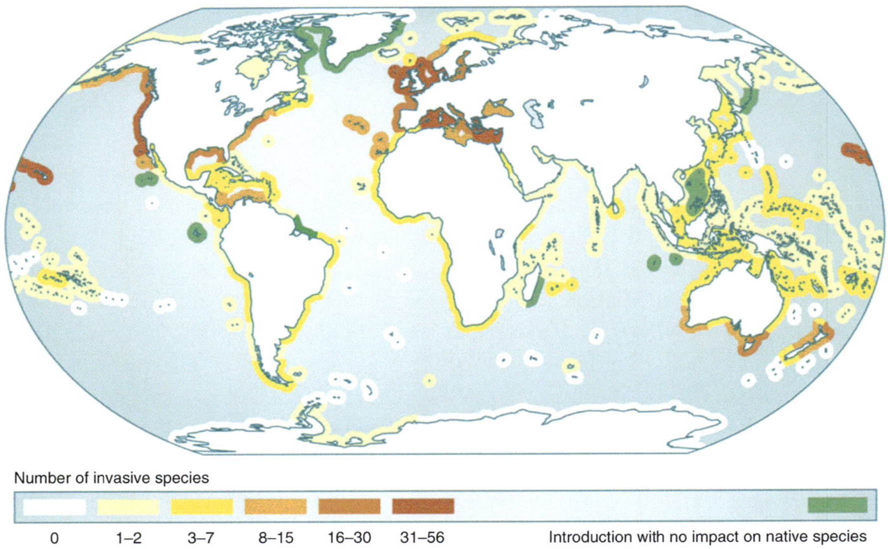

Fig. 7.3 Map showing the number of NIS by coastal ecoregion. Darker colors indicate a greater number of species with a large ecological impact. (World Ocean Review 2010)

In some ports, the community is dominated by NIS. In San Francisco Bay, for example, virtually every coastal habitat is dominated by NIS. In habitats as diverse as in- and epifaunal soft bottom benthos, fouling communities, brackish water zooplankton and fish fauna, NIS may account for 40–100% of the common species and make up to 99% of the biomass and account for 97% of the total species (Cohen and Carlton 1998; Bax et al. 2003). In North America, as much as 75% of NIS transport can be attributed to shipping. Seventy-four to ninety percent of NIS in the Hawaiian islands, 42% of accidental introductions to Japan, 69% of NIS to New Zealand, 78% of introduced species in Port Philip Bay, Australia, more than half of North Sea introductions and the majority of introductions in Britain and Europe were transported as biofouling (Jackson 2008; Ruiz et al. 2015).

For an organism to successfully invade a novel habitat, it must first survive a series of events culminating in establishment and spread in its new habitat (Fig. 7.4). This process involves four steps: (1) Entrainment, (2) Transport, (3) Establishment and (4) Spread. Because of strong selective pressures, only a small proportion of organisms survives each stage despite a large potential propagule pool at the source location (Sakai et al. 2001; Piola et al. 2009; Hopkins and Forrest 2010). The first two stages, entrainment and transport represent areas where the invasion process can be prevented, whereas the latter two become an eradication issue (Sakai et al. 2001). Prevention is always preferred because of the high cost and near impossibility of eradicating an established invader (Bax et al. 2003; Molnar et al. 2008; Westphal et al. 2008). In fact, there is only one proven incident of a successful eradication which occurred in Australia in 1999. This was the successful elimination of several species of invasive mussel from several marinas in Darwin at the cost of 2.2 M Au

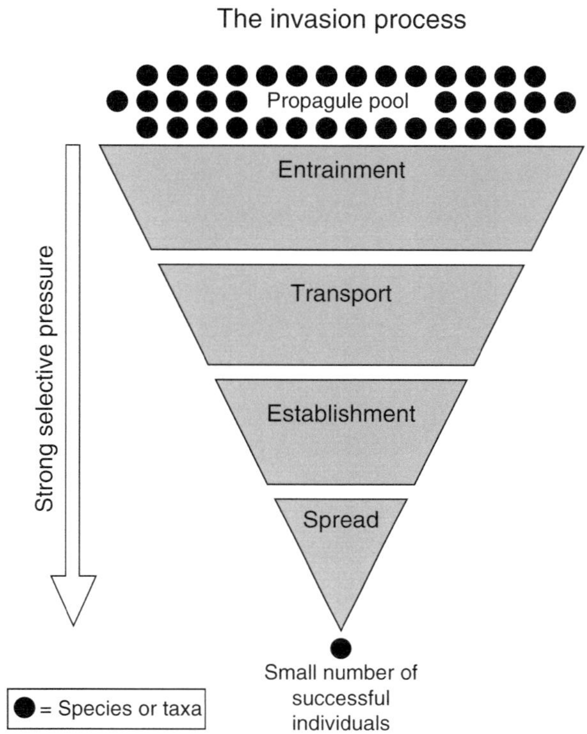

Fig. 7.4 Graphical representation of the invasion process. (Piola et al. 2009, www.tandfonline.com)

(Hutchings et al. 2002; Bax et al. 2003). Once an organism is introduced to a new location, it must become established. Often NIS are favored over native species because they can tolerate a highly polluted harbor and prefer to settle on anthropogenic structures (Tyrrell and Byers 2007; Piola and Johnston 2008; Dafforn et al. 2009b; Crooks et al. 2011). Once an NIS is established, it may begin to expand its range after some lag period (Sakai et al. 2001; Piola et al. 2009). This range expansion or spread is often facilitated by the movement of recreational vessels (Apte et al. 2000; Wasson et al. 2005; Wyatt et al. 2005; Piola et al. 2009). Recently, much work has been done recently to predict the risk of invaders on a ship hull. Factors including number of regions visted, time in previous port of call and time since last antifouling application were found to predict potential propagule pressure (Sylvester et al. 2011).

7.2 Hull Husbandry

The risk of accumulating biofouling can be minimized through optimizing a ships operations and scheduling, however, slow operating speeds, and long durations in port greatly increase the risk of accumulating biofouling. Ship owners minimize the risk associated with ship hull fouling through hull husbandry, the general maintenance and upkeep of the underwater portions of a ship hull. Hull husbandry techniques to combat biofouling include ship hull cleaning and grooming.

7.2.1 Ship Hull Cleaning

Mechanical removal of fouling through in-water hull cleaning is a reactive approach to help reduce the impact due to fouling on ship hulls. In-water cleaning allows for the removal of fouling without drydocking, thus minimizing costs. Properly done hull cleanings can extend the life of a hull coating system, and return the performance of the ship to a condition similar to a clean hull (Schultz et al. 2011). Cleaning cycles will be dependent on the ship duty cycle, coating type, coating area, and biofouling coverage. Slow operating speeds and long durations in port and between scheduled maintenance greatly increase the risk of accumulating biofouling. One such example are Naval ships, which sit pier side for long periods of time, giving them a high risk for biofouling and invasive species accumulation. Routine underwater hull inspections are performed to determine the extent of fouling and coating damage, as set forth in the US Navy's Naval Ships' Technical Manual (NSTM) (US Navy 2006). Divers will record the Fouling Rating (FR) on several locations along the ship hull, struts, rudders, propellers, and seachests. Fouling density and composition will vary depending on ship class, port of call, coating type, and coating condition. Because of this, the Navy does not specify routine hull cleaning. Instead the decision to clean is based on the regular in-water hull inspections (US Navy 2006). The NSTM calls for a full hull cleaning when a ship has FR 40 over 20% of the hull (antifouling coatings) or a FR 50 over 10% of the hull (fouling release coatings) (US Navy 2006). FR40 refers to the growth of small calcareous fouling (or weed) less than 6 mm in diameter or height. FR 50 refers to calcareous fouling in the form of barnacles less than 6.4 mm in diameter or height. In addition to regular inspections, the NSTM also outlines several performance indicators for the presence of biofouling accumulation which call for an underwater hull inspection.

1. The loss of 1kt of speed with shaft rpm set to standard speed
2. An increase in excess of 5% fuel required to maintain a specific speed
3. An increase in shaft RPM of 5% to maintain a given speed

In-water hull cleaning on both Navy and commercial ships is conducted by commercial divers or remotely operated vehicles (ROVs). The most common method to clean is by the use of hydraulically driven brushes, however, high pressure and cavitating water jets may also be used (Fig. 7.5). The brushes are stiff and often made of polypropylene or wire, to be able to dislodge attached fouling organisms from the hull. Divers or ROV pilots must try to delicately balance the forces imparted by the cleaning vehicle to remove the maximum amount of fouling while minimizing damage to the antifouling coating and ship hull.

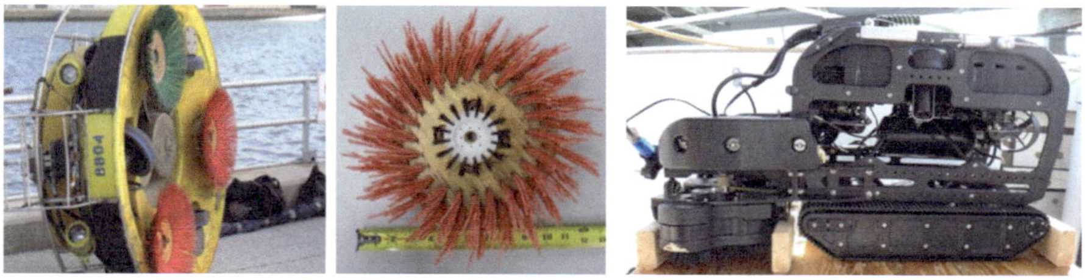

Fig. 7.5 Hull Husbandry Tools. A SCAMP vehicle (left) is commonly used to clean ship hulls. Brushes are often stiff (larger red brush, middle) compared to grooming brushes which are small and gentle on the surface (smaller black brush, middle). Grooming is conducted using a ROV which has several brushes mounted to the front (right)

7.2.2 *Robotic Ship Hull Grooming*

Grooming or the gentle, habitual, and mechanical maintenance of a ship hull, has been proposed as a viable method to reduce biofouling on ship hulls (Tribou and Swain 2010, 2015, 2017). Grooming maintains a ship hull free of extraneous material such as fouling and particulate debris, without damaging the coating. This is done by way of an underwater vehicle that wipes the surface with gentle brushes to dislodge potential fouling species before they have the time to become fully established (Fig. 7.5). The majority of the cost associated with fouling is the increased frictional drag and the resultant fuel consumption (Schultz et al. 2011). By controlling the macrofouling, grooming decreases a ship's fuel consumption, allowing for increased time at sea. It is estimated that active grooming of a ship hull to maintain a light biofilm layer has the potential to save 12M USD per ship over a fifteen-year period (Schultz et al. 2011). In addition, grooming disturbs the surface of the ship hull, thus removing the fouling at the juvenile and settlement stages, thus eliminating recruitment and the potential of invasive species transport.

Several long-term studies have proven grooming to be effective at controlling the macrofouling growth on antifouling and fouling release coatings (Hearin et al. 2015, 2016). A grooming regiment of once per week has been found to keep the macrofouling community from developing, on a range of commercially available coatings in an aggressive fouling environment.

7.3 Large Scale Test Facility for the Development of In-Water Hull Maintenace

A large-scale seawater testing facility (LSTF) was developed in Port Canaveral, Florida for the testing and development of hull husbandry tools, methods and technology (Hearin et al. 2015, 2016). The LSTF has four large steel test assemblies which are coated with Navy qualified coatings to represent a ship hull near the water line. The large test assemblies are fabricated from a 4.6 m by 3.4 m plate welded to a

.76 m diameter pipe for floatation. Two assemblies are bolted end to end to provide a large surface area, 9.2 m by 2.4 m, for testing tools and ROVs. The LSTF is located in an area of high fouling activity, and thus provides intensive year-round conditions to which coatings and grooming regimes are subjected.

7.3.1 A Case Study of Hull Cleaning

A test section from the LSTF coated with an ablative copper antifouling coating was allowed to freely foul for 23 months, to achieve a FR100 over 90% of the surface. The test section was then cleaned using a hand-held tool which included a polypropylene brush, to determine the impact the cleaning would have on coating. Figure 7.6 shows underwater photographs of the coating before and after cleaning. Fouling before cleaning was severe, consisting of calcareous organisms (barnacles and tube worms) over 90% of the coating (Fig. 7.6, left). In cases where divers applied enough force with the brush to remove all fouling organisms, damage also resulted to the antifouling coating. This can be seen in Fig. 7.6 (middle) in which the topcoat has been worn away, exposing the epoxy anti-corrosion coating. When divers applied just enough force with the brush to remove fouling while preserving the coating, some calcareous growth (barnacles and barnacle baseplates) remained, (Fig. 7.6, right). The barnacles present after cleaning are the striped acorn barnacle (*Balanus amphitrite*), an invasive species which is known to be copper tolerant (Weiss 1947).

7.3.2 A Case Study of Ship Hull Grooming

In 2015, a long-term grooming study was undertaken to assess the performance of several commercially available ship hull coatings. One fouling release coating (FR) and one antifouling coating (AF) were applied to replicate 15.2 × 30.5 cm panels, which were then attached to one of the large test assemblies. Grooming was conducted using a nine-brush system mounted to a ROV, which allowed for a groomed swath of 55.9 cm (Fig. 7.5). The polypropylene brushes rotate at

Fig. 7.6 Cleaning Results. (Left) Fouling before cleaning FR 100100%, (middle) anti-fouling coating damage, (right) fouling remaining after cleaning

approximately 500 RPM and a grooming regime of once per week was chosen, based on previous testing at the LSTF (Hearin et al. 2015, 2016). The same coatings were applied to an additional panel set, which were immersed and remained ungroomed throughout the course of the experiment.

Grooming maintained surfaces had significantly less fouling than the ungroomed treatments, throughout the course of the experiment (Fig. 7.7). After 2 years, all ungroomed panels had a significantly greater coverage of macrofouling organisms (Figs. 7.8 and 7.9). The groomed antifouling coating had minimal biofilm (less than 10%). The groomed fouling release coating were maintained in a fouling free condition, but at times had tubeworms and encrusting bryozoans present (Fig. 7.7). These tests demonstrated that a grooming regime of once per week is sufficient to keep antifouling coatings free of fouling, but the fouling release coatings may need a more frequent grooming cycle or a higher RPM to prevent settlement of organisms such as tubeworms and encrusting bryozoans, especially in an aggressive fouling environment.

7.3.3 Hull Maintenance and Ivasive Species

Table 7.1 provides a list of the benthic invaders found the LSTF, all of which have a potential to foul ship hulls. The presence of the ship hull coatings alone, was able to deter the settlement of certain organisms. For example, no settlement of the Asian Green Mussel *(Perna viridis)* was seen throughout the course of the study on the test

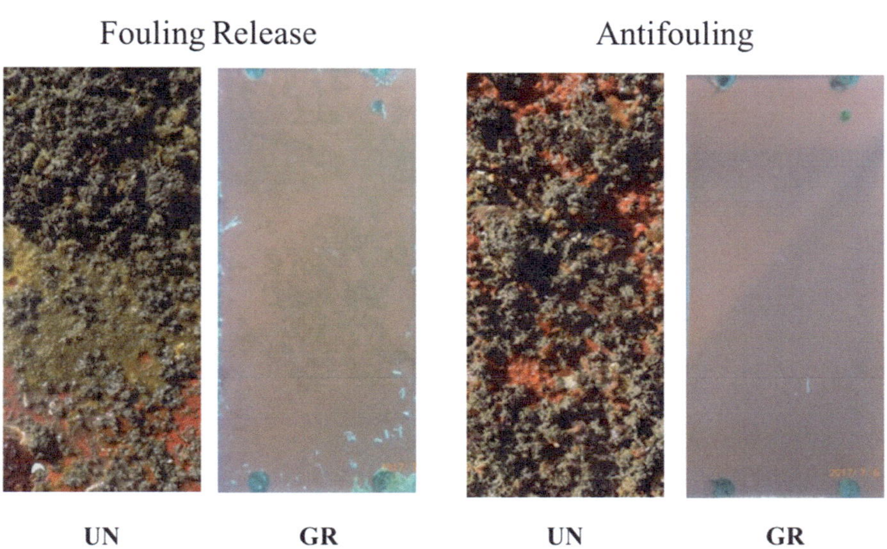

Fig. 7.7 Representative photographs of fouling release and antifouling test coatings after 2 years of immersion. Coatings were either ungroomed (UN) or groomed (GR)

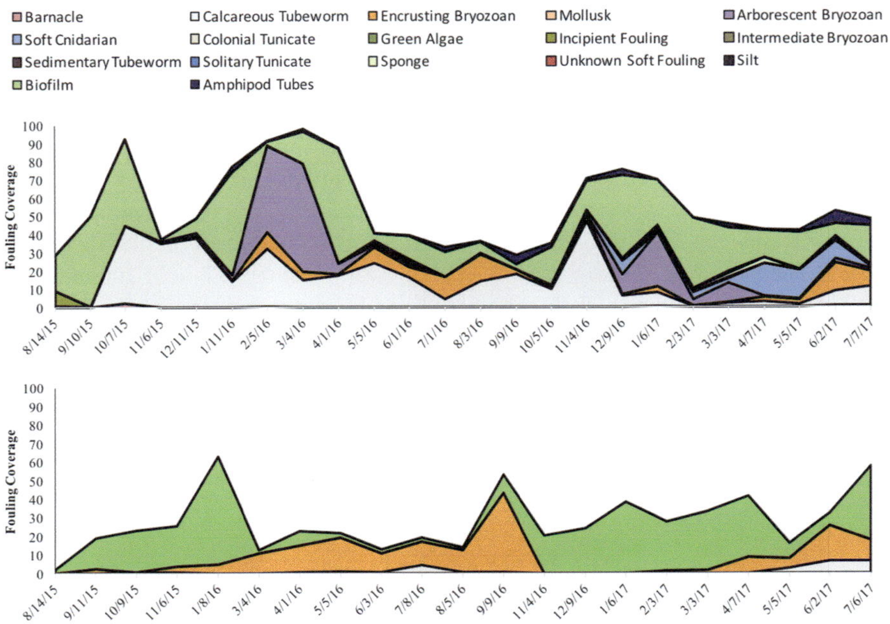

Fig. 7.8 Monthly fouling on a fouling release coating over a 2 year period on ungroomed (top) and groomed (bottom)

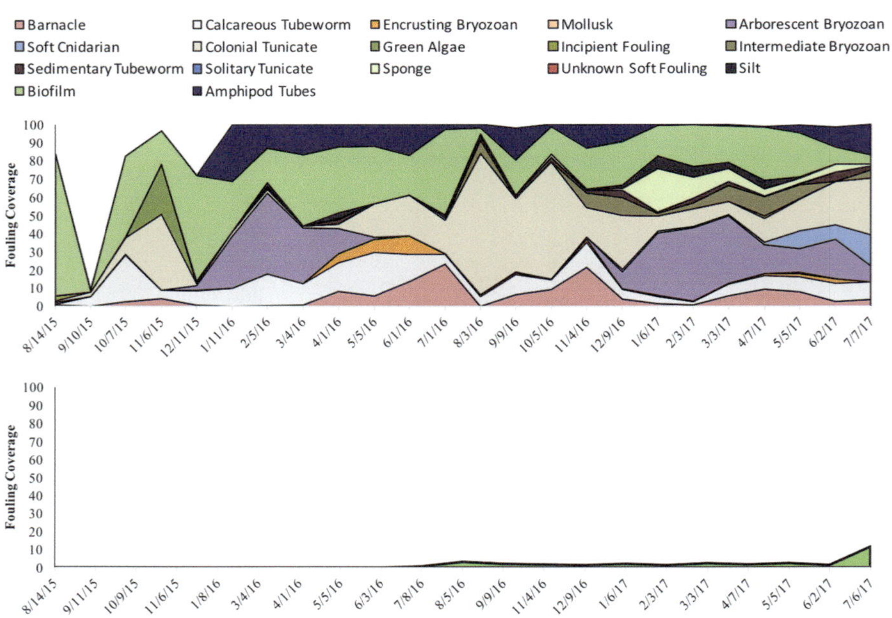

Fig. 7.9 Monthly fouling on an antifouling coating over a 2 year period on ungroomed (top) and groomed (bottom)

panels. (Note: The fouling in Port Canaveral is commonly dominated by colonial tunicates, which can often be a sign of an invaded community. However, without the proper genetic techniques it is often hard to provide a positive identification. Thus the list provided could be even larger.)

After 2 years of immersion, the abundance of the invasive organisms on the groomed versus ungroomed coatings was recorded (Table 7.1). Grooming was able to decrease the overall density of the invasive species on the coatings, and on the antifouling coating prevent them from becoming established. The ungroomed

Table 7.1 Invasive species present in Port Canaveral on a fouling release and antifouling coating, with their average abundance and standard deviation on the ungroomed and groomed treatments

		Fouling release		Antifouling	
Species name	Common name	Ungroomed	Groomed	Ungroomed	Groomed
Amphibalanus reticulatus	Reticulated Barnacle				
Balanus amphitrite	Striped Acorn Barnacle	0.33 ± 0.82		0.5 ± 0.84	
Balanus trigonus	Triangle Barnacle				
Botryllus schlosseri	Star Tunicate				
Bugula neritina	Arborescent bryozoan	0.50 ± 1.2		0.17 ± 0.41	
Diadumene lineata	Striped Sea Anemone				
Diplosoma listerianum	Colonial tunicate				
Hydroides elegans	Serpulid Worm	4.8 ± 1.6	1.7 ± 1.5	35.0 ± 13.4	
Lyrodus medilobatus	Shipworm				
*Lyrodus pedicellatus***	Blacktip Shipworm				
Megabalanus coccopoma	Titan Acorn Barnacle				
Mytella charruana	Charru Mussel				
*Paradella dianae***	Shipworm				
Perna viridis	Asian Green Mussel				
Styela plicata	Pleated Sea Squirt				
*Teredo bartschi***	Shipworm				
*Teredo furcifera***	Shipworm				
Teredo navalis	Naval Shipworm				
*Victorella pavida***	Bryozoan				
*Watersipora subtorquata complex***	Encrusting Bryozoan	9.0 ± 3.5	15.3 ± 17.6	1.00 ± 1.3	
*Zoobotryon verticillatum***	Bryozoan			0.8 ± 2.0	

**denotes cryptogenic species

antifouling coating had the presence of several invaders: *Balanus amphitrite, Bugula neritina, Hydroides elegans, Watersipora subtorquata complex,* and *Zoobotryon verticillatum.* Whereas the panels exposed to weekly grooming did not have any of the above, and remained free of macrofouling species altogether. The ungroomed fouling release coatings also accumulated several invasive species: *B. amphitrite, B. neritina, H. elegans,* and *W.subtorquata.* Grooming was able to eliminate the *B. amphitrite, B.neritina,* and reduce *H.elegans.* This demonstrates that a grooming regime of once per week is sufficient to keep antifouling coatings free of fouling.

7.4 Discussion

Biofouling is problematic for the shipping industry because it leads to many functional and financial setbacks. One of the major concerns with intact biofouling on ship hulls, is the transport of invasive species. While there are many ship hull coatings on the market, there is no perfect system. The antifouling coatings have been widely used (especially copper based coatings) and organisms have been able to adapt to the toxins, often proliferating on these coatings which are meant prevent fouling. Fouling release coatings offer an environmentally friendly alternative to putting biocides in the marine environment. Organisms are able to settle on the surface, but due to low surface energy, the organisms have a weaker attachment. This would allow organisms to be removed via a cleaning process or once a ship is underway. However, fouling release coatings are often softer and damage can occur through in-water cleanings and via organisms undercutting the top coating layer. In addition, studies have begun to show that many organisms remain attached to the fouling release coatings even at high speeds, leading to the transport of invasive species. Grooming, or the frequent gentle wiping of the ship hull, offers a protective approach that works in synergy with both antifouling and fouling release coatings. While disturbing the biofouling species in the juvenile or settlement stage, grooming prevents them from becoming established. Cleaning is another alternative, but as discussed below, has some consequences that may enhance the transport of non-indigenous species.

Hull cleaning has the potential to completely remove biofouling from a ship hull, retuning the condition to that similar of a clean hull, thus eliminating the potential for invasive species transport. Hull cleaning can also save money, as it reduces the hull husbandry costs associated with drydocking. However, it is a reactive measure and often times a complete hull cleaning is not performed. The NSTM recommends several types of cleanings to restore ship performance: a full cleaning (complete fouling removal from hulls, propellers, shafts, struts, rudders, all openings), an interim cleaning (fouling removal from propellers, shafts, struts, and rudders), and a partial cleaning (selected sections of the ship hull) (US Navy 2006, Schultz et al. 2011). Costs for cleanings alone are not inexpensive. Schultz et al. (2011) estimated costs for a full cleaning to range from $26,200 to $34,200 and $15,000 to $2500 for an interim cleaning. Conversely, grooming could save the Navy US$6.2M to US $12M over cleaning if a hull is maintained at an FR-20 or FR-10, respectively (Schultz et al. 2011). At present there are no cost estimates for grooming.

Hull cleaning is reactive and may damage hull coatings, which will increase the surface roughness and add increased frictional drag. Damaged areas are also prime locations for biofouling settlement and recruitment. Small areas of damage (0.5 cm wide) in a toxic antifouling coating quickly become colonized by a large variety of organisms, potentially leading to translocation of NIS (Piola and Johnston 2008; Piola et al. 2009). Fouling organisms, once established, may not be easy to remove once a ship gets up to speed. At Port Canaveral, the invasive barnacle (*B. Amphitrite)* and encrusting bryozoan *(W. subtorquata)* remained on antifouling coatings even after speeds of 10 m/s (Hunsucker et al. 2017). On fouling release coatings, the invasive encrusting bryozoan *(W. subtorquata*) and calcareous tubeworm (*H. elegans*) also remained intact after similar speeds. A study by Coutts et al. (2010) found encrusting or hard organisms had a high survival rate on a hydrodynamic keel subjected to speeds up to 9 m/s. Even prolonged exposure to high speeds may not remove fouling organisms. Survival of the invasive arborescent bryozoan. *B. neritina*, was not influenced by voyage duration (2–8 days) or voyage speed (6 and 3 and 9 m/s) (Schimanski et al. 2016).

Grooming, when performed at a frequency appropriate for the coating and environmental conditions, may prevent the growth of macrofouling organisms on the ship hull. In water cleaning may leave behind a diverse fouling community, particularly if fouling was heavy before cleaning occurred (Floerl et al. 2005; Davidson et al. 2008; Hopkins and Forrest 2008). This was seen above, in the cleaning case study. In order to preserve the integrity of the coating, a lighter cleaning was applied, thus resulting in barnacles and barnacle base plates remaining. In addition, surfaces that have been cleaned may experience enhanced, selective &/or more rapid recolonization (Floerl et al. 2005; Ralston and Swain 2014). The most common cleaning involves the use of rotating brushes or hand held scrapers, depending on size and vessel type. These are most effective at removing low to moderate fouling from flat surfaces, but tend to be less effective with more advanced fouling or curved surfaces. The fouling that is removed is generally crushed by the brushes, but a diverse range of fouling organisms and fragments may be viable after removal (Hopkins and Forrest 2008; Hopkins et al. 2010). The survival of defouled fragments was found to vary depending on species, fragment size, sedimentation and turbidity conditions and presence of predators, in subsequent experiments (Hopkins et al. 2011).

Copper-based antifouling coatings are still one of the most popular coating types on the market, but unfortunately many groups of marine organisms have been observed to be copper tolerant. In several studies, NIS were found to recruit more readily to surfaces coated with copper antifouling paints. This was attributed to reduced competition from native species, which suffered reduced recruitment in the presence of copper and to an increase in areal coverage of the NIS (Dafforn et al. 2008; Piola and Johnston 2009; Canning-Clode et al. 2011; Crooks et al. 2011). Perhaps the best studied are encrusting bryozoans in the genus *Watersipora*.

Watersipora has been observed to settle directly on copper coated surfaces (Floerl et al. 2004; Piola et al. 2009). These bryozoans may then act as a settlement substrate for organisms that may not be copper tolerant, facilitating transport of additional NIS (Floerl et al. 2004). In addition to *Watersipora*, the arborescent bryozoan *Bugula neritina*, the calcareous tube worm *Hydroides elegans* and the barnacle *Amphibalanus amphitrite* have also been reported to be copper tolerant (Weiss 1947; Floerl et al. 2004; Piola et al. 2009). The source population is critical to successful establishment by potentially tolerant species, as those from heavily polluted sources are more likely to be tolerant (Dafforn et al. 2008; Piola and Johnston 2009; Piola et al. 2009; Crooks et al. 2011; McKenzie et al. 2011). Grooming copper coatings, once a week, has the ability to prevent the settlement of these copper tolerant fouling organisms (Hearin et al. 2015, 2016).

7.5 Conclusions

Over the past decade, the Office of Naval Research (ONR) has been investigating the efficacy of grooming as a proactive method to keep ship hulls free of fouling. Eliminating fouling will allow for reductions in fuel consumption, frictional drag, greenhouse gas emissions, and invasive species transport. Research has been underway to develop an autonomous underwater hull crawler which will push brushes along the ship hull. These soft brushes rotate at speeds up to 500 RPM to provide the forces needed to disturb the recruitment of biofouling species. Long-term studies have shown that a grooming frequency of once per week is enough to prevent macrofouling buildup on antifouling and fouling release coatings. The frequency and brush rotational speed will vary depending on the ship hull coating and environmental conditions. The shipping industry is attributed to being one of the primary vectors for the transport of non-indigenous species. A marriage of grooming along with ship hull coatings, can act in synergy to eliminate fouling and thus prevent the distribution of potential harmful invasive species worldwide.

References

Apte S, Holland B, Godwin L, Gardner J (2000) Jumping ship: a stepping stone event mediated transfer of a non-indigenous species via a potentially unsuitable environment. Biol Invasions 2:75–79

Bax N, Williamson A, Aguero M, Gonzalez E, Geeves (2003) Marine invasive alien species: a threat to global biodiversity. Mar Policy 27:313–323

Brock R, Bailey-Brock J, Goody J (1999) A case study of efficacy of freshwater immersion in controlling introduction of alien marine fouling communities: the USS Missouri. Pac Sci 53:223–231

Campbell M, Hewitt C (2011) Assessing the port to port risk of vessel movements vectoring non-indigenous marine species within and across domestic Australian borders. Biofouling 27:631–644

Canning-Clode J, Fofonoff P, Riedel G, Torchin M, Ruiz G (2011) The effects of copper pollution on fouling assemblage diversity: a tropical-temperate comparison. PLoS One 6:e18026. https://doi.org/10.1371/journalpone.0018026

Chambers LD, Stokes KR, Walsh FC, Wood RJK (2006) Modern approaches to marine antifouling coatings. Surf Coat Technol 201:3642–3652

Cohen A, Carlton J (1998) Accelerating invasion rate in a highly invaded estuary. Science 279:555–558

Coutts A, Taylor M (2004) A preliminary investigation of biosecurity risks associated with biofouling on merchant vessels in New Zealand. N Z J Mar Freshwat Res 38:215–229

Coutts ADM, Piola RF, Hewitt CL, Connell SD, Gardner JPA (2010) Effect of vessel speed on survival of biofouling organisms: implications for translocation of non-indigenous marine species. Biofouling 26(1):1–13

Crooks J, Chang A, Ruiz G (2011) Aquatic pollution increases the relative success of invasive species. Biol Invasions 13:165–176

Dafforn K, Glasby T, Johnston E (2008) Differential effects of tributyltin and copper antifoulants on recruitment of non-indigenous species. Biofouling 24:23–33

Dafforn K, Johnston E, Glasby T (2009a) Shallow moving structures promote marine invader dominance. Biofouling 25:277–287

Dafforn K, Glasby T, Johnston E (2009b) Links between estuarine condition and spatial distributions of marine invaders. Divers Distrib 15:807–821

Dafforn K, Lewis J, Johnston E (2011) Antifouling strategies: history and regulation, ecological impacts and mitigation. Mar Pollut Bull 62:453–465

Davidson I, McCann L, Sytsma M, Ruiz G (2008) Interrupting a multi-species bioinvasion vector: the efficacy of in-water cleaning for removing biofouling on obsolete vessels. Mar Pollut Bull 56:1538–1544

Davidson I, Brown C, Sytsma M, Ruiz G (2009) The role of containerships as transfer mechanisms of marine biofouling species. Biofouling 25:645–655

Drake JM, Lodge DM (2007) Hull fouling is a risk factor for intercontinental species exchange in aquatic ecosystems. Aquat Invasions 2:121–131

Ferreira C, Goncalves J, Coutinho R (2004) Ship hulls and oil platforms as potential vectors to marine species introduction. J Coast Res 39:1340–1345

Floerl O, Inglis G (2005) Starting the invasion pathway: the interaction between source populations and human transport vectors. Biol Invasions 7:589–606

Floerl O, Pool T, Inglis G (2004) Positive interactions between nonindigenous species facilitate transport by human vectors. Ecol Appl 14:1724–1736

Floerl O, Inglis G, Marsh H (2005) Selectivity in vector management: an investigation of the effectiveness of measures used to prevent transport of non-indigenous species. Biol Invasions 7:459–475

Glasby T, Connell S, Holloway M, Hewitt C (2007) Nonindigenous biota on artificial structures: could habitat creation facilitate biological invasions. Mar Biol 151:887–895

Halpern B et al (2013) Cumulative human impact on the ocean as a result of shipping. National Center for Ecological Analaysis and Synthesis. https://www.nceas.ucsb.edu/globalmarine/panel/impactbyactivity

Hearin J, Hunsucker KZ, Swain G, Stephens A, Gardner H, Lieberman K, Harper M (2015) Analysis of long-term mechanical grooming on large-scale test panels coated with an antifouling and a fouling release coating. Biofouling 31(8):625–638

Hearin J, Hunsucker KZ, Swain G, Gardner H, Stephens A, Lieberman L (2016) Analysis of mechanical grooming at various frequencies on a large scale test panel coated with a fouling-release coating. Biofouling 32(5):561–569

Hewitt C, Campbell M (2010) The relative contribution of vectors to the introduction and translocation of marine invasive species. Report commissioned by The Australian Department of Agriculture, Fisheries and Forestry, p 42
Hopkins G, Forrest B (2008) Management options for vessel hull fouling: an overview of risks posed by in-water cleaning. ICES J Mar Sci 65:811–815
Hopkins G, Forrest B (2010) Challenges associated with pre-border management of biofouling on oil rigs. Mar Pollut Bull 60:1924–1929
Hopkins G, Forrest B, Coutts A (2010) The effectiveness of rotating brush devices for management of hull fouling. Biofouling 26:555–566
Hopkins G, Forrest B, Piola R, Gardner J (2011) Factors affecting survivorship of defouled communities and the effect of fragmentation on establishment success. J Exp Mar Biol Ecol 396:233–243
Hunsucker KZ, Gardner H, Swain G (2017) Using hydrodynamic testing to assess the performance of five fouling control coatings immersed at two field sites along the East Coast of Florida. American Towing Tank Conference, Bethesda
Hutchings P, Hilliard R, Coles S (2002) Species introductions and potential for marine pest invasions into tropical marine communities, with special reference to the indo-Pacific. Pac Sci 56:223–233
International Maritime Organization (2001) International convention on the control of harmful antifouling systems on ships. London, 18 October 2001
International Maritime Organization (2006) International shipping and world trade. Facts and figures. See http://www.imo.org/
Jackson L (2008) Marine biofouling: an assessment of risks and management initiatives. Report for global invasive species programme, 68 pp
Mack R, Simberloff D, Lonsdale W, Evans H, Clout M, Bazzaz F (2000) Biotic invasions: causes, epidemiology, global consequences, and control. Ecol Appl 10:689–710
McKenzie L, Brooks R, Johnston E (2011) Heritable pollution tolerance in a marine invader. Environ Res 111:926–932
Molnar J, Gamboa R, Revenga C, Spalding M (2008) Assessing the global threat of invasive species to marine biodiversity. Front Ecol Environ 6:485–492
Moser C, Wier T, Grant J, First M, Tamburri M, Ruiz GM, Miller AW, Drake LA (2016) Quantifying the total wetted surface area of the world fleet: a first step in determining the potential extent of ships' biofouling. Biol Invasions 18:265–277
Omae I (2003) General aspects of tin-free antifouling paints. Chem Rev 103:3431–3448
Piola R, Johnston E (2008) The potential for translocation of marine species via small-scale disruptions to antifouling surfaces. Biofouling 24:145–155
Piola R, Johnston E (2009) Comparing differential tolerance of native and non-indigenous marine species to metal pollution using novel assay techniques. Environ Pollut 157:2853–2864
Piola R, Dafforn K, Johnston E (2009) The influence of antifouling practices on marine invasions. Biofouling 25:633–644
Ralston E, Swain G (2014) The ghost of fouling communities past: the effect of original community on subsequent recruitment. Biofouling 30:458–471
Ruiz G, Carlton J, Grosholz E, Hines A (1997) Global invasions of marine and estuarine habitats by non-indigenous species: mechanisms, extent, and consequences. Am Zool 37:621–632
Ruiz GM, Rawlings TK, Dobbs FC, Drake LA, Mullady T, Huq A, Colwell RR (2000) Global spread of microorganisms by ships. Nature 408:49–50
Ruiz G, Fofonoff P, Steves B, Carlton J (2015) Invasion history and vector dynamics in coastal marine ecosystems: a North American perspective. Aquat Ecosyst Health Manag 18:299–311
Sakai A, Allendorf F, Holt J, Lodge D, Molofsky J, With KA, Baughman S, Cabin RJ, Cohen JE, Ellstrand NC, McCauley DE, O'Neil P, Parker IM, Thompson JN, Weller SG (2001) The population biology of invasive species. Annu Rev Ecol Syst 32:305–332
Sammarco P, Atchison A, Boland G (2004) Expansion of coral communities within the Northern Gulf of Mexico via offshort oil and gas platforms. Mar Ecol Prog Ser 280:129–143

Schimanski KB, Piola RF, Goldstein SJ, Floerl O, Grandison C, Atalah J, Hopkins GA (2016) Factors influencing the en route survivorship and post-voyage growth of a common ship biofouling organism, Bugula neritina. Biofouling 32(8):969–978
Schultz MP, Bendick JA, Holm ER, Hertel WM (2011) Economic impact of biofouling on a naval surface ship. Biofouling 27(1):87–98
Swain G (1999) Redefining antifouling coatings. J Prot Coating Lining 16(9):26–35
Swain G (2010) The importance of ship hull coatings and maintenance as drivers for environmental sustainability. In: Proceedings of Ship Design Operation Environmental Sustainability. RINA, London, 10–11 March, 2010
Swain GW, Kovach B, Touzot A, Casse F, Kavanagh CJ (2007) Measuring the performance of today's antifouling coatings. J Ship Production 23(3):164–170
Sylvester F, Kalaci O, Leung B, Lacoursiere-Roussel A, Murray C, Choi FM, Bravo MA, Therriault TW, HJ MI (2011) Hull fouling as an invasion vector: can simple models explain a complex problem? J Appl Ecol 48:415–423
Thomas KV, Brooks S (2010) The environmental fate and effects of antifouling paint biocides. Biofouling 26(1):73–88
Tribou M, Swain G (2010) The use of proactive in-water grooming to improve the performance of antifouling coatings. Biofouling 26(1):47–56
Tribou M, Swain G (2015) Grooming using rotating brushes as a proactive method to control ship hull fouling. Biofouling 31(4):309–319
Tribou M, Swain G (2017) The effects of grooming on a copper ablative coating: a six year study. Biofouling 33(6):494–504
Tyrrell M, Byers J (2007) Do artificial substrates favor nonindigenous fouling species over native species? J Exp Mar Biol Ecol 342:54–60
US Navy (2006) Naval ships' technical manual chapter 081:waterborne underwater hull cleaning of Navy ships (5th rev). US Naval Sea Systems Command. Washington (DC). Pub#S9086-CQ-STM-010/CH-081
Wanless R, Scott S, Sauer W, Andrew T, Glass J, Godfrey B, Griffiths C, Yeld E (2010) Semi-submersible rigs: a vector transporting entire marine communities around the world. Biol Invasions 12:2573–2583
Wasson K, Fenn K, Pearse J (2005) Habitat differences in marine invasions of Central California. Biol Invasions 7:935–948
Weiss C (1947) The comparative tolerances of some fouling organisms to copper and mercury. Biol Bull 93:56–63
Westphal M, Browne M, McKinnon K, Noble I (2008) The link between international trade and the global distribution of invasive alien species. Biol Invasions 10:391–398
World Ocean Review (2010) Living with the oceans. A report on the state of world's oceans. Maribus, Hamburg, 232p
Wyatt A, Hewitt C, Walker D, Ward T (2005) Marine introductions in the Shark Bay World Heritage Property, Western Australia: a preliminary assessment. Divers Distrib 11:33–44

Part II
Localized Effects of Individual Coastal Invasives

Chapter 8
Feeding Habits of *Pterois volitans*: A Real Threat to Caribbean Coral Reef Biodiversity

Arturo Acero P., Diana Bustos-Montes, P. Pabón Quintero, Carlos Julio Polo-Silva, and A. Sanjuan Muñoz

Abstract Lionfish consume at least 250 fish and crustaceans prey species in the western Atlantic. Main taxa eaten include grunts (*Haemulon aurolineatum)*, wrasses (*Thalassoma bifasciatum* and *Halichoeres* spp.), damselfishes *(Stegastes partitus* and *Chromis cyanea*), gobies (*Coryphopterus personatus)*, labrisomids (*Malacoctenus triangulatus*) and *Pterois volitans*. Because lionfish prey on such a long list of Caribbean reef fauna it should be considered a generalist invasive species that even threats commercially and ecologically important species such as grunts, groupers, snappers, triggerfishes, parrotfishes, surgeonfishes, gobies, lobsters, and cleaner shrimps. Four richness estimators indicate that lionfish may consume around 300 species. Stable isotopes analysis ratifies that most prey eaten by lionfish are reef dwellers. Lionfish diets from the Colombian Caribbean appear distant from the Bahamas and Cayman Island diets in a cluster analysis. Research and monitoring of this dangerous invading species should be maintained.

8.1 Introduction

The negative effects of invasive species have been widely documented. They include threats to biological diversity (Mooney and Cleland 2001; Muñoz et al. 2011) and transformation of marine habitats which may lead to financial losses (Molnar et al. 2008). In marine ecosystems, environmental degradation of habitats, changes in community structure, spread of new diseases and parasites, and decrease of fishery

A. Acero P. (✉) · D. Bustos-Montes · P. Pabón Quintero
Instituto de Estudios en Ciencias del Mar (CECIMAR). Universidad Nacional de Colombia sede Caribe, Santa Marta, Colombia
e-mail: aacerop@unal.edu.co; dmbustosmo@unal.edu.co; pepabonq@unal.edu.co

C. J. Polo-Silva · A. S. Muñoz
Facultad de Ciencias Naturales e Ingeniería, Universidad Jorge Tadeo Lozano, Santa Marta, Colombia
e-mail: carlosj.polos@utadeo.edu.co; adolfo.sanjuan@utadeo.edu.co

C. Makowski, C. W. Finkl (eds.), *Impacts of Invasive Species on Coastal Environments*, Coastal Research Library 29,
https://doi.org/10.1007/978-3-319-91382-7_8

resources have been observed (Morris et al. 2009; Rilov 2009). In addition, bioinvasions occur more frequently (Mooney and Cleland 2001), so invading species establish themselves in new areas where they proliferate, distribute and persist in detriment of native species and ecosystems (Mack et al. 2000).

Invasions carried out by predator species are considered one of the causes of decrease and even local extinction of native species in various parts of the world (Mack et al. 2000). Generalist invasive species can potentially affect a large number of taxa, but it is considered that the risk of them leading to local extinction of a species is much lower than the risk posed by a specialist species that focuses on one or a few prey (Rilov 2009). Nonetheless, recently several studies have shown that generalist species can affect prey populations in a similar way to that of specialist species. Specifically, regarding the reduction and possible extinction of prey populations as generalist species may present trophic specialization of individuals, with subgroups specialized in small niches (Araújo et al. 2011; Clavel et al. 2011).

The red lionfish *Pterois volitans*, and its close relative the devil firefish *Pterois miles*, has invaded the western tropical and temperate Atlantic Ocean since at least 1985 (Morris and Akins 2009; Schofield 2009) (Fig. 8.1). By the end of last century, it was proposed that the invasion was directly related to Hurricane Andrew, which hit Miami in 1994, supposedly destroying some fish tanks close to the sea shore and releasing lionfish individuals. However, it is clear nowadays that the phenomenon is due to numerous releases, intentional or not, of lionfish individuals throughout more than a decade in the Miami area (Betancur-R. et al. 2011). Recently, it has been estimated that the minimum number of lionfish specimens that may have originated the invasion in the Atlantic is about 118 (54–514, 95% HPD), assuming a balanced sex ratio and no Allee effects (Selwyn et al. 2017).

In the western Atlantic, lionfish has a higher density and reaches larger sizes than in its natural distribution area (Darling et al. 2011). Green and Côté (2009) evaluated the abundance of lionfish in reefs of the southeast coast of New Providence (Bahamas) which exceeded the highest documented density for both the Indo-Pacific (~80 adult lionfish/km^2) and the Atlantic Ocean and the Caribbean Sea. The 393.3 ± 144.4 fish/km^2 density is 8–10 times higher than the one found by Whitfield et al. (2007) for the North Carolina coast and five times the one observed by Fishelson (1997) for the Indo-Pacific. Densities found in the Colombian Caribbean do not differ too much from those figures; González-Corredor et al. (2014) reported densities over 379 ± 220 specimens/hectare in the western Caribbean island of San Andrés.

Lionfish also uses hunting strategies different to those of any other predator in the Atlantic (Côté and Maljković 2010; Green et al. 2011; Albins and Lyons 2012; Rojas-Vélez et al. in press) and ingests entire prey that can be as big as half its own size (Fig. 8.2), so it can significantly reduce the recruitment, biomass and richness of native coral fishes (Albins and Hixon 2008; Green et al. 2012a; Albins 2015). In both its original habitat and the invaded one, *P. volitans* is a demersal generalist mesocarnivore. The ecological impacts that the invading species can potentially trigger in the tropical and temperate western Atlantic ecosystems have not been totally determined. For this purpose, it should be understood how the lionfish

Fig. 8.1 Lionfish stalking a school of bigtooth cardinalfish (*Paroncheilus affinis*) in the Magdalena area (Colombia). Photograps by Santiago Estrada R

interacts with other species and how its arrival and establishment affects community structure (Kalogirou et al. 2007; Rilov 2009; Côté and Maljković 2010; Côté et al. 2013a). Arias-González et al. (2011), using trophic models, show that lionfish may have a widespread effect on the coral reef trophic nets, influencing the biomass of native species directly and indirectly. According to those authors, models suggest an increase from 10% to 65% of the fish biomass in absence of lionfish. Through simulations of different management strategies scenarios, the eradication and the recuperation ability of the species was analyzed, showing that without a constant fishing pressure as control measure, not only including adult individuals, the invading fish would have a considerable and irreversible impact on the trophic nets of coral reefs in the Caribbean.

Fig. 8.2 Lionfish captured in Santa Marta locality, regurgitating a peppermint bass (*Liopropoma rubre*, Serranidae). Photograph by Santiago Estrada R

Most studies have focused on identifying lionfish stomach contents, based on traditional taxonomic techniques, with few examples of modern techniques, such as DNA barcoding. Complementarily, Stable Isotopes Analyses (SIA), joint to recently developed analytical tools, represents a commonly employed methodology for the study of trophic interactions in aquatic ecosystems (Pinnegar and Polunin 2000), and a powerful way to depict the trophic characteristics of an organism, upon which niche research relies (Newsome et al. 2007). This technique offers long-term estimates of the prey that have been incorporated into predator tissue, with promising applications in invasion biology (Cucherousset et al. 2007). Stable isotope composition provides an indication of the origin and transformations of organic matter to trace the use and fluxes of resources from the individual to the community level (Newsome et al. 2012).

In this chapter, we review the available literature about feeding habits of lionfish in the Greater Caribbean, including our own work of more than five years studying the biology and the ecology of the invading scorpaenid in the southern and western Caribbean. We emphasize the importance of using all available tools to adequately evaluate and provide elements to mitigate the effects of this deleterious phenomenon.

Fig. 8.3 Colombian Caribbean areas where lionfish was collected. The cluster analysis shows the relationship between areas using Dice index and linkage UPMGA

8.2 Methods

Lionfish material was collected in five Colombian Caribbean regions using spear guns and Hawaiian spears. In the continental Colombian Caribbean specimens were captured in three regions (La Guajira GUA, Magdalena MAG, and Bolívar BOL) from the border with Venezuela on the northeast to the southwestern Colombian Caribbean. The fourth region was the oceanic western Colombian Caribbean (Archipiélago de San Andrés, Providencia y Santa Catalina SAI), from San Andrés in the south to Serrana in the north (Fig. 8.3). An additional region only very recently sampled is the Deep Reef Ecosystem (DRE) located at relatively deep water (50–70 m) in the continental Colombian Caribbean.

Each lionfish specimen was measured and weighted before stomach extraction. Stomach contents were extracted and kept in 70% ethanol. Morphologically, fishes were identified using mainly Carpenter (2002) and Robertson (2015) and crustaceans were determined with Williams (1984) and Carpenter (2002). Additionally, some material was identified using DNA barcoding following Ivanova et al. (2007) and Côté et al. (2013b). To quantify stomach content three methods were used: frequency (%F), number (%N) and weight (%G) (Hyslop 1980). Pinkas et al. (1971)

Index of Relative Importance (IRI) was used to estimate the importance of each item. All published information about lionfish diets was used to compare our findings and to elaborate a complete list of prey items.

Curves of species accumulation were performed using the software Estimates 9.1.0 (Colwell 2013), to calculate how adequate the data is to the sampling effort and to estimate the diet richness by the use of the non-parametric estimators Chao 2, Jacknife 1, and Bootstrap with a confidence interval of 95%. The sampling effort sufficiency was determine plotting the Coefficient of Variation (CV), where a CV 5% indicates the curve stabilization. The estimation was made based on preys contained in 3000 lionfish stomachs that have been examined in the invaded region by Albins and Hixon (2008), Morris and Akins (2009), Côté and Maljković (2010), Jud et al. (2011), Muñoz et al. (2011), Abril (2012), Valdez-Moreno et al. (2012), Côté et al. (2013b), Gómez-Pardo (2014), Villaseñor-Derbez and Herrera-Pérez (2014), Rocha et al. (2015), Eddy et al. (2016), Pabón and Acero P (2016), Pantoja et al. (2017), Romero (2017), and our unpublished data.

For the SIA, approximately 5 g of white muscle were removed from the dorsal anterior section of each fish. The samples were freeze-dried for 24 h, followed by lipid- and urea-extraction following the protocol proposed by Kim and Koch (2012). Analyses were carried out at the Stable Isotope Laboratory of the Estación Experimental del Zaidín (CSIC, Granada, Spain). Results are expressed in delta notation where $d^hX = 1000 \times [(R_{sample}/R_{standard}) - 1]$, where h is the high mass of element X and R_{sample} and $R_{standard}$ are the $^{13}C/^{12}C$ or $^{15}N/^{14}N$ ratios of the sample and standard, respectively. Units are expressed as parts per thousand (‰). The within and between-run standard deviations of an acetanilide standard was $\leq$0.2‰ for both d ^{13}C and d^{15}N values. The weight-percent C: N ratios of each sample were in the expected range for protein (3.1–3.5), indicating that lipid and urea extraction were effective (Post et al. 2007).

8.3 Results and Discussion

8.3.1 A Review of Lionfish Diet in the Greater Caribbean

Studies about lionfish diet based on stomach content have been carried out in several western Atlantic regions such as Bermuda, USA (Carolina and Florida), Bahamas, Cuba, Cayman Islands, Mexico, Belize, and Colombia. Those countries comprise the marine ecoregions defined by Spalding et al. (2007): Bermuda, Bahamian, Southwestern Caribbean, Western Caribbean, Southern Gulf of Mexico, and Floridian. On the other hand, a total of 1292 lionfish stomachs from the Colombian Caribbean, containing 141 prey items (Fig. 8.4), have been examined by us.

At least 250 taxa among fishes, crustaceans, mollusks, echinoderms, and other groups have been registered as lionfish prey (Appendix). The amount of taxa found in every study varied between less than ten to almost one hundred (Table 8.1), depending on the number of analyzed lionfishes (minimal = 52, maximum = 1508)

Fig. 8.4 Fishes, crustaceans, and mollusks identified as lionfish stomach contents in the Colombian Caribbean. In all cases, black bar corresponds to 1 cm. (Photographs by María Camila Rodríguez, Humberto Gómez-Pardo and Josselyn Bryan)

and the specific objectives and methods of every research, showing that lionfish is a generalist and opportunistic predator. It has been determined that lionfish is a mesotrophic predator with an ample trophic niche [Layman and Allgeier 2012: standard ellipse areas = 0.91 (CI: 0.67–1.18); Gómez-Pardo 2014: Hulbert

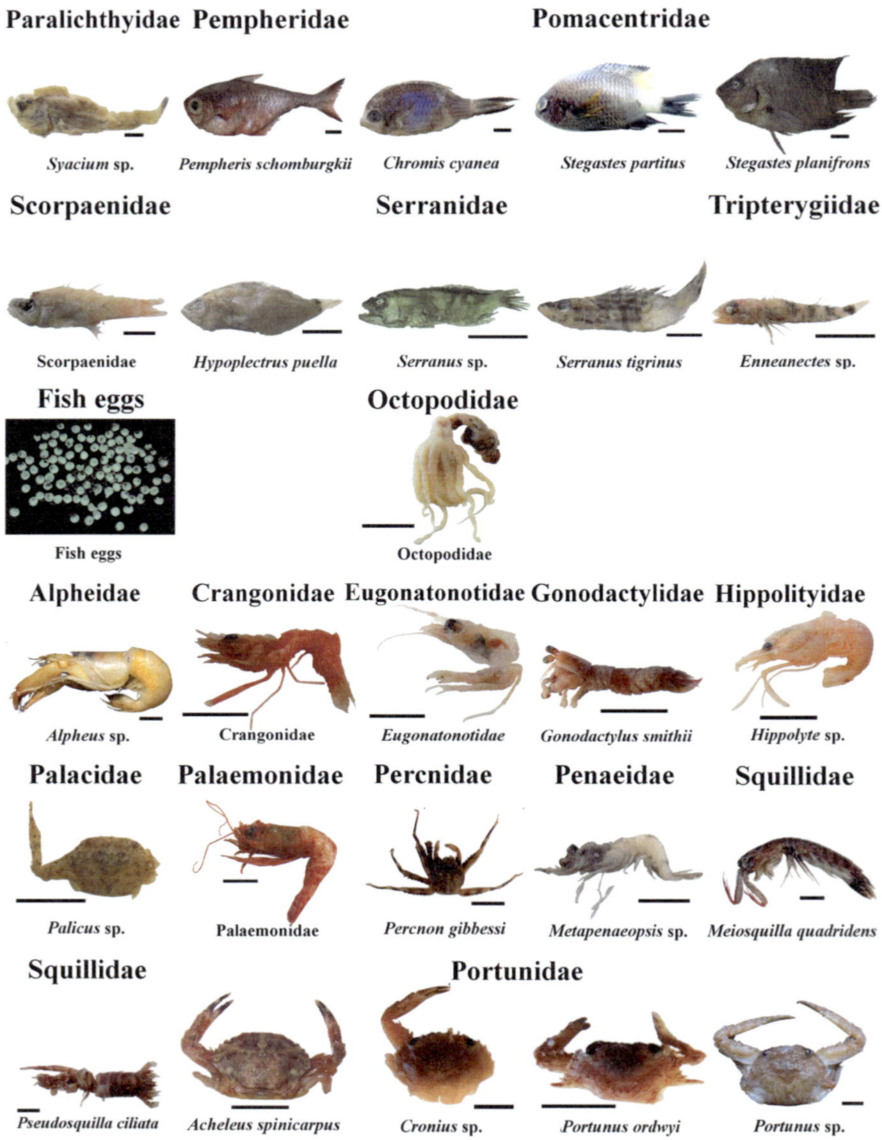

Fig. 8.4 (continued)

Index = 0.91 (CI: 0.84–0.97); Pabón and Acero 2016: Levins Index = 0.97] even though it has a high level of individual specialization (Layman and Allgeier 2012: Similarity Index = 0.34).

The collected data and the four richness estimators show that the curve stabilizes around 300 species. Because for the four estimators the curve stabilizes (CV <5%) with different sampling efforts (Fig. 8.5), it seems that the total sampling effort

Table 8.1 List of diet studies made on lionfish in the western Atlantic, including localities, prey richness and abundances, and lionfish sizes

Author (year)	Lionfish analized	Prey identified	Taxa richness						
			Familes	Total taxa	Fishes	Crustaceans	Mollusks	Equinoderms	Unidentified
Albins and Hixon (2008)	52	12		9	9				
Morris and Akins (2009)	1069	1876	26	50	41	7	2	–	
Côté and Maljković (2010)	192	–	9	20	21	21			
Jud et al. (2011)	71			15					
McCleery (2011)	70	57							
Muñoz et al. (2011)	226	392	16	18	24	3	2	1	
Abril (2012)	181	590		36	17	18	1		
Cure et al. (2012)			12	14	13	1			2
Green et al. (2012)	567		16	42					
Layman and Allgeier (2012)	122			22					
Muñoz-Escobar and Gil-Agudelo (2012)	65	16	6	10	8	2			
Valdez-Moreno et al. (2012)	157	330	14	34	34	20	–	–	
Côté et al. (2013b)	130	397	16	37	37	–	–	–	
Pimiento et al. (2013)	20	24	10	11	6	5	–	–	–
Dahl and Patterson III (2014)	934			88	36	38	14	–	
Gómez-Pardo (2014)	810	273	24	47	27	20	–	–	
Sandel et al. (2014)	373	924	14 (only fishes)	24	17	5	–	–	1
Villaseñor-Derbez and Herrera-Pérez (2014)	109	92	15	18	14	4	–	–	
García-Rodríguez (2015)	269	–	18	54	39	11	2	–	2
Rocha et al. (2015)	68	105	8	16	16	–	–	–	
Eddy et al. (2016)	1508	2703	39	63	44	17	2		7
Pabón and Acero (2016)	144	97	19	26	17	9	–	–	
Pantoja-Echevarría et al. (2017)	899		30	81	62	16	3	–	
Romero (2017)	568	394	29	56	41	15		–	

(continued)

Table 8.1 (continued)

Own data (Colombia)	328	481	26	46	29	15	2		
	393	599	26	43	30	13			
	27	65	10	16	11	5			
	538	747	31	66	45	21			
	6	6	1	1					

Author (year)	Relative abundance (n) Fishes (%)	Crustaceans (%)	Others (%)	Locality (Country)	TL (cm) Min-Max	TL (cm) Mean
Albins and Hixon (2008)	100.0			Bahamas	11.8–28.5	16.5
Morris and Akins (2009)	71.2	28.5	0.3	Bahamas	6.2–42.4	21.7 ± 0.2 (SE)
Côté and Maljković (2010)	100.0	–	–	New Providence Island (Bahamas)	13.0–39.0	
Jud et al. (2011)				Loxahatchee River, Florida (USA)	2.3–18.5	9.2 ± 3.4 (SD)
McCleery (2011)	75.0	25.0	–	Bonaire (Netherland Antilles)	6.1–34.0	
Muñoz et al. (2011)	91.6	7.4	1.02	Onslow Bay – North Carolina (USA)	15.0–34.9	30.9 ± 0.53 (SE, 2004) 28.4 ± 0.74 (SE, 2006)
Abril (2012)	62.5	37.1	0.4	San Andrés Island (Colombia)	12.2–42.4	26.9 ± 0.46 (SE)
Cure et al. (2012)				Cayman Island and Bahamas (United Kingdom)		Cayman Island: 15.5 ± 5.6 (SD) Bahamas: 22.7 ± 5.4 (SD)

Green et al. (2012)				New Providence Island (Bahamas)		
Layman and Allgeier (2012)	91.0	9.0		Abaco (Bahamas)	6.0–20.8 (SL)	14.3 (SL)
Muñoz-Escobar and Gil-Agudelo (2012)				Santa Marta (Colombia)	6.4–26.5	16.3 ± 0.6 (SE)
Valdez-Moreno et al. (2012)	74.4	25.6	–	Mexican Caribbean (México)		
Côté et al. (2013b)	100.0	–	–	New Providence Island (Bahamas)	12.2–37.2	
Pimiento et al. (2013)	–	–	–	San Salvador (Bahamas)	9.9–24.8	17.5 ± 0.8 (SE)
Dahl and Patterson III (2014)				Gulf of Mexico (México)	6.7–37.7	
Gómez-Pardo (2014)	69.5	30.5	–	Santa Marta, Cartagena and Capurganá (Colombia)	4.5–41.0	23.8 ± 0.2 (SE)
Sandel et al. (2014)	41.2	58.3	0.5	Parque Nacional Cahuita y Refugio Nacional de Vida Silvestre Gandoca-Manzanillo (Costa Rica)		18.7 ± 0.3 (SE)
Villaseñor-Derbez and Herrera-Pérez (2014)	32.61	67.39		Quintana Roo (México)	5.13–29.28 (midpoint of classes)	
García-Rodríguez (2015)	63.2	33.0	3.8	La Habana (Cuba)		22.0 ± 8.0 (SD)
Rocha et al. (2015)	100.0	–	–	Belize barrier reef (Belize)	5.0–35.0	
Eddy et al. (2016)	55.5	43.0	3.8	Bermuda	12.4–46.7	33.5 ± 0.2 (SE)

(continued)

Table 8.1 (continued)

Author (year)	Relative abundance (n) Fishes (%)	Crustaceans (%)	Others (%)	Locality (Country)	TL (cm) Min-Max	TL (cm) Mean
Pabón and Acero (2016)	62.9	37.1	–	Santa Marta and San Andrés Island (Colombia)	4.9–41.0	26.1 ± 0.36 (SE)
Pantoja-Echevarría et al. (2017)	84.3	15.4	0.3	Cayo las Brujas, Guanahacabibes and La Habana (Cuba)	10.1–35.0	
Romero (2017)	83.5	16.5	–	Santa Marta and Cartagena (Colombia)	4.5–43.2	24.0 ± 0.3 (SE)
Own data (Colombia)	83.3	16.1	0.6	Archipiélago de San Andrés, Providencia and Santa Catalina(	12.9–41.5	28.67 ± 0.30 (SE)
	87.7	12.3		Barú, Bolívar	8.5–42	25.38 ± 0.30 (SE)
	97.8	2.2		Cabo de la vela, La Guajira	14.1–40	28.27 ± 1.34 (SE)
	95.2	4.8		Taganga, Magdalena	4.3–43.7	23.30 ± 0.32 (SE)
	100.0	0.0		Deep reef ecosystem	18.2–39.5	30.51 ± 3.28 (SE)

TL Total length, *Min* Minimum, *Max* Maximum, *SL* Standard length *SD* Standard deviation, *SE* Standard error

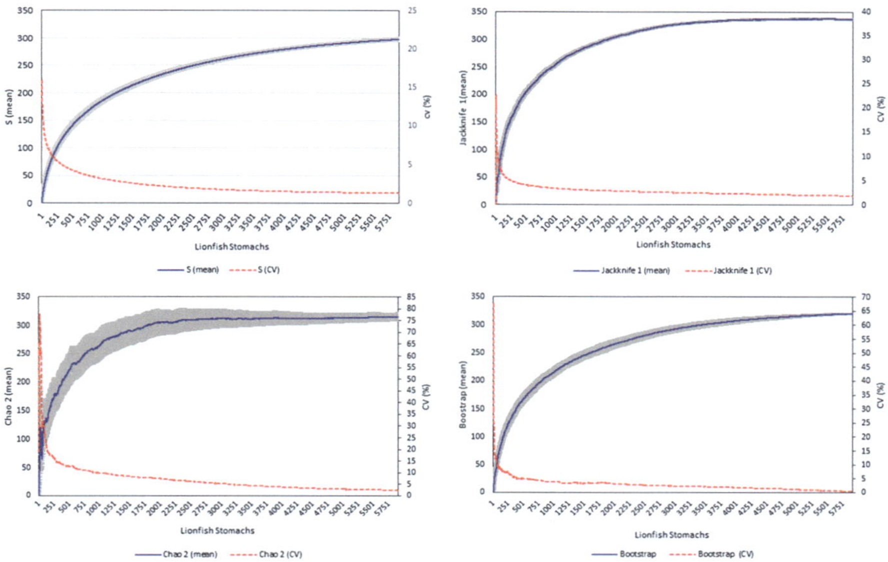

Fig. 8.5 Mean curves of species accumulation in *Pterois volitans* diet and coefficients of variation (CV). (**a**) Curve of total number of observed species in all pooled stomachs (S), (**b**) Richness estimation by Chao 2. (**c**) Richness estimation by Jacknife 1. (**d**) Richness estimation by Bootstrap

(5911) has been sufficient. Nevertheless, as stated by Eddy et al. (2016), a large quantity of well-digested items could not be identified in the stomachs, so it may be possible that lionfish diet is broader than characterized until now.

Lionfish consumes mainly fish and crustaceans and sporadically mollusks and echinoderms (Table 8.1). In most studies, fish is the preferred food, followed by crustaceans although sometimes the latter can be consumed in relatively high proportions (Sandel et al. 2014; Villaseñor-Derbez and Herrera-Pérez 2014; Eddy et al. 2016). Studies show that lionfish feed upon juveniles and smallbodied adults of many species (Morris and Akins 2009; Muñoz et al. 2011; Dahl and Patterson III 2014; Gómez-Pardo 2014; Eddy et al. 2016). Green and Côté (2014) showed that having a shallow body is a trait correlated with an increased vulnerability to predation by lionfish. Nonetheless, the great opportunism shown by lionfish allows it to access prey that lack characteristics that increased their vulnerability, like some crustaceans that have greater abundance than teleosts at certain times of the day (e.g. crepuscular periods) or in certain locations (e.g. deep crevices and caves), as proposed by Eddy et al. (2016). Pantoja et al. (2017) found that in La Habana even large sized lionfish eat mainly penaeid shrimps; their explanation for the situation was that shrimps are extremely abundant in the Cuban seagrass beds where they collected.

Some studies have found that small-sized lionfish (<15–20 cm) tend to consume more invertebrates, mainly crustaceans, and that they become piscivores as they grow (Morris and Akins 2009; Gómez-Pardo 2014; García-Rodríguez 2015;

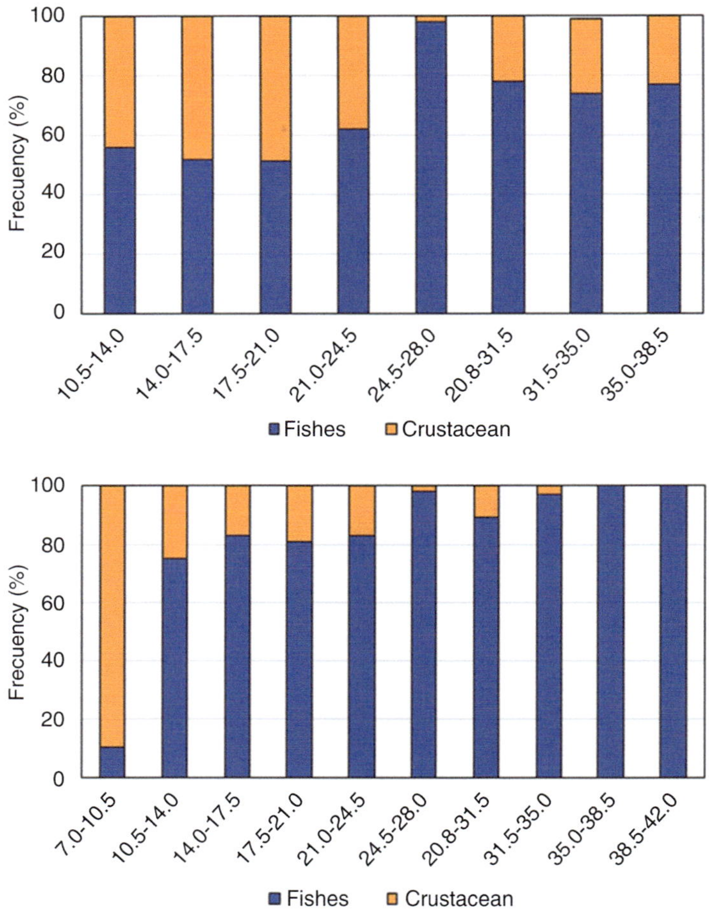

Fig. 8.6 Lionfish size-classes (cm) showing relationship between crustacean and fish prey. Proportion individuals (**a**) and biomass (**b**) fishes and crustacean consumed by lionfish

Romero 2017) (Fig. 8.6). As discussed by Eddy et al. (2016), a cause for this phenomenon could be that small lionfish need to spend more time in caves, rifts and holes protecting themselves of potential predators (among them bigger specimens of their own species). This leads to a high frequency of encounters with small invertebrates that also take refuge in those spaces. Furthermore, as crustaceans have less mobility than fish they are easier to catch for small lionfish. The higher consumption of fish by larger lionfish is probably due to the fact that a larger sized fish gives lionfish a higher supply food with one capture effort. Larger sized lionfish can patrol around the reef looking for prey with a lower risk of being predated, so they can find bigger prey (Gómez-Pardo 2014). Additionally, according to Cabrera-Guerra (2014), crustaceans have an exoskeleton that is too complex for digestion and very difficult to utilize as a digestible item in contrast to fish.

Using the IRI, it is evident that the three more important families in lionfish diets in the western Atlantic are Haemulidae, Labridae (including Scaridae), and Pomacentridae. However, it seems that the relative importance of each prey family varies depending on the locality. Haemulids were by far the most important item in the southeast US; Muñoz et al. (2011) reported an IRI of 72.8% for the family, made mainly by the tomtate *Haemulon aurolineatum*. However, no other work found such a high IRI for this family; reported values in Colombia varied between 5.8% (Gómez-Pardo 2014) and 16.2% (Pabón and Acero P 2016). Labridae (*sensu lato*) was found to be very important in Colombia; reported values varied between 26.2% (Romero 2017) and 50.7% (Gómez-Pardo 2014). No high IRI was found in any other locality for this family; nevertheless, authors reporting only %N found high values in Belize (Rocha et al. 2015), the Mexican Caribbean (Valdez-Moreno et al. 2012), and the Bahamas (Côté et al. 2013b). Pomacentridae was also quoted as having high %IRI in Colombia, with figures going from 22.4% (Gómez-Pardo 2014) to 50.9% (Romero 2017). Only Côté et al. (2013b) reported an important %N for damselfishes in the Bahamas. Those three families share several features, including relatively high abundance in western Atlantic reefs, recruiting on shallow reefs, and a relatively large number of species (over 10 species per family). It should be mentioned that Gobiidae, a very rich family of comparatively small species, does not usually reach high %IRI values. However, when the %N is considered by itself, it presents values varying between 18.3% and 48.0% (Côté and Malkjović 2010; Valdez-Moreno et al. 2012; Rocha et al. 2015); those relatively high numbers have led Linardich et al. (2017) to consider eight western Atlantic gobiid species as threatened. Another point that should be highlighted is that three relatively conspicuous families of reef fishes have not yet been detected in lionfish stomach contents: Muraenidae, Kyphosidae, and Sciaenidae. Kyphosidae and Sciaenidae share one feature, a relatively low richness in western Atlantic reefs that does not surpasses five species. Other than that, their life histories are different; kyphosids recruit offshore, on floating *Sargassum*, arriving to reefs at relatively large sizes to be preyed upon by lionfish. A hypothesis that may help to explain the situation of sciaenids is that they are usually not very abundant, which, by chance, has precluded researchers from finding them in examined lionfish stomachs. Muraenids, on the other hand, are highly secretive species, mainly at undeveloped stages, living deep within the reef frame; which may difficult lionfish predation over muraenid recruits.

Because all authors reported data based on counting prey items (%N), it is of interest to check the list of species against those data. Eight of ten papers reviewed present one labrid species as the most important preyed fish species; usually that species is *Thalassoma bifasciatum*, but Rocha et al. (2015) report the Belizean endemic *Halichoeres socialis* and Villaseñor-Derbez and Herrera-Pérez (2014) found *Halichoeres garnoti*. The second most recorded top prey species (as stated in four papers) is the gobiid *Coryphopterus personatus*, already discussed; it is worth mentioning that those four researches were carried on in the northern part of the Greater Caribbean: Bahamas, Belize, and Cuba. Other species reported with high %N figures in at least one paper are the damselfishes

Stegastes partitus and *Chromis cyanea*, the haemulid *Haemulon aurolineatum*, the labrisomid *Malacoctenus triangulatus*, and even the lionfish itself.

A suggestive indicator of the negative impact of the invader in San Andrés island was presented by Abril (2012), who calculated that the lionfish average daily intake was 1.1 ± 0.5 g. Considering that González-Corredor (2014), using different capture models and multiple recaptures, estimated that the lionfish population fluctuated between one and six million lionfish in that western Caribbean area, our conservative calculation is that lionfish can remove annually between 410 and 2409 t of biomass in San Andrés reefs.

8.3.2 *Prey Origin*

We analyzed 78 muscle sample from lionfish (46 from Magdalena and 32 from Bolívar). The mean values for $\delta^{13}C$ (−16.6 ± 1.2‰; −16.2 ± 0.8‰) and $\delta^{15}N$ (10.7 ± 0.7‰; 10.2 ± 0.6‰) were similar in both areas. A multiple comparison test revealed differences in isotope values of $\delta^{15}N$ between locations (Kruskal Wallis test: $\delta^{15}N$, H = 12.3, P = 0.002). In Bolívar, Barú and Tierra Bomba (Fig. 8.7) presented higher values (10.5 ± 0.5‰; 10.4 ± 0.3‰ respectively) than Salmedina (9.9 ± 0.5‰). The mean values for $\delta^{13}C$ and $\delta^{15}N$ in Magdalena localities did not show significant differences (Fig. 8.8). Those values of $\delta^{13}C$ are similar to those reported by Muñoz et al. (2011) in the Bahamas (−16.6 ± 0.2‰) and by Arredondo (2016) in the Mexican Caribbean (−16.2‰); therefore, those $\delta^{13}C$ values seem to be typically found in reef areas.

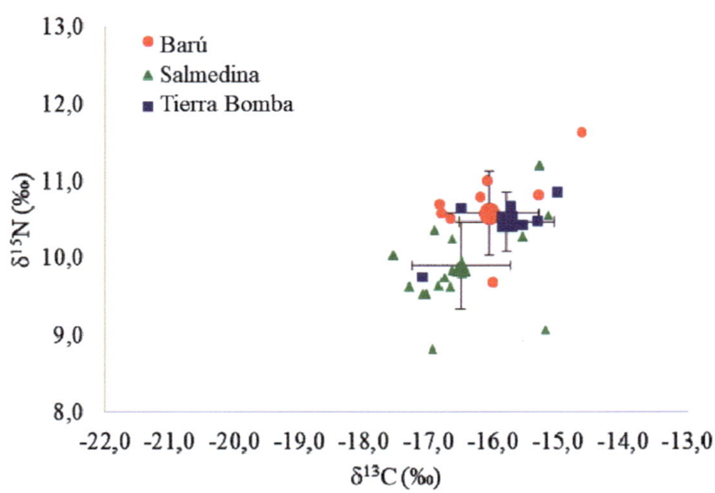

Fig. 8.7 Muscle $\delta^{13}C$ and $\delta^{15}N$ values (small symbols) and mean ± SD (large symbols and error bars) for Bolívar locations

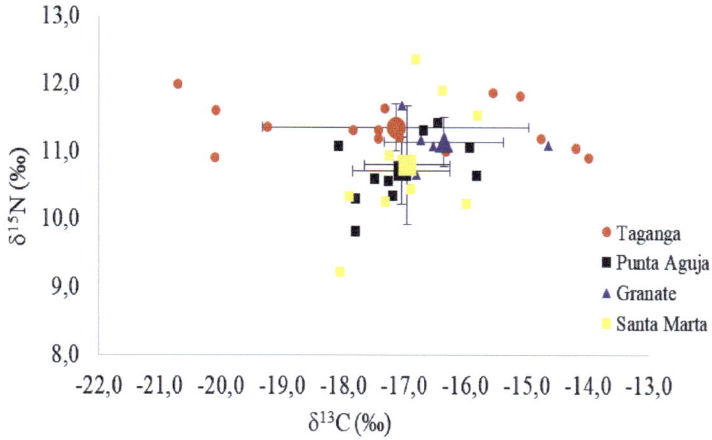

Fig. 8.8 Muscle $\delta^{13}C$ and $\delta^{15}N$ values (small symbols) and mean ± SD (large symbols and error bars) for Magdalena locations

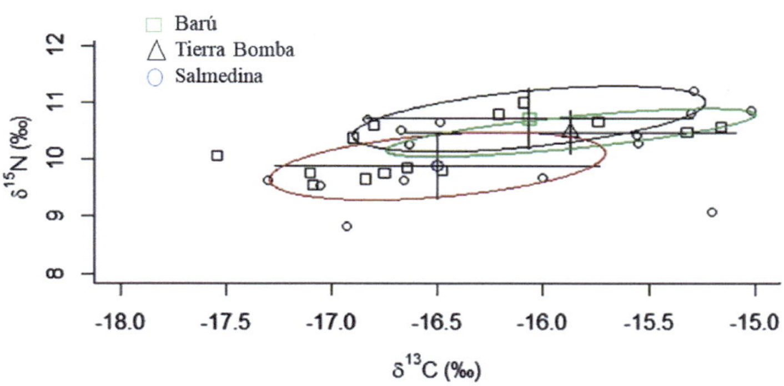

Fig. 8.9 $\delta^{13}C$ and $\delta^{15}N$ values and mean ± SD of each location in Bolívar. Ellipses represent the estimated standard area determined by SIBER analysis (Stable Isotope Bayesian Ellipses in R)

The ellipse comparison in the same area showed high niche overlap (more than 70%) among individuals from Barú with Tierra Bomba, but no overlap with individuals from Salmedina (Fig. 8.9). In Magdalena, the overlap was also low among Punta Aguja, Granate and Santa Marta, with 70% of values under 0.5. The highest overlap (82%) was between Punta Aguja and Santa Marta (Fig. 8.10). Individual niche width (Standard Ellipses Area, SEAc) of lionfish from Bolívar has a range between 0.05 and 1.20, whereas Magdalena has a wider range, between 0.29 and 5.30.

For $\delta^{15}N$, differences were found among Bolívar locations, which could be related to the prey consumed in each area. Perhaps prey from Salmedina came mainly from lower trophic levels and/or were smaller compared to those consumed by lionfish in Barú and Tierra Bomba. Layman and Allgeier (2012) found lionfish values of $\delta^{15}N$ of 9.1 ± 0.38‰ in the Bahamas, which were lower than in other

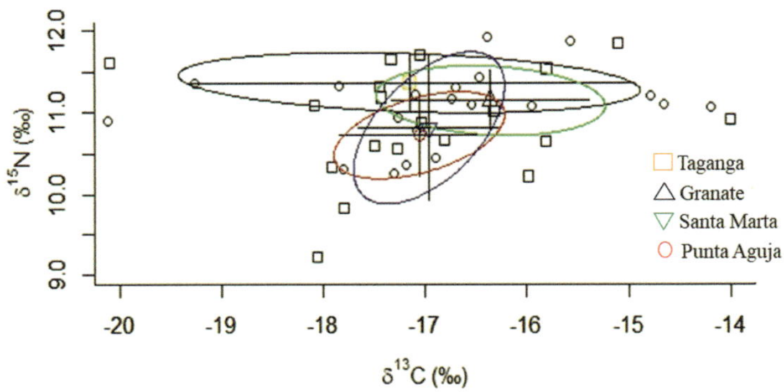

Fig. 8.10 δ^{13}C and δ^{15}N values and mean ± SD of each location in Magdalena. Ellipses represent the estimated standard area determined by SIBER analysis (Stable Isotope Bayesian Ellipses in R).

sympatric species. That was attributed to the fact that lionfish in that area feed on small fishes and crustaceans, while the other species consumed a greater proportion of benthic crustaceans, which usually have larger amounts of δ^{15}N.

Arredondo (2016) found similar values (9.6 ± 0.5‰) in lionfish caught in Puerto Morelos, Mexican Caribbean. However, those values were low compared to native species such as *Lutjanus apodus* (10.5 ± 0.58‰). Those differences may be attributed to consumption of different prey, because lionfish feeds on small crustaceans and fish, while *L. apodus* preys on large sized items. Although no differences were found in the Magdalena material, it is notorious that lionfish from Taganga showed a high level of individual variation in δ^{13}C (~7‰) (Fig. 8.10). These differences suggest that lionfish individuals showing lower levels of δ^{13} C (−20.7‰) may have fed in offshore areas, because those values are typical of offshore food webs (De Niro and Epstein 1978). The overlap between Barú and Tierra Bomba could be explained by the geographical closeness of the locations. The SEAc values were larger in the individuals from Magdalena when compared to those from Bolívar. Those differences could be associated to the higher prey richness in lionfish caught in the Magdalena area (Appendix, Table 8.1).

8.3.3 Geographic and Environmental Differences

Variations in habitat preferences (Fig. 8.11) (Whitfield et al. 2002; Green and Côté 2009; Barbour et al. 2010; Arbeláez and Acero 2011; Claydon et al. 2012; Pimiento et al. 2013; García-Rodríguez 2015; García-Urueña et al. 2015; Hernández-Abello et al. 2015) as well as seasonal and depth distributions of prey species affect lionfish diet (Pimiento et al. 2013; MacNeil and Connolly 2015; Eddy et al. 2016). A comparison of the prey species richness in the available Caribbean studies

Fig. 8.11 Lionfish inhabits almost all kinds of hard bottom environments. Left, Santa Marta; right Cartagena. (Photographs by Santiago Estrada R. and Adolfo Sanjuan-Muñoz)

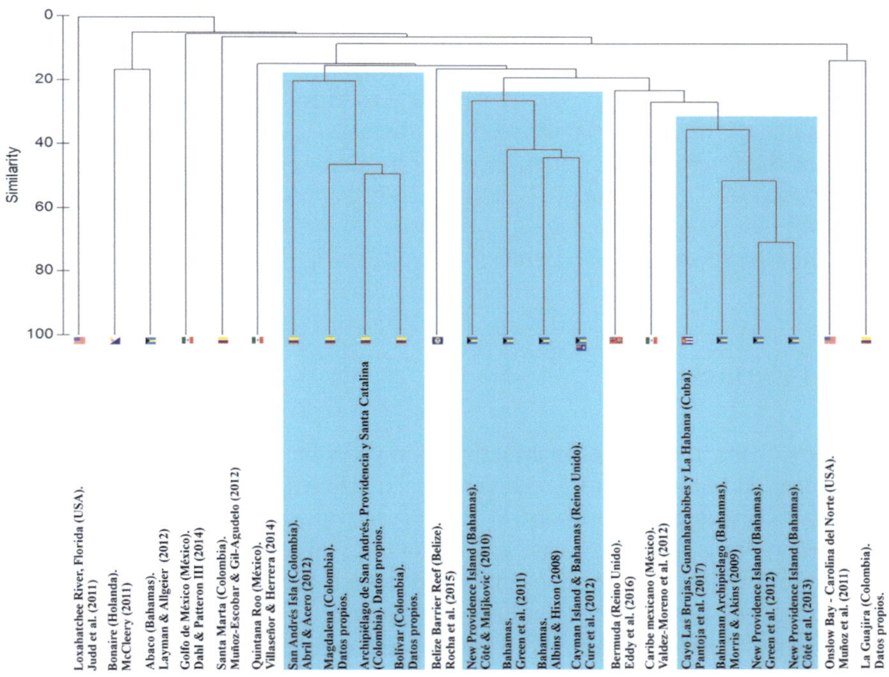

Fig. 8.12 Cluster analysis showing the relationship between western Atlantic regions where lionfish have been studied using Dice index and linkage UPMGA. The three main groups (two for the Bahamas, Cuba, and Cayman Islands and one for the southern Caribbean) are blue colored

(Fig. 8.12) shows differences between regions. A cluster analysis between diets show three main groups, two of them made by studies carried upon in the Bahamas and Cayman islands. The other main group includes only southern Caribbean (Colombian studies). Diet composition of lionfish from the eastern coast of the

United States (North Carolina and Florida) and the Gulf of Mexico and the Mexican Caribbean appear to not be closely related to any of the main groups. Diet composition of lionfish from the Guajira Peninsula (Colombia), an area heavily influenced by a rather strong upwelling (Fajardo 1979; Andrade and Barton 2005; Bernal et al. 2006), is also separated from the main groups. Those results ratify that the success of lionfish invasion to the western Atlantic is strongly influenced by its opportunistic feeding habits.

A more detailed analysis using only Colombian Caribbean data shows significant differences between the upwelling area (GUA), deep reefs (DRE), and shallow environments (MAG, SAI and BOL) (Fig. 8.3). The stomach contents of sixty-six lionfish captured in continental Colombian deep reefs (50–70 m) using Closed Circuit Rebreather (CCR) equipment, Remote Operated Vehicles-ROV and commercial fishermen catches have been examined. 17% of them were empty, and a total of 19 prey groups (8 fishes and 11 crustaceans, unpublished data) were identified in the other 55 lionfish specimens (size-range 15–45 cm, 91% > 25 cm). The number of prey items per stomach ranged from 0 to 6 (mean $\pm$ DE: 1.31 $\pm$ 1.15). Although lionfish were relatively large, the crustaceans prey abundance may indirectly reflect the dominance of that group in deep environments and prove the opportunistic predator character of the invader, as found by Eddy et al. (2016) and similar to what Pantoja et al. (2017) report. Because most fish prey (i. e. *Acanthurus coeruleus, Stegastes planifrons* and *Serranus* sp.) at those depths were juveniles, the recruitment of those species may be significantly impacted by lionfish.

8.4 Conclusions

Pterois volitans, a generalist invasive species, preys on a wide array of Caribbean reef critters and simultaneously represents an extinction threat to several fish species. We have compiled an inventory of 259 species (73 crustaceans and 186 fish) included in 81 families (37 crustaceans and 44 fish); this extensive list may evidence that lionfish is one of the reef predators exploiting biodiversity more efficiently. Haemulidae, Labridae, and Pomacentridae are the most important fish families consumed by lionfish in the invaded range. The wrasses *Thalassoma bifasciatum* and *Halichoeres* spp., the gobiid *Coryphopterus personatus*, the damselfishes *Stegastes partitus* and *Chromis cyanea*, the haemulid *Haemulon aurolineatum*, the labrisomid *Malacoctenus triangulatus*, and *P. volitans* have been reported among the preferred food items of lionfish in the Greater Caribbean. Among the various fishes consumed by lionfish are some that are commercially important and therefore usually threatened species, such as the grunt *Haemulon plumierii,* the groupers *Mycteroperca venenosa* and *Epinephelus striatus,* the snappers *Ocyurus chrysurus,*

Rhomboplites aurorubens, and *Lutjanus synagris,* and the triggerfish *Balistes vetula*. Key ecological groups that control algal growth in coral reefs (i. e. parrotfishes and surgeonfishes) as well as extremely abundant minute species (i.e. *Coryphopterus* spp.) that may be important food items for piscivorous species are also heavily preyed. There are also crustaceans with an important commercial value (e. g. *Panulirus argus*) and ecological value (e. g. cleaner shrimps). To date, at least 83% of the statistically estimated prey richness of around 300 species have been identified from lionfish stomach content; in any case it may be possible that the figure of prey items will increase when deep reefs and other undersampled environments are studied. A stable isotope ($\delta^{13}C$ and $\delta^{15}N$) analysis study of lionfish stomach contents coming from the Colombian continental Caribbean have shown that most consumed prey are of reef origin. A cluster analysis between lionfish diets in the invaded range show three main groups: two for the Bahamas and Cayman Islands, and another for the Colombian Caribbean. Lionfish diets from the eastern coast of the United States, the Gulf of Mexico, the Mexican Caribbean, and the Colombian Guajira Peninsula appear separated from the three main groups. Finally, we recommend that research covering unexplored subjects, such as temporal dynamics of the lionfish diet, would allow the inference of recruitment patterns, generally unknown for many reef species, and how they can be altered by the invading scorpaenid. The former without reducing the permanent monitoring of the abundance, feeding habits, and reproductive biology of this highly deleterious species.

Acknowledgments We deeply thank our Colombian institutions (Universidad Nacional de Colombia Sede Caribe and Universidad Jorge Tadeo Lozano), which let us face coordinately this unexpected and powerful menace. Funds for our research came from COLCIENCIAS (Project code 1361-521-28271), Universidad Nacional de Colombia (project codes 10498, 12397, 16265, 34831, 36649, and 39149) and Universidad Jorge Tadeo Lozano (project codes 558-10-13, 639-11-14 and 694-12-15). Santiago Estrada R. (Reef Shepperd Center) loaned the underwater photographs and collaborated with the Santa Marta field trips. Officers from the Parque Nacional Natural Corales de Profundidad, especially Alejandro Henao, donated deep reef lionfish specimens. Humberto Gómez-Pardo, María Camila Rodríguez and Josselyn Bryan helped us with the photographs of some stomach contents items. Many other graduate and undergraduate students from our institutions generously helped us to gather data from Colombian environments. Luisa María Sanjuan Agudelo checked English grammar and style. Contribution No. 470 of the Instituto de Estudios en Ciencias del Mar, Cecimar, Universidad Nacional de Colombia sede Caribe.

Appendix: Historical List of Taxa Found in Lionfish Stomach Contents in the Western Atlantic

Family	Specie/taxon	Albins and Hixon (2008)	Morris and Akins (2009)	Côté and Maljković (2010)	Green etal. (2011)	Jud etal. (2011)	McCleery (2011)	Muñoz etal. (2011)	Abril (2012)	Cure etal. (2012)	Green etal. (2012)	Layman and Allgeier (2012)	Muñoz – Escobar and Gil – Agudelo (2012)	Valdez – Moreno etal. (2012)	Côté etal. (2013a, b)	Gómez – Pardo (2014)	Dahl and Patteron III (2014)	Villaseñor – Derbez and Herrera – Pérez (2014)	Rocha etal. (2015)	Pabón and Acero (2016)	Eddy etal. (2016)	Romero (2017)	Pantoja etal. (2017)	Own Data. Archipiélago de San Andrés, Providencia y Santa Catalina	Own Data. Bolívar	Own Data. La Guajira	Own Data. Magdalena
Fishes																											
Acanthuridae	*Acanthurus chirurgus*				Y																Y		Y				
Acanthuridae	*Acanthurus coeruleus*				Y																		Y				
Acanthuridae	*Acanthurus tractus*		Y		Y																Y		Y				
Acanthuridae	*Acanthurus* sp.																				Y	Y	Y	Y	Y		
Antennariidae																					Y						
Apogonidae	*Apogon aurolineatus*	Y																									
Apogonidae	*Apogon binotatus*		Y		Y										Y				Y		Y						
Apogonidae	*Apogon lachneri*													Y													

Apogonidae	*Apogon maculatus*													Y		Y					Y	Y	Y				Y
Apogonidae	*Apogon mosavi*													Y													
Apogonidae	*Apogon phenax*														Y												
Apogonidae	*Apogon planifrons*											Y															
Apogonidae	*Apogon pseudomaculatus*							Y									Y				Y						
Apogonidae	*Apogon quadrisquamatus*															Y											
Apogonidae	*Apogon townsendi*		Y								Y		Y								Y						
Apogonidae	*Apogon* spp.																Y	Y				Y	Y	Y			Y
Apogonidae	*Astrapogon puncticulatus*													Y													
Apogonidae	*Paroncheilus affinis*																					Y					Y
Apogonidae	*Phaeoptyx pigmentaria*										Y				Y												
Apogonidae																			Y		Y	Y					
Atherinidae	*Atherinomorus* sp.										Y																
Aulostomidae	*Aulostomus maculatus*								Y		Y				Y					Y	Y		Y	Y			
Balistidae	*Balistes vetula*																			Y			Y				
Blenniidae	*Entomacrodus nigricans*																				Y						
Blenniidae	*Hypleurochilus geminatus*							Y																			
Blenniidae	*Hypleurochilus springeri*																									Y	
Blenniidae	*Ophioblennius macclurei*																	Y									
Blenniidae	*Parablennius marmoreus*																		Y		Y						
Blenniidae																	Y		Y		Y						
Bothidae	*Bothus lunatus*													Y						Y							Y

(continued)

Family	Specie/taxon	Albins and Hixon (2008)	Morris and Akins (2009)	Côté and Maljković (2010)	Green etal. (2011)	Jud etal. (2011)	McCleery (2011)	Muñoz etal. (2011)	Abril (2012)	Cure etal. (2012)	Green etal. (2012)	Layman and Allgeier (2012)	Muñoz – Escobar and Gil – Agudelo (2012)	Valdez – Moreno etal. (2012)	Côté etal. (2013a, b)	Gómez – Pardo (2014)	Dahl and Patteron III (2014)	Villaseñor – Derbez and Herrera – Pérez (2014)	Rocha etal. (2015)	Pabón and Acero (2016)	Eddy etal. (2016)	Romero (2017)	Pantoja etal. (2017)	Own Data. Archipiélago de San Andrés, Providencia y Santa Catalina	Own Data. Bolívar	Own Data. La Guajira	Own Data. Magdalena
Bothidae	*Bothus ocellatus*														Y												
Bothidae	*Bothus* sp.																					Y				Y	Y
Bothidae																					Y						
Carangidae	*Decapterus punctatus*																Y					Y			Y		Y
Carangidae	*Decapterus tabl*																					Y					Y
Carangidae	*Decapterus* spp.																Y				Y						
Carangidae	*Selar crumenophthalmus*							Y																			
Carangidae	*Trachurus lathami*																Y										
Carangidae																	Y					Y					

Chaenopsidae	*Acanthemblemaria aspera*										Y				Y												
Chaenopsidae	*Acanthemblemaria betinensis*																										Y
Chaenopsidae	*Acanthemblemaria* sp.																						Y			Y	
Chaenopsidae	*Emblemaria* sp.																								Y		
Chaenopsidae	*Lucayablennius zingaro*		Y								Y												Y				
Chaetodontidae	*Chaetodon capistratus*																				Y						
Chaetodontidae	*Chaetodon ocellatus*																				Y						
Chaetodontidae	*Chaetodon* sp.																						Y				
Chaetodontidae																					Y						
Cirrhitidae	*Amblycirrhitus pinos*		Y																			Y	Y				Y
Clupeidae	*Harengula* sp.																					Y			Y		
Clupeidae	*Jenkinsia lamprotaenia*																				Y						
Fistulariidae	*Fistularia* sp.																							Y			
Gerreidae						Y																					
Gobiidae	*Coryphopterus dicrus*		Y		Y																		Y				
Gobiidae	*Coryphopterus eidolon*		Y		Y						Y																
Gobiidae	*Coryphopterus glaucofraenum*	Y	Y		Y					Y	Y		Y						Y		Y	Y	Y				Y
Gobiidae	*Coryphopterus hyalinus*		Y	Y										Y	Y												
Gobiidae	*Coryphopterus personatus*		Y	Y					Y						Y	Y			Y		Y		Y	Y			Y
Gobiidae	*Coryphopterus thrix*													Y													

(continued)

Family	Specie/taxon	Albins and Hixon (2008)	Morris and Akins (2009)	Côté and Maljković (2010)	Green etal. (2011)	Jud etal. (2011)	McCleery (2011)	Muñoz etal. (2011)	Abril (2012)	Cure etal. (2012)	Green etal. (2012)	Layman and Allgeier (2012)	Muñoz – Escobar and Gil – Agudelo (2012)	Valdez – Moreno etal. (2012)	Côté etal. (2013a, b)	Gómez – Pardo (2014)	Dahl and Patteron III (2014)	Villaseñor – Derbez and Herrera – Pérez (2014)	Rocha etal. (2015)	Pabón and Acero (2016)	Eddy etal. (2016)	Romero (2017)	Pantoja etal. (2017)	Own Data. Archipiélago de San Andrés, Providencia y Santa Catalina	Own Data. Bolívar	Own Data. La Guajira	Own Data. Magdalena
Gobiidae	*Coryphopterus tortugae*													Y													
Gobiidae	*Coryphopterus venezuelae*										Y			Y	Y												
Gobiidae	*Coryphopterus* sp.																					Y	Y	Y	Y		Y
Gobiidae	*Elacatinus prochilos*																	Y									
Gobiidae	*Gnatholepis thompsoni*	Y			Y						Y				Y						Y		Y			Y	
Gobiidae	*Lythrypnus spilus*										Y				Y												
Gobiidae	*Microgobius* sp.																					Y			Y		

Gobiidae	*Priolepis hipoliti*		Y		Y						Y			Y	Y								Y				
Gobiidae																	Y		Y		Y		Y				
Grammatidae	*Gramma loreto*	Y	Y	Y	Y				Y	Y	Y			Y	Y			Y		Y			Y	Y			
Grammatidae	*Gramma melacara*		Y																				Y				
Haemulidae	*Anisotremus virginicus*																					Y					
Haemulidae	*Haemulon aurolineatum*				Y			Y								Y	Y			Y	Y					Y	Y
Haemulidae	*Haemulon flavolineatum*				Y											Y					Y						
Haemulidae	*Haemulon melanurum*	Y																									
Haemulidae	*Haemulon plumierii*	Y																									
Haemulidae	*Haemulon sciurus*																						Y				
Haemulidae	*Haemulon* sp.																					Y		Y	Y	Y	Y
Haemulidae	*Haemulon vittatum*										Y				Y												
Haemulidae																					Y						
Holocentridae	*Holocentrus adscensionis*																				Y						
Holocentridae	*Holocentrus rufus*														Y	Y						Y					Y
Holocentridae	*Holocentrus* sp.																							Y			
Holocentridae	*Myripristis jacobus*																						Y				
Holocentridae	*Neoniphon marianus*																						Y				
Holocentridae	*Sargocentron coruscum*										Y			Y	Y						Y						
Holocentridae	*Sargocentron vexillarium*		Y																		Y		Y				
Holocentridae																					Y	Y	Y	Y	Y		Y
Labridae	*Bodianus rufus*		Y								Y	Y				Y											
Labridae	*Clepticus parrae*		Y								Y				Y	Y				Y	Y	Y	Y	Y	Y		Y
Labridae	*Cryptotomus roseus*																	Y									

(continued)

Family	Specie/taxon	Albins and Hixon (2008)	Morris and Akins (2009)	Côté and Maljković (2010)	Green etal. (2011)	Jud etal. (2011)	McCleery (2011)	Muñoz etal. (2011)	Abril (2012)	Cure etal. (2012)	Green etal. (2012)	Layman and Allgeier (2012)	Muñoz – Escobar and Gil – Agudelo (2012)	Valdez – Moreno etal. (2012)	Côté etal. (2013a, b)	Gómez – Pardo (2014)	Dahl and Patteron III (2014)	Villaseñor – Derbez and Herrera – Pérez (2014)	Rocha etal. (2015)	Pabón and Acero (2016)	Eddy etal. (2016)	Romero (2017)	Pantoja etal. (2017)	Own Data. Archipiélago de San Andrés, Providencia y Santa Catalina	Own Data. Bolívar	Own Data. La Guajira	Own Data. Magdalena
Labridae	*Halichoeres bathyphilus*																Y										
Labridae	*Halichoeres bivittatus*	Y	Y		Y					Y	Y	Y			Y		Y		Y		Y		Y				
Labridae	*Halichoeres garnoti*	Y	Y		Y										Y						Y		Y	Y			Y
Labridae	*Halichoeres maculipinna*		Y								Y				Y						Y		Y				
Labridae	*Halichoeres pictus*	Y																									
Labridae	*Halichoeres socialis*																		Y								

Labridae	*Halichoeres* sp.																				Y	Y	Y			Y
Labridae	*Scarus iseri*		Y											Y					Y				Y				
Labridae	*Scarus taeniopterus*													Y							Y	Y			Y		
Labridae	*Scarus* sp.																					Y		Y	Y		Y
Labridae	*Sparisoma atomarium*																		Y				Y				
Labridae	*Sparisoma aurofrenatum*				Y					Y	Y			Y	Y				Y	Y	Y		Y	Y			
Labridae	*Sparisoma chrysopterum*															Y											
Labridae	*Sparisoma radians*							Y																			
Labridae	*Sparisoma viride*		Y		Y									Y					Y								
Labridae	*Sparisoma* sp.																					Y			Y		Y
Labridae	*Thalassoma bifasciatum*	Y	Y	Y	Y			Y	Y	Y	Y	Y		Y	Y			Y		Y	Y	Y	Y	Y	Y	Y	Y
Labridae	*Xyrichthys martinicensis*																				Y						
Labridae	*Xyrichthys novacula*																Y										
Labridae	*Xyrichthys* sp.		Y		Y																						
Labridae																	Y				Y	Y	Y				
Labrisomidae	*Gobioclinus haitiensis*										Y				Y												
Labrisomidae	*Labrisomus nuchipinnis*															Y											
Labrisomidae	*Labrisomus* sp.																		Y								
Labrisomidae	*Malacoctenus boehlkei*		Y								Y				Y												
Labrisomidae	*Malacoctenus gilli*	Y			Y																						
Labrisomidae	*Malacoctenus macropus*											Y															

(continued)

Family	Specie/taxon	Albins and Hixon (2008)	Morris and Akins (2009)	Côté and Maljković (2010)	Green etal. (2011)	Jud etal. (2011)	McCleery (2011)	Muñoz etal. (2011)	Abril (2012)	Cure etal. (2012)
Labrisomidae	*Malacoctenus triangulatus*		Y							
Labrisomidae	*Malacoctenus* sp.									
Labrisomidae	*Starksia langi*									
Labrisomidae	*Starksia occidentalis*									
Labrisomidae	*Starksia ocellata*									
Labrisomidae										
Lutjanidae	*Lutjanus synagris*									
Lutjanidae	*Ocyurus chrysurus*		Y							

Family	Specie/taxon	Green etal. (2012)	Layman and Allgeier (2012)	Muñoz – Escobar and Gil – Agudelo (2012)	Valdez – Moreno etal. (2012)	Côté etal. (2013a, b)	Gómez – Pardo (2014)	Dahl and Patterson III (2014)	Villaseñor – Derbez and Herrera – Pérez (2014)	Rocha etal. (2015)
Labrisomidae	*Malacoctenus triangulatus*								Y	
Labrisomidae	*Malacoctenus* sp.									
Labrisomidae	*Starksia langi*				Y					
Labrisomidae	*Starksia occidentalis*									Y
Labrisomidae	*Starksia ocellata*				Y					
Labrisomidae										Y
Lutjanidae	*Lutjanus synagris*									
Lutjanidae	*Ocyurus chrysurus*									

Family	Specie/taxon	Pabón and Acero (2016)	Eddy etal. (2016)	Romero (2017)	Pantoja etal. (2017)	Own Data. Archipiélago de San Andrés, Providencia y Santa Catalina	Own Data. Bolívar	Own Data. La Guajira	Own Data. Magdalena
Labrisomidae	*Malacoctenus triangulatus*				Y		Y		
Labrisomidae	*Malacoctenus* sp.				Y	Y			
Labrisomidae	*Starksia langi*								
Labrisomidae	*Starksia occidentalis*								
Labrisomidae	*Starksia ocellata*								
Labrisomidae									
Lutjanidae	*Lutjanus synagris*						Y		
Lutjanidae	*Ocyurus chrysurus*				Y				

Lutjanidae	*Rhomboplites aurorubens*																Y										
Monacanthidae	*Aluterus schoepfii*																				Y						
Monacanthidae	*Aluterus scriptus*																				Y						
Monacanthidae	*Cantherhines* spp.																				Y						
Monacanthidae	*Monacanthus ciliatus*							Y											Y		Y						
Monacanthidae	*Monacanthus tuckeri*		Y								Y			Y	Y						Y	Y	Y	Y	Y		Y
Monacanthidae	*Monacanthus* sp.																Y										
Monacanthidae																			Y			Y	Y	Y	Y		Y
Mullidae	*Mulloidichthys martinicus*																				Y		Y				
Mullidae	*Pseudupeneus maculatus*		Y								Y										Y		Y				
Opistognathidae	*Opistognathus aurifrons*	Y																					Y				
Opistognathidae																						Y					
Paralichthyidae	*Citharichthys* sp.																								Y		
Paralichthyidae	*Paralichthys albigutta*																Y										
Paralichthyidae	*Syacium micrurum*															Y											
Paralichthyidae	*Syacium* sp.																					Y					Y
Paralichthyidae																	Y										
Pempheridae	*Pempheris schomburgkii*															Y					Y						Y
Pleuronectiformes																	Y										
Pleuronectidae																	Y										
Poeciliidae	*Gambusia* sp.																						Y				
Pomacanthidae	*Holacanthus tricolor*																						Y				
Pomacanthidae																							Y				

(continued)

Family	Specie/taxon	Albins and Hixon (2008)	Morris and Akins (2009)	Côté and Maljković (2010)	Green etal. (2011)	Jud etal. (2011)	McCleery (2011)	Muñoz etal. (2011)	Abril (2012)	Cure etal. (2012)	Green etal. (2012)	Layman and Allgeier (2012)	Muñoz – Escobar and Gil – Agudelo (2012)	Valdez – Moreno etal. (2012)	Côté etal. (2013a, b)	Gómez – Pardo (2014)	Dahl and Patteron III (2014)	Villaseñor – Derbez and Herrera – Pérez (2014)	Rocha etal. (2015)	Pabón and Acero (2016)	Eddy etal. (2016)	Romero (2017)	Pantoja etal. (2017)	Own Data. Archipiélago de San Andrés, Providencia y Santa Catalina	Own Data. Bolívar	Own Data. La Guajira	Own Data. Magdalena
Pomacentridae	*Abudefduf saxatilis*								Y			Y		Y									Y				
Pomacentridae	*Chromis cyanea*		Y								Y				Y	Y							Y				Y
Pomacentridae	*Chromis enchrysura*							Y									Y										
Pomacentridae	*Chromis flavicauda*																				Y						
Pomacentridae	*Chromis insolata*		Y																		Y						
Pomacentridae	*Chromis multilineata*		Y								Y				Y			Y				Y	Y				Y
Pomacentridae	*Chromis scotti*																Y										
Pomacentridae	*Chromis* sp.																					Y		Y	Y		

Pomacentridae	*Stegastes adustus*																Y										
Pomacentridae	*Stegastes diencaeus*				Y				Y																		
Pomacentridae	*Stegastes leucostictus*		Y																								
Pomacentridae	*Stegastes partitus*		Y		Y				Y	Y	Y			Y	Y	Y		Y		Y		Y	Y	Y	Y		Y
Pomacentridae	*Stegastes planifrons*															Y						Y		Y	Y		Y
Pomacentridae	*Stegastes xanthurus*		Y								Y				Y				Y		Y						
Pomacentridae	*Stegastes* spp.																				Y	Y	Y				
Pomacentridae																	Y		Y								
Priacanthidae	*Heteropriacanthus cruentatus*																						Y				
Scorpaenidae	*Pterois volitans*													Y	Y			Y									
Scorpaenidae																	Y					Y					Y
Serranidae	*Baldwinella vivanus*																Y										
Serranidae	*Centropristis ocyurus*																Y										
Serranidae	*Centropristis* sp.																Y										
Serranidae	*Cephalopholis cruentata*										Y			Y	Y	Y							Y				
Serranidae	*Cephalopholis* sp.																										Y
Serranidae	*Diplectrum bivittatum*																					Y					Y
Serranidae	*Diplectrum* sp.							Y									Y										
Serranidae	*Epinephelus* spp.																	Y									
Serranidae	*Epinephelus striatus*		Y								Y																
Serranidae	*Hypoplectrus puella*																		Y		Y	Y			Y		
Serranidae	*Hypoplectrus* sp.		Y								Y												Y				

(continued)

Family	Specie/taxon	Albins and Hixon (2008)	Morris and Akins (2009)	Côté and Maljković (2010)	Green etal. (2011)	Jud etal. (2011)	McCleery (2011)	Muñoz etal. (2011)	Abril (2012)	Cure etal. (2012)	Green etal. (2012)	Layman and Allgeier (2012)	Muñoz – Escobar and Gil – Agudelo (2012)	Valdez – Moreno etal. (2012)	Côté etal. (2013a, b)	Gómez – Pardo (2014)	Dahl and Patteron III (2014)	Villaseñor – Derbez and Herrera – Pérez (2014)	Rocha etal. (2015)	Pabón and Acero (2016)	Eddy etal. (2016)	Romero (2017)	Pantoja etal. (2017)	Own Data. Archipiélago de San Andrés, Providencia y Santa Catalina	Own Data. Bolívar	Own Data. La Guajira	Own Data. Magdalena
Serranidae	*Liopropoma carmabi*																	Y									
Serranidae	*Liopropoma rubre*		Y								Y			Y	Y						Y						
Serranidae	*Liopropoma* spp.																				Y						
Serranidae	*Mycteroperca venenosa*																	Y									
Serranidae	*Paranthias furcifer*																				Y						
Serranidae	*Schultzea beta*							Y																			
Serranidae	*Serranus baldwini*							Y																			
Serranidae	*Serranus flaviventris*																		Y								

Serranidae	*Serranus phoebe*							Y																			
Serranidae	*Serranus subligarius*							Y									Y										
Serranidae	*Serranus tabacarius*										Y																
Serranidae	*Serranus tigrinus*		Y					Y			Y		Y		Y	Y				Y		Y	Y	Y	Y		Y
Serranidae	*Serranus* sp.																						Y				
Serranidae																	Y		Y			Y					
Sparidae	*Diplodus bermudensis*																				Y						
Syngnathidae																				Y							
Synodontidae	*Saurida normani*							Y																			
Synodontidae	*Synodus intermedius*										Y					Y											
Synodontidae	*Synodus saurus*										Y																
Synodontidae	*Synodus synodus*										Y				Y		Y				Y		Y		Y		
Synodontidae	*Synodus* sp.																						Y		Y		Y
Synodontidae																	Y				Y	Y					
Tetraodontidae	*Canthigaster rostrata*		Y		Y											Y		Y					Y				
Tetraodontidae	*Sphoeroides spengleri*																				Y						
Triglidae	*Bellator brachychir*																Y										
Triglidae	*Prionotus carolinus*																Y										
Triglidae																	Y			Y							
Tripterygidae	*Enneanectes altivelis*													Y													
Tripterygiidae	*Enneanectes* sp.																							Y			
Crustaceans																											
Alpheidae	*Alpheus* sp.															Y											
Alpheidae																	Y										

(continued)

Family	Specie/taxon	Albins and Hixon (2008)	Morris and Akins (2009)	Côté and Maljković (2010)	Green etal. (2011)	Jud etal. (2011)	McCleery (2011)	Muñoz etal. (2011)	Abril (2012)	Cure etal. (2012)	Green etal. (2012)	Layman and Allgeier (2012)	Muñoz – Escobar and Gil – Agudelo (2012)	Valdez – Moreno etal. (2012)	Côté etal. (2013a, b)	Gómez – Pardo (2014)	Dahl and Patteron III (2014)	Villaseñor – Derbez and Herrera – Pérez (2014)	Rocha etal. (2015)	Pabón and Acero (2016)	Eddy etal. (2016)	Romero (2017)	Pantoja etal. (2017)	Own Data. Archipiélago de San Andrés, Providencia y Santa Catalina	Own Data. Bolívar	Own Data. La Guajira	Own Data. Magdalena
Anomura																							Y				
Axiidae																	Y										
Brachyura																	Y						Y				
Calappidae	*Acanthocarpus* sp.																							Y			
Calappidae	*Calappa gallus*																				Y						
Calappidae	*Calappa ocellata*																				Y						
Calappidae	*Calappa sulcata*																Y										
Calappidae	*Calappa* sp.								Y							Y				Y							Y
Calappidae	*Cryptosoma bairdii*																				Y						
Calappidae	*Hepatus* sp.								Y																		

Calappidae																	Y				Y		Y				
Caridea																	Y						Y	Y	Y		Y
Cirripedia								Y																			
Corallanidae			Y																								
Crangonidae	*Crangonidae*																								Y		
Cumacea																	Y										
Decapoda																	Y										
Dendrobranchiata	*Dendrobranchiata*																								Y		Y
Eugonatonotidae	*Eugonatonotus crassus*																					Y					
Eugonatonotidae																								Y	Y		
Eumunididae	*Eumunida* sp.																Y										
Euphausiacea																	Y										
Euphausiidae	*Euphausia americana*													Y													
Euphausiidae									Y																		
Gammaridae																	Y										
Goneplacidae									Y							Y											
Gonodactylidae	*Gonodactylus smithii*																	Y									Y
Gonodactylidae	*Neogonodactylus oerstedii*																				Y			Y			
Gonodactylidae									Y											Y							
Grapsidae	*Percnon gibbesi*																				Y						
Grapsidae	*Plagusia depressa*																				Y						
Grapsidae	*Planes minutus*																				Y						
Hippoidea																	Y										
Hippolytidae	*Hippolyte* sp.																										Y
Hyppolytidae	*Lysmata grabhami*																				Y						
Hyppolytidae	*Lysmata* sp.					Y												Y									
Hyppolytidae	*Thor* sp.													Y		Y											
Hyppolytidae									Y							Y						Y				Y	Y

(continued)

Family	Specie/taxon	Albins and Hixon (2008)	Morris and Akins (2009)	Côté and Maljković (2010)	Green etal. (2011)	Jud etal. (2011)	McCleery (2011)	Muñoz etal. (2011)	Abril (2012)	Cure etal. (2012)	Green etal. (2012)	Layman and Allgeier (2012)	Muñoz – Escobar and Gil – Agudelo (2012)	Valdez – Moreno etal. (2012)	Côté etal. (2013a, b)	Gómez – Pardo (2014)	Dahl and Patteron III (2014)	Villaseñor – Derbez and Herrera – Pérez (2014)	Rocha etal. (2015)	Pabón and Acero (2016)	Eddy etal. (2016)	Romero (2017)	Pantoja etal. (2017)	Own Data. Archipiélago de San Andrés, Providencia y Santa Catalina	Own Data. Bolívar	Own Data. La Guajira	Own Data. Magdalena
Inachidae	*Stenorhynchus seticornis*															Y	Y										
Inachoididae	*Anasimus latus*																Y										
Isopoda																							Y				
Lophogastrida																	Y										
Lysiosquilla	*Lysiosquilla scabricauda*																				Y						
Majidae	*Mithrax* spp.						Y					Y															
Majidae			Y					Y									Y										
Menippidae	*Menippe mercenaria*																Y										

Mithracidae	*Mithraculus forceps*																				Y						
Mithracidae	*Mithraculus* sp.																					Y					Y
Munididae	*Munida simplex*																				Y						
Mysida																	Y						Y				
Mysidae	*Mysidium* spp.																	Y									
Paguridae	*Pagurus* sp.																Y										
Paguridae									Y																Y		
Palaemonidae	*Brachycarpus biunguiculatus*															Y					Y						
Palaemonidae	*Palaemon northropi*																				Y						
Palaemonidae	*Periclimenes* sp.																						Y				
Palaemonidae									Y					Y								Y		Y	Y		Y
Palicidae	*Palicus* sp.																					Y			Y		
Pandalidae																						Y					
Panopeidae						Y																					
Parthenopidae																	Y										
Pasiphaeidae																								Y			
Penaeidae	*Farfantepenaeus duorarum*																Y										
Penaeidae	*Litopenaeus setiferus*																Y										
Penaeidae	*Metapenaeopsis* sp.																					Y		Y	Y		Y
Penaeidae	*Metapenaeus* sp.								Y																		
Penaeidae	*Penaeus* sp.															Y											
Penaeidae	*Rimapenaeus* sp.															Y				Y				Y			
Penaeidae	*Trachypenaeus similis*																Y										
Penaeidae												Y					Y				Y	Y		Y	Y		Y
Penaeoidea																							Y				
Percnidae	*Percnon gibbesi*																										Y

(continued)

Family	Specie/taxon	Albins and Hixon (2008)	Morris and Akins (2009)	Côté and Maljković (2010)	Green etal. (2011)	Jud etal. (2011)	McCleery (2011)	Muñoz etal. (2011)	Abril (2012)	Cure etal. (2012)	Green etal. (2012)	Layman and Allgeier (2012)	Muñoz – Escobar and Gil – Agudelo (2012)	Valdez – Moreno etal. (2012)	Côté etal. (2013a, b)	Gómez – Pardo (2014)	Dahl and Patteron III (2014)	Villaseñor – Derbez and Herrera – Pérez (2014)	Rocha etal. (2015)	Pabón and Acero (2016)	Eddy etal. (2016)	Romero (2017)	Pantoja etal. (2017)	Own Data. Archipiélago de San Andrés, Providencia y Santa Catalina	Own Data. Bolívar	Own Data. La Guajira	Own Data. Magdalena
Pleocyemata																	Y										
Porcellanidae	*Petrolisthes* sp.															Y											
Porcellanidae	*Porcellana sigsbeiana*																Y										
Porcellanidae						Y																					
Portunidae	*Callinectes sapidus*																	Y									
Portunidae	*Charybdis hellerii*															Y											
Portunidae	*Cronius* sp.																			Y		Y					Y
Portunidae	*Portunus anceps*																				Y						
Portunidae	*Portunus ordwayi*																					Y					

Portunidae	*Portunus sayi*																Y										
Portunidae	*Portunus* sp.								Y							Y						Y	Y	Y		Y	Y
Portunidae	*Portunus spinicarpus*																Y					Y		Y			
Portunidae												Y					Y						Y				
Processidae									Y							Y											
Pseudosquillidae	*Pseudosquilla* sp.													Y													
Pseudosquillidae	*Pseudosquilla ciliata*																								Y		Y
Pseudosquillidae														Y													
Rhynchocinetidae	*Cinetorhynchus rigens*						Y														Y						
Scyllaridae	*Scyllarides nodifer*																				Y						
Scyllaridae																	Y										
Sicyoniidae	*Sicyonia brevirostris*															Y											
Sicyoniidae	*Sicyonia* sp.																							Y		Y	
Squillidae	*Meiosquilla quadridens*																					Y					Y
Squillidae	*Squilla empusa*																Y										
Squillidae	*Squilla* sp.															Y							Y				
Squillidae									Y								Y					Y			Y		Y
Stenopodidae	*Stenopus hispidus*								Y									Y		Y	Y		Y	Y			
Stenopodidae	*Stenopus scutellatus*																						Y				
Stenopodidae	*Stenopus* sp.																										Y
Stenopodidae																	Y										
Stomatopoda																							Y				
Thalassinidea																				Y							Y
Thychidae	*Pitho* sp.											Y															
Thychidae												Y															
Xanthidae									Y							Y	Y				Y						

(continued)

Family	Specie/taxon	Albins and Hixon (2008)	Morris and Akins (2009)	Côté and Maljković (2010)	Green etal. (2011)	Jud etal. (2011)	McCleery (2011)	Muñoz etal. (2011)	Abril (2012)	Cure etal. (2012)	Green etal. (2012)	Layman and Allgeier (2012)	Muñoz – Escobar and Gil – Agudelo (2012)	Valdez – Moreno etal. (2012)	Côté etal. (2013a, b)	Gómez – Pardo (2014)	Dahl and Patteron III (2014)	Villaseñor – Derbez and Herrera – Pérez (2014)	Rocha etal. (2015)	Pabón and Acero (2016)	Eddy etal. (2016)	Romero (2017)	Pantoja etal. (2017)	Own Data. Archipiélago de San Andrés, Providencia y Santa Catalina	Own Data. Bolívar	Own Data. La Guajira	Own Data. Magdalena
Xanthoidea																						Y		Y	Y		Y
Mollusks																											
Loliginidae	*Loligo* sp.																Y										
Loliginidae	*Sepioteuthis sepioidea*																				Y						
Marginellidae	*Volvarina albolineata*																				Y						
Marginellidae																				Y				Y			
Octopodidae	*Octopus* spp.																				Y			Y			
Octopodidae								Y									Y						Y				
Bivalvia								Y																			
Cephalopoda									Y														Y				
Mollusca																	Y										

References

Abril A (2012) Aspectos de la biología y ecología trófica y reproductiva del pez león, *Pterois volitans* (Linnaeus 1758), en San Andrés Isla, Reserva de la Biosfera-Seaflower, Caribe colombiano. Bachelor Thesis. Universidad Nacional de Colombia, Bogotá, 33 p

Albins M (2015) Invasive Pacific lionfish *Pterois volitans* reduce abundance and species richness of native Bahamian coral-reef fishes. Mar Ecol Prog Ser 522:231–243

Albins MA, Hixon MA (2008) Invasive Indo-Pacific lionfish *Pterois volitans* reduce recruitment of Atlantic coral-reef fishes. Mar Ecol Prog Ser 367:233–238

Albins MA, Lyons PJ (2012) Invasive red lionfish *Pterois volitans* blow directed jets of water at prey fish. Mar Ecol Prog Ser 448:1–5

Andrade CA, Barton ED (2005) The Guajira upwelling system. Cont Shelf Res 25(9):1003–1022

Araújo M, Bolnick DI, Layman CA (2011) The ecological causes of individual specialisation. Ecol Lett 14(9):948–958

Arbeláez N, Acero PA (2011) Presencia del pez león *Pterois volitans* (Linnaeus) en el manglar de la bahía de Chengue, Caribe colombiano. Bol Invest Mar Cost 40(2):431–435

Arias-González JE, González-Gándara C, Luis Cabrera J, Christensen V (2011) Predicted impact of the invasive lionfish *Pterois volitans* on the food web of a Caribbean coral reef. Environ Res 111 (7):917–925

Arredondo AT (2016) Comparación trófica del pez león, *Pterois volitans* (Linnaeus, 1758) con tres especies de peces arrecifales del Caribe mexicano: a través del uso de isótopos estables y contenido estomacal. Master Thesis. Centro de Investigación Científica de Yucátan, Posgrado en Ciencias del Agua, Cancún, Quintana Roo, 108 p

Barbour A, Montgomery M, Adamson A, Díaz-Ferguson E, Silliman B (2010) Mangrove use by the invasive lionfish *Pterois volitans*. Mar Ecol Prog Ser 401:291–294

Bernal G, Poveda G, Roldán P, Andrade C (2006) Patrones de variabilidad de las temperaturas superficiales del mar en la costa Caribe colombiana. Cienc Tierra 30(115):195–208

Betancur-R R, Hines A, Acero PA, Ortí G, Wilbur AE, Freshwater DW (2011) Reconstructing the lionfish invasion: insights into Greater Caribbean biogeography. J Biogeogr 38(7):1281–1293

Cabrera-Guerra D (2014) Caracterización de la dieta del pez león (Teleostei: Scorpaenidae: *Pterois* sp.) en cuatro localidades de Cuba. Bachelor thesis, Universidad de La Habana. La Habana

Carpenter KE (ed) (2002) The living marine resources of the Western Central Atlantic. FAO species identification. Guide for Fishery Purposes and American Society of Ichthyologists and Herpetologists Special Publication No. 5. FAO. Rome. 2055 p

Clavel J, Julliard R, Devictor V (2011) Worldwide decline of specialist species: toward a global functional homogenization? Front Ecol Environ 9(4):222–228

Claydon JAB, Calosso MC, Traiger SB (2012) Progression of invasive lionfish in seagrass, mangrove and reef habitats. Mar Ecol Prog Ser 448:119–129

Colwell RK (2013) Estimates version 9.1 statistical estimation of species richness and shared species from samples (software and user's guide). Freeware for Windows and Mac OS, http://viceroy.eeb.uconn.edu/estimates/

Côté IM, Maljković A (2010) Predation rates of indo-pacific lionfish on Bahamian coral reefs. Mar Ecol Prog Ser 404:219–225

Côté IM, Green SJ, Hixon MA (2013a) Predatory fish invaders: insights from Indo-Pacific lionfish in the western Atlantic and Caribbean. Biol Conserv 164:50–61

Côté IM, Green SJ, Morris JA, Akins JL, Steinke D (2013b) Diet richness of invasive Indo-Pacific lionfish revealed by DNA barcoding. Mar Ecol Prog Ser 472:249–256

Cucherousset J, Aymes JC, Santoul F, Céréghino R (2007) Stable isotope evidence of trophic interactions between introduced brook trout *Salvelinus fontinalis* and native brown trout *Salmo trutta* in a mountain stream of south-west France. J Fish Biol 71(Suplemment D):210–223

Cure K, Benkwitt CE, Kindinger TL, Pickering EA, Pusack TJ, McIlwain JL, Hixon MA (2012) Comparative behavior of red lionfish *Pterois volitans* on native Pacific versus invaded Atlantic coral reefs. Mar Ecol Prog Ser 467:181–192

Dahl KA, Patterson WF III (2014) Habitat-specific density and diet of rapidly expanding invasive red lionfish, *Pterois volitans*, populations in the northern Gulf of Mexico. PLoS One 9(8):1–13
Darling ES, Green SJ, O'Leary JK, Côté IM (2011) Indo-Pacific lionfish are larger and more abundant on invaded reefs: a comparison of Kenyan and Bahamian lionfish populations. Biol Inv 13(9):2045–2051
DeNiro MJ, Epstein S (1978) Influence of diet on the distribution of carbon isotopes in animals. Geochim Cosmochimica Acta 42:495–506
Eddy C, Pitt J, Morris JA, Smith S, Goodbody-Gringley G, Bernal D (2016) Diet of invasive lionfish (*Pterois volitans* and *P. miles*) in Bermuda. Mar Ecol Prog Ser 558:193–206
Fajardo EGG (1979) Surgencia costera en las proximidades de la península colombiana de La Guajira. Bol Cient CIOH 2:7–19
Fishelson L (1997) Experiments and observations on food consumption, growth and starvation in *Dendrochirus brachypterus* and *Pterois volitans* (Pteroinae, Scorpaenidae). Environ Biol Fish 50(4):391–403
García-Rodríguez A (2015) Principales relaciones ecológicas del pez león (*Pterois volitans/miles*) en arrecifes de La Habana, Cuba. Master Thesis. Centro de Investigaciones Marinas, Universidad de La Habana, La Habana, Cuba, 107 p
García-Urueña R, Acero PA, Coronado-Carrascal V (2015) Reef fish community in presence of the lionfish (*Pterois volitans*) in Santa Marta, Colombian Caribbean. Rev MVZ Córdoba 20:4989–5003
Gómez-Pardo HM (2014) Ecología trófica del pez león *Pterois volitans* (Linnaeus, 1758) en tres formaciones coralinas del Caribe colombiano. Bachelor Thesis, Universidad Jorge Tadeo Lozano, Santa Marta, 51 p
González-Corredor J (2014) El pez león *Pterois volitans* (Scorpaenidae) como estudio de caso de invasiones biológicas marinas: aspectos biológicos y ecológicos en el Caribe occidental insular colombiano. Master thesis. Universidad Nacional de Colombia. Santa Marta, Colombia, 75 p
Green SJ, Côté IM (2009) Record densities of Indo-Pacific lionfish on Bahamian coral reefs. Coral Reefs 28(1):107
Green SJ, Côté IM (2014) Trait-based diet selection: prey behaviour and morphology predict vulnerability to predation in reef fish communities. J Anim Ecol 83(6):1451–1460
Green SJ, Akins JL, Côté IM (2011) Foraging behaviour and prey consumption in the Indo-Pacific lionfish on Bahamian coral reefs. Mar Ecol Prog Ser 433:159–167
Green SJ, Akins JL, Maljković A, Côté IM (2012) Invasive lionfish drive Atlantic coral reef fish declines. PLoS One 7(3):1–3
Hernández-Abello J, García-Urueña R, Acero PA (2015) Estructura de tallas y preferencia al sustrato del pez león (*Pterois volitans*) (Scorpaeniformes: Scorpaenidae) en Santa Marta (Colombia). Acta Biol Colomb 20(2):175–182
Hyslop EJ (1980) Stomach contents analysis-a review of methods and their application. J Fish Biol 17:411–429
Ivanova NV, Zemlak TS, Hanner RH, Hebert PDN (2007) Universal primer cocktails for fish DNA barcoding. Mol Ecol Notes 7(4):544–548
Jud ZR, Layman CA, Lee JA, Arrington DA (2011) Recent invasion of a Florida (USA) estuarine system by lionfish *Pterois volitans/P. miles*. Aq Biol 13(1):21–26
Kalogirou S, Corsini M, Kondilatos G, Wennhage H (2007) Diet of the invasive piscivorous fish *Fistularia commersonii* in a recently colonized area of the eastern Mediterranean. Biol Inv 9:887–896
Kim SL, Koch PL (2012) Methods to collect, preserve, and prepare elasmobranch tissues for stable isotope analysis. Environ Biol Fish 95(1):53–63
Layman CA, Allgeier JE (2012) Characterizing trophic ecology of generalist consumers: a case study of the invasive lionfish in the Bahamas. Mar Ecol Prog Ser 448:131–141
Linardich C, Ralph G, Carpenter K, Cox N, Robertson DR, Harwell H, Acero P A, Anderson Jr W, Barthelat F, Bouchereau L, Brown JJ, Buchanan J, Buddo D, Collette B, Comeros-Raynal M, Craig M, Curtis M, Defex T, Dooley J, Driggers III W, Elfes-Livsey C, Fraser T, Gilmore Jr R,

Grijalba Bendeck L, Hines A, Kishore R, Lindeman K, Maréchal JP, McEachran J, McManus R, Moore J, Munroe T, Oxenford H, Pezold F, Pina Amargós F, Polanco-Fernández A, Polidoro B, Pollock C, Robins R, Russell B, Sayer C, Singh-Renton S, Smith-Vaniz W, Tornabene L, Van Tassell J, Vié JC, Williams JT (2017) The conservation status of marine bony shorefishes of the Greater Caribbean. IUCN, Gland., 75 p

Mack RN, Simberloff D, Lonsdale WM, Evans H, Clout M, Bazzaz FA (2000) Biotic invasions: causes, epidemiology, global consequences, and control. Ecol Appl 10(3):689–710

MacNeil MA, Connolly SR (2015) Multi-scale patterns and processes in reef fish abundance. In: Mora C (ed) Ecology of fishes on coral reefs. Cambridge University Press, Cambridge, pp 116–124

McCleery C (2011) A comparative study of the feeding ecology of invasive lionfish (*Pterois volitans*) in the Caribbean. Physis J Mar Sci 9:38–43

Molnar JL, Gamboa RL, Revenga C, Spalding MD (2008) Assessing the global threat of invasive species to marine biodiversity. Front Ecol Environ 6(9):485–492

Mooney HA, Cleland EE (2001) The evolutionary impact of invasive species. Proc Natl Acad Sci U S A 98(10):5446–5451

Morris JA, Akins JL (2009) Feeding ecology of invasive lionfish (*Pterois volitans*) in the Bahamian archipelago. Environ Biol Fish 86(3):389–398

Morris JA, Akins JL, Barse A, Cerino D, Freshwater DW (2009) Biology and ecology of the invasive lionfishes, *Pterois miles* and *Pterois volitans*. Proceedings of the 61st Gulf Carib Fish Inst 61: 1–6

Muñoz RC, Currin CA, Whitfield PE (2011) Diet of invasive lionfish on hard bottom reefs of the Southeast USA: insights from stomach contents and stable isotopes. Mar Ecol Prog Ser 432:181–193

Muñoz-Escobar L, Gil-Agudelo DL (2012) Composición dietaria del pez león, *Pterois volitans* (Pisces: Scorpaenidae), en Santa Marta y El Parque Nacional Natural Tayrona. Bol Invest Mar Cost 41(2):471–477

Newsome SD, Martínez del Río C, Bearhop S, Phillips DL (2007) A niche for isotopic ecology. Front Ecol Environ 5(8):429–436

Newsome SD, Yeakel JD, Wheatley PV, Tinker MT (2012) Tools for quantifying isotopic niche space and dietary variation at the individual and population level. J Mamm 93(2):329–341

Pabón P, Acero PA (2016) Ecología trófica del invasor pez león *Pterois volitans* en el Caribe colombiano: Impacto sobre familias ícticas de Santa Marta y San Andrés. In: Campos-Campos-N, Acero PA (eds) Contribuciones en Ciencias del Mar. Universidad Nacional de Colombia, Santa Marta, pp 50–74

Pantoja-Echevarría LM, Chevalier-Monteagudo PP, Cabrera-Guerra D, Corrada-Wong RI, Cobián-Rojas D, Caballero-Aragón H, García-Rodríguez A, Fernández-Osoria RM (2017) Superposición de la dieta del pez león *Pterois volitans* (Teleostei: Scorpaenidae) con la de peces nativos de nivel trófico similar en Cuba. Bol Invest Mar Cost 46:115–134

Pimiento C, Nifong JC, Hunter ME, Monaco E, Silliman BR (2013) Habitat use patterns of the invasive red lionfish *Pterois volitans*: a comparison between mangrove and reef systems in San Salvador, Bahamas. Mar Ecol 36:28–37

Pinkas L, Oliphant MS, Iverson LK (1971) Food habits ofalbacore, bluefin tuna and bonito in Californian waters. Calif Fish Game 152:1–105

Pinnegar J, Polunin N (2000) Contributions of stable-isotope data to elucidating food webs of Mediterranean rocky littoral fishes. Mar Biol 122:399–409

Post DM, Layman CA, Arrington DA, Takimoto G, Quattrochi J, Montaña CG (2007) Getting to the fat of the matter: models, methods and assumptions for dealing with lipids in stable isotope analyses. Oecologia 152(1):179–189

Rilov G (2009) Predator-prey interactions of marine invaders. In: Rilov G, Crooks JA (eds) Biological invasions in marine ecosystems. Ecological studies (analysis and synthesis) vol. 204. Springer, Berlin/Heidelberg, pp 261–285

Robertson DR, Peña EA, Posada JM, Claro R (2015) Peces costeros del Gran Caribe: Sistema de información en línea. Version 1.0. Instituto Smithsonian de Investigaciones Tropicales, Balboa, República de Panamá. http://biogeodb.stri.si.edu/caribbean/es/pages

Rocha LA, Rocha CR, Baldwin CC, Weigt LA, McField M (2015) Invasive lionfish preying on critically endangered reef fish. Coral Reefs 34(3):803–806

Rojas S, Tavera y JJ, Acero P. A (In press) Unraveling lionfish invasion: Is Pterois volitans truly a morphologically novel predator in the Caribbean? Biol Inv

Romero D (2017) Dieta del pez león *Pterois volitans* en dos ambientes del Caribe colombiano. Bachelor Thesis, Universidad Jorge Tadeo Lozano, Santa Marta, 58 p

Sandel V, Martínez-Fernández D, Wangpraseurt D, Sierra L (2014) Ecology and management of the invasive lionfish *Pterois volitans/miles* complex (Perciformes: Scorpaenidae) in southern Costa Rica. Rev Biol Trop 63(1):213–221

Schofield PJ (2009) Geographic extent and chronology of the invasion of the non-native lionfish (*Pterois volitans* [Linnaeus 1758] and *P.miles* [Bennett 1828]) in the western North Atlantic and Caribbean Sea. Aq Inv 4(3):473–479

Selwyn JD, Johnson JE, Downey-Wall AM, Bynum AM, Hamner RM, Hogan JD, Bird CE (2017) Simulations indicate that scores of lionfish (*Pterois volitans*) colonized the Atlantic Ocean. PeerJ 5:e3996. https://doi.org/10.7717/peerj.3996

Spalding MD, Fox HE, Allen GR, Davidson N, Ferdaña ZA, Finlayson M, Robertson J (2007) Marine ecoregions of the world: a bioregionalization of coastal and shelf areas. Bioscience 57 (7):573–583

Valdez-Moreno M, Quintal-Lizama C, Gómez-Lozano R, García-Rivas MDC (2012) Monitoring an alien invasion: DNA barcoding and the identification of lionfish and their prey on coral reefs of the Mexican Caribbean. PLoS One 7(6):1–8

Villaseñor-Derbez JC, Herrera-Pérez R (2014) Brief description of prey selectivity and ontogenetic changes in the diet of the invasive lionfish *Pterois volitans* (Actinopterygii, Scorpaenidae) in the Mexican Caribbean. Pan-Am J Aq Sci 9(2):131–135

Whitfield P, Gardner T, Vives S, Gilligan M, Courtenay W, Ray G, Hare J (2002) Biological invasion of the Indo-Pacific lionfish *Pterois volitans* along the Atlantic coast of North America. Mar Ecol Prog Ser 235:289–297

Whitfield PE, Hare JA, David AW, Harter SL, Muñoz RC, Addison CM (2007) Abundance estimates of the Indo-Pacific lionfish *Pterois volitans/miles* complex in the western North Atlantic. Biol Inv 9(1):53–64

Williams AB (1984) Shrimps, lobsters and crabs of the Atlantic coast of the Eastern United States, Maine to Florida. Smithsonian Institution Press, Washington

Chapter 9
Environmental Impact of Invasion by an African Grass (*Echinochloa pyramidalis*) on Tropical Wetlands: Using Functional Differences as a Control Strategy

Hugo López Rosas, Eduardo Cejudo, Patricia Moreno-Casasola, Luis Alberto Peralta Peláez, María Elizabeth Hernández, Adolfo Campos C., and Gustavo Aguirre León

Abstract Tropical wetlands are commonly used for cattle ranching and have been modified either by draining them or introducing non-native species that are palatable to cattle. Some of these introduced species have become wetland and dune invaders. In Mexico, the introduction of antelope grass (*Echinochloa pyramidalis*) and its effects are being documented. This grass species is highly appreciated by cattle ranchers and is invading natural wetlands. It has C4 photosynthesis, high biomass production and high vegetative propagation, is tolerant to grazing and able to grow in both flooded and dry conditions. It is reducing plant biodiversity by increasing its own aerial coverage, changing wetland hydrology, reducing faunal habitat and causing soil physicochemical changes (e.g. vertical accretion). Reducing its dominance and increasing the density of native wetland species is difficult, expensive and time-consuming. We began a restoration project in a coastal wetland in central Veracruz, Gulf of Mexico, which included using shade to control the invader. This strategy reduced *E. pyramidalis* cover and increased the cover of native species, highlighting the importance of understanding the functional differences between native and invasive species when developing strategies for the control and eradication of problematic species.

H. López Rosas · P. Moreno-Casasola (✉) · M. E. Hernández · A. Campos C. · G. Aguirre León
Instituto de Ecología A.C, Xalapa, Veracruz, Mexico
e-mail: patricia.moreno@inecol.mx; elizabeth.hernandez@inecol.mx; adolfo.campos@inecol.mx; gustavo.aguirre@inecol.mx

E. Cejudo
Centro de Investigación Científica de Yucatán A.C, Cancún, Quintana Roo, Mexico
e-mail: eduardo.cejudo@cicy.mx

L. A. Peralta Peláez
Instituto Tecnológico de Veracruz, Veracruz, Veracruz, Mexico
e-mail: peralta@itver.edu.mx

C. Makowski, C. W. Finkl (eds.), *Impacts of Invasive Species on Coastal Environments*, Coastal Research Library 29,
https://doi.org/10.1007/978-3-319-91382-7_9

Keywords C4 photosynthesis · Ecological restoration · Ecosystem services · Invasive grasses · Mexico · Shade treatments

9.1 Introduction

Human populations are concentrated along coasts, and consequently coastal ecosystems are some of the most impacted and altered worldwide. They are exposed to various hazards such as coastal flooding, hurricanes and transmission of marine-related infectious diseases. Hazards in coastal areas often become disasters through the erosion of resilience, driven by environmental change and by human action (Adger et al. 2005). At the same time, estuarine and coastal ecosystems are some of the most heavily used and threatened natural systems globally (Halpern et al. 2008; Lotze et al. 2006; Worm et al. 2006). They have suffered biodiversity losses, reduction of coastal vegetation cover and ecosystem functions, which may have contributed to biological invasions, declining water quality, and decreased coastal protection from flooding and storm events (Barbier et al. 2011; Cochard et al. 2008; Koch et al. 2009). Human dependence on marine and coastal ecosystems services is intensifying, even so these systems still are increasingly being degraded or destroyed (Lau 2013).

Natural and even transformed ecosystems provide goods and services that have been called natural capital and ecosystem services (Costanza et al. 1997, 2017; Farber et al. 2002). Most of the time, however, they are neither valued nor considered in the context of their economic impact, due to the scarcity of information (Duncan et al. 2004). Wetlands are ecosystems that provide many environmental services (Barbier 1994; Costanza et al. 1997; Pérez-Maqueo et al. 2007) and are a fundamental part of the natural environment. In tropical areas, mangroves have received most of the attention, but freshwater wetlands have been neglected or poorly studied. Liquete et al. (2013) reviewed and summarized the existing scientific literature related to marine and coastal ecosystem services aiming to extract and classify indicators used to assess and map ecosystem services. Food provision, particularly fisheries, had the largest body of literature; water purification and coastal protection were the most frequently studied regulation and maintenance services; and under the cultural services, recreation and tourism were relatively well assessed, but the majority of the case studies were carried out on mangroves and coastal wetlands, mainly in Europe and North America.

Freshwater wetlands are ecologically and economically recognized as systems of social and cultural importance (Mitsch and Gosselink 2007; Turner and Harrison 1983). However, wetland degradation continues almost all over the world. Mitsch and Gosselink (2007) indicate that the United States has lost 53% of its wetlands, Australia around 50%, China 60%, New Zealand and Europe over 90%. In Mexico, Landgrave and Moreno-Casasola (2012) estimated that 62% of the Mexican wetlands have disappeared or deteriorated.

In coastal areas, population increase in recent decades has affected coastal, freshwater, brackish and saline wetlands. All wetlands are subjected to an increasing degree of human pressure through water abstraction, changes in the natural flood regime, land reclamation, pollution, over-utilization of natural resources, and poaching, but also owing to extensive cattle grazing on the wetlands, mainly in the tropics (Gomes Pinto et al. 2010; Junk et al. 2006; Junk and Nunhes da Cunha 2012; López Rosas et al. 2006; Melgarejo Vivanco 1980; Moreno-Casasola et al. 2012; Travieso-Bello et al. 2005). Wetland transformation to flooded pastures for cattle rearing is frequently coupled with the introduction of exotic grass species to increase pasture productivity. Some of these species have become invaders in tropical wetlands, such as *Pennisetum purpureum* (Williams and Baruch 2000), *Echinochloa pyramidalis* (López Rosas et al. 2006), *Spartina alterniflora* (Li and Zhang 2008), and *Phalaris arundinacea* (Maurer et al. 2003). Some of these were introduced from Africa (Parsons 1972, Williams and Baruch 2000), with severe consequences to the biological diversity of tropical wetlands and ecosystem functioning (D'Antonio and Vitousek 1992).

9.2 Exotic Grass Introduction

A biological invasion is the introduction and naturalization of non-indigenous species and has been recognized as a major threat to natural ecosystems (Vitousek 1990). Non-indigenous invasive plants have a variety of negative effects on the structure and functioning of the invaded ecosystems, such as altering the food webs (O'Connor et al. 2000), changing the rates of erosion and sedimentation in streams (Lacey et al. 1989), altering the hydrological cycle (Loope and Sánchez 1988, Vitousek 1990), modifying nutrient cycles (Ramakrishnan and Vitousek 1989; Vitousek 1990) and causing considerable disturbance (MacDougall and Turkington 2005).

Specifically, invasion by non-indigenous plants is a concern because of the biological and ecological traits of these species, which make them better competitors than the plants of other native families. Among the trends found in invasive grasses are increased litter production, dense root systems, highly efficient water and soil nutrient acquisition, a vigorous capacity for vegetative reproduction, and tolerance to drought (D'Antonio and Vitousek 1992, Chapman 1996). Among grasses, the C4 perennial African species are very successful invaders in tropical and subtropical regions outside of Africa because their adaptations give them a strong competitive advantage over native plants such as tolerance to grazing, fire, low soil nutrient requirements and tolerance to both droughts and flooding (Chapman 1996; D'Antonio and Vitousek 1992; López Rosas et al. 2006; López Rosas and Moreno-Casasola 2012; Parsons 1972; Williams and Baruch 2000). Table 9.1 show some examples of C4 perennial African grasses that have invaded regions outside of Africa.

Table 9.1 Some examples of C4 perennial African grasses that have invaded regions outside of Africa

Species, common name and invaded regions	Problems	References
Andropogon gayanus – Gamba grass; South American savannas, and Tropical Australia	Introduced as forage to other regions[a]. Displacement of native grasses; invasion of natural parks and reserves; high production of deep roots and potential change in the dynamics of soil C and N; alteration of the fire regime, with higher char height	Fisher et al. (1994), Lonsdale (1994), Rossiter-Rachor et al. (2009) and Setterfield et al. (2010)
Urochloa brizantha – Palisade grass; Tropical South America	Introduced as forage to other regions[a]. Displacement of native plants and impoverishment of the associated fauna; alteration of C:N litter ratio in soils	Piccolo et al. (1994)
Urochloa decumbens – Signal grass; Brazilian savannas, and Tropical Australia	Introduced as forage to other regions[a]. Displacement of native grasses and vegetaton	Lonsdale (1994), Pivello et al. (1999) and Weeds of Australia (2015)
Urochloa humidicola – Creeping signal grass; South American savannas	Introduced as forage to other regions[b]. High production of deep roots and potential change in the dynamics of soil C	Fisher et al. (1994)
Urochloa mutica – Para grass; Wetlands of Australia, Tasmania, and Tropical America	Introduced as forage to other regions[a]. Aggressive colonization and displacement of native plants; facilitation of the invasion by other non-indigenous grasses; formation of large floating mats in ponds and other wetlands	Lonsdale (1994) and Parsons (1972)
Eragrostis lehmanniana – Lehmann lovegrass, United States, Northern Mexico	Introduced for range restoration in southwest USA[a]. Displacement of native plants and decrease of the associated birds and insects species	Bock et al. (1986) and COTECOCA (1991)
Hyparrhenia rufa – Jaragua grass, Tropical America	Introduced as forage to other regions[a]. Displacement of native grasses in the Venezuelan and Brazilian savannas; impedes the regeneration of trees at riverine forest edges; alteration of the fire regime	Nepstad et al. (1991) and Rodríguez (2001)
Melinis minutiflora – Molasses grass; Tropical America and Hawaii	Introduced as forage to other regions[a]. Displacement of native grasses in the	D'Antonio et al. (2000), Hoffmann et al. (2004), Hoffmann, and Haridasan (2008),

(continued)

Table 9.1 (continued)

Species, common name and invaded regions	Problems	References
	Venezuelan and Brazilian savannas; enhancement of the fire cycle; successful light competition; modification of N cycle, reduced tree regeneration	Nepstad et al. (1991), Pivello et al. (1999) and Rodríguez (2001)
Panicum maximum – Guinea grass; Tropical America and South of United States	Introduced as forage to other regions[a]. Alteration of the fire regime and displacement of native plants	Lonard and Judd (2002), Nepstad et al. (1991) and Rodríguez (2001)
Cenchrus polystachios (Pennisetum polystachyum) – Mission grass; North of Australia	Introduced as forage to other regions[c]. Alteration of the fire regime	Lonsdale (1994)
Andropogon glomeratus var. *pumilus (Schizachyrium condensatum* –- Colombian or Bush bluegrass; Hawaii	Reasons for introduction unknown[d]. Promotes the spread of fire	D'Antonio et al. (1998)

Modified from López Rosas et al. (2006)
[a]Invasive Species Compendium (CABI)
[b]Assessment of non-native plants in Florida's Natural Areas
[c]Invasive pasture grasses in Northern Asutralia
[d]Hawaii Invasive Species Council

9.2.1 The Introduction of Echinochloa pyramidalis

Echinochloa pyramidalis (Lam.) Hitchc. & A. Chase is abundant in the floodplains of Tropical Africa where it grows in dense, pure stands (Denny 1993; Howard-Williams and Walker 1974). Also known as antelope grass, it is a very productive C4 perennial grass, tolerant to both drought and flooding (Skerman and Riveros 1990), and is used by local fauna (Lewinson and Carter 2004) such as hippopotamus in its native Africa.

In south-eastern Mexico it is commonly found in coastal freshwater wetlands or at the boundary between wetlands and terrestrial ecosystems where it has been introduced to sustain cattle during the dry season, when the wetlands are still productive (COTECOCA 1991). Several authors have reported that the introduction and expansion of *E. pyramidalis*, together with grazing practices, has caused a decrease in plant diversity (López Rosas et al. 2006; Thomas and Reid 2007; Travieso-Bello et al. 2005; this study). The approximate dates of introduction are known for a few countries., In Mexico and Guyana it was introduced during the 1960s and 1970s respectively, specifically introduced as forage (Tapia et al. 1962), and was likely introduced much earlier in other countries such as India (Bor 1960). In Guyana it was introduced in 1984 and has become a serious weed in sugar cane fields

(Bushundial 1991; Bushundial et al. 1997; EPA Guyana 2011). Dates obtained from the earliest herbarium specimens are found in Australia (1931), St. Helena (1974), Argentina (1977), Costa Rica (1982) and Peru (1992; GBIF 2012). In Brazil, it has been cultivated in Pará for at least 20 years (López Rosas 2007). In all cases, it is believed that the introduction was deliberate, and perhaps earlier than the documented dates. There are no confirmed instances of accidental introduction (Invasive Species Compendium 2017). This grass occurs as a weed of rice in Cote d'Ivoire (Kent et al. 2001) and is also spreading as a grassy weed over lowland forest zones in Australia, India and the Philippines (Pancho 1991). It is listed as a serious weed in Nigeria, Swaziland, Sudan and Madagascar (Holm et al. 1979). Pancho (1991) mentioned that it was introduced to the Philippines "recently" and is now abundant in rice crops in Laguna.

9.3 Description of *Echinochloa pyramidalis* (Antelope Grass)

Echinochloa pyramidalis belongs to the Poaceae family. It is known as antelope grass, limpopo grass, sil grass, ahidrano (English, US), and pasto or zacate alemán (Mexico). Synonyms according to Tropicos include: *Echinochloa guadeloupensis* (Hack.) Wiegand, *Echinochloa holubii* (Stapf) Stapf, *Echinochloa pyramidalis* var. *guadeloupensis* (Hack.) Stehlé, *Echinochloa quadrifaria* var. *atroviolacea* (A. Rich.) Chiov., *Echinochloa senegalensis* Mez, *Echinochloa verticillata* Berhaut, *Panicum atroviolaceum* A. Rich., *Panicum crus-galli* var. *molle* Pilg. ex Peter, *Panicum crusgalli* var. *polystachyum* Asch. & Schweinf., *Panicum frumentaceum* Benth., *Panicum holubii* Stapf, *Panicum plicatum* Lam., *Panicum pyramidale* Lam, *Panicum pyramidale* var. *hebetatum* Stapf, *Panicum pyramidale* var. *quadrifarium* (Hochst. ex A. Rich.) Chiov., *Panicum pyramidale* var. *spadiceum* Peter, *Panicum quadrifarium* Hochst. ex A. Rich., *Panicum setarioides* Peter, *Panicum setarioides* Steud, *Panicum spadiciferum* Peter, and *Panicum spectabile* var. *guadeloupense* Hack. (http://www.tropicos.org/Name/25520876).

This grass is a robust, rhizomatous, reed-like perennial, erect up to 300 cm high with solid stems, rarely to 450 cm, glabrous or short-haired, with longer hairs on the veins. Its leaves are up to 60 cm long, 2 cm wide, glabrous when growing in the water, but erect plants can have sharp, irritating hairs on the leaf and sheath. Inflorescence is 15–30 cm long with racemes up to 8 cm long having purplish, acute, awnless spikelets 3–4 mm long with many overlapping racemes (Napper 1965). The upper glume as long as the spikelet; the lower glume less than half as long. Lower lemma awnless or with an occasional awn up to 1 or 2 mm long. Upper glume 2–3 mm (Chippindall 1955; Clayton 1989; Clayton and Renvoize 1982; Invasive Species Compendium 2017; Fig. 9.1). This grass establishes along the edges of water (Fig. 9.2), but also develops spongy horizontal stems spreading for many meters across the water's surface, or open mud, rooting at the nodes, with the ability to produce tillers easily at the nodes (Yabuno 1968).

Fig. 9.1 Antelope grass (*Echinochloa pyramidalis)* flowering

Fig. 9.2 *Echinochloa pyramidalis* can establish along the edges of water courses and develop spongy horizontal stems spreading across the water surface or unconsolidated mud

9.3.1 Range Description

Echinochloa pyramidalis is native to Africa and widely distributed across the continent; whether it is native or introduced in Madagascar is uncertain. Kuri cattle

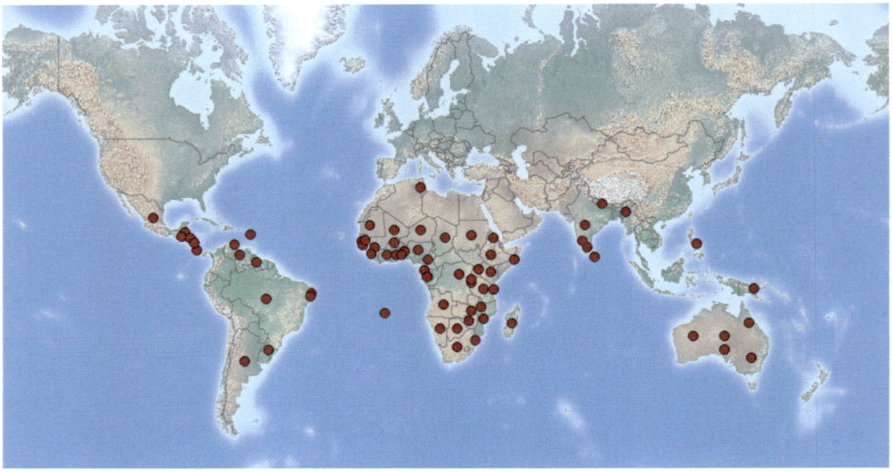

Fig. 9.3 World distribution of the antelope grass, *Echinochloa pyramidalis*

graze on the inundated antelope grass around Lake Chad (Purseglove 1976). Although it is extremely coarse, indigenous animals graze it readily to ground level at the end of the dry season. This grass is tolerant to intensive grazing (John et al. 1993). *E. pyramidalis* and *Echinochloa scabra* (previously *E. stagnina*) are the dominant species in the great floating meadows of the Niger and Lake Chad and in the major part of the Sudd at the head-waters of the Nile (FAO 2012). Its distribution is associated with wetlands, rivers and irrigation systems. It has been introduced to many other countries outside Africa as a fodder grass (Fig. 9.3). It is not known to have naturalized in most of those countries. Is has become a problem in Mexico (López Rosas et al. 2015; Invasive Species Compendium 2017) where it has been found in the coastal states of Campeche, Chiapas, Nayarit, Oaxaca, Tabasco, Veracruz, Yucatán and in three non-coastal states, Guanajuato, Puebla and Mexico (SNIB 2017). World reports indicate it is distributed from sea level to 1500 m a.s.l.

9.3.2 Environmental Requirements

Echinochloa pyramidalis is considered an aquatic species associated with wetlands, rivers and irrigation systems, flood plain grasslands and lake shores, floating meadows, swamps and marshes, wherever rainfall accumulates. Hence, these are more important factors in distribution than climate type, though its range is limited by sensitivity to frost. It withstands a wide range of water depths and duration of flooding (FAO 2012; Howard-Williams and Walker 1974; Scholte 2007; Skerman and Riveros 1990). It is also tolerant to moderate salinity and drought conditions. In Brazil, *E. pyramidalis* was among the more productive species tested in a mangrove area, suggesting some tolerance to salinity (Nascimento et al. 1988). Work in

Cameroon confirmed that it was tolerant to salinity levels of up to 6.5 ppt under both drained and flooded conditions for a 100 day period (Pare et al. 2011). It can succeed where there is seasonal flooding up to 1 m depth (Scholte 2007); however, its growth is suppressed by prolonged flooding (16 months) after cutting to soil level (Zuloaga and Morrone 2003). It is usually associated with badly drained black clays ("black cotton" soils) of alluvial nature, which become very sticky when wet but when dry are very hard and deeply fissured or have a blocky structure, often alkaline with a pH as high as 9.2 (when drained soil is normally slightly acid at pH 6; Vesey-Fitzgerald 1963). Phosphorus addition can increase yield (Smith et al. 1991).

9.3.3 Reproductive Biology and Physiology

This species is a C4 grass (Howard-Williams and Walker 1974), indifferent to day length (Evans et al. 1964). Reproduction occurs from seed but also by the vegetative spread of both rhizomes and stolons. It is a heavy seed producer that sheds its seeds during the rains, but sometimes seed germination is low (Fig. 9.4). No information has been found on the conditions for seed germination (Invasive Species Compendium 2017). In our working area on the coast of Veracruz we have not seen it germinate in the field (López Rosas, *pers. obs.*). It can be propagated by cuttings. Flowering occurs about half way through the wet season and seeds are shed before

Fig. 9.4 Infrutescence of *Echinochloa pyramidalis*

the end of the rains, when translocation of nutrients below ground starts and the subaerial parts dry off even though the site may still be flooded. Under optimum conditions the previous season's accumulation of dry matter rots away in the water after nutrient translocation, and the new growth is very vigorous. However, node shoots remain green and there is some secondary flowering later.

9.3.4 Economical Relevance

Antelope grass is widely used, both in Africa and in other parts of the world where it was introduced for forage, fodder, hay or silage (Fig. 9.5). It is mainly grown for fodder in pure, dense stands with a leafy table at 1.2–2 m. It is relished by livestock and able to withstand heavy grazing (Animal Feed Resources Information System 2017; Ecocrop 2010; FAO 2012; Quattrocchi 2006). Although no quantitative figures have been cited, it is an important type of dry-season grazing fodder throughout tropical Africa. After fires, vigorous growth from ground level occurs in the absence of rain, and this provides a green dry- season pasture, which may remain available until the next rainy season begins. The young growth is very palatable after the old material has been burnt off (Vezey-Fitzgerald 1963). In Cheffa plain, Ethiopia, local people perceive Antelope grass as high quality feed (Tamene et al. 2000).

Under cultivation *Echinochloa pyramidalis* can achieve dry matter (DM) yields of 15.3 t/ha/year of DM when grown in permanently flooded mangrove areas (Nascimento et al. 1988). In Guyana, DM production increased from 21.3 t/ha DM

Fig. 9.5 Antelope grass forms monodominant stands, making it a very productive forage

to 27.6 t/ha DM when harvested after 21 or 35 days (Seaton et al. 1994). In a semi-arid climate, harvest at 42 days gave better quality hay than at 56 or 98 days (Braga et al. 2008). In the Amazon Basin and under rotational grazing, much lower yields of 0.39–0.76 t/ha were obtained with a 50% decrease between cuts, suggesting that this management was inadequate (Abreu et al. 2006). In order to maximize forage use efficiency and to prevent losses due to leaf senescence, a 25-day regrowth interval has been recommended, as leaf senescence increases strongly after 25 days (Andrade et al. 2008; Heuze et al. 2016).

In Guyana, heifers (young cows that have not yet had a calf) continuously grazing antelope grass had a daily weight gain of 300 g day^{-1} at an optimum stocking rate of 1.1 head ha^{-1} (Seaton et al. 1994). Pasture containing *Echinochloa pyramidalis*, *Cynodon dactylon*, *Digitaria eriantha* and *Digitaria swazilandensis* with a 28-day (or less) grazing, was able to maintain up to 5 cows ha^{-1} year^{-1} (Cunha et al. 1975).

Echinochloa pyramidalis has a variable composition that depends on the season and age of regrowth. Heuze et al. (2016) reported average crude protein 6.5% DM (± 3.9); however, values as high as 15–20% have been reported (Abreu et al. 2006; Bogdan 1977). At 6 weeks, the concentrations of crude protein, structural and non-structural carbohydrates were 9.8%, 44.3% and 51.2% DM respectively (Adebowale 1988). Crude protein content, P, K, Na, Fe, Zn and Cu were higher during periods of rainfall, while the Ca, Mg, S, Mn and B were higher in drier periods. Sulfur is the only element that was below the ideal level for the nutrition of cattle (0.2 to 0.5 g/kg; Abreu et al. 2006). The crude protein content of hay decreased from 6.6% to 4.7% between days 42 and 98 (sampled at 42, 56, 70, 84 and 98 days; Braga et al. 2008). Nutritional tables provide information for both fresh grass and hay on different components important to nutrition (Dougall and Bogdan 1965). It is highly palatable, especially at the early growth stage, and was among the most palatable (with *Digitaria eriantha*) of 15 species evaluated for their feeding potential in Venezuela (Cunha and Bryan 1974). It becomes less palatable at later stages of regrowth, even at the early flowering stage (Bogdan 1977).

It has been used to rehabilitate floodplains in North Cameroon (Scholte et al. 2000). Since the 1979 construction of a dam, annual inundations have decreased, reducing perennial vegetation as an important grazing source for nomadic herds and wildlife during the dry season. With the release of excess water, floodplain rehabilitation has occurred. Perennial grass cover, most notably that of *Echinochloa pyramidalis* and *Oryza longistaminata*, increased from 41% to 61% in the reflooded zone. Should the observed conversion rate of annual into perennial grassland was extrapolated, recovery towards a 100% perennial state would have been expected after the 2003 flooding season.

Echinochloa pyramidalis is also mentioned in relation to food in Africa. The grain is used as human food in some parts of Africa (Chippindall and Crook 1976). Field trials in Kenya demonstrated that the forage grass, *Sorghum vulgare sudanense* (Sudan grass) attracted greater oviposition by stemborers than cultivated maize, resulting in a significant increase in maize yield. It is believed that these wild grasses can act as trap plants and control stemborers naturally (Khan et al. 1997). Its dense, tangled, floating stems, rooting at the nodes, provide efficient protection against wave action on the walls of earth dams, or flood-induced erosion of riverbanks (FAO 2012; Rose-Innes 1977).

9.4 Invasiveness and Impact

Echinochloa pyramidalis has decidedly invasive characteristics with its fast, vigorous shoot and rhizome growth, long life, asexual reproduction, and abundant seed production. It is very abundant in its extensive native range. As an aquatic, it also has the potential to be very damaging to sensitive aquatic habitats. It has a high degree of genetic variability and is highly adaptable to different environments; these characteristics have allowed it to be invasive both within its native range and outside of it. It is also a pioneer in disturbed areas (Invasive Species Compendium 2017).

In many wet and regularly flooded situations, this grass species can come to dominate the vegetation, as can be seen in many parts of Africa, where it may be a virtually climax type of vegetation, desirable for local livestock or wildlife (Denny 1993; John et al. 1993). Wherever it is an exotic species, it may be quite undesirable, replacing local native species. López Rosas et al. (2005) suggested that it is more efficient at using water than the native plants are, and has greater biomass; both characteristics that can change the hydrological pattern of the wetland. In Mexico, there is clear evidence that *E. pyramidalis* has replaced local flora including *Sagittaria lancifolia, Laportea mexicana, Pontederia sagittata and Typha domingensis* (López Rosas et al. 2010; Invasive Species Compendium 2017).

Competition and rapid growth are the main mechanisms that allow it to have such a great impact and to monopolize resources. This results in ecosystem changes (see Sect. 9.5). It has a negative impact on agriculture in which the cultivar is flooded, such as rice and sugar cane. In addition to its competitive growth, Wells et al. (1986) note its tendency to obstruct water flow. Additionally, in aquaculture and fisheries it invades water bodies and changes their physico-chemical conditions, i.e. light and decomposing organic matter, among others. It also disrupts transportation in channels and waterways.

Holm et al. (1979) record it as a major weed in its native area in Nigeria, Swaziland, Sudan and Madagascar. In Guyana, after being introduced and cultivated for some years, it was noticed as a weed in sugar cane in the early 1980s and increased rapidly to become one of the most troublesome weeds in the Guyana Sugar Corporation's aquatic system (Bishundial et al. 1997). In Mexico, also after introduction as a fodder grass, it became widely invasive in wetlands, tending to reduce native wetland species (López Rosas et al. 2010).

The risk of introduction is relatively high as *E. pyramidalis* is widely valued as a fodder grass, for soil conservation, and as a mean of alleviating pollution from heavy metals and sewage. Accidental introduction as a contaminant of seed lots is possible but rather unlikely, because it is not common as a weed of crops (Invasive Species Compendium 2017). In the United States of America it is highly ranked as a potential invasive weed of the future (Parker et al. 2007) and has been identified as a species 'not authorized (for introduction) pending pest risk analysis' (NAPPRA) (USDA-APHIS 2012). The IUCN Red List of Threatened Species indicates that there are no known significant past, ongoing or future threats to this species (Lansdown 2013). Thus, the pathways for the introduction of this species are animal production, breeding and propagation, crop production, forage, habitat restoration and improvement, and water treatment (Invasive Species Compendium 2017).

9.5 Alterations to Wetland Structure and Functions Caused by *Echinochloa pyramidalis*: A Study Case in the Gulf of Mexico

According to Costanza et al. (2006), freshwater wetlands provide the greatest value of all ecosystem services (\$23 billion ha^{-1} $year.^{-1}$in New Jersey). Furthermore, the most valuable services were disturbance regulation, water filtration, water supply and waste treatment. There is evidence of the impact of invasive species on ecosystems (Vitousek 1990; Van der Velde et al. 2006) and of how invasive species can hinder ecosystem services. In this chapter, we present experimental evidence of how the invasive grass *E. pyramidalis* negatively alter wetlands, modifying critical processes and leading to the loss of ecosystem services. We present data relative to the alterations of wetland structure and functions caused by *E. pyramidalis* in a freshwater marsh in the central area on the Gulf of Mexico, a region where its introduction as forage was common practice. These alterations are considered proxies for the loss of ecosystem services, namely biodiversity (sustaining indigenous biota), disturbance regulation, carbon storage and sequestration, maintaining water quality and the regulation of atmospheric gases (Costanza et al. 2006; Dise 2009). The experimental evidence provided includes a decrease in plant and animal biodiversity, a reduction of soil water holding capacity, a decrease in denitrification potential, a reduction in organic carbon accumulation and a reduced capacity to remove xenobiotic compounds. Methods not previously described elsewhere are presented in the corresponding section; otherwise, the references that describe methods in detail are cited in the text. We evaluate the consequences of invasion by this grass relative to the economic production of cattle ranching in today's tropical wetlands in order to understand the true cost of interfering with these essential natural systems.

Research undertaken since the late 1970s at the La Mancha field station (Ramsar site 1336 La Mancha-El Llano) has documented changes in the wetlands resulting from several causes, one of the most important being the introduction of the African grass *E. pyramidalis*. The La Mancha field station is located in the municipality of Actopan, state of Veracruz (19°36′N, 96°22′W), Mexico. It is located in the central coastal area on the Gulf of Mexico (Fig. 9.6). It belongs to the Palma Sola volcanic unit, the only landscape where the Trans-Mexican Volcanic Belt dips into the coastal

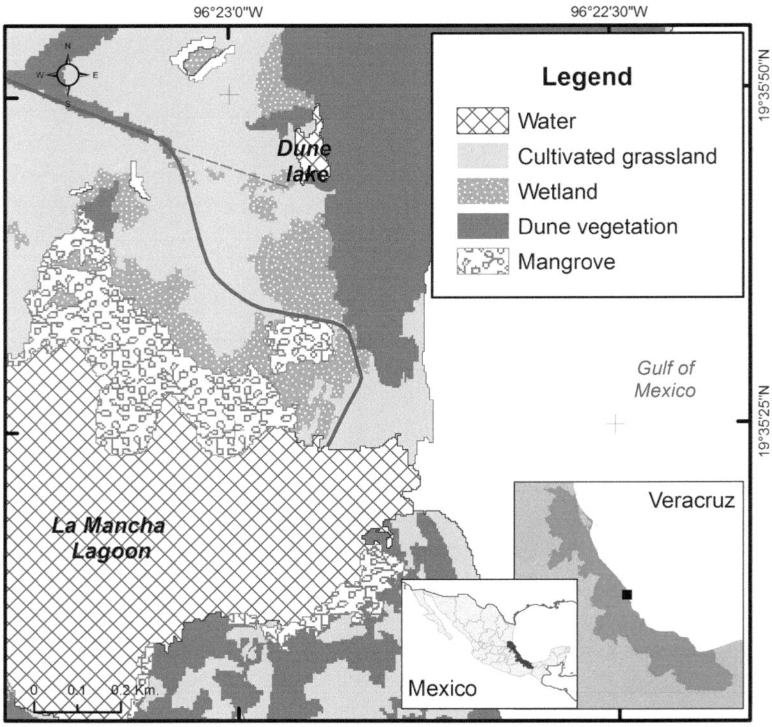

Fig. 9.6 Location of "La Mancha" wetland, Veracruz, central Gulf of Mexico coast. Cultivated grassland represents *Echinochloa pyramidalis*-dominated area. Wetland represents the native freshwater marsh. In this site, a wetland dominated by Antelope grass is being restored

plains, as a result of effusive volcanic activity during the mid- and upper Miocene (15 to 8 million years BP), which produced spills of basaltic andesite and dacitic breccias (Geissert 2006 and references therein). Recent geomorphology (Pliocene and Quaternary) is hills with erosion processes, depositional low plains and eroded mountains shaped by exogenic processes such as fluvial erosion, sedimentation and karstification (Geissert 1999). There are at least 10 types of soils; the wetland is dominated by molic gleysol and fibric histosol (Travieso Bello and Campos 2006). The region receives an average annual rainfall of 1160 mm, with markedly seasonal rains from June to September, and strong northern winds from October to February with some winter storms, followed by the dry season in March to the end of May; the mean annual temperature is 24.5 °C (CNA-ERIC III 2007). The region also receives rainfall from hurricanes and has endured severe droughts (e.g. 1998). This is a very important catchment area and drainage for rainwater into the Gulf of Mexico, and receives additional water from groundwater discharge (Yetter 2004).

During the late 1970s, part of the original wetland (located in a dune depression) was used for agricultural projects such as aquaculture ponds and poultry farms, which modified the topography and plant community mainly through soil removal

Fig. 9.7 During the 1980s the wetland (now under restoration) was transformed into a vegetable and fish farm, using the traditional production system called *chinampa*, (elevated terraces). See text for explanation

(Ortiz-Espejel and Hernandez-Trejo 2006). Five ponds were built for aquaculture by excavating the soil and sediments; nine elevated wetland farming plots (locally known as *chinampas*) were set up to grow vegetables in an area of 729 m^2 (López Martínez 1985). To build the chinampas, ground level needs to be raised and canals built to simultaneously reduce flooding and maintaining fish production, as shown in the photo of the area taken in 1988 (Fig. 9.7). The chinampas were abandoned in 1989. This historical land use resulted in a microsite with a high degree of heterogeneity, with areas ranging from permanently inundated to permanently saturated with freshwater. Before the introduction of the grass, the original physiognomy of this wetland was herbaceous vegetation dominated by *Typha domingensis* and *Sagittaria lancifolia* (Novelo 1978). With the invasion of *E. pyramidalis*, the wetland was transformed and the grass became dominant; however, the original wetland plant species *S. lancifolia, Laportea mexicana,* and *T. domingensis* (López Rosas et al. 2005) remained in small patches.

Currently, the wetland is still under restoration, surrounded by sugar cane fields and grasslands used to feed cattle. The introduced African grasses were *Panicum maximum, Cynodon dactylon* and *Echinochloa pyramidalis.* Some areas were used several years for pasture, then abandoned approximately 20 years ago. However, significant increases in grass cover (and biomass) were observed in former wetland zones where the water table drops down below ground level. Alteration of topography and the consequent change in the hydroperiod are likely two of the main causes of wetland degradation in the study site. By compiling the historical land use over the last 40 years, and the consequences of land use in terms of the topography and hydrology of the area, it was possible to understand and reconstruct the invasion patterns of the African grass introduced into neighboring cattle ranches. Yetter (2004) found that sub-superficial groundwater flow year round fed the wetland, and the flow direction was from the sugarcane fields (northwest) and the dune lake through the dune slack (east), toward the west, into the mangrove and the coastal lagoon (Fig. 9.8). In addition, a small dam was built on a neighbouring ranch (south-

Fig. 9.8 Study site in La Mancha, Veracruz, Mexico. Dune systems, mangroves and freshwater wetlands are indicated in the aerial photograph (2009, elaborated by R. Monroy). The wetland in the dune depression (**W** -wetland under restoration) is surrounded by stabilized dunes, mangrove, and the coastal lagoon of La Mancha. Other types of wetlands surround the site. On the north side, there was originally a dune slack marsh that was seasonally flooded, but was subsequently excavated and converted into a permanent lake and is now used for watering cattle. It is bordered by a swamp dominated by *Pachira aquatica* and *Annona glabra* (**W1**). Behind it, there are sugar cane crops on old dunes. The wetland on the west side geomorphologically belongs to the same slack of the study site, but the owner uses it as pasture land; thus, vegetation is dominated by the Antelope grass *Echinochloa pyramidalis* and the southern cattail *Typha domingensis* (**W2**). Elevated terraces (**Ch** -chinampas) were located inside the wetland. To the west the stabilized dune system continues, covered with grassland for cattle ranching. On the south side there is a road that separates the dune slack from a mangrove swamp and the coastal lagoon. A concrete draining pipe facilitates the water flow between these two systems

western border of the wetland) to contain run-off. Because of the latter the water accumulates, forming a dune lake 1.2 ha in area. A sluice is used to control surface water depth, which spills into the rest of the wetland; however, there is enough groundwater discharge flow to maintain the water above ground level in the lower lying areas (Yetter 2004). Whenever the inundation has been maintained, the freshwater marsh maintains the same surface area. With the flood control structures, three areas have become clearly differentiated: the first is a distinctive wetland community dominated by the southern cattail *Typha domingensis*, the second is also a wetland dominated by broadleaf arrowhead *Sagittaria lancifolia* and the third is dominated by *Echinochloa pyramidalis*, which forms a flooded grassland (see López Rosas et al. 2005 for details).

9.5.1 Biodiversity – Decreased Ability to Sustain Biota

9.5.1.1 Plants

Between May 1998 and November 1999 evenly spaced vegetation plots (0.7 × 0.7 m) were monitored in areas dominated by *E. pyramidalis* and *S. lancifolia*. In each plot the percentage cover was estimated per species (Kent and Coker 1992). During the study, a total of 112 observations were made (56 plots in *E. pyramidalis* sites and 56 in *S. lancifolia* sites). With this information diversity was estimated (richness, evenness, Shannon's index) and compared among communities (*t* test modified by Hutchinson to compare two indices of diversity Zar 1996), as was similarity between communities (percentage similarity, Quanta software). Similar procedures were performed later (March 2007 and February 2010) while monitoring plots used as control sites in restoration experiments (see López Rosas et al. 2010 for details). It was observed that plant species richness was higher but evenness and Shannon's diversity index were lower in the flooded *E. pyramidalis* grassland, compared to the *S. lancifolia* wetland (Table 9.2). Lower diversity was expected

Table 9.2 Richness (S), evenness (E), Shannon's diversity index (H′) and percent similarity between the flooded *Echinochloa pyramidalis* grassland and *Sagittaria lancifolia* wetland over two periods

Vegetation area	Index or coefficient							
	(1998–1999)				(2007–2010)			
	S	E	H′	Similarity (%)	S	E	H′	Similarity (%)
S. lancifolia	13	0.75	0.84(a)	40.5	18	0.75	0.94(a)	21.4
E. pyramidalis	24	0.53	0.73 (b)		18	0.48	0.60(b)	

Different letters for Shannon's index indicate significant differences (t0.01(2), 12,117 = 13.5 in 1998–1999; t0.01(2), 4128 = 21.3 in 2007–2010)

The 1998–1999 period represents a time when *Echinochloa pyramidalis* dominated and grazing had recently stopped. The 2007–2010 period represents the beginning of the restoration activities

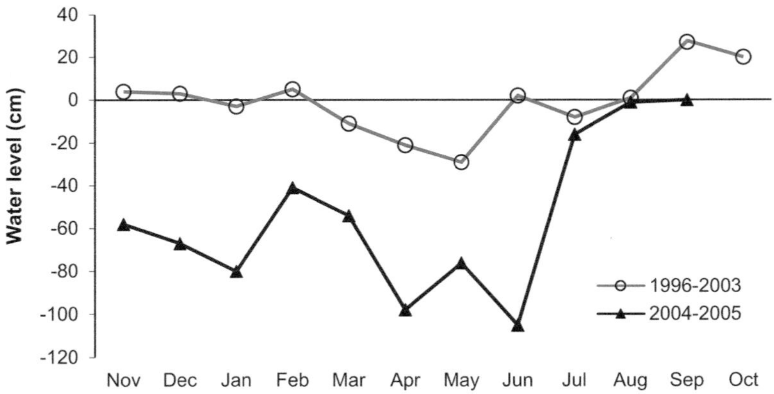

Fig. 9.9 Fluctuation of water level throughout the year in the freshwater marsh invaded and partially dominated by the African grass *Echinochloa pyramidalis*. 0 represents ground level. Historical data plotted as monthly averages. Dry years (*triangles*) did not have water above ground level

as a result of the invasion of the exotic grass; but higher richness was not expected, since cattle grazing had stopped prior to 1998–99 and no restoration activities had been carried out at that time. It was hypothesized that the increased richness could be due to *E. pyramidalis* functioning as a facilitator to other secondary species and creating favourable conditions for facultative hydrophytes by causing accretion and soil compaction during a period of reduced flooding. The increased species richness is not reflected in a higher diversity because *E. pyramidalis* is strongly dominant. This is reflected by its low evenness value and significantly lower diversity value than the values recorded for it in the *S. lancifolia* wetland. Finally, water levels for areas dominated by *E. pyramidalis* changed substantially over time (Fig. 9.9), facilitating its expansion into saturated and temporarily flooded areas.

From 2007 to 2010, while restoration activities were underway, species richness in the flooded *E. pyramidalis* grassland was lower than richness in the *S. lancifolia* dominated wetland (compared to the observation on 1998–99). This may have been a result of restoration efforts (2007 to date), which increased the flood level as well as flood duration and perennial water spaces for strict free-floating hydrophytes (such as *Lemna* and *Salvinia*) and emergent plants (sedges, such as *Pontederia sagittata*). Wherever *E. pyramidalis* had dominated previously, conditions were no longer suitable for facultative or ruderal species and these were displaced by wetland native species. This shift reduced the similarity between the marsh and grassland communities (21.4%). Evenness and diversity values maintained the same trend as in the previous period, with higher values in the *S. lancifolia* wetland than in the flooded *E. pyramidalis* grassland.

Changes in richness, evenness, diversity and invader/native ratio (*Inv/Nat*) after the introduction of *E. pyramidalis* in experimental plots dominated by *S. lancifolia* at the La Mancha freshwater marsh are described in Table 9.3. A value of 0 in the *Inv/Nat* ratio means no *E. pyramidalis* is present, 1 means *E. pyramidalis* is equal to the

Table 9.3 Changes in richness (S), evenness (E), Shannon's diversity index (H´) and invader/native ratio (Inv/Nat) after introducing *Echinochloa pyramidalis* in experimental plots dominated by *Sagittaria lancifolia* at the La Mancha freshwater marsh, Veracruz (May, July, October 1998, and February, June, November 1999)

	Treatment				
Index	Control (*S. lancifolia*)	Herbicide	*E. pyramidalis* transplants	*E. pyramidalis* + herbicide	$F_{3,12}$
S	6.2 ± 0.2 a	7.2 ± 0.1 a	6.6 ± 0.2 a	6.6 ± 0.2 a	1.8 (*n.s.*)
E	0.83 ± 0.01 a	0.72 ± 0.02 a	0.70 ± 0.03 a	0.27 ± 0.03 b	21.8***
H′	0.66 ± 0.01 a	0.61 ± 0.02 a	0.57 ± 0.02 a	0.22 ± 0.02 b	21.1***
Inv/Nat	0.02 ± 0.02 b	0 b	1.3 ± 0.46 b	11.7 ± 4.5 a	6.0**

Mean ± 1 SE is shown for five replicates. Different letters indicate significant differences between treatments (two-way ANOVA; **$P < 0.01$, ***$P < 0.001$). For the purpose of analysis, the S data were square root transformed, while Inv/Nat ratios were arcsine square root transformed

sum of the biomass of the native species and values >1 indicate that the biomass of the invading grass is greater than the biomass of native species. For the plot where *E. pyramidalis* was intentionally transplanted, the value obtained is slightly greater than 1, but it is not different and cannot be distinguished from the control plot or from the herbicide-control treatments, where there was no *E. pyramidalis*. We propose that the experimental plots with grass transplants simulated the beginning of the invasion process. *E. pyramidalis* was dominant but did not displace native species. In treatments with *E. pyramidalis* + herbicide, the biomass of *E. pyramidalis* was 11.7 times greater than that of native species, indicating complete dominance by the invasive grass, which displaced almost all the native species. This could have been the situation at advanced stages of invasion. In the study case, we think that the invasive species is likely the direct cause of the decline in biodiversity, the so-called *driver* (*sensu* Thomas and Reid 2007), given that the previous management strategy was based on the intentional dispersal of the invasive species and the removal of the original vegetation.

To verify whether the conditions of the *S. lancifolia* dominated wetland were suitable for *E. pyramidalis* survival, ramets *of E. pyramidalis* were transplanted into 0.7 × 0.7 m plots in the *S. lancifolia* dominated area. A randomized block experimental design was used, with random allocation of four treatment levels nested within each of five replicate plots. Each block measured 0.7 × 5.5 m and was separated from the others by a 1 m buffer area. Blocks were further divided into seven 0.7 × 0.7 m permanent plots (experimental units) that were 0.5 m apart. The size of the experimental units was determined by the herbaceous nature of the plants (Barbour et al. 1999). In this wetland, the height of the vegetation is between 0.5 and 1.5 m (when in bloom), while the root zone is no deeper than 0.5 m. In each plot, one of the following disturbance or transplant treatments was randomly applied: (i) Control (the plot was left in a natural condition); (ii) Herbicide (all vegetation was sprayed once with the systemic herbicide glyphosate, Round Up®); (iii) *E. pyramidalis* transplants (three 15-cm-long individuals of *E. pyramidalis*), and (iv) Herbicide + *E. pyramidalis* (all vegetation was sprayed with Round Up®, removed when dead and three 15-cm-long individuals of *E. pyramidalis* were transplanted). In treatments 3 and 4, the transplanted individuals had a similar number of leaves, underground stems and root biomass. The borders of all experimental units were protected by a buffer area where the roots and rhizomes were regularly cut along the plot edges with a straight shovel to a depth of 37 cm. Twenty-one months after the experiment, aerial biomass production was measured by species. All plots were harvested to soil level; the vegetation contained in each was separated by species and oven-dried for 120 h at 65 °C to obtain the dry aerial biomass of each species per plot. The relative aerial biomass of invasive and native species, richness, evenness, Shannon's diversity index and the invasive/native ratio were analyzed using a two-way Analysis of Variance (ANOVA; Zar 1996) to identify differences between treatments with SigmaPlot 11.0 software. As observed in Table 9.4, vegetation biomass changed almost 2 years after the introduction of the grass. The biomass production was significantly higher; the invasive grass grew vigorously and came to represent almost half of the total biomass of the plots, compared to its biomass in the control plots. These results suggest that when *E. pyramidalis* appears on a wetland, it is a strong competitor that can displace native species and gain dominance very quickly when an open space is available (simulated by the removal of plants with the herbicide). This shows the capacity of non-indigenous grass species for invading and transforming wetlands.

Grazing was also observed to have an influence upon the diversity of the plant community in areas where *E. pyramidalis* was present. Experimental plots in patches dominated by the grass were studied. Plots with grazing, without grazing, canopied (under the canopy of isolated trees of *Salix humboldtiana* and *Tabebuia rosea)* and in open areas were compared to address plant species composition and the number of individuals (culms for *E. pyramidalis* and leaves for *Hydrocotyle umbellata*) (Fig. 9.10). We estimated the Shannon-Wiener index H', evenness (E), species richness (S) and abundance, and applied a Kruskal-Wallis one-way analysis of variance on ranks, pair-wise multiple comparison was applied following the Student-Newman-Keuls method. We found that grazing was associated with higher values of species richness, abundance, diversity and evenness (Table 9.5), probably because it promotes heterogeneity and reduces the dominance of grasses. Differences in species richness, abundance and diversity between open and canopied area (isolated trees) were observed in the wetland exposed to grazing, but evenness was the same in both treatments. The four indicators of diversity did not differ in the wetlands in the absence of grazing, in open areas or under isolated trees.

We interpret these results to mean that grazing increases diversity, especially when cleared areas are compared with canopied ones. For grazed pastures, these results are similar to those reported by Guevara et al. (1986) and Laborde (1996), where plant diversity was higher under isolated trees in pastures, given that canopied sites provide a favourable microclimate and edaphic conditions for plants that cannot tolerate the conditions of cleared sites, and also due to the fact that trees provide

Table 9.4 Relative aerial biomass (%) of plant species on experimental plots in the *Sagittaria lancifolia* wetland

	Treatment				
Species	Control (*S. lancifolia*)	Herbicide	*E. pyramidalis* transplants	*E. pyramidalis* + herbicide	$F_{3,12}$
E. pyramidalis	1.7 ± 1.7 c	0 c	48.3 ± 10.0 b	88.0 ± 3.3 a	65.27***
S. lancifolia	23.3 ± 6.0 ab	42.8 ± 8.4 a	12.0 ± 3.7 bc	4.5 ± 1.7 c	9.58**
T. domingensis	20.1 ± 6.7 a	14.6 ± 5.1 a	9.2 ± 7.0 a	1.8 ± 0.9 a	2.31

Mean ± 1 SE is shown for five replicates. Different letters indicates significant differences between treatments (two-way ANOVA; **$P < 0.01$, ***$P < 0.001$). For the purpose of analysis, all data were arcsine square root transformed

Fig. 9.10 *Echinochloa pyramidalis* grassland with standing trees of *Tabebuia rosea* in a wetland transformed for cattle grazing. The broadleaf plants in the front are *Pontederia sagittata* and *Sagittaria lancifolia*

Table 9.5 Diversity index for the flooded Echinochloa pyramidalis grassland with and without grazing, comparing the vegetation under the tree canopy of standing isolated trees and in open areas at the La Mancha freshwater marsh

	With grazing		Without grazing		
Index	*Under isolated trees*	*Open areas*	*Under isolated trees*	*Open areas*	*H*
Richness	5 (a)	4 (a)	2 (b)	2 (b)	58.1***
	(4–6)	(2.5–5)	(1–2)	(1–2)	
Abundance	140 (a)	116 (ab)	90 (c)	100 (bc)	32.4***
	(117.5–170.0)	(97.5–142.5)	(76.5–112.5)	(90.0–110.0)	
Diversity (H′)	0.62 (a)	0.55 (a)	0.09 (b)	0.190 (b)	65.0***
	(0.55–0.72)	(0.33–0.63)	(0–0.26)	(0–0.26)	
Evenness (E)	0.91 (a)	0.91 (a)	0.31 (b)	0.58 (b)	42.9***
	(0.85–0.95)	(0.85–0.95)	(0–0.75)	(0–0.71)	

Values are the median of 24 quadrats (first – third quartile). Different letters indicate significant differences (Kruskal-Wallis one-way analysis of variance on ranks; ***$p < 0.001$)

shelter or stopover sites for seeds dispersers (namely birds and bats). The effectiveness of grass invasion could be facilitated by the activity of the cattle and the physiological characteristics of the plant, helping it to successfully invade an area. Having a high tolerance to grazing due to its large biomass production (common in

C4 species) and the intrinsic ability of the Poaceae family to succeed at invasion would also help (Goodwin et al. 1999; Scott and Panetta 1993; Williams and Baruch 2000).

9.5.1.2 Amphibians

The assemblage of amphibians in the study area was monitored on a monthly basis (2008 to 2009) in order to measure changes in diversity, composition and abundance over time, and to assess the differences between the *Sagittaria lancifolia* wetland and the adjacent *E. pyramidalis* dominated grassland (Valdez Lares 2010). Strip transects of 50 × 4 m were set up throughout the wetland and the grassland. Amphibians were sampled using three techniques: visual encounter surveys, audible recordings and by offering artificial shelters made of PVC pipes (Boughton et al. 2000). Individuals were recorded for six consecutive nights per month from July 2008–March 2009. We used non-parametric tests to detect differences in richness and abundance (Kolmogorov-Smirnov test), and in diversity (1/D Simpson's index; t-test proposed by Solow 1993). A total of eight species of amphibians were observed in the study: *Bolitoglossa platydactyla*, *Incilius valliceps*, *Hypopachus variolosus*, *Lithobates berlandieri*, *Scinax staufferi*, *Leptodactylus melanonotus*, *Trachycephalus venulosus*, and *Smilisca baudinii* (Fig. 9.11). Only the latter three were recorded in the grassland. The most abundant species was *L. melanonotus* in both the *E. pyramidalis* and *S. lancifolia* habitats. The two species of hylids (*T. venulosus* and *S. baudinii*) observed in the grassland were only seen inside the artificial shelter pipes, which functioned as temporary refuges. Richness and species composition were different between the wetland and the flooded grassland

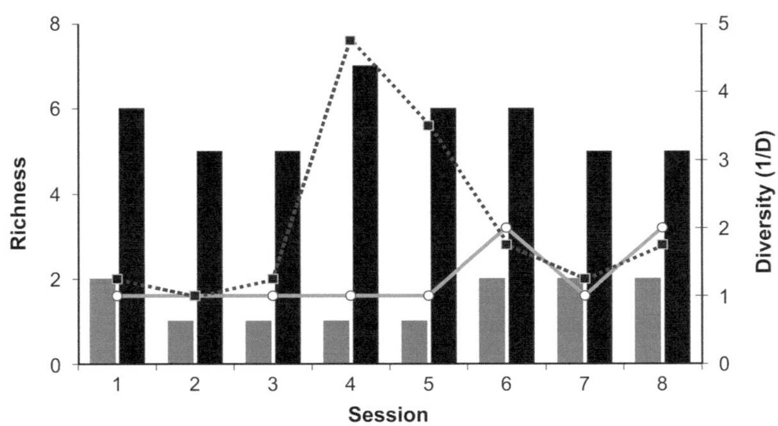

Fig. 9.11 Richness and diversity of amphibians in the study area for eight sampling events from July 2008 through March 2009. Bars represent richness; lines represent diversity (1/D). Black: *Sagittaria lancifolia* dominated wetlands. Grey: flooded *Echinochloa pyramidalis* grassland

($p < 0.05$), as was diversity, with the *S. lancifolia* dominated wetland the most diverse ($p < 0.01$).

In general, diversity, richness and abundance were much lower in the grassland than in the wetland. This might be an indication of the influence of open spaces on amphibian populations (Babbitt et al. 2009), and more directly, the results show that the majority of the species of amphibians in the area avoid the flooded *E. pyramidalis* grassland. This is possibly due to its low water level and dense vegetation. At the time of sampling, *E. pyramidalis* grew up to two meters high and might be considered a poor community that eliminates a variety of microhabitats that different species of amphibians could use (Tews et al. 2004). Soil compaction (*see below*) can have a negative effect on burrowers like *H. variolosus*, due to the compact matrix that the roots and stems formed, preventing the presence of surface water. This could hinder the presence of aquatic species (such as *L. berlandieri*). Also, the absence of trees could be a factor that lessens the presence of arboreal species such as *S. baudinii* and *T. venulosus* (these two species were only recorded in the grassland in the shelter pipes). In fact, this observation indicates that flooded grasslands are possibly used by those species (and perhaps others) for dispersion from or to the wetland, but considering the low values of diversity, richness and abundance observed, this habitat could represent more of a barrier for other species that do not have the same dispersal ability (Kolozsvary and Swihart 1999). The artificial shelters may have served as stepping stones, allowing these two species to move from less suitable to more suitable habitats. A low water level is potentially problematic for some amphibian species, particularly in ephemeral wetlands (Chandler et al. 2017). Connectivity between areas would appear to be very important for facilitating the movement and dispersal of amphibians.

9.5.1.3 Insects

In order to assess changes in the insect community, shallow ponds in the freshwater marsh and in the flooded grassland were sampled following the method described by Merritt and Cummins (1996, "D" net, 350 cm^2, 500 μm mesh). Ponds were identified by numbers based on vegetation composition and sampled once during the rainy season and once during the dry season, samples were always collected at the same location and at approximately the same time of day: Site 1 was bordered and partially covered by *E. pyramidalis*; Site 2 was dominated by *S. lancifolia*, and Site 3 was co-dominated by *Pontederia sagittata* and *T. domingensis*. Five samples were taken per pond at each site, at a depth between 10 and 50 cm, during the dry and the rainy seasons. All biological material was collected in plastic bottles and preserved *in situ* with ethanol (96%). In the laboratory, leaves and organic material were discarded, and insects were separated by high taxa (Order) and preserved in 70% ethanol. Specimens were identified to family using specialized keys (Borror et al. 1989; Contreras-Ramos 1997; McCafferty et al. 1997; Merritt and Cummins 1996; Novelo-Gutiérrez 1997a, b). The Shannon diversity index (H'), evenness (E) and richness (S, based on families) were obtained per site. A Kruskal-Wallis analysis and

Dunn's multiple pair comparisons test were used to compare diversity among sites. Finally, a Principal Components Analysis (PC-ORD 5) was used to explain any differences in insect composition between ponds. From a total of 30 samples collected, we obtained 1385 organisms: 665 in the dry season and 720 in the rainy season, belonging to 28 families and six orders (Table 9.6). During the rainy season, there was a greater number of families at all sites compared to the dry season. During the dry season, the ponds dominated by P. sagittata and T. domingensis remained partially inundated; thus, likely having higher diversity, with families such as Hydrophilidae, Scirtidae, Ceratopogonidae, Stratiomidae, Aeshnidae and

Table 9.6 Presence-absence of families of aquatic insects collected at the La Mancha freshwater marsh

Order	Family	Dry season			Rainy season		
		Site 1	*Site 2*	*Site 3*	*Site 1*	*Site 2*	*Site 3*
Coleoptera	Curculionidae			x		x	
	Dytiscidae	x	x	x	x	x	x
	Elmidae		x	x			
	Hydrophilidae	x	x	x	x	x	x
	Noteridae					x	x
	Scirtidae			x	x	x	x
	Staphylinidae		x	x			
Diptera	Ceratopogonidae	x	x	x	x	x	x
	Chironomidae	x	x	x	x		x
	Culicidae	x	x	x	x	x	x
	Scirphidae					x	x
	Stratiomidae	x	x	x	x	x	x
	Tabanidae				x	x	
	Tipulidae	x	x	x			x
Ephemeroptera	Baetidae	x					
Hemiptera	Belostomatidae	x		x	x	x	x
	Hydrometridae	x		x			
	Naucoridae	x	x	x			
	Nepidae		x	x			x
	Vellidae		x				
	Notonectidae					x	x
	Plaidae				x		x
Lepidoptera	Pyrelidae						x
Odonata	Aeshnidae		x	x	x	x	x
	Coenagrionidae	x		x	x	x	
	Libellulidae	x			x	x	x
	Perilestidae				x	x	
	Protoneuridae	46	x	x	x	x	x

Site 1 was bordered and partially covered by *Echinochloa pyramidalis*, Site 2 was dominated by *Sagittaria lancifolia*, Site 3 was co-dominated by *Pontederia sagittata* and *Typha domingensis*. Dry season: March 2003; Rainy season: September 2003

Libelulidae predominating. In contrast, the ponds dominated by *E. pyramidalis* lost almost all of their surface water, with water aboveground only around the root system, where Chironomidae and Baetidae dwell and dominated.

In Fig. 9.12 the ordination shows that during the rainy season, sites are grouped (1b, 2b and 3b), suggesting similarities among them. However, during the dry season, the ponds bordered and partially covered by *E. pyramidalis* (1a) are separated from the other ponds. Batzer and Wissinger (1996) mentioned that insect communities in marshlands preferentially use submerged and emergent macrophytes as habitats. In our study, aquatic plants may play an important role by providing dwelling, nursery and adequate conditions for the development of resources (Hann

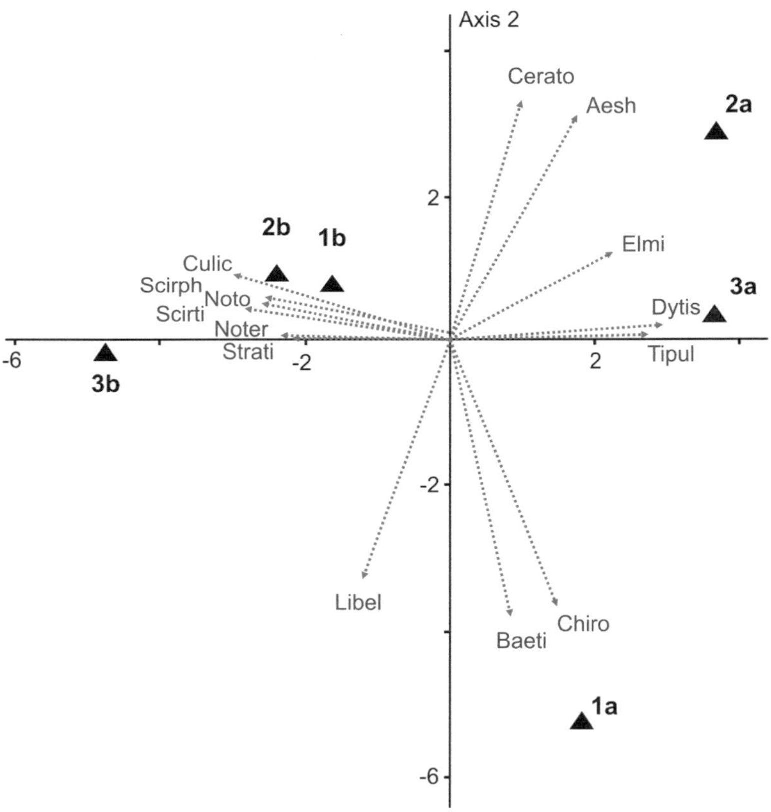

Fig. 9.12 PCA explaining 65% of insect variability (families) between ponds in La Mancha freshwater wetland. Site 1 was bordered and partially covered with *Echinochloa pyramidalis*; Site 2 was dominated by *Sagittaria lancifolia*, Site 3 was co-dominated by *Pontederia sagittata* and *Typha domingensis*, 18 samples collected (1385 organism). Letters represent (*a*): dry season (March 2003), (*b*): rainy season (September 2003). Note that site 1 is distinguished by the presence of Baetidae (*Baeti*) and Chironomidae (*Chiro*). Insect Families are: Aeshnidae (*Aesh*), Ceratopogonidae (*Cerat*), Culicidae (*Culic*), Dytiscidae (*Dytis*), Elmidae (*Elmi*), Libelulidae (*Libel*), Noteridae (*Noter*), Notonectidae (*Noto*), Scirphidae (*Scirph*), Scirtidae (*Scirti*), Stratiomidae (*Strati*), Tipulidae (*Tipul*)

Table 9.7 Family richness (*S*), evenness (*E*) and Shannon's Diversity index (*H*′) for aquatic insects in three ponds inside the La Mancha freshwater marsh during the 2003 dry (March to June) and rainy (July to October) seasons

	Dry season			Rainy season			Annual		
	S	E	H′	S	E	H′	S	E	H′
E. pyramidalis	14	0.14	0.16	15	0.17	0.2	19	0.27	0.35
Sagittaria lancifolia	14	0.14	0.16	17	0.23	0.3	24	0.38	0.52
Pontederia sagittata – Typha domingensis	18	0.25	0.32	18	0.25	0.3	24	0.38	0.52

1991). Although we did not quantify microalgae and decaying biomass, according to Batzer and Resh (1991) midge larval abundance (grazers) is correlated with periphyton biomass (producers). Also, an increase in midge larvae (Chironomidae) was related to the absence of beetle larvae (predators), such as Hydrophilidae and Dytiscidae. The latter was observed in the dry season in the monitored ponds. In temporary ponds, habitat preference between these two groups is sometimes important. Ephemeral habitats (such as temporarily flooded freshwater marshes) tend to be dominated by beetles and mosquitoes, whereas midges and dragonflies predominate in permanently flooded habitats (Batzer and Wissinger 1996).

The seasonal differences in the insect community suggest that different strategies of colonization could have occurred (Hargeby 1990), especially when some families disappeared completely. Finally, the differences observed in insect diversity due to the dominant vegetation might have also influenced trophic networks (Table 9.7). For instance, Odonata are frequently reported as the top predatory insects in fishless habitats (Blois-Heulin et al. 1990) and the presence or absence of top predators would likely impact prey (community structure) to the point where compensating effects by other predators such as Coleoptera and Hemiptera (Wissinger and McGrady 1993) are observed to promote or increase changes in insect diversity.

9.5.2 Disturbance Regulation- Soil Water Holding Capacity

In this section, we take soil water holding capacity as a proxy for the ecosystem service of disturbance regulation, which is related to flood buffering (Costanza et al. 2006). Soil parameters were measured in three areas dominated by *T. domingensis, S. lancifolia* and *E. pyramidalis*, respectively. Soil bulk density (g cm^{-3}), electrical conductivity (mS cm^{-1}), moisture (%), pH of interstitial water and oxidation-reduction potential (mV) of the upper horizon (0–15 cm) were measured during the rainy season. To assess the soil water holding capacity, the procedure described by Dane and Hopmans (2002) was used in soil of areas dominated by *S. lancifolia* and *E. pyramidalis*, respectively. Five randomly chosen sites were sampled to collect undisturbed soil from the upper layer (10 cm), to determine water retention properties at field capacity and bulk density. Briefly, the soil sample is placed on a plate

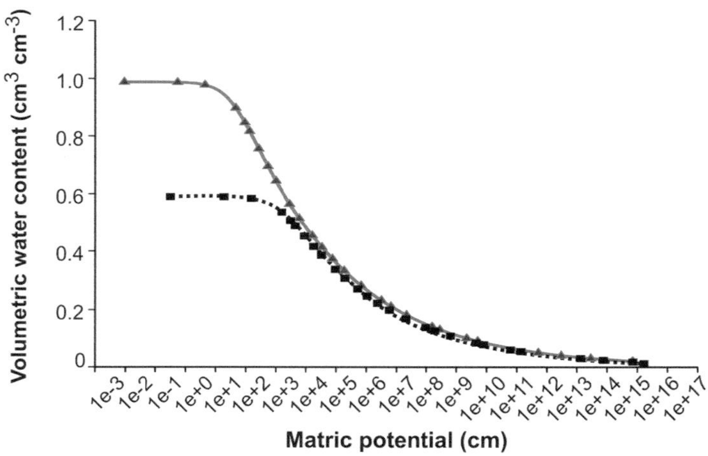

Fig. 9.13 Soil water retention curves: *Sagittaria lancifolia*-dominated wetland (triangles) and *Echinochloa pyramidalis* flooded grassland (squares). The water retention curve reflects the relationship between soil water content and pressure potential

with water, allowed to saturate for 24 h, and then placed in pots with pressure membranes and subjected to 0.1, 0.3, 1.0, 5.0 and 15.0 bar of pressure. Soil samples were dried (105 °C for 24 h) for water content determination and a water retention curve was obtained. Average water retention for the two types of wetlands is showed in Fig. 9.13. The water retention curve reflects the relationship between soil water content and pressure potential. Mean values of soil water content were higher in the *S. lancifolia* dominated marshland (0.98 cm3 cm^{-3}) than in the flooded grassland dominated by *E. pyramidalis* (0.58 cm3 cm^{-3}). Similar freshwater marshes in the state of Veracruz had a soil water holding capacity of 687 to 880 L m^{-2}, in which the thickness of the organic layer was assumed to be the major factor affecting water storage capacity (Campos et al. 2011). The differences in water holding capacity between the marshland and grassland (approximately 59% at the saturation point) suggested that differences in soil among sites could be related to the extent of degradation of the organic substrate, given that drainage and intensive use of organic soils leads to a decrease in water content and a consequent increase in air content (Päivänen 1982). In organic soils, the relationship between pressure potential and water content depends on the decomposition rate and the quality and quantity of the organic matter (Boelter 1969), which in our site is 23.4 (±8.2) % (Cejudo-Espinosa et al. 2009). Soil in the freshwater marshes in Veracruz had around 18% organic carbon content (Campos et al. 2011); thus, large volumes of water can be stored in organic soils (as can carbon, estimated at 31 kg C m^{-2}, Campos et al. 2011).

Additionally, increase in bulk density was also attributed to the invasion of *E. pyramidalis* (Table 9.8), which could have reduced the vertical migration of water. Bulk density results are consistent with data from water retention curves. The *S. lancifolia* marshland had a significantly lower bulk density than the *E. pyramidalis* grassland, probably due to a lower degree of decomposition of the

Table 9.8 Physicochemical characteristics of the soil in the three wetland areas in the La Mancha freshwater marsh, Veracruz

	Vegetation area			
Characteristic	*Typha domingensis*	*Sagittaria lancifolia*	*Echinochloa pyramidalis*	*F*
Soil bulk density (g cm^{-3})	0.1 ± 0.005 (b)	0.1 ± 0.002 (b)	0.4 ± 0.1 (a)	12.8***
Soil moisture (%)	89.3 ± 0.3 (a)	89.4 ± 0.7 (a)	59.1 ± 6.9 (b)	21.2***
Soil ORP (mV)	51.8 ± 10.9 (b)	69.2 ± 5.5 (b)	106.0 ± 6.8 (a)	11.7**
Interstictial water conductivity (mS cm^{-1})	1.4 ± 0.1 (a)	0.8 ± 0.02 (a)	1.5 ± 0.6 (a)	32.4***
Interstitial water pH	6.9 ± 0.01 (a)	7.0 ± 0.01 (a)	7.1 ± 0.1 (a)	2.4 (n.s.)

The species name in the column header is the dominant species. Interstitial water measurements were made at a depth of 15 cm. Values are the mean of 10 replicates ±1 SE. Significant differences are indicated with different letters (one-way ANOVA, ** $p < 0.01$; *** $p < 0.001$). Modified from López Rosas et al. (2005)

organic substrate; decomposition of the organic substrate increases because of drainage, thus, bulk density is higher (Boelter 1969). The bulk density data show that the *S. lancifolia* soil was more porous than that of the *E. pyramidalis* grassland; therefore, it has a greater water storage capacity. To summarize, soil water holding capacity in both sites is related to bulk density, which influences the amount of water held at low retention values. Based on the characteristics of the water retention curves, we suggest that the *S. lancifolia* marshland and the *E. pyramidalis* grassland have a common origin; however, the growth of *E. pyramidalis* affected the water holding capacity, diminishing the amount of water potentially retained by about 40%. López Rosas et al. (2005) found that after almost 30 years of transformation, the soil of the grass-dominated area was significantly different in conductivity, moisture and ORP compared to wetland soil (Table 9.8).

9.5.3 Maintaining Water Quality

9.5.3.1 Nutrient Removal

Nitrification potential was used as a proxy for nutrient removal in wetland soils during 2008 and 2009. Intact soil cores (10 cm deep × 4 cm diameter) were sampled with PVC pipes in five 1m^2 plots located in zones where *S. lancifolia* was dominant and in zones where the invasive *E. pyramidalis* was dominant. PVC pipes with soil cores were placed in plastic bags, sealed with a rubber band, placed on ice and transported to the laboratory. Two samples were used as a mixed sample for measuring nitrification potential and to quantify soluble organic carbon. Triplicate 20 g portions of soil at field water capacity were mixed with 100 ml substrate (phosphate buffer solution pH 7.2 and NH_4Cl 50 mg l^{-1}), incubated at 25 °C in 100 ml wide mouth plastic bottles, and placed on a horizontal shaker. The bottles were covered with aluminum foil with pinholes to reduce evapotranspiration. Every 8 h, 10 ml of the slurry was sampled from each flask, centrifuged (2500 g, 15 min) and the supernatant was stored, frozen, until nitrate analysis (salicylic acid method, Robarge et al. 1983). Nitrification potential was calculated from the production of nitrates (following Ambus 1998) and estimated to be 283% greater in *E. pyramidalis* than in *S. lancifolia* soils.

Soluble organic carbon was determined by extracting soil with cold water and quantifying organic carbon in the extracts (Hernández and Mitsch 2007). Briefly, 8 g of fresh soil were placed in 50 ml centrifuge tubes and 15 ml of distilled water was added. Samples were placed in a horizontal shaker for 24 h, then the tubes were centrifuged (2500 g, 15 min) and the supernatant separated and filtered (0.45 μm). Organic carbon in the supernatant was quantified using the Moebius method (García and Ballesteros 2005). Soluble organic carbon was 233% greater in the *S. lancifolia* plots than in the *E. pyramidalis* plots.

Denitrification potential was also measured as a proxy for nutrient removal following the sampling process described above (*nitrification potential*). Three intact soil cores were used to quantify denitrification potential. The PVC pipes were sealed at the bottom with a tightly fitting rubber stopper and silicone. After the silicone had dried, 100 ml of distilled water was added to each core, and they were left for 24 h to saturate the soil completely. Prior to adding the nitrate, a 0.5 ml water sample was taken from each core to establish the background nitrate level in the floodwater column (Maltby et al. 1998). The supernatant liquid in each column was amended with 10 ml of a concentrated aqueous potassium nitrate solution to give a final concentration of 10 mg nitrate l^{-1} in the supernatant. Immediately after the amendment, and every 8 h thereafter, the water column was sampled and the nitrate concentration was quantified using the salicylic acid method (Robarge et al. 1983). The average nitrate removal (denitrification) potential was 1200% greater in the *S. lancifolia* plots than in the *E. pyramidalis* plots (Fig. 9.14).

One of the important ecosystem services of wetlands is functioning as a sink for terrestrially derived nitrogen. This prevents the release of nutrients, which could lead to eutrophication in adjacent surface waters (Comin et al. 1997; Mitsch et al. 2001). Changes in this cycle could negatively affect the potential of wetlands to decrease the input of nitrogen to surface and groundwater. The nitrogen attenuation function in wetlands depends on the capacity of vegetation and microbial communities to intercept and process pollutants (Bowden 1987; Hill and Cardaci 2004). Denitrification is a key process resulting in the net removal of N from the biosphere into the atmosphere; thus, it is an important pathway for ecosystem losses of N (Groffman et al. 2002).

Our results show that *E. pyramidalis* changes nitrogen transformation in coastal wetlands, decreasing the denitrification and increasing the nitrification potential of the soils. This may have a negative impact on the capacity of the coastal wetlands to improve water quality. Denitrification is controlled by oxygen availability, temperature, nitrogen, and organic carbon supply (Hernández and Mitsch 2007; Hortan-Hartwing et al. 2002). The decrease in denitrification in this coastal wetland might be related to the physical changes in the soils caused by *E. pyramidalis*, such as an

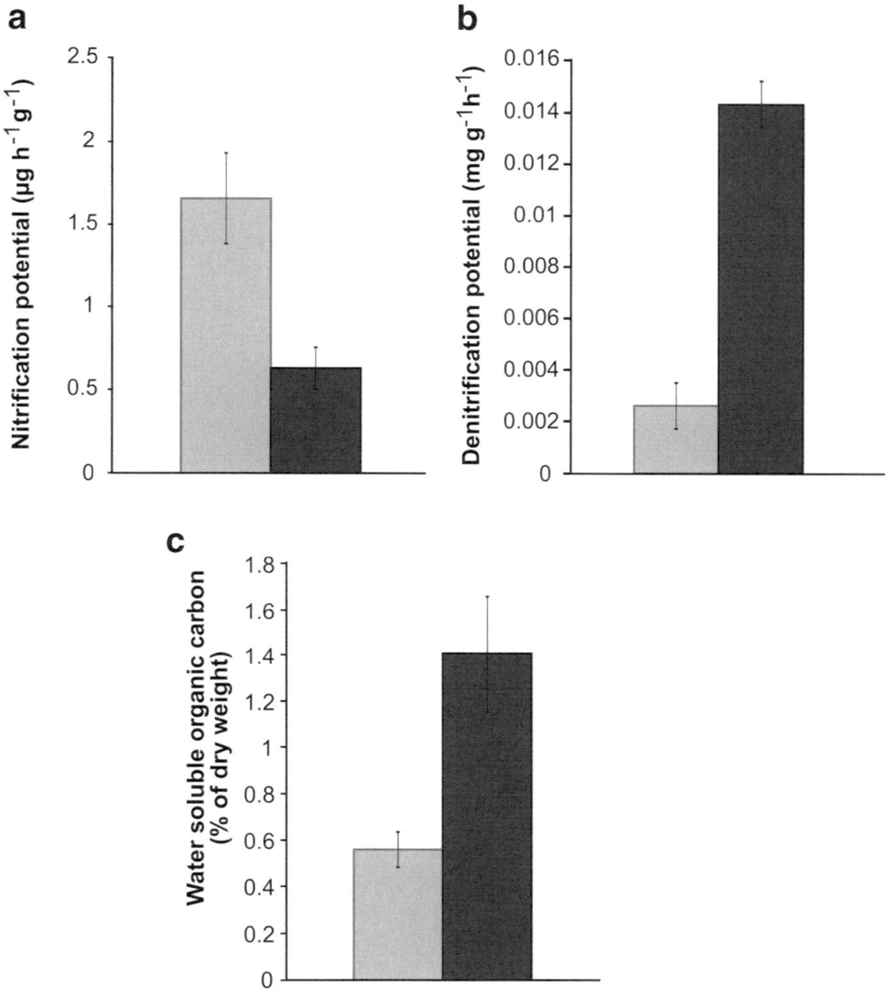

Fig. 9.14 Comparison of the nitrification potential (**a**), denitrification potential (**b**) and water Soluble Organic Carbon (**c**) between the *Echinochloa pyramidalis* dominated flooded grassland (white) and *Sagittaria lancifolia* wetland (gray). Values are annual average at 10 cm depth. Bars are standard error SE

increase in the redox potential and shorter hydroperiods (López Rosas et al. 2010). Also, this decrease seems to be influenced by changes in the quality of carbon input (Ehrenfeld 2003). Heterotrophic denitrification is performed by facultative anaerobic soil bacteria, which under anoxic conditions are able to use nitrates as a final electron acceptor, and organic carbon is required as an essential electron donor for heterotrophic denitrifiers (Beauchamp et al. 1989). It has been demonstrated that wetland plants provide a different quality of organic carbon, which strongly influences denitrification (Bastviken et al. 2007; Hernández and Mitsch 2007; Hume et al.

2002). We found that available (water soluble) organic carbon was higher in soils with native plants than in zones invaded by *E. pyramidalis*, indicating that in the latter plots, there were fewer available electron donors for denitrification.

Decreasing denitrification in wetlands soils causes a negative effect on the environmental services provided by wetlands. Denitrification is a key biogeochemical process to remove nitrogen from the water column in wetlands (Song et al. 2014), therefore the invasion of *E. pyramidalis* decreases wetland capacity to improve water quality. Changes in nitrogen transformation by invasive exotic plants have been described for temperate salt and freshwater marshes. Angeloni et al. (2006) found that plots invaded by *Typha glauca* had higher nutrient concentrations and higher *nir*S genotype denitrifiers than plots with native vegetation in the Cheboygan Marsh in Michigan. Fickbohm and Zhu (2006) investigated the effects of the invasion by the exotic emergent plant purple loosestrife (*Lythrum salicaria*) in a freshwater wetland dominated by native cattail (*Typha latifolia*) in New York. They found that net nitrification rates were significantly higher in *L. salicaria* plots than in *T. latifolia* and mixed plots.

9.5.3.2 Improving Water Quality

Herbicide removal can be considered a proxy for improving water quality. Three plant species commonly found in wetlands (*T. domingensis, S. lancifolia* and *E. pyramidalis)* were used in a greenhouse experiment to assess their potential of removing the herbicide atrazine (2-chloro-4-(ethylamine)-6-(isopropylamine)-s-triazine) from soil and water (Cejudo-Espinosa et al. 2009). This herbicide is commonly used in sugar cane fields, a very important crop on the coast of the Gulf of Mexico. The herbicide accumulated in the root systems was reported as mg of atrazine per kilogram of root tissue. The accumulation of atrazine in the roots of the plants differed, the roots of *S. lancifolia* accumulated greater amounts of atrazine (680.4 mg atrazine kg^{-1} root DW) than the African grass *E. pyramidalis* did (Fig. 9.15); arguably less efficient in accumulating the herbicide and removing it from the environment. It is important to say that plants of the genera *Typha*, *Phragmites*, *Eichhornia*, *Azolla*, *Lemna, Eleocharis* and other aquatic macrophytes have been used for pollutant removal (Rai 2008; Vymazal 2013), consequently, they seem to be more suitable than introduced grasses for phytoremediation and improving water quality.

Changes in land use from wetlands to agriculture and cattle ranching, together with an increase in the use of agrochemicals has led to the degradation of wetlands in the tropics. The most common problem associated with agriculture and cattle ranching is the large load of agrochemicals (fertilizers and herbicides) in the run-off water, with the consequent impacts on flooded lowlands, wetlands, rivers, lagoons and the sea. Pesticides and their transformation products are frequently detected in groundwater (Steele et al. 2008). In fact, the groundwater from an artesian well at the La Mancha field station had atrazine in concentrations of 0.65 mg L^{-1} (January 2006) and 0.44 mg L^{-1} (March 2006, unpublished data).

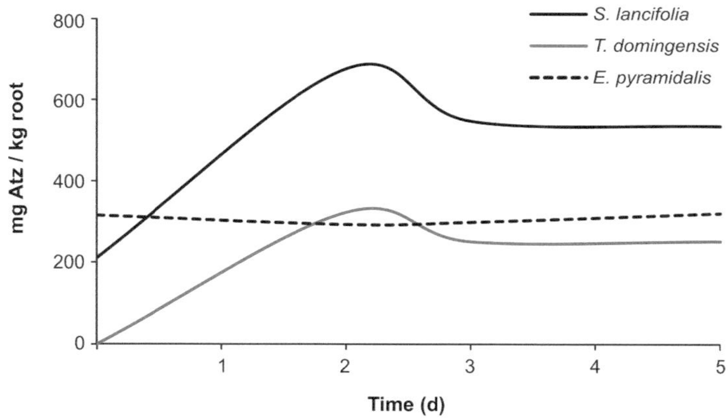

Fig. 9.15 Accumulation of Atrazine in roots of three wetland plant species (*Sagittaria lancifolia, Typha domingensis, Echinochloa pyramidalis*) after herbicide application (as mg Atrazine kg^{-1} root, dry weight). Lines correspond to treatments of 30 mg atrazine L^{-1}, single application. Time 0 is 15 min after application under greenhouse conditions

The occurrence of pesticides in groundwater or sediments depends mainly on physicochemical properties, such as solubility, pK_a, the K_{ow} partition coefficient (Costa Paraíba et al. 2003; Neurath et al. 2004), amount and rate of use (Steele et al. 2008), agricultural management practices, the vegetation associated with the crop, and the amount and intensity of rainfall and irrigation (Bouldin et al. 2006; Moore et al. 2006). Considering that the soil in the study area has a high sand content and precipitation is over 1000 ml $year^{-1}$, the detection of atrazine in the groundwater suggests an inappropriate use of herbicides in the area, and that the plants of the undisturbed wetland could play an important role in removing xenobiotic compounds from the water. Wetlands (natural or engineered) have been used as mitigation or attenuation methods for herbicides discharged into surface waters (Anderson et al. 2002; Maddison et al. 2005, Vymazal and Brezinová 2015) and have promising applications in tropical wetlands: however, the selection of species largely depends on the targeted compound to be removed.

9.5.4 Economics vs. Ecoservices: The Cost of Interfering with Natural Systems

As a result of the introduction of the invasive African grass *E. pyramidalis* in the freshwater marshes of central Veracruz, we observed less diversity but higher richness in areas where the grass has become dominant. This observation suggests a negative effect where fewer species can colonize or establish in areas where *E. pyramidalis* is present; however, more individuals are able to colonize, likely the result of intermittent flooding and habitats that are less stressful environments because of water logging. Increased nitrification and decreased denitrification also suggest a negative effect on net N removal, as denitrification is not occurring at the same rates between marshes with indigenous species compared to more terrestrial environments dominated by grasses.

As a foraging species, *E. pyramidalis* is used in Africa for feeding cattle and also has decorative uses (Tamene et al. 2000). It was also reported to be economically important in Cameroon, where its price as forage can range from USD $0.1 to $0.3 per kilogram (fresh weight), and nutritional values were acceptable for being sold as animal feed in local markets (Pare et al. 2012). *E. pyramidalis* has a high nutrition value compared with other *Echinochloa* species (Abreu et al. 2006); however, proper management and grazing intensity was necessary to ensure adequate production for cattle grazing.

Engineered wetlands with *E. pyramidalis* (horizontal surface flow) successfully reduced total suspended solids, nitrate, orthophosphates and fecal indicators (coliforms and streptococci) when used to treat domestic effluent. Lekeufack et al. (2012) reported that removal efficiency was best in the second year, arguably due to the fact that the wetland was well established, had a greater bacterial biomass and root network, indicating that this grass is useful for domestic water treatment whenever conventional wastewater treatment plants cannot be built and operated. Fonkou et al. (2010) ran trials in a yard scale subsurface flow wetland system for treating wastewater from a distillery and found that this grass contributed to nutrient removal. Similarly, a small-scale dewatering operation observed that *E. pyramidalis* growth was enhanced by sludge application, estimating that approximately 150–180 t DW ha^{-1} $year^{-1}$ could be harvested and used as a forage plant (Kengne et al. 2008, 2009, 2011), representing a net removal of 366 g N and 97 g P m^{-2}. Other studies have measured phosphorus release rather than net removal in engineered wetlands with *E. pyramidalis*, suggesting that more efficient removal was achieved with the cyperaceous *Cyperus papyrus* (Bojcevska and Tonderski 2007). It is important to mention that the studies performed in Africa can be safely conducted on that continent due to the fact that it is the natural range of the grass.

Those using *E. pyramidalis* for applications and engineered wetlands in other regions should keep in mind that the grass is an invasive species in Mexico. Vymazal (2013) reviewed the most common macrophytes used in constructed wetlands and found that the most commonly used genera are *Typha, Scirpus (Schoenoplectus), Phragmites, Juncus* and *Eleocharis*; with *Typha latifolia* and *Typha domingensis* the most common in North, Central, and South America, respectively. Thus, the use of *Echinochloa pyramidalis* in engineered wetlands should be weighed against all of the scientific information available, under rigorous impact assessment, and considering that indigenous species should also be evaluated for their potential application given their null risk of invasion.

This also applies to its use as a fodder species. While it may be attractive because of its nutritional value and tolerance to flooding, its impact on native wetlands and their environmental services should be taken into account.

Additionally, *E. pyramidalis* has been reported to have a high density of stemborers (Muyekho et al. 2005: Le Ru et al. 2006), thus it also is a potential host plant for exotic borer species and possibly other exotic insects.

We recognize that economic estimates in monetary units are useful to highlight the importance and magnitude of ecosystem services (Costanza et al. 2014); however, they still have no straightforward application or influence in decision making or improving regulations in Mexico. This is an aspect of governance that we think should be addressed with all possible haste.

9.6 Controlling *Echinochloa pyramidalis*

Echinochloa pyramidalis has been considered one of the most troublesome invaders in the aquatic system of the Guyana Sugar Corporation (Bushundial et al. 1997) and in the freshwater wetlands of the Mexican tropics (López Rosas et al. 2006). This invading species reduces biodiversity by replacing native species. To eliminate this species or at least reduce its presence considerably, it is necessary to recreate the natural topography and hydrology of the wetland and to select control mechanisms that disrupt the growth characteristics (e.g. rapid propagation from rhizomes and horizontal expansion via tillers) that make this grass more competitive than native species (Lopez Rosas et al. 2006).

Reducing the dominance of this species and increasing the density of native wetland species is a difficult, expensive and time-consuming process. In 2007, using shade to control the invader, we began a restoration project in a coastal wetland in central Veracruz, Gulf of Mexico, in a freshwater marsh that had been partially invaded by *E. pyramidalis* in La Mancha (Figs. 9.1 and 9.6). Before initiating the restoration project, some of the authors of this chapter applied a series of disturbance treatments aimed at eliminating the invading grass and recovering the plant species native to the marsh. The most successful treatments for eliminating or reducing the invader included soil disking or using herbicide on the grass. Plant richness increased with soil disking. The first native hydrophytes to colonize after soil was disked were *Pontederia sagittata* (with shade) and *Fuirena simplex* and other sedges (without shade). The use of shade mesh gradually reduced the cover of *E. pyramidalis* to the point of its complete elimination 15 months after the beginning of the experiment. The treatment that included cutting all the vegetation to soil level and transplanting individuals of native species such as *S. lancifolia* only reduced the cover of *E. pyramidalis* during the first 7 months, after which the invader increased its cover to the same levels as those in the control plot (see López Rosas et al. 2005, 2006). The first general results of the restoration project were published in López Rosas et al. (2010) and López Rosas et al. (2015).

Treatments included the use of shade mesh (López Rosas et al. 2015), based on the positive results of other authors (Belsky 1994; Bunn et al. 1998; Cole and Weltzin 2005) who used shade mesh to control C4 species. Used this way, shade also provided satisfactory results, reducing the cover of the invader and increasing the cover of native species. In this project, the site invaded by this grass had significantly lower richness and diversity than shaded and reference sites. Shading with mesh decreased the presence of the grass and increased the cover of the native species. Our results highlight the importance of understanding the functional differences between native and invasive species when developing strategies for the control and eradication of problematic species.

The results of the shade mesh experiments helped us to design and test these restoration techniques in the field in the spots in the freshwater marsh still invaded by *E. pyramidalis*. The main objective was to use the shade mesh to control small areas of invasion where there were many native plant species that needed protection to produce their seeds and grow vegetatively in the restored areas, and to prevent the spontaneous reintroduction of the invasive species.

At the beginning of the restoration activities, in March 2007, López Rosas et al. (2010) set up groups of five 0.7 × 0.7 m quadrats, following a strategy of cluster sampling to monitor vegetation before and during the restoration process (Ratti and Garton 1996). One quadrat cluster was placed in an area dominated by the invasive *E. pyramidalis*, which also had some *Typha domingensis* cover (the grassland wetland site). The second quadrat cluster was located on a border area with an equal abundance of *E. pyramidalis* and native hydrophytes, on the edge of the wetland and close to the dune. In this area, larger patches of *E. pyramidalis* were clipped by hand and, in September 2007 this area was covered 45 × 25 m (the shaded wetland site) with a shade mesh (no. 50) to make a shade house with each side approximately two meters above the ground (Fig. 9.16). In October 2008 the

Fig. 9.16 A large patch of *Echinochloa pyramidalis* was clipped by hand and the area was covered with a shade mesh (no. 50) to make a shade house with each side approximately two meters above the ground

mesh was removed. The third quadrat cluster was placed in an area dominated by *S. lancifolia* and other desirable native hydrophytes but no invasive species (the reference wetland site). The grassland and reference sites were used as controls because they were not directly managed.

As part of restoration techniques, at the beginning of the rainy season (May 2007), the drainpipe leading from the marsh to the mangroves was blocked with a rustic gate to flood the wetland for 16 continuous months, and ensure that the antelope grass stems, which had been cut, were drowned. This flood affected all wetland sites (see López Rosas et al. 2010), including the controls. Between March 2007 (at the beginning of restoration activities) and May 2010, we sampled plant species composition and percentage cover per species eight times (March and October 2007, March, July, September and December 2008, March 2009 and May 2010). With the exception of September 2008, on all other sampling dates and at the same time we sampled the plants, and measured the physicochemical characteristics of the water and soil (water level, electrical conductivity and pH of interstitial water, soil humidity, Eh and soil bulk density). The bulk density (oven dry weight by volume) was determined after drying 198 cm^3 soil samples at 105 °C for 24 h (Hausenbuiller 1972).

We calculated the cov*I*/cov*N* ratio and diversity indicators (richness (S), Shannon' index (H), evenness (E)). The cov*I*/cov*N* ratio, dominant species cover, S, H and E, and the physicochemical characteristics obtained on each sampling date were analyzed with an RM-ANOVA to detect differences between wetland sites, sampling dates, and any interaction between the two. To meet the assumptions of normality and homogeneity of variance required for the ANOVA, we transformed cov*I*/cov*N*, percentage cover and relative soil moisture values using the arcsine square root, and richness using the square root. When there was a significant effect by wetland areas, sampling dates, or the interaction, we used the SNK test to identify significant differences among means (Zar 1996). The dominant species during the restoration process were the invasive grass *E. pyramidalis* in the grassland site; the native hydrophyte *Typha domingensis* in the grassland and reference sites; the native hydrophyte *Sagittaria lancifolia* in the reference and shade sites, and *Pontederia sagittata*, and *Hydrocotyle umbellata* in the shaded site (López Rosas et al. 2010).

Invader:Natives Ratio and Species Cover. The RM-ANOVA detected significant sampling date x site interactions for the cov*I*/cov*N* ratio ($F_{14,56} = 13.5$; $P < 0.001$) and the aerial cover of the five dominant species ($F_{14,56} = 21.8$, 4.4, 2.2, 50.2, and 8.2; $P < 0.001$, 0.001, 0.05, 0.001, and 0.001, respectively for *E. pyramidalis*, *S. lancifolia*, *T. domingensis*, *P. sagittata*, and *H. umbellata*). The invader *E. pyramidalis* was absent from the reference wetland site, so its cov*I*/cov*N* ratio was always significantly lower (Fig. 9.17a). At the beginning of the restoration, in March 2007, the cov*I*/cov*N* ratio was significantly higher in the shaded wetland site, followed by the grassland site; in October 2007 there was no difference between the values obtained for these two sites; but from March 2008 to the last sampling date in May 2010, *E. pyramidalis* was eliminated from the shaded site, so the cov*I*/cov*N* ratio was significantly higher in the grassland site (Fig. 9.17a). This pattern was also observed for the aerial cover of *E. pyramidalis* (Fig. 9.17b). *Sagittaria lancifolia* was

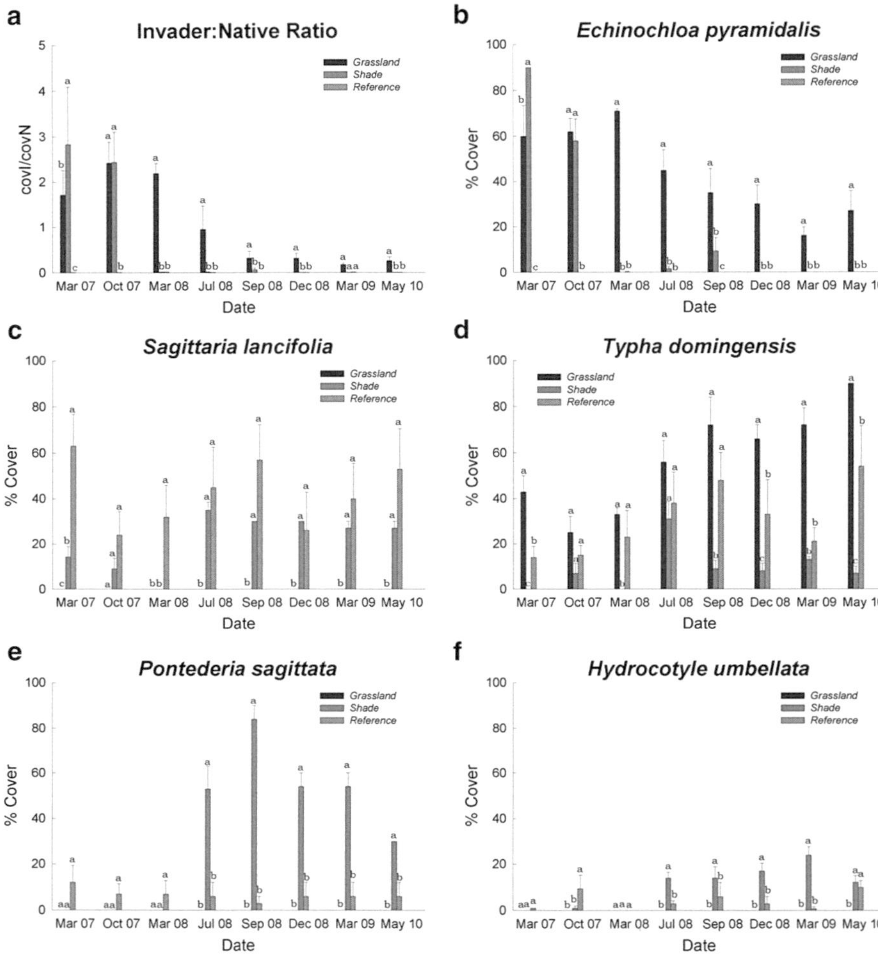

Fig. 9.17 Ratio response for cov*I*/cov*N* *(a)* and the percent cover response (mean + 1SE; $n = 5$) of five dominant hydrophytes over time during the restoration process in three wetland sites (*b* to *f*, see text for description). Different letters indicate significant differences among means for each sampling date (two-way RM-ANOVA; $P < 0.05$). For the analysis, all data were *arcsine square root* transformed

only present in the reference and shaded sites (Fig. 9.17c). Only at the beginning of the restoration, *S. lancifolia* cover in the reference site was significantly higher than at the shaded site, but from October 2007 to May 2010 there was no difference between the two sites (Fig. 9.17c). *Typha domingensis* was present in the three wetland sites (Fig. 9.17d). In the grassland site it maintained its cover throughout the first year but in July 2008 it began to increase and stayed high from then onward. The highest value was recorded in May 2010. *T. domingensis* was absent from the shaded site until October 2007, after which its cover values were low until July 2008 when it attained its highest degree of cover and from then onwards its cover decreased. It was always present at the reference sites, attaining its highest degree of cover in September 2008 and then again in May 2010. From December 2008 to May 2010, *T. domingensis* cover was significantly higher at the grassland site than at the reference site (Fig. 9.17d). *Pontederia sagittata* was absent from the grassland site (Fig. 9.17e) and appeared in the shaded site in July 2008. It reached its maximum cover in September 2008 and was present until the last measurement in May 2010, but with a lower degree of cover. At the reference site it was always present but with very low cover values (Fig. 9.17e). This same pattern, but with approximately one quarter the cover, was obtained for *H. umbellata* (Fig. 9.17f).

Species Richness and Diversity. The RM-ANOVA detected significant sampling date x site interactions for all diversity indicators ($F_{14,\ 56} = 7.7$, 6.8, and 5.1 for *S*, *H* and *E* respectively; all with $P < 0.001$; Table 9.9). In March and October 2007 the species richness (*S*) and *H* values were significantly higher in the reference site and lower in the grassland site; values were intermediate or high at the shaded site. No differences were detected in equitability (*E*). In March 2008 all of the study sites were flooded, and the quadrats in the shaded site were only covered by a bloom of an unidentified alga; thus the shaded site had the lowest values of S, H and E, and values were higher for the reference and grassland sites. From July 2008 to the last measurement in May 2010, no changes were detected in equitability between the three wetland sites. In July 2008 plant vegetation recovered at the shaded site so, from that date on, this site had the highest values of *S* and *H*, while the reference site had high or intermediate values. In July 2008 the grassland site has the lowest *S* and *H* values; from September 2008 to March 2009 these values did not differ significantly from those of the shaded or reference sites. In May 2010 the grassland site had the significantly lowest values for *S* and *H* (Table 9.9).

Physicochemical Parameters. The RM-ANOVA did not detect any significant sampling *date x site* interactions for water level ($F_{12,48} = 1.9$; $P > 0.05$). For interstitial pH, electrical conductivity, soil moisture, soil bulk density and soil Eh (redox potential) there were significant sampling date x site interactions ($F_{12,48} = 3.7$, 8.5, 7.1, 7.1 and 4.6 respectively; all with $P > 0.001$). Water level changed significantly with wetland site ($F_{5,30} = 5.6$; $P < 0.001$), with the highest value recorded at the *grassland* site (25.5 ± 2.3 cm, $n = 70$), an intermediate value at the *reference* site (17.4 ± 2.5 cm, $n = 70$), and the lowest value at the *shaded site* (8.3 ± 2.4 cm, $n = 70$). Water level measurements taken between October 2007 and July 2008 were affected by artificially blocking the drainpipe from the marsh to the mangrove, resulting in the highest water levels. Significantly lower levels were obtained in March 2007, during the dry season before blocking the pipe, and in March 2009 during the dry season after the blocking the pipe (Fig. 9.18). Interstitial pH changed in October 2007, and in March and December 2008 among wetlands (Table 9.10). In October 2007 and March 2008 this value was significantly higher for the *grassland* site, while in December 2008 it was higher for the *reference* site. Electrical conductivity was significantly different among wetlands in March and October 2007, March and December 2008, and March 2009 (Table 9.10). In March 2007, December 2008 and March 2009 conductivity was highest in the *grassland*

Table 9.9 Species richness (*S*), Diversity (*H*), and Equitability (*E*) at different wetland sites (grassland, shaded, reference; see text for description) on eight sampling dates during a wetland restoration process at La Mancha, Veracruz

	Wetland site		
	Grassland	*Shaded*	*Reference*
March 2007			
Richness (*S*)	1.8 ± 0.2 c	4.0 ± 0.4 b	5.8 ± 0.2 a
Diversity (*H*)	0.48 ± 0.12 c	0.84 ± 0.13 b	1.21 ± 0.11 a
Equitability (*E*)	0.69 ± 0.17 a	0.63 ± 0.10 a	0.69 ± 0.06 a
October 2007			
Richness (*S*)	2.0 ± 0.0 b	3.6 ± 0.2 a	4.8 ± 0.4 a
Diversity (*H*)	0.60 ± 0.03 b	0.86 ± 0.09 b	1.29 ± 0.12 a
Equitability (*E*)	0.87 ± 0.05 a	0.68 ± 0.07 a	0.82 ± 0.05 a
March 2008			
Richness (*S*)	1.8 ± 0.2 b	1.0 ± 0.0 c	3.0 ± 0.6 a
Diversity (*H*)	0.51 ± 0.13 a	0.00 ± 0.00 b	0.67 ± 0.21 a
Equitability (*E*)	0.73 ± 0.18 a	0.00 ± 0.00 b	0.60 ± 0.10 a
July 2008			
Richness (*S*)	4.2 ± 0.2 b	7.2 ± 0.7 a	6.0 ± 0.3 a
Diversity (*H*)	0.95 ± 0.08 c	1.71 ± 0.09 a	1.36 ± 0.11 b
Equitability (*E*)	0.66 ± 0.04 a	0.88 ± 0.02 a	0.76 ± 0.05 a
September 2008			
Richness (*S*)	4.6 ± 0.2 a	5.4 ± 0.7 a	5.4 ± 0.5 a
Diversity (*H*)	1.24 ± 0.11 a	1.37 ± 0.08 a	1.46 ± 0.08 a
Equitability (*E*)	0.81 ± 0.05 a	0.83 ± 0.04 a	0.88 ± 0.04 a
December 2008			
Richness (*S*)	5.0 ± 0.7 a	5.2 ± 0.5 a	3.8 ± 0.8 b
Diversity (*H*)	1.15 ± 0.10 ab	1.46 ± 0.07 a	1.03 ± 0.29 b
Equitability (*E*)	0.74 ± 0.03 a	0.90 ± 0.03 a	0.69 ± 0.18 a
March 2009			
Richness (*S*)	5.2 ± 0.4 a	6.2 ± 0.6 a	5.6 ± 0.4 a
Diversity (*H*)	1.06 ± 0.08 b	1.61 ± 0.10 a	1.29 ± 0.14 ab
Equitability (*E*)	0.65 ± 0.04 a	0.89 ± 0.02 a	0.74 ± 0.06 a
May 2010			
Richness (*S*)	3.2 ± 0.2 c	9.0 ± 0.5 a	6.6 ± 0.7 b
Diversity (*H*)	0.81 ± 0.05 c	2.07 ± 0.05 a	1.47 ± 0.08 b
Equitability (*E*)	0.70 ± 0.03 a	0.95 ± 0.01 a	0.79 ± 0.03 a

Means ±1 SE shown are for the five replicate plots. Different letters between columns indicate significant differences (two-way RM-ANOVA, $P < 0.05$). For the analysis, species richness data were square root transformed

site. In October 2007 the *grassland* site had the lowest value, while in March 2008 the *shaded* site had the lowest value. Soil moisture did not differ among wetlands in March or October 2007. In March and July 2008 soil moisture was higher in the *grassland* and *reference* sites and lower in the *shaded* area. For the last three

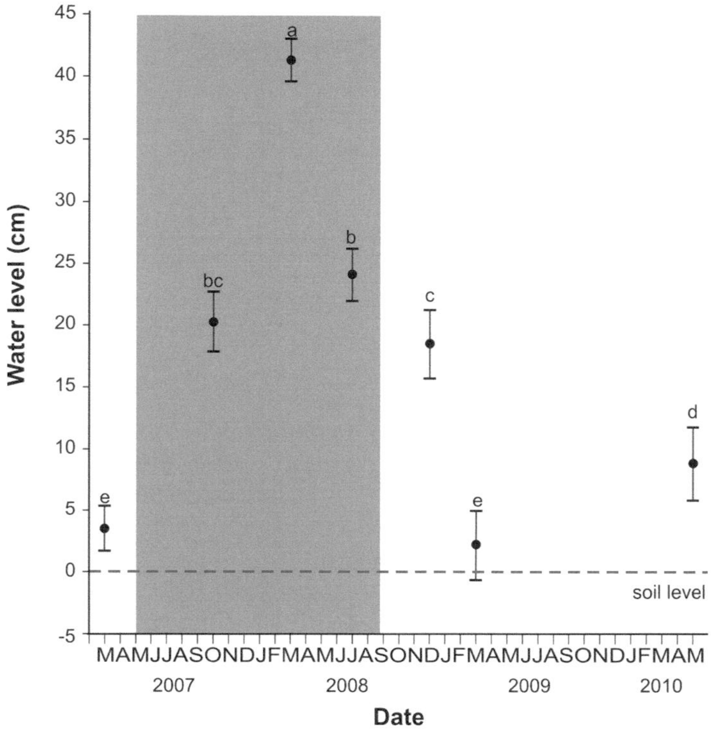

Fig. 9.18 Water level for seven sampling events (mean ± 1SE; $n = 15$) during the restoration process. Different letters indicate significant differences (two-way RM-ANOVA; $F_{6,24} = 85.4$ $P < 0.001$). Shaded area indicates the time affected by artificially blocking the draining pipe from the marsh to the mangrove

measurements (December 2008 to May 2010) values were high for the *reference* site and lowest in the *grassland*. For soil bulk density the pattern was the inverse of that of soil moisture. This parameter did not differ among wetlands in March or October 2007. In March and July 2008 bulk density was higher in the *shaded* site. In December 2008 and March 2009 the *grassland* site had the highest value, while in May 2010 there were no differences among wetlands. Eh at the *grassland* site was always significantly lower, values at the *reference* site were higher, and those of the *shaded* site were high or intermediate. The reduced soil Eh at the *grassland* site was because the treatment coincided with the more flooded area in the wetland.

Effects of shade on hydrophytes. The results show that shade was efficient at eliminating the invading grass, which only survived in the *grassland* site. Desirable natives such as *S. lancifolia*, *P. sagittata* and other forbs and herbs increased under shaded conditions and are now dominant. Thus, in this wetland area applying shade helped eliminate an undesirable species and increase plant diversity, maintaining the native species that would have been destroyed by other treatments. After 3 years, these native species had high cover values. At the *grassland* site there was a decrease

Table 9.10 Physiochemical characteristics of soil and water at different wetland sites (grassland, shaded, reference; see text for description) for seven sampling dates in the wetland of La Mancha, Veracruz, during the restoration process

	Wetland site		
	Grassland	*Shaded*	*Reference*
March 2007			
pH	7.5 ± 0.05 a	7.4 ± 0.07 a	7.5 ± 0.15 a
Conductivity (mS cm^{-1})	1.24 ± 0.02 a	0.82 ± 0.04 b	0.84 ± 0.07 b
Soil moisture (%)	74.7 ± 1.07 a	70.4 ± 0.55 a	73.6 ± 0.72 a
Bulk density (g cm^{-3})	0.10 ± 0.016 a	0.11 ± 0.008 a	0.10 ± 0.005 a
Eh (mV)	43.8 ± 6.6 c	115.3 ± 3.3 b	139.0 ± 9.2 a
October 2007			
pH	7.6 ± 0.09 a	7.3 ± 0.04 b	7.2 ± 0.02 b
Conductivity (mS cm^{-1})	0.95 ± 0.09 b	1.30 ± 0.07 a	1.20 ± 0.05 a
Soil moisture (%)	69.9 ± 1.55 a	66.7 ± 1.61 a	70.8 ± 0.64 a
Bulk density (g cm^{-3})	0.14 ± 0.033 a	0.21 ± 0.034 a	0.12 ± 0.004 a
Eh (mV)	112.9 ± 2.0 b	131.7 ± 5.2 a	141.5 ± 8.1 a
March 2008			
pH	7.9 ± 0.10 a	7.6 ± 0.04 b	7.6 ± 0.05 b
Conductivity (mS cm^{-1})	0.96 ± 0.14 b	1.40 ± 0.05 a	1.16 ± 0.05 b
Soil moisture (%)	72.5 ± 0.49 a	63.1 ± 4.23 b	71.5 ± 0.72 a
Bulk density (g cm^{-3})	0.09 ± 0.011 b	0.26 ± 0.080 a	0.10 ± 0.005 b
Eh (mV)	119.7 ± 8.9 b	139.5 ± 3.0 a	141.6 ± 9.9 a
July 2008			
pH	7.2 ± 0.09 a	7.1 ± 0.02 a	7.2 ± 0.05 a
Conductivity (mS cm^{-1})	0.96 ± 0.15 a	1.02 ± 0.06 a	1.08 ± 0.05 a
Soil moisture (%)	74.4 ± 0.96 a	67.2 ± 2.18 b	71.5 ± 1.33 ab
Bulk density (g cm^{-3})	0.05 ± 0.007 b	0.18 ± 0.045 a	0.12 ± 0.025 ab
Eh (mV)	99.9 ± 5.6 c	121.1 ± 3.2 b	147.1 ± 4.6 a
December 2008			
pH	7.3 ± 0.02 b	7.2 ± 0.04 b	7.5 ± 0.04 a
Conductivity (mS cm^{-1})	1.22 ± 0.10 a	0.95 ± 0.05 b	0.81 ± 0.15 b
Soil moisture (%)	50.5 ± 4.74 c	63.4 ± 1.77 b	70.0 ± 0.77 a
Bulk density (g cm^{-3})	0.44 ± 0.072 a	0.24 ± 0.027 b	0.12 ± 0.008 c
Eh (mV)	100.5 ± 7.1 b	133.0 ± 2.2 a	147.7 ± 2.5 a
March 2009			
pH	7.3 ± 0.05 a	7.3 ± 0.04 a	7.3 ± 0.15 a
Conductivity (mS cm^{-1})	1.41 ± 0.10 a	0.93 ± 0.01 b	0.75 ± 0.02 b
Soil moisture (%)	51.2 ± 2.71 b	68.5 ± 1.08 a	71.4 ± 0.96 a
Bulk density (g cm^{-3})	0.42 ± 0.041 a	0.16 ± 0.016 b	0.12 ± 0.015 b
Eh (mV)	127.5 ± 3.5 b	176.1 ± 4.1 a	185.4 ± 4.5 a
May 2010			
pH	6.9 ± 0.04 a	6.8 ± 0.02 a	6.9 ± 0.02 a
Conductivity (mS cm^{-1})	1.16 ± 0.003 a	0.98 ± 0.02 a	0.97 ± 0.06 a
Soil moisture (%)	65.6 ± 5.49 b	70.7 ± 0.54 ab	73.2 ± 0.64 a

(continued)

Table 9.10 (continued)

	Wetland site		
	Grassland	*Shaded*	*Reference*
Bulk density (g cm^{-3})	0.20 ± 0.077 a	0.16 ± 0.005 a	0.13 ± 0.005 a
Eh (mV)	80.7 ± 1.3 c	125.2 ± 3.6 b	144.8 ± 1.9 a

Means ±1 SE shown are for the five replicate plots. Different letters between columns indicate significant differences (two-way RM-ANOVA, $P < 0.05$). For the analysis, soil moisture data were arcsine square root transformed

in invader cover and an increase in that of *T. domingensis*. This is probably because at the beginning of the restoration, flooded conditions were maintained over 16 consecutive months and that was followed by a normal hydroperiod (9–10 months of flooding) (López Rosas et al. 2010). This limited the invader and favored *T. domingensis*, which can tolerate permanent flooding (López Rosas and Moreno-Casasola 2012; Miao and Zou 2012; Osland et al. 2011). *Sagittaria lancifolia* dominated the *reference* site, followed by *T. domingensis*. Many of the species attained their highest cover in September 2008, probably because we had let the water out of the marsh, the surface of the soil had dried, and the reduced vegetation cover allowed many species to grow. After this date, a balance was established in which the water level was not manipulated, and species began competing among themselves, many of them decreasing in cover, but remaining more or less stable in the plots.

After an overall analysis of the effects of the different treatments and management types on vegetation, and based on previous studies, we can conclude that the high cover and biomass production of *E. pyramidalis* in the wetlands is favored by its C4 photosynthesis and its high capacity for vegetative propagation by rhizomes and stems in a warm climate with an unlimited water supply. Once established, it can survive and reproduce under drier conditions. The high cover of this invader inhibits the germination of most native wetland species. The exclusion of native vegetation coupled with the thick growth of the grass, contributes to its success as an invader. Shading would contribute to the elimination of the invader in the first stage of restoration, and with monitoring it would be possible to assess whether the system will shift into a passive restoration process or whether, as suggested by Osland (2009) and Middleton et al. (2010), it is necessary to select and plant native species that are able to compete with invading species and resist invasion.

9.6.1 Other Methods of Control

9.6.1.1 Physical/Mechanical Control

E. pyramidalis has been controlled by various combinations of disking, shading, cutting to soil level, and prolonged flooding (López Rosas and Moreno-Casasola 2012; López Rosas et al. 2010). But Bushundial et al. (1997) caution that great care is needed to dispose of cut material safely, well away from the cropped area, or buried 1 m deep because of the high capacity of *E. pyramidalis* to regrow from small canes.

9.6.1.2 Chemical Control

Among a range of herbicides tested by Bushundial (1991), the best results were obtained with a mixture of Asulam and Dalapon. Later work by the same team (Bushundial et al. 1997) reported that Asulam and Asulam mixtures with Dalapon, Paraquat, Diuron or Atrazine gave what they referred to as acceptable control. Hexazinone plus Paraquat also gave good control when used as a basal and foliar treatment, but none of these prevented eventual regrowth. For complete kill in non-crop areas, Imazapyr was effective at 120 g ha^{-1}. In sugar cane, the recommendation was the repeated use of Asulam when the weed is about 30 cm tall. López Rosas et al. (2010) refer to tests with Glyphosate, but this was less successful than soil disking and shading.

9.7 Discussion

As summarized in Table 9.11, the invasion and the subsequent dominance of *E. pyramidalis* in natural wetlands has led to changes in the hydrological regime and a reduction of the water holding capacity of the wetland soil, changes in the biogeochemical cycles, changes in plant, amphibian and insect biodiversity, and a reduction in herbicide retention capacity.

Table 9.11 Summary of changes resulting from the invasion and dominance of the exotic grass *Echinochloa pyramidalis* in a native tropical broad leave freshwater marsh in La Mancha, Veracruz, Mexico.

Environmental service	Potential changes attributed to *Echinochloa pyramidalis*
Hydrology regime and flood control	Decreased bulk density and water holding capacity of the soil, thus reducing flood control. Accumulates organic matter from roots and leaves
Biogeochemical cycles	Increase in ORP, from near nitrate and manganese reduction to near iron reduction, increase in
	Nitrification and decrease in denitrification and in soluble organic carbon in the soils
Biological control	Unbalanced population of vector species (Diptera) and predators (Coleoptera and Odonata)
Diversity and genetic resources	Lower richness and diversity at invaded grassland, in both plants and macroinvertebrates (insects)
Water depuration (herbicide retention)	Roots retain less atrazine (herbicide) than other native wetland plants
Cultural and recreational	Reduced biodiversity, poor panorama potential for ecotourism

Kolar and Lodge (2001) carried out an interesting review of the particular characteristics that might predispose species to become invaders in non-indigenous regions. This category is applied to species introduced outside of their native range by human activity. Some generalizations were made in order to predict the species' invasiveness, i.e. characteristics that are important to their establishment and common among invaders. They found that invasive plants have a history of invasion (at the genus or family level), vegetative reproduction and low seed variability. In addition, they tend to have small seeds and spend only a short period of time as juveniles. Some studies (e.g. Goodwin et al. 1999) found that there is a relationship between biological characteristics and propensity for invasiveness.

It has been argued that plant metabolism can enhance the intensity of plant invasion into natural ecosystems, but Kao-Kniffin and Balser (2007) found that even in an atmosphere with elevated CO_2, there were no visible effects on above-ground plant biomass, and that the increased CO_2 had large impacts on soil microbial biomass and might be playing an important role in invasion. According to the evidence presented above (see section on Biomass production and competition), vigorously growing species may be dominant under a variety of hydrological conditions, when nutrients are not scarce (Kercher and Zedler 2004). These authors describe the behaviour of 16 species of Poaceae, Apocynaceae, Typhaceae, Cyperaceae and Asteraceae under flooding conditions, and state that their ecological strategies are valuable for predicting invasion.

There has been some work on the role of perceptions and cultural background on charismatic species. Fischer and van der Wal (2007) suggest that applying cultural values and attitudes toward biodiversity would lead us to a better and more effective policy for nature conservation, by considering for example, aggregated values rather than just typical uses. In our case, the typical use would be cattle ranching and exclusion of the exotic would mean limiting production to only one head of cattle per hectare where there are native wetland grasses (Rodríguez-Medina and Moreno-Casasola 2013). Productive activities such as rustic aquaculture, plant nurseries, handcrafts and nature tourism all linked to the conservation of wetlands are excellent alternatives for rural zones, because these activities can increase the income of the local inhabitants while preserving wetlands.

Studies carried out in the La Mancha wetlands (currently under restoration) contribute to our understanding of the consequences of invasion by an exotic grass and the environmental modification it causes, such as the degradation and loss of ecosystem services; a significant effect considering that the central coastal plain in the state of Veracruz is the most important catchment area and drainage for rainwater in the Gulf of Mexico. Other case studies (Ogden and Rejmànek 2005) have shown that small-scale experimental work reflects the conditions of the larger surrounding landscape. Our results suggest that a combination of management (such as grazing or manual cutting), better practices and restoration techniques will help ecosystems recover, maintain the environmental services that benefit us all, and allow for the integral conservation of basins.

9.7.1 Coastal Ecosystems and Environmental Services

Resilience reflects the degree to which a complex adaptive system is capable of self-organization and the degree to which the system can build capacity for learning and adaptation. Part of this capacity lies in the regenerative ability of ecosystems and their capability in the face of change to continue to deliver resources and ecosystem services that are essential for human livelihoods and societal development (Adger et al. 2005). López Rosas and Moreno-Casasola (2012) concluded that La Mancha's weed invasion was a consequence of the deliberate introduction of *E. pyramidalis* into the relatively dry areas of the wetland. This non-indigenous grass gradually advanced towards the wetter areas, changing the marsh's topography and hydroperiod in the process, and inducing changes in species composition. The capacity of *E. pyramidalis* to produce biomass (Andrade et al. 2008; Braga et al. 2008) modifies the hydrological regime by accumulating organic matter and raising the soil level (Table 9.11). Local cattle ranchers say that they like the species because it "builds soil and dries it out" (Melgarejo-Vivanco 1980). With time, this certainly modifies the wetlands character and functioning. Hydrological changes bring about soil transformations and reduction in biodiversity.

The success of this invader would eliminate less competitive species such as *Laportea mexicana*, *Ludwigia octovalvis*, and reduce the cover of others such as *Hydrocotyle umbellata, Sagittaria lancifolia, Pontederia sagittata*, and lead to a decrease in plant richness and evenness (Travieso-Bello et al. 2005; López Rosas et al. 2006; López Rosas and Moreno-Casasola 2012). A temporary reduction in the hydroperiod and would create an opportunity for mats or bushes of other facultative hydrophyte invasive species to establish in the marsh. Such species include *Dalbergia brownei* (Fabaceae) a native aggressive climber, the grass *Pennisetum purpureum,* and the creepers *Ipomoea sagittata* (Huerta-Ramos et al. 2015) and *I. tiliacea* all of which have already been sighted in the wetland. These species are frequently seen as invaders in other coastal wetlands (P. Moreno-Casasola, *pers. obs.*).

After 20 years of observation, field studies and experiments in the La Mancha marsh, none of the native species have been seen to displace *E. pyramidalis* and it is becoming increasingly common at a regional level. In rainy years, *Typha domingensis* and more recently *Leersia hexandra* increase their cover and reduce the invader cover, but this trend reverses in the dry season. In the medium term it can drastically alter regional wetland distribution, functions and biodiversity (López Rosas and Moreno-Casasola 2012). The cost of reversing these changes may be unsustainable. In Guyana alone, the federal government spent over U.S. $3.4 M to control this grass invader (Ministry of Agriculture of Guyana 2008).

From the farming perspective, it is a very attractive species, which has high productivity and beneficial impacts on cattle. Thus, its introduction remains very appealing to the productive sector. It can increase the number of animals that can be reared in flooded areas. Cattle ranching in wetlands is currently a common practice in many parts of tropical America (Junk et al. 2006; Junk and Nunhes da Cunha 2012; Rodríguez-Medina and Moreno-Casasola 2013) and can be a sustainable activity if the number of animals is kept low, one or two head per hectare (Rodríguez-Medina and Moreno-Casasola 2013; Rodríguez-Medina et al. 2017). It is necessary to further research and have clear economic data on the impact of *E. pyramidalis* and other forage species on wetlands and their environmental services to be able to argue against its introduction.

Wetlands are recognized for being highly effective sinks for atmospheric carbon dioxide, a characteristic they share with mangroves and swamps. Coastal wetlands are threatened by increasing rates of sea level rise and changes in freshwater superficial and subterranean flow, and it is essential to determine their sustainability, thus meeting the requirement of permanence of the carbon sink, as Chmura (2013) suggests. Natural and induced changes in hydrology and soil accretion may result either in the submergence or elevation of present day wetlands, favoring species invasion and modifying carbon accumulation processes.

Hurricanes, typhoons, and their related impacts affect societies throughout the world, both directly through acute damage on human settlements, and indirectly through their impact on coastal ecosystems such as coral reefs, seagrass beds, mangroves that support local societies and economies (Adger et al. 2005), including freshwater wetlands. Responses to hurricanes and tropical storms and their effectiveness depend on social and ecological resilience. Soil compaction and loss of organic matter reduce water storage capacity. Thus invasion by *E. pyramidalis* in coastal freshwater wetlands will reduce their water holding capacity, reducing the flood mitigation service. The cost to society of losing this service is high: property, infrastructure and potentially lives during hurricane season.

Loss of biodiversity is another consequence of *E. pyramidalis* wetland invasion. Decreased species numbers reduces both material goods and ecosystem services. Myers (2005) states "that the environmental services of biodiversity are certainly significant, probably much more so than the direct benefits of biodiversity in the form of material goods. Biodiversity conservation is complementary to, rather than competitive with, other pursuits of human well-being. The time has come when biodiversity cannot be safeguarded primarily in protected areas". Tropical herbaceous wetlands are important biodiversity reservoirs. Myers (2005) argues that it is far from true that all forms of biodiversity can contribute all ecosystemic services or that similar forms of biodiversity can perform similar tasks with similar efficiency. Recent research suggests that they are highly resilient to some loss of species and they can keep on supplying their services even in highly modified states. For example a wetland dominated by *E. pyramidalis* can be very efficient at producing organic material, maybe more so than the natural vegetation it replaced, and even with its low biodiversity might have a high capacity to fix carbon, but there is a significant downside: cattle trampling and soil accretion dries the wetland resulting in the loss of water depuration services, as well as flood mitigation.

Freshwater coastal wetlands play an important role in protecting areas vulnerable to the impact of both seasonal climate events and climate change, and preserving their functions makes ecological and economic sense. Acquiring knowledge about how invading species affect these wetlands is of the utmost importance in the new scenarios facing the coasts of the Gulf of Mexico.

9.8 Conclusions

Several changes in the structure and function of the wetland hereby used as study case were observed through time. Such changes were assessed and interpreted as proxies for the loss of ecosystem services. We observed that the introduction of the African grass *Echinochloa pyramidalis* decreased plant, insect and amphibian biodiversity; impaired soil with reduced capacity for regulating flooding; modified the carbon storage and sequestration, and reduced the ability of maintaining water quality. The best methods of control of *E. pyramidalis* are those that attack subterranean structures or that affect their photosynthetic rate, such as the use of shade mesh. Collectively, the evidence suggested a negative effect on coastal native wetlands and their ecosystem services that should be weighted relative to the use of the grass as fodder species, since it is a preferred species due to its tolerance to flooding and rapid growth.

Acknowledgements This research was made possible by funding from Conacyt-SEP (106451), CONABIO (HH024), the Instituto de Ecología A.C. (Red Ecología Funcional 902-16). Special thanks to all the staff at the Instituto de Ecología A.C., especially Roberto Monroy for his help with the figures and Gerardo Sánchez Vigil for the photos, and the CICOLMA field station for their invaluable support, especially Javier Tolome and Anastacio P. García Rodríguez. Bianca Delfosse for revising the English.

References

Adebowale EA (1988) Nutritive evaluation of *Echinochloa pyramidalis* using West African Dwarf goats. In: Goat production in the humid tropics. Proceedings of a workshop at the University of Ife, Ife, Nigeria, 20–24 July 1987

Adger WN, Hughes TP, Folke C, Carpenter ST, Rockström J (2005) Social-ecological resilience to coastal disasters. Science 309:1036–1039

Ambus P (1998) Nitrous oxide production by denitrification and nitrification in temperate forest, grassland and agricultural soils. Eur J Soil Sci 49:495–502

Anderson K, Wheeler K, Robinson J, Touvinen O (2002) Atrazine mineralization potential in two wetlands. Water Resour 36:4785–4794

Andrade AC, Rodrigues BHN, Azevedo DMMR, Magalhaes JA, de Carvalho KS (2008) Morphogenetic characteristics of canarana (*Echinochloa pyramidalis* Lam.) at different ages of regrowth. Rev Cient Prod Anim 10(1):37–41

Angeloni NL, Jankowski KJ, Tuchman NC, Kelly JK (2006) Effects of an invasive cattail species (*Typha* × *glauca*) on sediment nitrogen and microbial community composition in a freshwater wetland. FEMS Microbiol Lett 263(1):86–92

Animal Feed Resources Information System (2017) https://www.feedipedia.org/node/448. Accessed 18 Sept 2017

Assessment of non-native plants in Florida's Natural Areas. https://assessment.ifas.ufl.edu/site/assets/files/24212/urochloa_humidicola_wra.pdf. Accessed 10 July 2017

Babbitt KJ, Baber MJ, Childers DL, Hocking D (2009) Influence of agricultural upland habitat type on larval anuran assemblages in seasonally inundated wetlands. Wetlands 29(1):294–301

Barbier E (1994) Valuing environmental functions: tropical wetlands. Land Econ 70(2):155–173

Barbier EB, Hacker SD, Kennedy C, Koch EW, Stier AC, Silliman BR (2011) The value of estuarine and coastal ecosystem services. Ecol Monogr 81(2):169–193

Barbour M, Burk JH, Pitts WD, Gilliam FS, Schwartz MW (1999) Terrestrial plant ecology. Benjamin/Cummings, Menlo Park

Bastviken KS, Eriksson PG, Ekstrom A, Tonderski K (2007) Seasonal denitrification potential in wetland sediments with organic matter from different plant species. Water Air Soil Pollut 183:25–35

Batzer DP, Resh VH (1991) Trophic interactions among a beetle predator, a chironomid grazer, and periphyton in a seasonal wetland. Oikos 60(2):251–257

Batzer DP, Wissinger SA (1996) Ecology of insect communities in nontidal wetlands. Annu Rev Entomol 41(1):75–100

Beauchamp EG, Trevors JT, Paul JW (1989) Carbon sources for bacterial denitrification. Adv Soil Sci 10:113–134

Belsky AJ (1994) Influences of trees on savanna productivity: tests of shade, nutrients, and tree grass competition. Ecology 75:922–932

Blois-Heulin C, Crowley P, Arrington M, Johnson D (1990) Direct and indirect effects of predators on the dominant invertebrates of two freshwater littoral communities. Oecologia 84(3):295–306

Bock CE, Bock JH, Jepson KL, Ortega JC (1986) Ecological effects of planting African lovegrasses in Arizona. Natl Geogr Res 2:456–463

Boelter DH (1969) Physical properties of peats as related to degree of decomposition. Soil Sci Soc Am Proc 33:606–609

Bogdan AV (1977) Tropical pasture and fodder plants. Longman, London

Bojcevska H, Tonderski K (2007) Impact of loads, season, and plant species on the performance of a tropical constructed wetland polishing effluent from sugar factory stabilization ponds. Ecol Eng 29(1):66–76

Bor NL (1960) The grasses of Burma, Ceylon, India and Pakistan (excluding Bambuseae). Pergamon Press, Oxford

Borror DJ, Triplehorn C, Johnson N (1989) An introduction to the study of insects. Sanders College Publishing, Philadelphia

Boughton RG, Staiger J, Franz R (2000) Use of PVC pipe refugia as a sampling technique for Hylid treefrogs. Am Midl Nat 144(1):168–177

Bouldin JL, Farris JL, Moore MT, Smith JS, Cooper CM (2006) Hydroponic uptake of atrazine and lambda-cyhalothrin in *Juncus effusus* and *Ludwigia peploides*. Chemosphere 65(6):1049–1057

Bowden WB (1987) The biogeochemistry of nitrogen in freshwater wetlands. Biogeochemistry 4:313–348

Braga AP, Braga ZCA da C, AH d R, de LJDM, Maciel M do V (2008) Green mass production and effect of cut age on chemical composition of canarana erecta hay *Echinochloa pyramidalis* hitch. Caatinga 21(4):1–5

Bunn SE, Davies PM, Kellaway DM, Prosser IP (1998) Influence of invasive macrophytes on channel morphology and hydrology in an open tropical lowland stream, and potential control by riparian shading. Freshw Biol 39:171–178

Bushundial DP (1991) Recent work on the control of *Echinochloa pyramidalis* (Lam.) Hitch. and Chase and *Echinochloa polystachya* (Kunth) Hitchc. In: Proceedings of the 24th West Indies sugar techologists conference, Sugar Association of the Caribbean, Kingston, Jamaica, 8–12 April 1991

Bushundial DP, Dasrat B, Rajkumar A, Kumar D (1997) Management of antelope grass (*Echinochloa pyramidalis*) in sugarcane water ways. In: Proceedings of the 26th conference West Indies sugar technologists, Sugar Association of the Caribbean, Bridgetown, Barbados, 22–26 September 1997
Campos CA, Hernández ME, Moreno-Casasola P, Cejudo Espinosa E, Robledo RA, Infante Mata D (2011) Soil water retention and carbon pools in tropical forested wetlands and marshes of the Gulf of Mexico. Hydrol Sci J 56(8):1388–1406
Cejudo-Espinosa E, Ramos-Valdivia AC, Esparza-García F, Moreno-Casasola P, Rodriguez-Vazquez R (2009) Short-term accumulation of atrazine by three plants from a wetland model system. Arch Environ Con Toxicol 56(2):201–208
Chandler HC, McLaughlin DL, Gorman TA, McGuire KJ, Feaga JB, Haas CA (2017) Drying rates of ephemeral wetlands: implications for breeding amphibians. Wetlands 37(3):545–557
Chapman GP (1996) The biology of grasses. CAB International, Wallingford
Chippindall LKA (1955) A guide to the identification of grasses in South Africa. Part 1. In: Meredith D (ed) The grasses and pastures of South Africa. Central News Agency 361, Cape Town
Chippindall LKA, Crook AO (1976) 240 grasses of southern Africa. Irwin Press (Pvt) Ltd, Salisbury
Chmura GL (2013) What do we need to assess the sustainability of the tidal salt marsh carbon sink? Ocean Coast Manag 83:25–31
Clayton WD (1989) Gramineae. XXIV. Paniceae R. Br. In: Launert E, Pope GV (eds) Flora Zambesiaca 10(3). Flora Zambesiaca Management Committee, London
Clayton WD, Renvoize SA (1982) Gramineae, Part I. In: Polhill RM (ed) Flora of tropical East Africa. Balkema, Rotterdam
CNA (Comisión Nacional del Agua) (2007). Extractor Rápido de Información Climatológica (ERIC III). México
Cochard R, Ranamukhaarachchi L, Shivakoti GP, Shipin OV, Edwards PJ, Seeland KT (2008) The 2004 tsunami in Aceh and Southern Thailand: a review on coastal ecosystems, wave hazards and vulnerability. Perspect Plant Ecol 10(1):3–40
Cole PG, Weltzin JF (2005) Light limitation creates patchy distribution of an invasive grass in eastern deciduous forests. Biol Invasions 7:477–488
Comin FA, Romero JA, Astorga V, García C (1997) Nitrogen removal and cycling in restored wetlands used filters of nutrients for agricultural runoff. Water Sci Technol 35:255–261
Contreras-Ramos A (1997) Clave para la determinación de los Megaloptera (Neuropterida) de México. Dugesiana 4:51–61
Costanza R, d'Arge R, Groot R, Farber S, Grasso M, Hannon B, Limburg K, Naeem S, O'Neill R, Paruelo J, Raskin R, Sutton P, van den Belt M (1997) The value of the world's ecosystem services and natural capital. Nature 387:253–260
Costanza R, Troy WM, Voinov A, Liu S, D'Agostino J (2006) The value of New Jersey's ecosystem services and natural capital. Report to New Jersey Department of Environmental Protection, Division of Science, Research, and Technology, Trenton, NJ
Costanza R, de Groot R, Sutton P, van der Ploeg S, Anderson SJ, Kubiszewski I, Farber S, Turner K (2014) Changes in the global value of ecosystem services. Glob Environ Chang 26:152–158
Costanza R, de Groot R, Kibszewski I, Fioramonti L, Sutton P, Farber S, Grasso M (2017) Twenty years of ecosystem services: how far have we come and how far do we still need to go? Ecosyst Serv 28:1–16
Costa-Paraiba L, Luiz-Cerdeira A, Fraga da Silva E, Souza-Martins J, Luiz da Costa Coutinho H (2003) Evaluation of soil temperature effect on herbicide leaching potential into groundwater in the Brazilian Cerrado. Chemosphere 53:1087–1095
COTECOCA (Comisión Técnica Consultiva de Coeficientes de Agostadero) (1991) Las Gramíneas de México III. Secretaría de Agricultura y Recursos Hidráulicos, México DF
Cunha ER, Bryan WB (1974) Preliminary study of pasture grasses in the Orinoco Delta. 2. Acceptability to the grazing animal. In: Sallette JE, Chenost M (eds) Colloque sur l'intensification de la

production fourragère en milieu tropical humide et son utilisation par les ruminants. Institut National de la Recherche Agronomique, 24–29 May 1971
Cunha E, Larez O, Bryan WB (1975) Pasture and livestock investigations in the humid tropics, Orinoco Delta, Venezuela. V. Milk production studies. Bulletn, IRI Research Institute 46
D'Antonio CM, Vitousek PM (1992) Biological invasions by exotic grasses, the grass/fire cycle, and global change. Annu Rev Ecol Syst 23(1):63–87
D'Antonio CM, Hughes RF, Mack MC, Hitchcock D, Vitousek PM (1998) Response of native species to the removal of non-native grasses in a seasonally dry Hawaiian woodland. J Veg Sci 9:699–712
D'Antonio CM, Tunison JT, Loh RK (2000) Variation in the impact of exotic grasses on native plant composition in relation to fire across an elevation gradient in Hawaii. Austral Ecol 25:507–522
Dane JH, Hopmans JW (2002) Pressure plate extractor. In: Dane JH, Topp GC (eds) Methods of soil analysis. Part 4. Physical methods. SSSA, Madison
de Abreu EMA, Fernandes AR, Martins ARA, Rodrigues TE (2006) Forage production and nutritive value of forage species under pasture conditions in lowland soils of the Guamá River. Acta Amazon 36(1):11–18
Denny P (1993) Wetlands of Africa. In: Whigham DF, Dykyjová D, Hejnÿ S (eds) Wetlands of the world I: inventory, ecology and management, Handbook of Vegetation Science. Kluwer Academic Publishers, Dordrcht
Dise NB (2009) Peatland response to global change. Science 326:810–811
Dougall HW, Bogdan AV (1965) The chemical composition of the grasses of Kenya- III. E Afr Agric For J 30(4):314–319
Duncan C, Jacheta J, Brown M, Carriethers V, Clark J, DiTomaso J, Lym R, McDaniel K, Renz M, Rice P (2004) Assessing the economic, environmental and societal losses from invasive plants on rangelands and wildlands. Weed Technol 18:1411–1416
Ecocrop (2010) Ecocrop database. FAO, Rome
Ehrenfeld JG (2003) Effects of exotic plant invasions on soil nutrient cycling. Ecosystems 6:503–523
EPA (The Environmental Protection Agency) Guyana (2011) Invasive alien species in Guyana: Assessment report, National Strategy and Action Plan. The Environmental Protection Agency (EPA) Ganges Street, Sophia, Georgetown
Evans LT, Wardlaw IF, Williams CN (1964) Environmental control of growth. In: Barnard C (ed) Grasses and grasslands. McMillan, London
FAO (2012) Grassland species profiles. Grassland species profiles. Rome, Italy: FAO. http://www.fao.org/ag/agp/agpc/doc/gbase/default.htm. Accessed 8 July 2017
Farber S, Costanza R, Wilson M (2002) Economic and ecological concepts for valuing ecosystems services. Ecol Econ 41(3):375–392
Fickbohm S, Zhu WX (2006) Exotic purple loosestrife invasion of native cattail freshwater wetlands: effects on organic matter distribution and soil nitrogen cycling. Appl Soil Ecol 32:123–131
Fischer A, van der Wal R (2007) Invasive plant suppresses charismatic seabird – the construction of attitudes towards biodiversity management options. Biol Conserv 135(2):256–267
Fisher MJ, Rao IM, Ayarza MA, Lascano CE, Sainz JI, Thomas RJ, Vera RR (1994) Carbon storage by introduced deep-rooted grasses in the South American Savannas. Nature 371:236–238
Fonkou T, Fonteh MF, Kanouo MD, Akoa A (2010) Performances of vegetated beds with *Echinochloa pyramidalis* in the purification of wastewater from distillery in sub-Saharan Africa. Tropicultura 28(2):69–76
García J, Ballesteros MI (2005) Evaluación de parámetros de calidad para la determinación de carbono orgánico en suelos. Rev Colomb Quím 34:2–7
GBIF (Global Biodiversity Information Facility) (2012) http://data.gbif.org. Accessed 8 July 2017
Geissert D (1999) Regionalización geomorfológica del estado de Veracruz. Invest Geog 1 (40):23–47

Geissert D (2006) Procesos y cambios. In: Moreno-Casasola P (ed) Entornos veracruzanos: la costa de La Mancha. Instituto de Ecología AC, Xalapa, pp 115–126
Gomes Pinto U, McManus C, Aparecida Santos S (2010) Cattle ranching, conservation and transhumance in the Brazilian Pantanal. Pastoralism 1(1):99–114
Goodwin B, McAllister A, Fahrig L (1999) Predicting invasiveness of plant species based on biological information. Conserv Biol 13(2):422–426
Groffman MP, Boulware NJ, Zipperer WZ, Pouyant RV (2002) Soil nitrogen cycle processes in urban riparian zones. Environ Sci Technol 36:4547–4552
Guevara S, Purata S, Van der Maarel E (1986) The role of remnant forest trees in tropical secondary succession. Vegetatio 66:77–84
Halpern BS, Walbridge S, Selkoe KA, Kappel CV, Micheli F, D'Agrosa C, Bruno JF, Casey KS, Ebert C, Fox HE, Fujita R, Heinemann D, Lenihan HS, Madin EMP, Perry MT, Selig ER, Spalding M, Steneck R, Watson R (2008) A global map of human impact on marine ecosystems. Science 319:948–952
Hann B (1991) Invertebrate grazer-periphyton interactions in a eutrophic marsh pond. Freshw Biol 26(1):87–96
Hargeby A (1990) Macrophyte associated invertebrates and the effect of habitat permanence. Oikos 57(3):338–346
Hausenbuiller RL (1972) Soil science principles and practices. Wm. C. Brown Publishers, Dubuque
Hawaii Invasive Species Council http://dlnr.hawaii.gov/hisc/info/species/bush-beardgrass/. Accessed 15 July 2017
Hernández ME, Mitsch W (2007) Denitrification potential and organic matter as affected by vegetation community, wetland age, and plant introduction in created wetlands. J Environ Qual 36:333–342
Heuzé V, Tran G, Boval M (2016) Antelope grass (*Echinochloa pyramidalis*). Feedipedia, a programme by INRA, CIRAD, AFZ, and FAO. https://www.feedipedia.org/node/448. Accessed 5 Sept 2017
Hill R, Cardaci M (2004) Denitrification and organic carbon availability in riparian wetland soils and subsurface sediments. Soil Sci Soc Am J 68:320–325
Hoffmann WA, Haridasan M (2008) The invasive grass, *Melinis minutiflora*, inhibits tree regeneration in a Neotropical savanna. Austral Ecol 33:29–36
Hoffmann WA, Lucatelli VMPC, Silva FJ, Azeuedo INC, Marinho MS, Albuquerque AMS, de Lopes A, Moreira SP (2004) Impact of the invasive alien grass *Melinis minutiflora* at the savanna-forest ecotone in the Brazilian Cerrado. Divers Distrib 10:99–103
Holm L, Pancho J, Herberger J, Plucknett D (1979) A geographical atlas of world weeds. Wiley, New York
Hortan-Hartwing L, Dorsh P, Bakken LR (2002) Low temperature control of soil denitrifying communities: kinetics of N2O production and reduction. Soil Biol Biochem 34:1797–1806
Howard-Williams C, Walker BH (1974) The vegetation of a tropical African lake: classification and ordination of the vegetation of Lake Chilwa (Malawi). J Ecol 62(3):831–853
Huerta-Ramos G, Moreno-Casasola P, Sosa V (2015) Wetland conservation in the Gulf of Mexico: the example of the salt marsh morning glory, *Ipomoea sagittata*. Wetlands 35(4):709–721
Hume NP, Fleming MS, Horne AJ (2002) Denitrification potential and carbon quality of four aquatic plants in wetland microcosms. Soil Sci Soc Am J 66(5):1706–1712
Invasive pasture grasses in Northern Asutralia. http://www.environment.gov.au/biodiversity/threatened/threat-abatement-advices/invasive-pasture-grasses-introduction. Accessed 2 July 2017
Invasive Species Compendium (2017) http://www.cabi.org/isc/datasheet/113990. Consulted on 8 Aug 2017
John EM, Lévêque C, Newton LE (1993) Western Africa. In: Whigham DF, Dykyjová D, Hejný S (eds) Wetlands of the world I: inventory, ecology and management. Kluwer Academic Publishers, Dordrecht, pp 47–78
Junk WJ, Nunes da Cunha C (2012) Pasture clearing from invasive woody plants in the Pantanal: a tool for sustainable management or environmental destruction? Wetl Ecol Manag 20:111–122

Junk WJ, Brown MB, Campbell IC, Finlayson M, Gopa B, Ramberg L, Warner BG (2006) The comparative biodiversity of seven globally important wetlands: a synthesis. Aquat Sci 68:400–414

Kao-Kniffin J, Balser T (2007) Elevated CO_2 differentially alters belowground plant and soil microbial community structure in reed canary grass-invaded experimental wetlands. Soil Biol Biochem 39:517–525

Kengne IM, Akoa A, Soh EK, Tsama V, Ngoutane MM, Dodane PH, Kone D (2008) Effects of faecal sludge application on growth characteristics and chemical composition of *Echinochloa pyramidalis* (Lam.) Hitch. and Chase and *Cyperus papyrus* L. Ecol Eng 34(3):233–242

Kengne IM, Dodane PH, Akoa A, Kone D (2009) Vertical-flow constructed wetlands as sustainable sanitation approach for faecal sludge dewatering in developing countries. Desalination 248 (2):291–297

Kengne IM, Kengne ES, Akoa A, Bemmo N, Dodane PH, Koné D (2011) Vertical-flow constructed wetlands as an emerging solution for faecal sludge dewatering in developing countries. J Water Sanit Hyg Dev 1(1):13–19

Kent M, Coker P (1992) Vegetation description and analysis. A practical approach. CRC Press, London

Kent R, Johnson DE, Becker M (2001) The influences of cropping system on weed communities of rice in Côte d'Ivoire, West Africa. Agric Ecosyst Environ 87(3):299–307

Kercher S, Zedler J (2004) Flood tolerance in wetland angiosperms: a comparison of invasive and noninvasive species. Aquat Bot 80:89–102

Khan ZR, Chiliswa P, Ampong-Nyarko K, Smart LE (1997) Utilisation of wild gramineous plants for management of cereal stemborers in Africa. Ecol Manage Trop Gramineous Stemborers 17 (1):143–150

Koch EW, Barbier E, Silliman BR, Reed DJ, Perillo GME, Hacker SD, Granek EF, Primavera JH, Muthiga N, Polasky S, Halpern BS, Kennedy CJ, Kappel CV, Wolanski E (2009) Non-linearity in ecosystem services: temporal and spatial variability in coastal protection. Front Ecol Environ 7(1):29–37

Kolar C, Lodge D (2001) Progress in invasion biology: predicting invaders. Trends Ecol Evol 16 (4):199–204

Kolozsvary MR, Swihart RK (1999) Habitat fragmentation and the distribution of amphibians: patch and landscape correlates in farmland. Can J Zool 77:1288–1299

Laborde J (1996) Patrones de vuelo de aves frugívoras en relación a los árboles en pie de los pastizales. Dissertation, Universidad Nacional Autónoma de México

Lacey JR, Clayton B, Marlow LJR (1989) Influence of spotted knapweed (*Centaurea maculosa*) on surface runoff and sediment yield. Weed Technol 3(4):627–631

Landgrave R, Moreno-Casasola P (2012) Evaluación cuantitativa de la pérdida de humedales en México. Investigación Ambiental 4(1):35–51

Lansdown RV (2013) *Echinochloa pyramidalis*. The IUCN red list of threatened species 2013:e. T164478A1053206. https://doi.org/10.2305/IUCN.UK.2013-1.RLTS.T164478A1053206.en. Accesed 22 Oct 2017

Lau WY (2013) Beyond carbon: conceptualizing payments for ecosystem services in blue forests on carbon and other marine and coastal ecosystem services. Ocean Coast Manag 83:5–14

Le Ru BP, Ong'amo GO, Moyal P, Ngala L, Musyoka B, Abdullah Z, Cugala D, Defabachew B, Haile TA, Matama TK, Lada VY (2006) Diversity of lepidopteran stem borers on monocotyledonous plants in eastern Africa and the islands of Madagascar and Zanzibar revisited. B Entomol Res 96(6):555–563

Lekeufack M, Fonkou T, Pamo Tedonkeng E, Akoa A (2012) Removal of faecal bacteria and nutrients from domestic wastewater in a horizontal surface flow wetland vegetated with *Echinochloa pyramidalis*. Afr J Environ Sci Technol 6(9):337–345

Lewison RL, Carter J (2004) Exploring behavior of an unusual megaherbivore: a spatially explicit foraging model of the hippopotamus. Ecol Model 171(1):127–138

Li H, Zhang L (2008) An experimental study on physical controls of an exotic plant *Spartina alterniflora* in Shanghai, China. Ecol Eng 32:11–21
Liquete C, Piroddi C, Drakou EG, Gurney L, Katsanevakis S, Charef A, Egoh B (2013) Current status and future prospects for the assessment of marine and coastal ecosystem services: a systematic review. PLoS One 8(7):e67737. https://doi.org/10.1371/journal.pone.0067737
Lonard RI, Judd FW (2002) Riparian vegetation of the lower Rio Grande. Southwest Nat 47:420–432
Lonsdale WM (1994) Inviting trouble: introduced pasture species in northern Australia. Aust J Ecol 19:345–354
Loope LL, Sánchez PG (1988) Biological invasions of arid land nature reserve. Biol Conserv 44:95–118
López Martínez A (1985) Estudios de productividad en áreas de chinampas: el caso de la Estación de Investigaciones sobre Recursos Bióticos "El Morro de La Mancha" Mpio. de Actopan,Ver. Dissertation. Colegio Superior de Agricultura Tropical, H. Cárdenas, Tabasco
López Rosas H (2007) Respuesta de un humedal transformado por la invasión de la gramínea exótica *Echinochloa pyramidalis* (Lam.) Hitchc. & A. Chase a los disturbios inducidos (cambios en el hidroperiodo, apertura de espacios y modificación de la intensidad lumínica), Veracruz, Mexico. Dissertation, Instituto de Ecología AC
López Rosas H, Moreno-Casasola P (2012) Invader versus natives: effects of hydroperiod on competition between hydrophytes in a tropical freshwater marsh. Basic Appl Ecol 13:40–49
López Rosas H, Moreno-Casasola P, Mendelssohn IA (2005) Effects of an African grass invasion on vegetation, soil and interstitial water characteristics in a tropical freshwater marsh in La Mancha, Veracruz (Mexico). J Plant Interact 1:187–195
López Rosas HL, Moreno-Casasola P, Mendelssohn IA (2006) Effects of experimental disturbances on a tropical freshwater marsh invaded by the African grass *Echinochloa pyramidalis*. Wetlands 26(2):593–604
López Rosas H, López-Barrera F, Moreno-Casasola P, Aguirre-León G, Cázares-Hernández E, Sánchez-Higueredo L (2010) Indicators of recovery in a tropical freshwater marsh invaded by an African grass. Ecol Restor 28(3):324–332
López Rosas H, Moreno-Casasola P, Espejel González VE (2015) Shade treatment affects structure and recovery of invasive C4 African grass *Echinochloa pyramidalis*. Ecol Evol 5:1327–1342
Lotze HK, Lenihan SH, Bourque BJ, Bradbury RH, Cooke RG, Kay MC, Kidwell SM, Kirby MX, Peterson CH, Jackson JBC (2006) Depletion, degradation, and recovery potential of estuaries and coastal seas. Science 312:1806–1809
MacDougall S, Turkington R (2005) Are invasive species the drivers or passengers of change in degraded ecosystems? Ecology 86(1):42–55
Maddison M, Soosaar K, Lohmus K, Mander U (2005) Cattail population in wastewater treatment wetlands in Estonia: biomass production, retention of nutrients and heavy metals in phytomass. J Environ Sci Health 40:1157–1166
Maltby E, Jewkes EC, Baker CJ, Mockler NJ, Flynn N, Gardner PJ (1998) A simple method for estimating nitrate removal from floodwaters in river marginal wetlands. Soil Use Manag 14:90–95
Maurer D, Lindig-Cisneros R, Werner K, Kercher S, Miller R, Zedler J (2003) The replacement of wetland vegetation by reed Canarygrass (*Phalaris arundinacea*). Ecol Restor 21(2):116–119
McCafferty C, Lugo-Ortíz R, Provosha AV, Wang TQ (1997) Los Efemerópteros de México: I Clasificación superior, diagnosis de familias y composición. Dugesiana 4:1–29
Melgarejo Vivanco J (1980) Historia de la ganadería en Veracruz. Ediciones del Gobierno de Veracruz, Xalapa
Merritt RW, Cummins KW (1996) An introduction to the aquatic insects of North America. Kendall/Hunt Publishing Company, Dubuque
Miao SL, Zou CB (2012) Effects of inundation on growth and nutrient allocation of six major macrophytes in the Florida Everglades. Ecol Eng 42:10–18

Middleton EL, Bever JD, Schultz PA (2010) The effect of restoration methods on the quality of the restoration and resistance to invasion by exotics. Restor Ecol 18:181–187

Ministry of Agriculture of Guyana (2008, Active July 2010) FAO plant specialist to support Agriculture Ministry in weed control. http://www.agriculture.gov.gy/Bulletins/January%202008/

Mitsch WJ, Gosselink J (2007) Wetlands. Wiley, New York

Mitsch WJ, Day JW, Wendell Gilliam J, Groffman PM, Hey DL, Randall GW, Wang N (2001) Reducing nitrogen loading to the Gulf of Mexico from the Mississippi River basin: strategies to counter a persistent ecological problem. Bioscience 51:373–388

Moore M, Bennett E, Cooper C, Smith S Jr, Farris J, Drouillard K, Schulz R (2006) Influence of vegetation in mitigation of methyl parathion runoff. Environ Pollut 142:288–294

Moreno-Casasola P, López Rosas H, Rodríguez Medina K (2012) From tropical wetlands to pastures on the coast of the Gulf of Mexico. Pastos 42(2):183–215

Muyekho FN, Barrion AT, Khan ZR (2005) Host range for stemborers and associated natural enemies in different farming systems of Kenya. Afr Crop Sci J 13(3):173–183

Myers N (2005) Environmental services of biodiversity. Proc Natl Acad Sci U S A 93:2764–2769

Napper DM (1965) Grasses of Tanganyika. Dar-es-Salaam, Ministry of Agriculture, Forestry and Wildlife

Nascimento CNB do, Carvalho LOD de M, Camarão AP, Salimos EP (1988) Evaluation of forage grasses on a mangrove area of Marajo Island. (Avaliacão de gramíneas forrageiras em área de mangue da Ilha de Marajó.) Boletim de Pesquisa – Centro de Pesquisa Agropecuária do Trópico Úmido, No. 93

Nepstad DC, Uhln C, Serrao EAE (1991) Recuperation of a degraded Amazonian landscape: forest recovery and agricultural restoration. Ambio 20:248–255

Neurath S, Sadeghi A, Shirmohammadi A, Isensee A, Torrents A (2004) Atrazine distribution measured in soil and leachate following infiltration conditions. Chemosphere 54:489–496

Novelo RA (1978) La vegetación de la estación biológica El Morro de la Mancha. Veracruz Biotica 3:9–23

Novelo-Gutiérrez R (1997a) Clave para la separación de familias y géneros de las náyades de Odonata de México, parte I Zygoptera. Dugesiana 4:1–10

Novelo-Gutiérrez R (1997b) Clave para la separación de familias y géneros de las náyades de Odonata de México, parte II Anisoptera. Dugesiana 4:31–40

O'Connor PJ, Covich AP, Scatena FN, Loope LL (2000) Non-indigenous bamboo along headwater streams of the Luquillo Mountains, Puerto Rico: leaf fall, aquatic leaf decay, and patterns of invasion. J Trop Ecol 16:499–516

Ogden J, Rejmànek M (2005) Recovery of native plant communities after the control of a dominant invasive plant species, *Foeniculum vulgare*: implications for management. Biol Conserv 125:427–439

Ortiz-Espejel B, Hernandez-Trejo H (2006) La historia socioambiental. In: Moreno-Casasola P (ed) Entornos veracruzanos: la costa de La Mancha. Instituto de Ecología AC, Xalapa, pp 25–34

Osland MJ (2009) Managing invasive plants during wetland restoration: the role of disturbance, plant strategies, and environmental filters. Dissertation, Duke University

Osland MJ, González E, Richardson CJ (2011) Coastal freshwater wetland plant community response to seasonal drought and flooding in northwestern Costa Rica. Wetlands 31:641–652

Päivänen J (1982) Physical properties of peat samples in relation to shrinkage upon drying. Silva Fenn 16:247–263

Pancho J (1991) Grassy weeds in the Philippines. In: Baker GP, Ferry PJ (eds) Tropical grassy weeds, CASAFA. Rept Series No 2. CAB International, Wallingford, pp 183–188

Pare MMN, Kone D, Kengne IM, Akoa A (2011) Growth and yield potential of *Echinochloa pyramidalis* (Lam.) Hitchc & Chase: a forage plant used in vertical-flow constructed wetlands in Cameroon. Afr J Environ Sci Technol 5(8):622–632

Pare MN, Dongo K, Kengne IM, Dodane PH, Akoa A, Kone D (2012) The economic potential of *Echinochloa pyramidalis* (Lam.) Hitchc & Chase forage plant used in liquid waste treatment in Cameroon: opportunity to link sanitation to food security. Int J Biol Chem Sci 6(1):210–236

Parker C, Caton BP, Fowler L (2007) Ranking nonindigenous weed species by their potential to invade the United States. Weed Sci 55(4):386–397

Parsons J (1972) Spread of African pasture grasses to the American tropics. J Range Manag 25 (1):12–170

Pérez-Maqueo O, Intralawan A, Martínez ML (2007) Coastal disaster and vulnerability assessment in the perspective of ecological economics. Ecol Econ 63:273–284

Piccolo MC, Neill C, Cerri CC (1994) Natural abundance of 15N in soils along forest-to pasture chronosequence in the western Brazilian Amazon Basin. Oecologia 99:112–117

Pivello VR, Shida CN, Meirelles ST (1999) Alien grasses in Brazilian savannas: a threat to the biodiversity. Biodivers Conserv 8:1281–1294

Purseglove JW (1976) Tropical crops: monocotyledons. Longmans, London

Quattrocchi U (2006) CRC world dictionary of grasses: common names, scientific names, eponyms, synonyms, and etymology. CRC Press, Boca Raton

Rai P (2008) Heavy metal pollution in aquatic ecosystems and its phytoremediation using wetland plants: an ecosustainable approach. Int J Phytoremediation 10(2):133–160

Ramakrishnan PS, Vitousek PM (1989) Ecosystem-level processes and the consequences of biological invasions. In: Drake JA, Mooney HA, di Castri F, Groves RH, Kruger FJ, Rejmánek M, Williamson M (eds) Biological invasions: a global perspective. Wiley, New York, pp 281–300

Ratti JT, Garton EO (1996) Research and experimental design. In: Bookhout TA (ed) Research and management techniques for wildlife and habitats. The Wildlife Society, Bethesda, pp 1–23

Robarge WP, Edwards E, Johnson B (1983) Water and waste water analysis for nitrate via nitration of salicylic acid. Commun Soil Sci Plant Anal 14:1207–1215

Rodríguez JP (2001) Exotic species introductions as a challenge for the conservation of South American biodiversity. Interciencia 26:479–484

Rodríguez-Medina K, Moreno-Casasola P (2013) Effect of livestock on soil structure and chemistry in the coastal marshes of the Central Gulf Coast of Mexico. Soil Res 51(4):341–349

Rodríguez-Medina K, Moreno-Casasola P, Yáñez-Arenas C (2017) Efecto de la ganadería y la variación estacional sobre la composición florística y la biomasa vegetal en los humedales de la costa central del Golfo de México. Acta Bot 119:79–99

Rose-Innes R (1977) A manual of Ghana grasses. Ministry Overseas Development, Land Resources Division. Surbiton, Surrey, UK

Rossiter-Rachor NA, Setterfield SA, Douglas MM, Hutley LB, Cook GD, Schmidt S (2009) Invasive *Andropogon gayanus* (Gamba grass) is an ecosystem transformer of nitrogen relations in Australian savanna. Ecol Appl 19(6):1546–1560

Scholte P (2007) Maximum flood depth characterizes above-ground biomass in African seasonally shallowly flooded grasslands. J Trop Ecol 23(1):63–72

Scholte P, Kirda P, Adam S, Kadiri B (2000) Floodplain rehabilitation in North Cameroon: impact on vegetation dynamics. Appl Veg Sci 3:33–42

Scott J, Panetta F (1993) Predicting the Australian weed status of Southern African plants. J Biogeogr 20(1):87–93

Seaton J, Cumberbatch N, Nedd S, Haynes A, Noble D, Knights H (1994) Liveweight gain of heifers continuously grazing *Echinochloa pyramidalis* pastures at three stocking rates. Proc. National Agricultural Research Institute with the Caribbean Agricultural Research and Development Institute (CARDI), October 20–23, 1992, Mon Repos, East Coast Demerara, Guyana, pp 102–105

Setterfield SA, Rossiter-Rachor NA, Hutley LB, Douglas MM, Williams RJ (2010) Turning up the heat: the impacts of *Andropogon gayanus* (gamba grass) invasion on fire behaviour in northern Australian savannas. Divers Distrib 16:854–861

Skerman P, Riveros F (1990) Tropical grasses. FAO Plant Production and Protection Series no. 23, Rome
Smith J, Seaton J, Osuji P, D'Aguiar P, Chesney P, McBean M, Haynes A, Harding C, Cumberbatch N (1991) Dry matter and crude protein yields of *Echinochloa pyramidalis* on coastal clay soil of Guyana. Turrialba 41(1):69–75
SNIB (Sistema Nacional de Información sobre Biodiversidad). Comisión Nacional para el Conocimiento y Uso de la Biodiversidad (CONABIO). http://www.enciclovida.mx/especies/6041655. Accessed 15 Aug 2017
Solow AR (1993) A simple test for change in community structure. J Anim Ecol 62(1):191–193
Song K, Hernández ME, Basto JA, Mitsch WJ (2014) Long-term denitrification rates in created riverine wetlands and their relationship with environmental factors. Ecol Eng 72:40–46
Steele G, Johnson H, Sandstrom M, Capel P, Barbash J (2008) Occurrence and fate of pesticides in four contrasting agricultural settings in the United States. J Environ Qual 37:1116–1132
Tamene B, Bekele T, Kelbessa E (2000) An Ethnobotanical study of the semi-wetland vegetation of Cheffa. Addis Ababa University, Department of Biology. Available at http://phe-ethiopia.org/resadmin/uploads/attachment-14-Hujub%20Sacred%20Forests.pdf
Tapia CJ, Carrera CM, Buller RE (1962) Zacate Alemán para terrenos inundables del trópico. Instituto Nacional de Investigaciones Agrícolas, Secretaría de Agricultura y Ganadería, México DF, Circular CIASE N° 6, 7
Tews J, Brose U, Grimm V, Tielbörger K, Wichmann MC, Schwager M, Jeltsch F (2004) Animal species diversity driven by habitat heterogeneity/diversity: the importance of keystone structures. J Biogeogr 31:79–92
Thomas M, Reid A (2007) Are exotic natural enemies an effective way of controlling invasive plants? Trends Ecol Evol 22(9):447–453
Travieso Bello AC, Campos A (2006) Los componentes del paisaje. In: Moreno-Casasola P (ed) Entornos veracruzanos: la costa de La Mancha. Instituto de Ecología AC, Xalapa, pp 139–150
Travieso-Bello A, Moreno-Casasola P, Campos A (2005) Efecto de diferentes manejos pecuarios sobre el suelo y la vegetación en humedales transformados a pastizales. Interciencia 30 (1):12–18
Tropicos. http://www.tropicos.org/Name/25520876
Turner B, Harrison P (eds) (1983) Pulltrouser swamp: ancient Maya habitat, agriculture and settlement in northern Belize. The University of Utah Press, Salt Lake City
USDA-APHIS (2012) Plant quarantine pest evaluation data sheets. Plant Quarantine pest evaluation data sheets. USA: US Department of Agriculture, Animal and Plant Health Inspection Service. http://www.aphis.usda.gov/plant_health/index.shtml
Valdez Lares R (2010) El ensamble de anfibios de un humedal en proceso de restauración (La Mancha, Veracruz, México): Diversidad y su relación con el ambiente. Dissertation. Instituto de Ecología, A. C. Xalapa, Veracruz
Van der Velde G, Rajagopal S, Kuyper-Kollenaar M, Bij de Vaate A, Thieltges DW, MacIsaac HJ (2006) Biological invasions: concepts to understand and predict a global threat. In: Bobbink R, Beltman B, Verhoeven JTA, Whigham DF (eds) Wetlands: functioning, biodiversity conservation, and restoration, Ecological studies, vol 191. Springer, Berlin, pp 61–90
Vesey-Fitzgerald DF (1963) Central African grasslands. J Ecol 57:243–274
Vitousek P (1990) Biological invasions and ecosystem processes: towards an integration of population biology and ecosystem studies. Oikos 57:7–13
Vymazal J (2013) The use of hybrid constructed wetlands for wastewater treatment with special attention to nitrogen removal: a review of a recent development. Water Res 47(14):4795–4811
Vymazal J, Březinová T (2015) The use of constructed wetlands for removal of pesticides from agricultural runoff and drainage: a review. Environ Int 75:11–20
Weeds of Australia (2015) http://keyserver.lucidcentral.org/weeds

Wells MJ, Balsinhas AA, Joffe H, Engelbrecht VM, Harding G, Stirton CH (1986) A catalogue of problem plants in South Africa. Memoirs of the botanical survey of South Africa No 53. Botanical Research Institute. Pretoria, South Africa

Williams D, Baruch Z (2000) African grass invasion in the Americas: ecosystem consequences and the role of ecophysiology. Biol Invasions 2:123–140

Wissinger S, McGrady J (1993) Intraguild predation and competition between larval dragonflies: direct and indirect effects on shared prey. Ecology 74(1):207–218

Worm B, Barbier EB, Beaumont N, Duffy JE, Folke C, Halpern BS, Jackson JBC, Lotze HK, Micheli F, Palumb SR, Sala E, Selkoe KA, Stachowicz J, Watson R (2006) Impacts of biodiversity loss on ocean ecosystem services. Science 314:787–790

Yabuno T (1968) Biosystematic studies of *Echinochloa stagnina* (Retz.) P. Beauv. and *E. pyramidalis* (Lamk.) Hitchc. et Chase. Cytologia 33:508–519

Yetter JC (2004) Hydrology and geochemistry of freshwater wetlands on the Gulf coast of Veracruz, Mexico. Dissertation, University of Waterloo. Available from: University of Waterloo Library; CA2ON UW420 2004Y47

Zar JH (1996) Biostatistical analysis, 4th edn. Prentice-Hall, Upper Saddle River

Zuloaga FO, Morrone O (2003) *Echinochloa*. Catalogue of new world grasses. Contrib US National Herbarium 46:215–224

Chapter 10
Environmental Impacts of an Alien Kelp Species (*Undaria pinnatifida*, Laminariales) Along the Patagonian Coasts

María Paula Bunicontro, Silvia C. Marcomini, and Graciela N. Casas

Abstract *Undaria pinnatifida* (Harvey) Suringar was recorded in Argentina for the first time in December 1992. Since then, it shows an invasive and competitive behavior, spreading quickly from the initial focus and increasing its population density. The first individuals were found attached to the port of Puerto Madryn, suggesting that the vector of introduction was the ballast water of cargo ships arriving to Golfo Nuevo from overseas. Since *U. pinnatifida* was recorded, it has been spreading along the Argentina coasts, extending its range within and outside Golfo Nuevo, along the coasts of Argentina from Puerto Deseado (Santa Cruz province) to Mar del Plata (Buenos Aires province) far as 1850 km from each other. The invasive condition of this species creates an awareness of the effects inflicted upon the indigenous biodiversity of this region, as well as, on the commercially-important benthic community structure. Moreover, the detachment of subtidal algae, especially during the summer, and its deposition upon the beach, have altered the sedimentary balance along the overall transverse beach profile. This effect, in addition with algal extraction by trucks mainly during tourist seasons, may increase beach erosion processes along Patagonian coasts.

Keywords *Undaria pinnatifida* · Marine bioinvasions · Invasive species · Argentine Patagonia

---

M. P. Bunicontro (✉) · S. C. Marcomini
Facultad de Ciencias Exactas y Naturales, Departamento de Geología, Geomorfología y Geología Costera, Universidad de Buenos Aires, Buenos Aires, Argentina

CONICET – Universidad de Buenos Aires – Instituto de Geociencias Básicas, Aplicadas y Ambientales de Buenos Aires (IGEBA), Buenos Aires, Argentina
e-mail: pbunicontro@gl.fcen.uba.ar; scm@gl.fcen.uba.ar

G. N. Casas
Laboratory of Marine Algae, Biology of Marine Organisms Institute (IBIOMAR) – CCT CENPAT – CONICET. Brown Boulevard 2915. (U9120ACD), Puerto Madryn, Argentina
e-mail: graciela@cenpat-conicet.gob.ar

C. Makowski, C. W. Finkl (eds.), *Impacts of Invasive Species on Coastal Environments*, Coastal Research Library 29,
https://doi.org/10.1007/978-3-319-91382-7_10

10.1 Introduction

Foreign seaweeds have been introduced along the coasts of the world since the Eighteenth and Nineteenth centuries. Exotic organisms began to appear in different ports around the world, transported by wooden sailing vessels. The community of organisms present in the dry ballast, as well as that growing attached to the hulls of vessels that used to be scrapped-out, allowed some of that organisms to settle and spread from this point of introduction while others will remain restricted without any threat to the native marine community.

Accidental human-mediated introductions have continued. Species are known to be spread internationally on ship's hulls, ballast water and fishing nets. Once introduced, the seaweeds may spread along the coasts by fishing vessels, artisanal fisheries, coastal or pleasure shipping and through the transport of aquaculture farming devices (South et al. 2017).

Macroalgae represent approximately 20% of the world's marine introduced species. The ability of foreign species to establish and spread in their new habitat and the risks they pose for the native flora and fauna varies greatly among different species.

Along the coast of Golfo Nuevo, a non-native algal species was observed settled onto the port of Puerto Madryn city on December 1992. The algae were identified as *Undaria pinnatifida* (Harvey) Suringar and it has not been previously registered in the local marine flora (Piriz and Casas 1994).

Undaria native distribution includes temperate regions. This species is native on Japanese, Korean and Chinese coasts and has spread to the Atlantic and Mediterranean coasts of Europe (Castric-Fey et al. 1993; Fletcher and Manfredi 1995; Curiel et al. 1998; Zenetos et al. 2005), to New Zealand (Hay and Luckens 1987; Russell et al. 2008) and Australia (Sanderson 1990; Campbell and Burridge 1998; Valentine and Johnson 2004). It was accidentally introduced to the argentine coasts, transported by shipping (Casas and Piriz 1996, 2001). After its introduction *Undaria pinnatifida* progressively spread along the coasts of argentine Patagonia. At the present time, with the exception of sandy bottoms, the coasts of the Golfo Nuevo are almost entirely colonized showing undesirable environmental and economic effects (Orensanz et al. 2002).

Environmental disturbances have significantly affected the world biota. Biological invasions are impacting ecosystems with dramatic effects on native species: changes in population structure and genetic modifications. The competitive relationships could determine the extinction of many indigenous species (Piriz and Casas 2001).

The invasive nature of *Undaria* also affects economic and touristic activities along Patagonian beaches, such as diving and recreation. During summer, the seaweed cleaning activity modifies the hydrodynamic of the beaches, by removing large volumes of algae including high percentages of sand. This sand will not return to the coastal system, increasing the beach erosion of Golfo Nuevo.

Urban growth during the last decades along the gulf has increased sewage to the sea, favoring *U. pinnatifida* expansion. The population grew rapidly, increasing the sewage water and domestic waste. The development of *Undaria* is not inhibited by residual discharges in the coast. Even, it was determined that the alga is capable of incorporating nutrients from the beach (Torres et al. 2004).

10.1.1 Golfo Nuevo Features

Golfo Nuevo is located in the eastern continental margin of Patagonia, Argentina (42° 42′S – 65°36′W) (Fig. 10.1). Golfo Nuevo is an elliptical basin with a surface of 2440 km^2 and a maximum depth of 184 m that connects to the continental shelf through a 17 km wide gap (Mouzo and Garza 1979). This gulf is 56.3 km long and 40.2 km wide, and its mouth has a width of 11.2 km, facing to the southeast with the Patagonian Shelf.

The main urban center is Puerto Madryn, located in the inner western section of the gulf with a coastal extension of about 12 km long. During the last decades urbanization increased and also the touristic activities. The population tripled between 1970 and 1980 from 6115 to 20,103 inhabitants after the installation of the aluminum company ALUAR in 1974 (Bunicontro 2018) (Graphic 10.1). This fact not only changed the demography of Puerto Madryn, but also the demand for services, housing and infrastructure. It is currently estimated that the population is approximately 100,000. The growth of urbanization between 1942 and 2012 is shown in Fig. 10.2. During that time, the coastline occupied by urban development expanded from 0.8 to 11.2 km. The rapid growth of the city, in addition to the deficient urban planning management of sewage discharge resulted in a high concentration of nutrients in shallow waters of Golfo Nuevo.

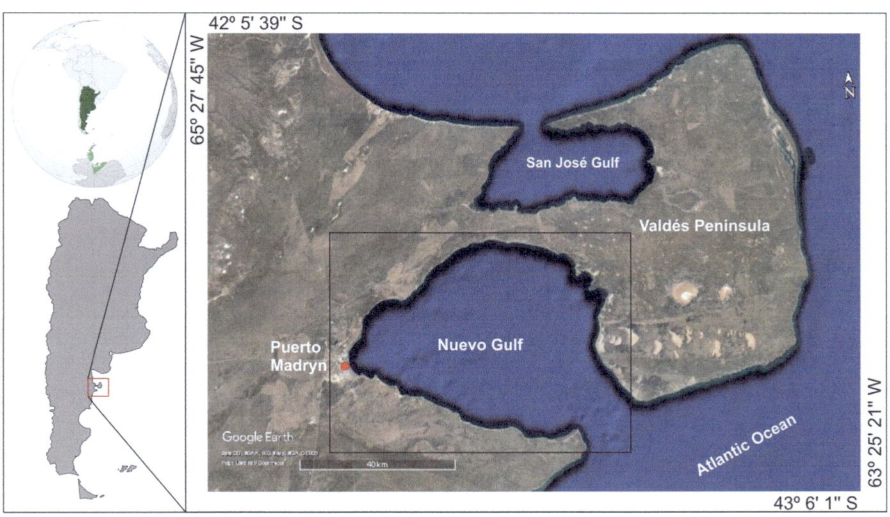

Fig. 10.1 Location map of Golfo Nuevo and Puerto Madryn city, Chubut province

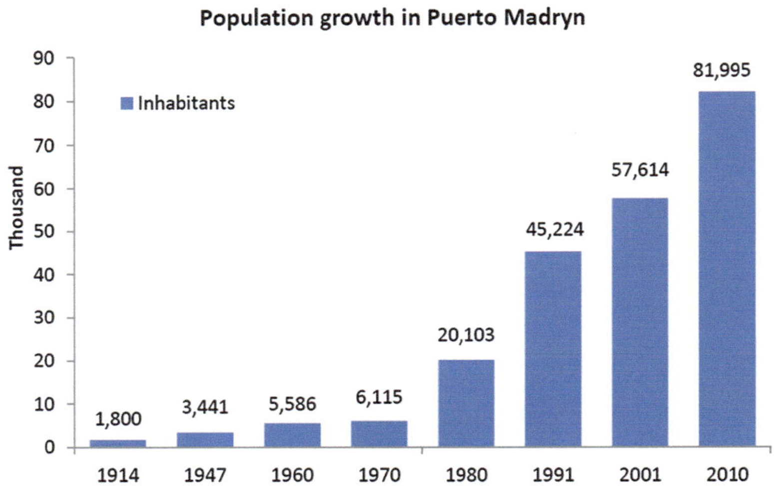

Graphic 10.1 Population growth in Puerto Madryn between 1914 and 2010. (Taken and modified from Bunicontro 2018)

This effects would have contributed to the algae species expansion. Also, *Undaria* is able to incorporate nutrients such as nitrates, ammonium and phosphate from the sewage (Torres et al. 2004).

The management of the sewage final disposal in Puerto Madryn has changed during the last decades. In 1980s wastewater was thrown to the seawater after primary and secondary treatment. Later, the treatment plant become deficient and most of the domestic wastewater outflow into the sea. About 8000 m^3/day of poorly treated sewage were discharged to the sea as far as 2001 (Sánchez et al. 2006). In consequence, some signs of eutrophication became evident in coastal water (Gil 2001 and Díaz et al. 2002). During the 2000s a new residual water treatment plant was built, reducing the amount of wastewater discharge to the coast. Since its installation the plant was operationally deficient and almost 80% of the wastewater continued to be thrown to the sea (Sánchez et al. 2006). At present, the treatment plant is operative and has been expanded, but there are still some deficiencies in the wastewater collection system. Nowadays, many storm drains discharging sewage directly to the beaches of Puerto Madryn, resulting in pollution and bad odors.

Puerto Madryn is also a very important destination for tourism. There are two high seasons during the year: (a) the summer season (between December and March) and (b) the "Right whales" season (between June and December). Every year the population doubles during each high season reaching between 200,000 and 250,000 tourists on average (Secretary of Tourism and Sports of Puerto Madryn, modified from Bunicontro 2018) (Graphic 10.2). This situation increases the demand for basic services, favoring the pollution of the coastal environment and the sewage discharges to the beaches.

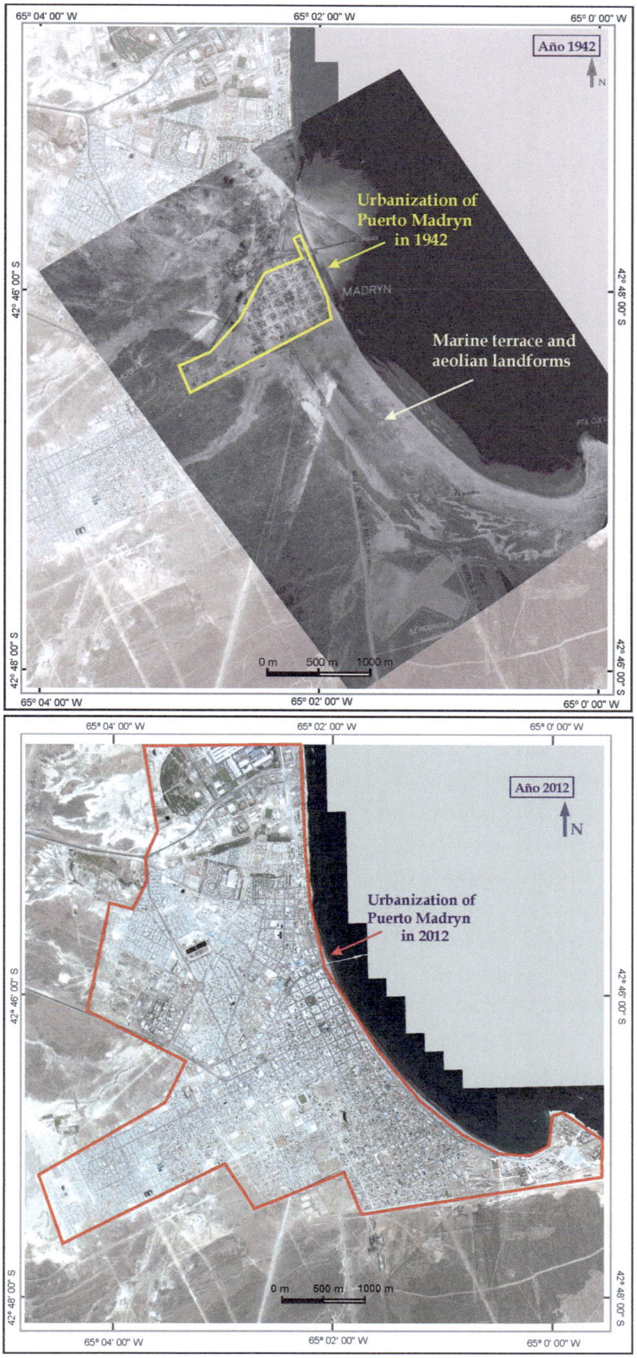

Fig. 10.2 Urban expansion between 1942 and 2012 of Puerto Madryn. The aerial photograph of 1942 shows a small village and original landscape. The satellite image of 2012 evidences urban expansion over original landforms. (Taken and modified from Bunicontro 2018)

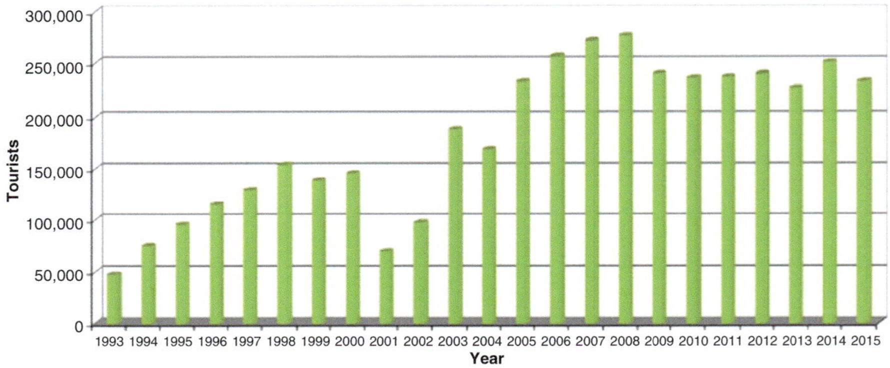

Graphic 10.2 Number of tourists per year arrived in Puerto Madryn during the last decades. (Source: Secretary of tourism and sports of Puerto Madryn. Taken and modified from Bunicontro 2018)

10.2 Taxonomy and Morphological Characteristics of Undaria

The genus *Undaria* (Phaeophyceae, Laminariales, Alariaceae) includes three species, *U. pinnatifida* (Harvey) Suringar, *U. undarioides* (Yendo) Okamura, and *U. peterseniana* (Kjellman) Okamura (Okamura 1915; Lee and Yoon 1998; Yoon and Boo 1999), although phylogenetic relationships are discussed (Yoon and Boo 1999; Kawai et al. 2016; Uwai et al. 2006) genetic studies suggest that the three species could be morphological variations of the same species (Uwai et al. 2006).

The analysis of its taxonomic position confirmed that the entity present in Argentina is the Japanese species *Undaria pinnatifida* with its forms *typica*, *distans* and the occasional occurrence of the form *narutensis*, as well as other phenotypic morphologies (Casas 2005).

The form *typica* is characterized by a short stipe where the sporophyll is closest to the blade. Blades are wide and their divisions pinnate. The form *distans* has a longer stipe and the sporophyll restricted to the low third of basal section, an elongated blade and deep pinnate divisions (Saito 1975; Stuart et al. 1999). A third form is called *narutensis* and sometimes it was observed in Golfo Nuevo in summer. It is characterized by a shortest stipe, a sporophyll always confluent with the blade, spreading on it. But the f. *narutensis*, could be an extreme form of f. *typica*, more than an independent form (Okamura 1915).

Although the different forms have a genetic base (Saito 1975), the expression of these forms could be modulated by environmental factors. Lee and Yoon (1998) suggested the use of *Undaria pinnatifida* var. *elongata* and var. *vulgaris* instead of those forms because of nomenclature precedent.

Within the local population, the observation of the forms *distans* and *typica* in Golfo Nuevo (Casas 2005) has led to suppose that both forms may have been introduced or that may have hybridized and combined their phenotypic characteristics, but later observations associated the f. *typica* with shallow waters and the f. *distans* to greater depths, showing an adaptive behavior. Even the f. *narutensis* was observed in structures near the sea surface and in tide pools.

In Argentine coasts *Undaria* has an annual life cycle with constant recruitment, being this characteristic related with the narrow range of seawater temperature (Casas and Piriz 1996). An inverse relationship was found between the sizes of sporophytes with seawater temperatures (i.e. larger plants associated to lower water temperatures). Highest densities were associated with recruiting peaks. The midrib width is related with the length of plants, and the larger sporophyll diameter is associated with summer conditions (Casas et al. 2008).

Undaria pinnatifida, is a species usually called "wakame" in Japan. The brownish-green large plants are the sporophytes and were of 1.45 m in length in our observations, with exceptional records of near 1.70 m, whereas in its native areas Undaria pinnatifida has usually 1.5–3 m in height. The kelps are attached to the marine rocky bottom by a strong fixing structure named holdfast. The blades of *Undaria* shows a conspicuous central midrib and smooth edges, groups of mucilage glands and criptostomata with hairs are observed scattered on the surface of the blade.

When individuals are mature a much undulated structure, the sporophyll, is developed between the holdfast and the blade. In Golfo Nuevo the sporophyll can reach 18 cm in wide and is able to release millions of spores that will develop in male and female microscopic filamentous gametophytes. Casas et al. (2008) mentioned the findings of dense brownish *Undaria* gametophytes settled on the crustose Corallinaceae *Synartrophyton* sp., on Ascidiaceae and on valves of *Aulacomya ater* Molina (Fig. 10.3).

In Argentina, *Undaria* grows reaching their biggest size at the end of winter and beginning of spring. In summer, they become damaged at the edges of the blades because of high temperatures of the sea water. Added to the strong swell produced by the winds and the marine currents, *Undaria* is thrown to the beaches. Damaged individuals with larger (near 18 cm in diameter) and healthier sporophylls were observed in the drift line on Puerto Madryn beaches in summer (G. Casas – personal observation) leading us to the conclusion that sporophytes are reproductive all year round in Southern Hemisphere (Casas et al. 2008).

10.3 Environmental Settings

The coastline of the gulf is an erosive coast emplaced in sandstones interbedded with volcaniclastic sediments. Active cliffs and wave cut platforms are the most frequent coastal landforms. *Undaria* lives in the subtidal wave cut platform settle in the rocky bottom up to 22 m depth, but algae mass are transported by littoral current to the intertidal zone (Fig. 10.4). Figure 10.5 shows a schematic coastal model to understand the site within the coastal profile where *Undaria* establishes and the site where they are thrown during summer when they are detached from the rocky bottom. It is

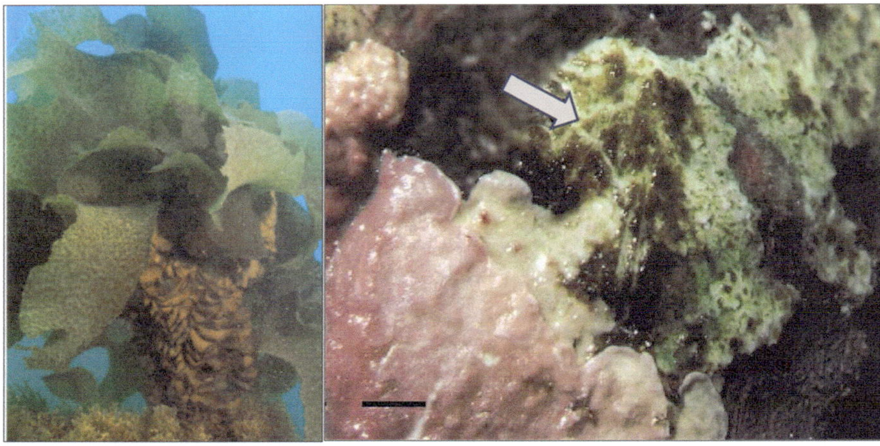

Fig. 10.3 Mature individual of *Undaria* in rocky shore of Golfo Nuevo (left); Gametophytes in the rocky bottom (arrow), on crustose Corallinaceae. Scale: 2 mm (Right). (Photos: G.Casas)

Fig. 10.4 Abrasion platforms along the erosive coast of Golfo Nuevo. This landscape of Patagonian coast and the rocky substrate represents the appropriated place for *Undaria*'s settlement. (Photos: P. Bunicontro)

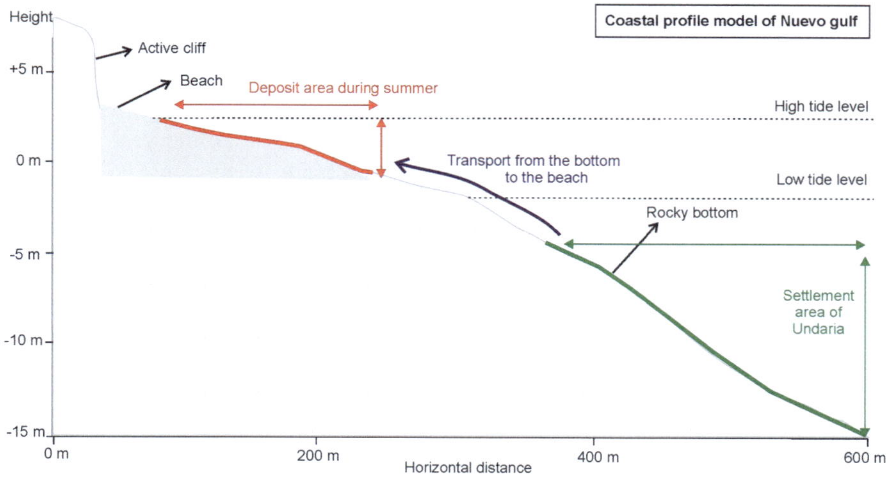

Fig. 10.5 Schematic model of a beach profile in Golfo Nuevo (the distances and depths are approximate). *Undaria* is settled on rocky bottom, up to 22 m depth depending on the light. It is moved by littoral currents from the bottom to the foreshore during summer, along a transport zone. The foreshore area occupied by the kelp depends specifically on the beach profile

casted ashore by tides from the bottom to the foreshore during summer, along transport zone. The foreshore area occupied by the kelp depends specifically on the beach profile.

The climate in Patagonia region is arid, characterized by extreme weather conditions, with sporadic precipitations and a mean annual temperature of 13 °C. Tidal regime is macrotidal (mean tidal range is 4.13 m) and the dominant tides are semidiurnal (Naval Hydrographic Service 2017). There are not morphological evidences of littoral drift inside the gulf. The currents are dominated by tides (not by wave action), making the gulf a place of relative calm waters.

The ecology of this invasive species and the phases of its life cycle are related with ***temperature***. In native areas of Japan, *Undaria* develops between 4–28 °C. In this wide range of temperatures *Undaria* has an annual life cycle. In the Northern Hemisphere seawater temperature has been also suggested to be the most important environmental factor affecting the life cycle of *U. pinnatifida* (Saito 1975; Stuart et al. 1999). In Golfo Nuevo, the superficial seawater temperature presents slight seasonal variations between 18 °C at the end of summer and 8 °C in spring (Rivas and Beier 1990). Recruitments in the *U. pinnatifida* population growing in Golfo Nuevo suggest that mature individuals can also reproduce throughout the year, in contrast to different patterns observed in Japanese populations where *U. pinnatifida* is native (Akiyama and Kurogi 1982).

Outside its native area where *Undaria* is present, as in mostly of the sites invaded by this species, the seawater temperatures are between 10 y 20 °C. This small temperature range in the invaded sites, could induce *Undaria* to have a constant recruitment. We experimentally confirmed the occurrence of juveniles throughout the cycle, indicating that the zoospores are released and germinate every month, even in the coldest months of July and August (11–10 °C), which means that viable zoospores are present in the column of water all year round (Casas et al. 2004). The presence of juvenile and mature sporophytes during almost all the year, could be indicating that when the canopy of mature sporophyte were removed, younger sporophytes were recurrently observed, suggesting that they were resting among large individuals and will develop when freed from the competitive "canopy effect" (Casas et al. 2008).

Although the ***salinity*** for the optimum growth of *Undaria* is over 27‰, the species might be observed in low salinities like in Venice (20‰), New Zealand (22–23‰) and Spain (27‰), showing tolerance to low salinities. The salinity of Golfo Nuevo water varies from 33.5 to 33.9‰ (Rivas and Ripa 1989), being a suitable condition for its development. A particular situation is that of the specimens developing in the tidal pools of the rocky intertidal where both the size and maturation of the sporophytes and the end of the life cycle are closely related to the abiotic conditions of the pools, showing a shorter life cycle and small size plants (Casas 2005).

The influence of ***light*** on growth of the kelps depends on transparency of marine water. The sporophytes of *Undaria* may grow deeper more than 20 m but in less transparent waters the sporophytes are observed attached near the water surface. In Golfo Nuevo *U. pinnatifida* has been found growing from the rocky intertidal up to 22 m in depth, showing tolerance to a wide range of radiation. This broad growing area indicates that if adequate substrates are available, the lower limit is determined by the quality of light. In coasts where the presence of *Undaria* has been reported, the percentage of light could be dramatically reduced by water turbidity. In these cases, although suitable substrates were available, the maximum depth where sporophytes were found was no more than 5 m. Physiologically, the juvenile samples of *Undaria* has better tolerance to low irradiation levels, becoming them in a good competitor with the native algae and the gametophytes or filamentous phase may become in resistance structures, able to survive in darkness for a long time (Saito 1975; Sanderson 1990; Brown and Lamare 1994; Casas et al. 2008).

10.4 Introduction of an Invasive Species

An introduced species could become an established one, which depends on both the environmental and ecological adaptive characteristics of the alga; if such a non-native species spreads from the point of introduction and becomes abundant, it is termed 'invasive species' (Kolar and Lodge 2001). The introduction of invasive species transported across biogeographical boundaries has been the cause of global ecological change in the past hundreds of years along with habitat destruction. Bioinvasions are considered the second cause of extinction of species (after habitat destruction), for this reason in 1992 the Convention of Biodiversity signed in Rio recommended its eradication or control. Living organisms that have been accidentally or intentionally transported by men who have crossed oceans, have led to an alteration in the diversity of many coastal marine communities (Carlton 1989, 1996, 1999, 2000).

The introduced species are able to affect the habitat profile by monopolizing space, acting as "ecosystem engineers" could significantly modify the composition of local communities by altering competitive interactions, trophic networks as well as ecologic and ecosystem processes (Wallentinus and Nyberg 2007).

Undaria pinnatifida showed a competitive profile, modifying the composition of local communities by altering competitive interactions and have led to economic and ecological consequences in their introduction areas range (Schaffelke et al. 2006; Schaffelke and Hewitt 2007; Williams and Smith 2007).

Opinions on the opportunistic and competitive features of *Undaria* (Battershill et al. 1998; Casas et al. 2004; Casas 2005) somewhat opposed to some indications about the non-aggressive characteristics of the kelp (Floc'h et al. 1991, 1996; Boudouresque and Verlaque 2002), the evidence suggests that *Undari*a should be considered as a successful species once established outside its native region (Casas et al. 2008).

The introduction of marine species can result in severe ecological disturbances on native communities. In algal communities, competition for light or substrate can be intense and might lead to partial exclusion or even total disappearance of native species. It was confirmed that the diversity of the associated biota is heavily impacted, affecting the richness and the biomass distribution of native algae (Raffo et al. 2012).

This could occur because the populations of this exotic species show a great development in height, especially by the end of the winter and in spring, added to the great number of individuals, would produce a significant shadowing on the seafloor. Also the holdfasts can overlap and entangle each other and covering the bottom, reducing the space available for the development of other native species living on it.

10.4.1 Dispersal of Undaria *Along Argentina Coasts*

In the Southern Hemisphere the first observations of *Undaria pinnatifida* were in New Zealand in 1987. In Argentina it was registered in 1992, being this the first occurrence in the coasts of Southeastern America (Hay and Luckens 1987; Piriz and Casas 1994).

A remarkable fact is that in all the sites where *Undaria* was introduced, the beginning of the invasion was in ports areas.

In Golfo Nuevo the dispersal rate of this kelp was about 17 km a year, *Undaria* spread here 172 km in 10 years (Fig. 10.6). Along with *Undaria* dispersion within the limits of Golfo Nuevo, populations increased their density, originating a "donor" area of structures (spores, gametophytes) that allowed the propagation of the species. In this way, *Undaria* could have dispersed to other locations far from Puerto Madryn, arriving 12 years after until Puerto Deseado (Santa Cruz province) far 760 km South from Golfo Nuevo and Mar del Plata (Buenos Aires province)

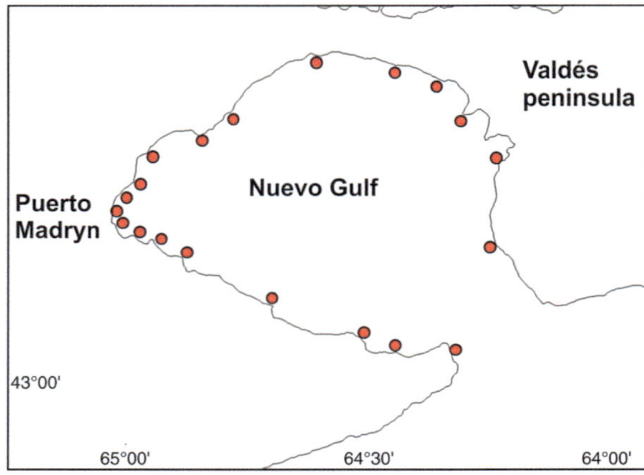

Fig. 10.6 Dispersal of *Undaria* in Golfo Nuevo between 1993 and 2003: near 172 km

1300 km Northwards from Golfo Nuevo (Martin and Cuevas 2006; Meretta et al. 2012; Casas 2013).

Undaria and its feasibility for dispersal around the world could be due to the latency state of its microscopic reproductive structures, which can remain for several months. This condition could have favored its resistance during transport within the ships' ballast water tanks (Hay 1990).

Using climatologic satellite sea-surface temperature (SST) data from several locations, a potential latitudinal thermal range extending between Puerto Deseado (Argentina, 47.75°S) and Cabo Torres (Brazil; 29.35°S) was predicted by Dellatorre et al. (2014). Natural dispersion of *Undaria* is unlikely given the presence of a major barrier formed by extensive exposed sandy beaches on the south coast of Río de la Plata estuary, so eventual introduction to the Uruguay coast would require human-mediated activities. Rocky outcrops along the Atlantic coast of Uruguay are vulnerable in terms of substrate suitability, salinity, and thermal regime. Cabo Torres (29.35°S, Brazil) is an isolated rocky outcrop, where the monthly SST range suits well within *Undaria*'s tolerance.

These northern locations might be colonized if gametophytes could reproduce within the range 17–20 °C, suggesting that vector control and surveillance are needed to prevent inoculation into new areas.

10.4.2 Dispersal Vectors: Vessels and Ballast Water

It was determined that the vector of introduction of *U. pinnatifida* to the Golfo Nuevo was the ballast water of ships. Vessels have been mentioned on several occasions as a vector of transport of exotic species (Carlton 1999; Casas 2005).

As a result of human movements throughout the world, by exploration, trade or colonization, species have been breaking natural biogeographical barriers between continents and between oceans. This process has gradually led to a profound alteration in the diversity and structure of many coastal marine communities.

Until about 1890 the boats were built entirely of wood. In this way, mammals, plants, birds, insects, dry algae and seeds were transported accompanying the sand and rocks that constituted the "dry ballast" that was used previously to give stability and balance to the boats.

After 1890 the vessels began to be constructed of metal and therefore the dry ballast was slowly replaced by ballast water, contained in special tanks. Not only the type of ballast of ships changed, but the shipping technology. The boats began to be faster, to travel in less time greater distances and therefore to visit a greater number of ports. Therefore, opportunities for new species introductions increased significantly.

Ballast water, beyond being necessary for the stability of the ship during navigation, constitutes a mechanism for transporting large volumes of water. The International Maritime Organization for January 2008 considered the world merchant fleet to be 50,525 vessels. These commercial vessels carrying more than 80% of the world's merchandise are moving globally about 7000 species in a volume of 3--10 trillion tons of ballast water. The ballast water is a non-selective transport, therefore the water pumped in each port contains not only animal and vegetable organisms and larvae, but also viruses and bacteria, many of them pathogenic like cholera and botulism.

Implementation of no-dispersal measures requires knowledge of the realized and potential range of the species in the region, and identification of natural barriers for natural dispersal within "internal borders" (Forrest and Taylor 2002; Forrest et al. 2009).

10.5 Environmental Impact of *Undaria* Invasion

Undaria invasion and its effects can be summarized in three serious consequences: ecological, economic and hydrodynamic. First of all, because of its large size (up to 2 m in length; Raffo et al. 2009) and invasive characteristics dense kelp beds can outcompete native macroalgal species (Casas et al. 2004) and induce changes in fish behavior, while positively affecting diversity and abundance of benthic macrofauna (Irigoyen et al. 2011).

Secondly, *Undaria* produces a negative visual impact on tourism and economy: (a) during diving activities, (b) when kelp arrivals spread along the beaches during summer and (c) on *Gracilaria gracilis* production (Casas et al. 2004).

Finally, it could be seen that those large amounts of algae on the beach during summer months and its treatment have a direct impact on beach sediment balance.

10.5.1 Ecological Impacts

Raffo et al. (2012) experimentally evaluated the effect of *Undaria* on the species richness, abundance, diversity and composition of the benthic macroalgae assemblages in the Golfo Nuevo in 2001 and 2008, at 8 and 15 years after the invasion. *Undaria* had a negative impact on macroalgae assemblages, reducing species richness and biomass, since in 2008 there were 18 species less than those found in 2001, which were not recovered at the end of 1 year of *Undaria* exclusion, suggesting that the changes occurred on the benthic macroalgal community are irreversible, with loss of native species.

Changes in macroalgal community structure caused by invasive seaweeds have strong impacts on the associated macrofauna due to the role of macroalgae as autogenic ecosystem engineers. The abundance and diversity of benthic macrofauna, species richness and diversity in Golfo Nuevo, were higher in plots covered by *Undaria* than when *Undaria* was removed. The abundance of two species of Crustaceans, one species of sea urchin, one species of Nemertinae and several species of Polychaetae was higher. Irigoyen et al. (2011) attribute these effects to the provision of new habitat structured by *Undaria*, a larger and structurally more complex species than the local native seaweeds.

Analyzing the shading effects of the exotic algae *Undaria pinnatifida* on the community of macroalgae in intertidal pools, Raffo et al. (2015) experimentally determined that the presence of *Undaria* diminishes the quantity and quality of light in the pools, affecting the diversity of the ephemeral macroalgae. Competition for light would then be one of the key processes in the growth and survival of the native macroalgae community.

Since the introduction of *Undaria* in the Golfo Nuevo, the underwater landscape has changed dramatically. *Undaria* has shown the ability to colonize both artificial substrates as natural reefs. On the Patagonian coast, the reefs consist of rocks that extend from the intertidal beach to the subtidal where they are never discovered by low tides. The biodiversity of these reefs are the habitat of fishes that could be impacted by the presence of *Undaria* producing changes in the abundance of fishes living there (Piriz and Casas 2001).

In summer, the algae are cast ashore to the beaches, phenomenon known as "*arribazón*" (Fig. 10.7). The large *Undaria* are removed by pulling out their holdfasts which carry with them marine fauna and flora, causing a removal and alteration in the rocky bottoms.

If *Undaria* is considered by native fauna as a new source of food, its introduction could generate other ecological changes. A small sea snail, the gastropod *Tegula patagonica* and the sea urchins *Arbacia dufresnii* and *Pseudechinus magellanicus* are recognized as potential herbivores and live habitually among the populations of *Undaria*. However, the true regulatory impact of these organisms on the kelp has not yet been studied, whether the populations of these invertebrates have increased considerably or not since the occurrence of *Undaria* in the Golfo Nuevo as a new and abundant resource (Teso et al. 2009).

Fig. 10.7 Significant amounts of sporophytes casted-off on beaches during high tides in summer: "Arribazón". (Photos: C. Eyras (with permission) and P. Bunicontro)

10.5.2 Economic Impact

An important aspect in the study of invasive species is the economic impact they can produce. The *Undaria pinnatifida* case has led to well-documented economic and ecological consequences in their introduction areas (Schaffelke et al. 2006; Schaffelke and Hewitt 2007; Williams and Smith 2007).

In Argentina the main risk is the dispersion of *Undaria* to other economically productive areas. In this sense, populations of this species have been observed growing on beds of the agarophyte *Gracilaria gracilis* (Casas et al. 2004) and recently *Undaria* was observed in a location characterized by the exploitation of bivalve mollusks collected by artisanal fishermen.

Another of the economically important activities that take place in the Golfo Nuevo is diving in submarine parks. Since the great size of *Undaria* individuals may cover the reefs, the diving activities are altered because of the cleaning activities- before and during- touristic diving season.

The greatest impact on the beaches of Golfo Nuevo produced by *Undaria* is in summer. Every day the immense algae mass deposited on the beach has to be removed. This is done to avoid re-entry of organic matter and reproductive structures into the sea. Also, beaches should be cleaned for hygiene reasons and visual impact in the tourism.

Fig. 10.8 Cleaning activities in Puerto Madryn beaches during summer, causing significant environmental and economic impacts. (Photo: C. Eyras (with permission))

Algae mass is removed by bulldozers and taken away is removed from the beach with trucks (Fig. 10.8). The daily work of the machines not only extracts the algae but also a large amount of sand with other organisms, increasing erosion.

10.5.3 Hydrodynamic Impact

The large arrivals of drift marine algae to Puerto Madryn during summer alter the transverse transport along the beaches profiles especially on sandy beaches. Algae arrivals acts as a carpet, avoiding the normal sand circulation along longshore and crosshore littoral currents. This effect result in the interruption of natural hydrodynamic balance (Fig. 10.9). During the time of algae arrivals to the foreshore, sediment transport is almost interrupted. The grains of sand are trapped within the algae biomass and transport from foreshore to backshore is stopped.

In addition to the alteration to sand transport, during the cleaning work, large amount of sand are removed with the algae. According to Piriz et al. (2003), between 1992 and 1994, there were removed between 2500 and 12,000 ton/year (wet weight) of algae biomass. According to these authors, between 100 and 400 m^3 of sand are removed within the algae every season. In their study, they showed that of 5 ton of algae in wet weight about 50% was sand (in dry weigh). Also, Eyras and Sar (2003) estimated that near 8000 ton of algae are removed from Puerto Madryn beaches every year and that about the 70% of it (dry weight) correspond to sand.

Fig. 10.9 *Undaria*'s arrivals on Puerto Madryn beaches. **a** and **b** show the seaweed deposit in sandy beaches on the foreshore, from foredune base to the low tide level (feeding place for coastal birds). **c** and **d** show the seaweed deposit in pebble beaches on the backshore during spring high tides or storms. In this type of beaches the area occupied by the seaweed is smaller depending on the beach slope. (Photos: P. Bunicontro)

These values are estimated since there are not official reports from the local authorities to assess the impact of the removal activities. The sand extracted from beaches does not return to the natural coastal system causing a deficit in sediment balance that is increasing every year.

10.6 Analysis of Beach Profiles

The configuration of the beach profile also control the dispersion and configuration of the algae arrivals. Along the coast of Golfo Nuevo two morphodynamic beach models regulate the final settlement of the algae arrivals. (Fig. 10.10).

Beach type 1 (Fig. 10.11a). It is a pebble and sand beach with poor sorted and bimodal sediments. Beach slope is 1° and 5° (Bunicontro 2018). The low intertidal zone is composed by fine sand (58%) and fine pebbles (42%) while the high intertidal zone is by fine pebble (52%) and fine sand (48%). It shows a slight increase in grain size from foreshore to backshore, where gravelly berms are common.

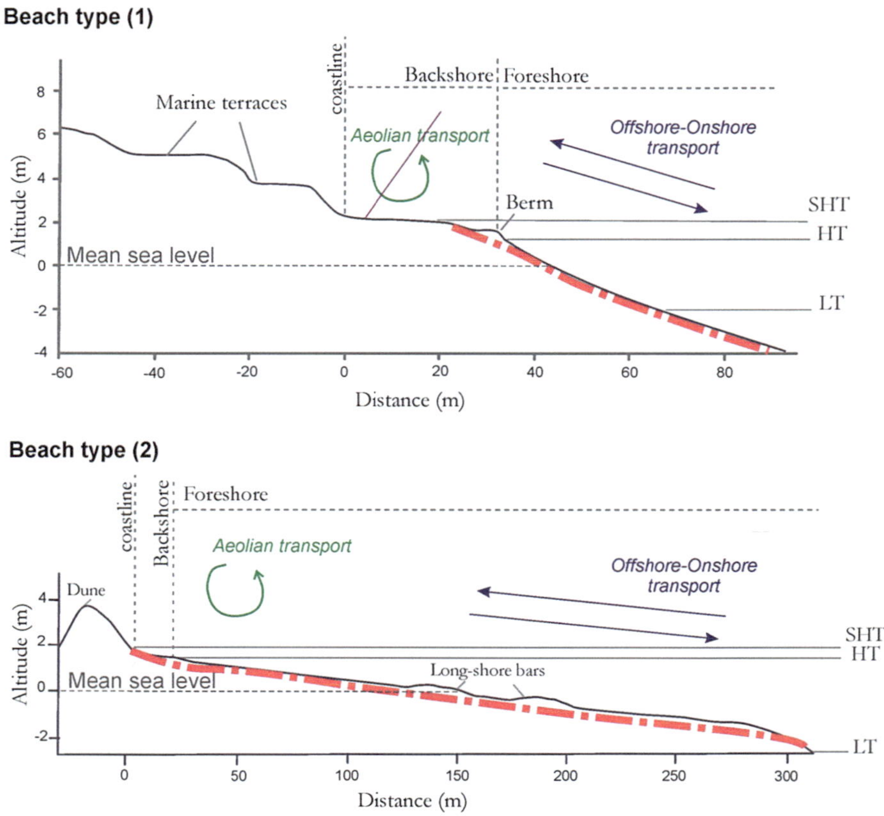

Fig. 10.10 Morphodynamic beach models of Golfo Nuevo coasts. References: LT (low tide level); HT (high tide level); SHT (storm high tide level); the red line represents the beach environment affected by algae arrivals. While in both models offshore-onshore transport of sediments by waves is frequent, only in the second one aeolian transport domain. (Taken and modified from Bunicontro (2018)

Beach type 2 (Fig. 10.11b) is a sandy beach (99% sand), with well sorted and unimodal sediments. Beaches slope is about 1° and they are typically represented in Puerto Madryn. Long-shore linear bars (between 10–30 m width) and berms are frequent but temporary (Bunicontro et al. 2017; Bunicontro 2018).

The aeolian transport is important in the second model, even though in both models the offshore-onshore wave transport domain. This is because in the first model the beach is dominated by pebbles and the wind is not very important in the movement of those sediments. In this way, the model 2 shows not only the sand is stuck by onshore-offshore transport intermixed with the biomass of the seaweed but also the sand moved by the wind from the shore stay trapped in the arrivals.

The cleaning activities during summer, in which all the seaweeds are removed from beaches and also great amounts of sand trapped on it, are concentrated in the

Fig. 10.11 Golfo Nuevo beaches where *Undaria* deposits during summer. (**a**) mixed (pebble and sand) beaches (Beach model 1). (**b**) fine sandy beaches (Beach model 2). (Photos: P. Bunicontro)

touristic beaches of Puerto Madryn (dominated by model 2) and not in those characterized by model 1. So, in those beaches where most of the sand is stuck in the seaweed and all the transport of sediment is interrupted, the erosion impact is higher due to the cleaning work.

The second model shows that almost all the beach is dominated by foreshore while backshore is much reduced. Conversely, the first model shows both sub-environments well developed, and in the case of an algae arrival only the half on the beach would be affected. Considering that foreshore is the main beach sub-environment affected by the algae arrivals, the coast dominated by model 1 is less susceptible that model 2. This last model is the most hydrodynamically vulnerable and sediment transport is by far more disrupted.

The entire coast can be zoned determining susceptible beaches to be affected by the seaweed detachment. In those touristic beaches, cleaning costs can be estimated and an effective coastal management plan can be applied.

10.7 Legislation and Management

After Undaria invasion and its rapid dispersal, the government of Chubut declared in 2008 (Decree N°1071/08) the environmental emergency in Golfo San José (northward from Golfo Nuevo) because of Undaria occurence. In 2009, the Promotion for Industrialization of the Undaria seaweed was established by Law N° 5796, based on the previous Law N° 1891 on algae exploitation and extraction of exotic algae.

At present, the efforts to control and eradicate invasive species worldwide have very poor results. In 2010, Irigoyen et al. established guidelines for the management of the invasive alga *Undaria* in coast of Chubut to prevent and restrict its dispersal. These authors encourage the use of *Undaria* for commercial purposes, the control of ballast water and ships cleaning, the identification of future colonization points and the monitoring of invaded sites.

It is necessary to continue with the biological investigations of *Undaria* impacts on native biota not only in time but also in space. It is recommended to stop or regulate at least the amounts and quality of the sewage thrown to the beaches. There should be more control over these discharges in order to reduce the eutrophication of the coastal water, and thus reduce the proliferation of unwanted species. To mitigate hydrodynamic effects, it is recommended to use new technologies for beach cleaning in order to diminish the extraction of sand from the coastal system. Besides, it is necessary to make frequent reports of the cleaning activities, detailing the area cleaned (foreshore, backshore, etc.), the amount of biomass extracted from beaches or trucks involved, the site of final deposition of waste and the frequency of the activities during the summer.

10.8 Conclusions

Since *Undaria* was introduced in Golfo Nuevo in 1992, its dispersion, favored by human vectors, seems to be inevitable. The speed with which it colonized different environments results in a high concern of scientist and authorities.

Puerto Madryn is one of the most important touristic cities in Patagonian coast and the kelp invasion results not only in ecological problems to native biota but also in an economic negative impact. The biodiversity of native algae is reducing due to the competitive behavior of *Undaria*, while other species take advantage of the seagrass beds as refuge environments. As a consequence, the marine ecosystems are affected in diverse and complex ways that need to be studied.

On the other hand, the biomass is detached and cast to the beaches every summer. The seaweed biomass spread in the beaches interrupts the natural sand transport from foreshore to backshore and along the beaches, causing a hydrodynamic impact. The algal biomass cast ashore also interferes with recreational uses of the beaches and therefore must be periodically collected by trucks during cleaning activities. This means a big but necessary expense for the local government. However, the trucks not only remove the seaweed biomass but also a lot of sand from the beaches, altering the natural sedimentary balance and inducing beach erosion.

In view of this, it is important to manage an integrated coastal strategy in order to control the dispersal vectors of the invasive species, promote the diffusion of the problematic to the local community and develop a plan for coastal management from a multidisciplinary research. In this sense, beach models proposed in this contribution will help to identify susceptible coastal areas to be affected by seaweed detachment along Patagonian coast.

References

Akiyama K, Kurogi M (1982) Cultivation of *Undaria pinnatifida* (Harvey) Suringar, the decrease in crops from natural plants following crop increase from cultivation. Bull Tohoku Reg Fish Res Lab 44:91–100

Battershill C, Miller K, Cole R (1998) The understorey of marine invasions. Seafood N Z 6 (2):31–33

Boudouresque CF, Verlaque M (2002) Biological pollution in the Mediterranean Sea: invasive versus introduced macrophytes. Mar Pollut Bull 44:32–38

Brown MT, Lamare MD (1994) The distribution of *Undaria pinnatifida* (Harvey) Suringer (sic) within Timaru harbour, New Zealand. Jpn J Phycol (Sôrui) 42:63–70

Bunicontro P (2018) Geomorfología, evolución y dinámica costera: implicancias en el manejo costero y ambiental en el Golfo Nuevo, provincia de Chubut. Tesis Doctoral Facultad de Ciencias Exactas y Naturales de la Universidad de Buenos Aires (inédito)

Bunicontro P, Marcomini S, Weiler N, López R, Quenardelle S (2017) Caracterización textural, composicional y análisis de procedencia de los sedimentos de playa del Golfo Nuevo, provincia de Chubut. Rev Asoc Geol Argent 74(2):68–83

Campbell SJ, Burridge TR (1998) Occurrence of *Undaria pinnatifida* (Phaeophyta: Laminariales) in Port Phillip Bay, Victoria, Australia. Mar Freshw Res 49:379–381

Carlton JT (1989) Man's role in changing the face of the ocean: biological invasions and implications for conservation of near-shore environments. Conserv Biol 3(3):265–273

Carlton JT (1996) Biological invasions and cryptogenic species. Ecology 77(6):1653–1655

Carlton JT (1999) The scale and ecological consequences of biological invasions in the world's oceans. In: Sandlund OT, Schei PJ, Viken A (eds) Invasive species and biodiversity management. Kluwer Academic Publishers, Dordrecht, pp 195–212

Carlton JT (2000) Global change and biological invasions in the oceans. In: Mooney HA, Hobbs RJ (eds) Invasive species in a changing world. Island Press, Covelo, pp 31–53

Casas GN (2005) Biology and ecology of *Undaria pinnatifida* (Phaeophyceae Laminariales) in Golfo Nuevo (Chubut, Argentina). PhD Thesis Universidad Nacional del Sur. Argentina. 234 p

Casas GN (2013) *Undaria* the Japanese Kelp in argentine coasts. Abstracts Invasive Seaweeds Symposium. X Argentine Meeting of Phycology. XXXIV Argentine Conference on Botany. La Plata. Argentina

Casas GN, Piriz ML (1996) Surveys on *Undaria pinnatifida* (Laminariales Phaeophyta) in Golfo Nuevo, Argentina. Hydrobiologia 326/327:213–215

Casas GN, ML Piriz (2001) The threat of bioinvasions on biodiversity of benthic communities: *Undaria pinnatifida* (Phaeophyta Laminariales) in Argentina. Applied Phycology Forum: 1–2. http://www.ib.usp.br/apf/argentina.html. Accessed 14 Mar 2008

Casas G, Schwindt E (2008) Un alga japonesa en la costa patagónica. Ciencia Hoy 18(107):31–39

Casas GN, Scrosati R, Piriz ML (2004) The invasive kelp *Undaria pinnatifida* (Phaeophyceae Laminariales) reduces native seaweed diversity in Nuevo gulf (Patagonia, Argentina). Biol Invasions 6:411–416

Casas GN, Piriz ML, Parodi ER (2008) Population features of the invasive kelp *Undaria pinnatifida* (Phaeophyceae Laminariales) in Nuevo Gulf (Patagonia, Argentina). J Mar Biol Assoc UK 88 (1):21–28

Castric-Fey A, Girard A, L'Hardy-Halos MT (1993) The distribution of *Undaria pinnatifida* (Phaeophyceae, Laminariales) on the coast of Saint Malo Brittany, France. Bot Mar 36:351–358

Curiel D, Bellemo G, Marzocchi M, Scattolin M, Parisi G (1998) Distribution of introduced Japanese macroalgae *Undaria pinnatifida*, *Sargassum muticum* (Phaeophyta) and *Antithamnion pectinatum* (Rhodophyta) in the lagoon of Venice. Hydrobiologia 385:17–22

Dellattorre FG, Amoroso R, Saravia J, (Lobo) Orensanz JM (2014) Rapid expansion and potential range of the invasive kelp *Undaria pinnatifida* in the Southwest Atlantic. Aquat Invasions 9 (4):467–478

Díaz P, López Gappa JJ, Piriz ML (2002) Symptoms of eutrophication in intertidal macroalgal assemblages of Nuevo Gulf (Patagonia, Argentina). Bot Mar 45:267–273

Eyras M, Sar E (2003) Arribazones estivales en Puerto Madryn, Argentina, como materiales para la obtención de compost. Bol Soc Argent Bot 38(1–2):105–111

Fletcher RL, Manfredi RL (1995) The occurrence of *Undaria pinnatifida* (Phaeophyceae, Laminariales) on the South Coast of England. Bot Mar 38:355–358

Floc'h JY, Pajot R, Wallentinus I (1991) The Japanese brown alga *Undaria pinnatifida* on the coast of France and its possible establishment in European waters. J Cons Int Explor Mer 47:379–390

Forrest BM, Taylor MD (2002) Assessing invasion impact: survey design considerations and implications for management of an invasive marine plant. Biol Invasions 4:375–386

Forrest B, Gardner J, Taylor MD (2009) Internal borders for managing invasive marine species. J Appl Ecol 46:46–54

Gil M (2001) Eutroficación: Rol del nitrógeno en ecosistemas marinos costeros. Tesis Doctoral en Ingeniería Química, Universidad Nacional del Sur Bahía Blanca (Bs As), 127 pp

Hay CH (1990) The dispersal of Sporophytes of *Undaria pinnatifida* by coastal shipping in New Zealand and implications for further dispersal of *Undaria* in France. Br Phycol J 25:301–313

Hay CH, Luckens PA (1987) The Asian kelp *Undaria pinnatifida* (Phaeophyta: Laminariales) found in a New Zealand harbour. N Z J Bot 25:329–332

Irigoyen AJ, Trobbiani G, Sgarlatta MP, Raffo MP (2011) Effects of the alien algae *Undaria pinnatifida* (Phaeophyceae, Laminariales) on the diversity and abundance of benthic macrofauna in Golfo Nuevo (Patagonia, Argentina): potential implications for local food webs. Biol Invasions 13:1521–1532

Floc'h JY, Pajot R, Mouret V (1996) *Undaria pinnatifida* (Laminariales, Phaeophyta) 12 years after its introduction into the Atlantic Ocean. Hydrobiologia 326/327:217–222

Kawai H, Hanyuda T, Uwai S (2016) Evolution and biogeography of Laminarialean kelps. In: Hu Z-M, Fraser C (eds) Seaweed phylogeography. Adaptation and evolution of seaweeds under environmental change. Springer, Berlin/Heidelberg, pp 227–250

Kolar CS, Lodge DM (2001) Progress in invasion biology: predicting invaders. Trends Ecol Evol 16(4):199–204

Lee Y-P, Yoon JT (1998) Taxonomy and morphology of *Undaria* (Alariaceae Phaeophyta) in Korea. Algae 13(4):427–446

Martin JP, Cuevas M (2006) First record of *Undaria pinnatifida* (Laminariales Phaeophyta) in Southern Patagonia, Argentina. Biol Invasions 8:1399–1402

Meretta PE, Matula CV, Casas G (2012) Occurrence of the alien kelp *Undaria pinnatifida* (Laminariales, Phaeophyceae) in Mar del Plata, Argentina. BioInvasions Rec 1:59–63

Mouzo F, Garza M (1979) Rasgos de la geología submarina del Golfo Nuevo (provincia del Chubut). *6° Congreso Geológico Argentino*. Actas:413–430 Buenos Aires

Naval Hydrography Service, 2017. Tides tables of Puerto Madryn, Chubut. Ministry of Defense. http://www.hidro.gov.ar/oceanografia/tmareas/form_tmareas.asp

Okamura K (1915) Undaria and its species. Bot Magaz (Tokyo) 29:269–281

Orensanz JM, Schwindt E, Pastorino G, Bortolus A, Casas G, Darrigran G, Elías R, López Gappa JJ, Obenat S, Pascual M, Penchaszadeh P, Piriz ML, Scarabino F, Spivak E, Vallarino EA (2002) No longer a pristine confine of the world ocean – a survey of exotic marine species in the southwestern Atlantic. Biol Invasions 4:115–143

Piriz ML, Casas G (1994) Occurrence of *Undaria pinnatifida* in Golfo Nuevo. Argent Appl Phycol Forum 10:4

Piriz ML, Casas G (2001) Introducción de especies y su impacto en la biodiversidad. El caso *Undaria pinnatifida*. In: Alveal K, Antezana T (eds) Sustentabilidad de la biodiversidad. Universidad de Concepción. Chile, pp 679–692

Piriz ML, Eyras MC, Rostagno CM (2003) Changes in biomass and botanical composition of beach-cast seaweeds in a disturbed coastal area from Argentine Patagonia. J Appl Phycol 15:67–74

Raffo MP, Eyras MC, Iribarne OO (2009) The invasion of *Undaria pinnatifida* to a *Macrocystis pyrifera* kelp in Patagonia (Argentina, south-west Atlantic). J Mar Biol Assoc UK 89 (8):1571–1580

Raffo MP, Irigoyen A J, Schwindt E, G N Casas (2012) Efectos del alga exótica *Undaria pinnatifida* sobre la comunidad de macroalgas bentónicas luego de 15 años de invasión (Golfo Nuevo, Chubut). Abstracts VIII Jornadas Nacionales de Ciencias del Mar (VIII National Conference on Marine Sciences): 69. Comodoro Rivadavia, Argentina

Raffo M P, Faleschini M, Casas G, E Schwindt (2015) Efecto de sombreado del alga exótica *Undaria pinnatifida* sobre la comunidad de macroalgas en pozas de marea (Patagonia, Argentina). Abstracts IX Jornadas Nacionales de Ciencias del Mar (IX National Conference on Marine Sciences): 287.Ushuaia, Argentina

Rivas A, Beier E (1990) Temperature and salinity fields in the North Patagonic Gulfs. Acta Oceanográfica Argentina 13(1):15–20

Rivas AL, Ripa P (1989) Variación estacional de la estructura termohalina del golfo Nuevo. Geofis Int 28(1):3–24

Russell LK, Hepburn CD, Hurd CL, Stuart MD (2008) The expanding range of *Undaria pinnatifida* in southern New Zealand: distribution, dispersal mechanisms and the invasion of wave-exposed environments. Biol Invasions 10:103–115

Saito Y (1975) *Undaria*. In: Tokida J and H Hirose (eds.) Advances of Phycology in Japan. VEB Gustav Fischer Verlag Jena: 304–320

Sánchez D, Acosta M, Jimenez L, Carabajal D (2006) La problemática del agua en Puerto Madryn: algunas consideraciones. En: GAEA Sociedad Argentina de Estudios Geográficos, *Congreso Nacional de Geografía 67° Semana de la Geografía, Contribuciones Científicas.* San Salvador de Jujuy. 235–244 pp.

Sanderson JC (1990) A preliminary survey of the distribution of the introduced macroalga, *Undaria pinnatifida* (Harvey) Suringer (sic) on the East Coast of Tasmania, Australia. Bot Mar 33:153–157

Schaffelke B, Hewitt CL (2007) Impacts of introduced seaweeds. Bot Mar 50:397–417

Schaffelke B, Smith JE, Hewitt CL (2006) Introduced macroalgae – a growing concern. J Appl Phycol 18:529–541

South PM, Floerl O, Forrest BM, MS Thomsen (2017) A review of three decades of research on the invasive kelp *Undaria pinnatifida* in Australasia: an assessment of its success, impacts and status as one of the world's worst invaders, Marine Environmental Research (in press)

Stuart MD, Hurd CL, Brown MT (1999) Effects of seasonal growth rate on morphological variation of *Undaria pinnatifida* (Alariaceae Phaeophyceae). Hydrobiologia 398(399):191–199

Teso VS, Bigatti G, Casas GN, Piriz ML, Penchaszadeh PE (2009) Do native grazers from Patagonia, Argentina, consume the invasive kelp *Undaria pinnatifida*? Rev Mus Argentino Cienc Nat 11(1):7–14

Torres AI, Gil MN, Esteves JL (2004) Nutrient uptake rates by the alien alga *Undaria pinnatifida* (Phaeophyta) (Nuevo Gulf, Patagonia, Argentina) when exposed to diluted sewage effluent. Hydrobiologia 520:1–6

Uwai S, Nelson W, Neill K, Wang WD, Aguilar-Rosas LE, Boo SM, Kawai KTH (2006) Genetic diversity in *Undaria pinnatifida* (Laminariales Phaeophyceae) deduced from mitochondria genes – origins and succession of introduced populations. Phycologia 45(6):687–695

Valentine JP, Johnson CR (2004) Establishment of the introduced kelp *Undaria pinnatifida* following dieback of the native macroalga *Phyllospora comosa* in Tasmania, Australia. Mar Freshw Res 55:223–230

Wallentinus I, Nyberg CD (2007) Introduced marine organisms as habitat modifiers. Mar Poll Bull 55:323–332

Williams SL, Smith JE (2007) Distribution, taxonomy and impacts of introduced seaweeds. Ann Rev Ecol Evol Syst 38:327–359

Yoon HS, Boo SM (1999) Phylogeny of Alariaceae (Phaeophyta) with special reference to *Undaria* based on sequences of the RuBisCo spacer region. Hydrobiologia 398(399):47–55

Zenetos A, C¸ i ME, Pancuci-Papadopoulou MA, Harmelin JG, Furnari G, Andaloro F, Bellou N, Streftaris N, Zibrowius H (2005) Annotated list of marine alien species in the Mediterranean with records of the worst invasive species. Mediterr Mar Sci 6(2):63–118

Chapter 11
Only the Strictest Rules Apply: Investigating Regulation Compliance of Beaches to Minimize Invasive Dog Impacts on Threatened Shorebird Populations

Grainne S. Maguire, Kelly K. Miller, and Michael A. Weston

Abstract In many countries domesticated dogs occur abundantly on coasts, where they may co-occur with and pose a threat to coastal wildlife such as threatened shorebirds. Dogs on beaches fit the ecological definition of invasive species. The management of dogs on coasts is controversial, with polarised debate surrounding dog access to public open spaces, and questions around the effectiveness of prevailing dog management regulations. We examined the levels of compliance with dog regulations (3516 checks, 69 ocean beaches) under six prevailing management regimes in Victoria, Australia. Compliance was low to moderate across all dog management 'types', but varied significantly. The highest compliance rates were associated with 'no dog' areas. Despite poor overall compliance, dog regulations appeared to be associated with different rates of occurrence and relative abundances of dogs, suggesting either they effectively displaced dog walkers or that dog area designations reflect usage patterns, or both.

Keywords Dog regulations · Coastal management · Compliance · Threatened species · Human-wildlife conflict · Domestic animal management

---

G. S. Maguire (✉)
BirdLife Australia, Carlton, VIC, Australia
e-mail: grainne.maguire@birdlife.org.au

K. K. Miller · M. A. Weston
Centre for Integrative Ecology, Faculty of Science, Engineering and the Built Environment, Deakin University, Burwood, VIC, Australia
e-mail: kelly.miller@deakin.edu.au; mweston@deakin.edu.au

C. Makowski, C. W. Finkl (eds.), *Impacts of Invasive Species on Coastal Environments*, Coastal Research Library 29,
https://doi.org/10.1007/978-3-319-91382-7_11

11.1 Introduction

Across the globe, domestic dogs *Canis familiaris* (those living closely with humans) and their management represent one of the most controversial coastal management issues (Gompper 2014; Weston et al. 2014). Dog ownership in many developed countries is extremely common, with >700 million dogs being owned worldwide, many of which enjoy frequent outdoor exercise (Gompper 2014; Hughes and Macdonald 2013; Ioja et al. 2011). Public open spaces are often designated to support dog exercise (Weston et al. 2014), and as the human population and their companion dogs increase in number, so does the demand for dog access to shared open spaces. This becomes a major management challenge for decision makers who must balance the needs of multiple users of public open spaces, as well as the potential for environmental impacts (Johnston et al. 2013; Le Corre et al. 2009; Walsh 2011).

In coastal areas in particular, views on the suitability and impacts of dogs on coasts are polarised (Maguire et al. 2011; Miller et al. 2014; Morgan 1999; Pereira et al. 2003). Coastal zoning to facilitate or limit dog access involves lengthy public consultation processes and public debates which sometimes evoke violent confrontations (Johnston et al. 2013; Gompper 2014). Critical issues include the impact that dogs have in terms of public health and safety, and impacts of dogs on wildlife including threatened species (Dowling and Weston 1999; Johnston et al. 2013; Maguire 2008). Dogs have a variety of effects on wildlife, such as causing disturbance, altering ecologies and spreading disease (Glover et al. 2011; Weston and Stankowich 2014). One faunal group especially affected by dogs are shorebirds, which are chased by dogs, and their behaviour (including breeding) is disrupted (Gompper 2014). Dogs are also a direct predator of eggs, young and even adult shorebirds (Burrell and Colwell 2012; Teunissen et al. 2008; Fig. 11.1). Figure 11.2 presents images which demonstrate some of the impacts of dogs on wildlife.

Fig. 11.1 Images showing the phases of the reproductive cycle when shorebirds are most vulnerable to dogs. While it varies between shorebird species, the egg phase (**a**) typically lasts a month, and the chick phase (**b**) lasts a month or more. At these stages, direct mortality (depredation or crushing) due to dogs is more likely, with reports of dogs killing flying birds rarer. Disturbance occurs during all life history phases. (Images: Hooded Plover *Thinornis cucullatus* nest and chick, M.A. Weston)

11.1.1 Dogs as Invasive Species

Dogs on coasts occur in a variety of contexts (Fig. 11.3). Traditional definitions of invasive species generally emphasise those species which spread geographically, are exotic and cause deleterious impacts (Lockwood et al. 2013; Valéry et al. 2008). More recently, ecologically coherent definitions have been based on the principles of invasion ecology and describe the spectrum of species and circumstances which are considered to constitute 'invasive species' (Blackburn et al. 2011). Dogs are rarely considered invasive species but fit the definition in many ways (see Table 11.1). On many coasts around the world, humans bring their dogs to the beach with them, or their presence is supported via the provision of food (e.g., in villages), thus their abundance and distribution in coastal environments is above that which would otherwise occur (Schlacher et al. 2015).

11.1.2 Managing Dogs on Coasts

Coastal managers around the world vary in awareness and acknowledgement of the range of deleterious impacts of dogs on beaches. Primarily, a variety of management strategies exist to reduce human-related impacts, providing opportunities for dog and dog-free use of coasts. Prominent among these are 'no dog' areas, on-leash areas (sometimes constrained seasonally or according to time of day), and dedicated off-leash areas (Weston and Stankowich 2014). These measures aim to constrain the occurrence of dogs on coasts temporally or spatially, thereby reducing any deleterious effects on beach goers or wildlife. Examples of managements implemented on the Victorian coast are provided in Fig. 11.4. The process of implementing and changing these designations usually involves balancing competing uses of beach space by engaging with identified stakeholders and including a public consultation phase. The resultant designations can thus be a compromise between mitigating impacts and public pressure. Along many coasts, these approaches form the foundation of dog management (Weston and Stankowich 2014). While no clear costings are known to us, dog management is likely to cost management authorities substantial amounts of money (Johnston et al. 2013), especially given that many strategies require ongoing resource requirements, such as signage installation and upkeep, patrols and enforcement (Dowling and Weston 1999).

Despite the management effort required, evidence suggests that compliance with dog regulations on the coast may be low in at least some circumstances (Miller et al. 2014). For example, in many coastal and wetland areas in Victoria, Australia, few dogs are leashed in areas where they are required to be so (Weston et al. 2009; Williams et al. 2009), while compliance with other managements aimed at pedestrians in the same areas, is much higher (Weston et al. 2011).

Fig. 11.2 Example images of selected dog impacts on shorebirds: (**a–b**) Actively chasing adult Hooded Plovers (**a**), Bev Wood; (**b**), Eric Woehler, BirdLife Tasmania; (**c–d**), Glenn Ehmke, BirdLife Australia); (**e**), Dog inspects chick shelter (subsequently urinates on it; BirdLife

Dog management regimes may work in two distinct ways, they may instil desirable on-site behaviour (e.g. leashing at a location), or displace behaviour which is undesirable at one site to a more appropriate location (e.g. where dog walkers travel to designated off leash areas to enjoy leash free dog exercise). The former effect will be evident by on-site levels of compliance with prevailing regulations. The latter effect would be evident if the occurrence of dogs differs between prevailing management regulations (although it is also likely that management designations have been made on the basis of demand for certain dog regulations). This study examines the relative abundance and levels of compliance associated with dogs in different prominent dog management regimes on ocean beaches in Victoria, Australia.

11.2 Materials and Methods

Data were collected as part of a threatened species monitoring program of Hooded Plover *Thinornis cucullatus*, a beach-nesting bird thought to be threatened by a range of processes including dogs (Dowling and Weston 1999). Over three consecutive breeding seasons (2006/07, 2007/08, 2008/09), 69 Hooded Plover (high energy, ocean) beaches were monitored over the breeding months (August – March). During each monitoring visit to a beach site ($n = 3516$, average visit duration 43.47 ± 0.88 mins), the presence or absence of dogs within a 250 metre radius was recorded, as well as the number on and off leash, and the presence of dog prints in the dune and/or nesting zone (see Maguire 2008 for details).

Figure 11.5 (a–c) provides a map of beach sites according to the prevailing dog regulations at that beach. Six 'types' of dog management regulations were evident across the Victorian coast. Four of these were suitable for an examination of compliance ('no dog' areas, time of day and seasonal restricted access areas, and on leash areas; see Table 11.2 for definitions of compliance). We excluded dogs under 'effective control' (under the control of their owner at all times even when off leash) and 'dogs off leash' management regulations, on the basis that compliance was not required or was not possible to measure.

Data were analysed on GenStat v.7 (VSN International 2005) using General Linear Mixed Modelling, incorporating a random factor of beach identity to account

Fig. 11.2 (continued) Australia and Deakin University); (**f–h**), Close encounters with Hooded Plover nests (wildlife cameras placed with active nests in the foreground; several indicated by red circles; nest in (**g**) was crushed by the dog; BirdLife Australia and Deakin University); (**i**), Dog preys upon Pied Oystercatcher *Haematopus longirostris* eggs (Priscilla Park, BirdLife Tasmania); (**j**), chick killed by male Labrador named "Jazz" (Glenn Ehmke, BirdLife Australia); (**k–l**), Radio-tracked Hooded Plover chick found dead in rubbish bin, in plastic bag containing dog faeces, with autopsy concluding dog bite most likely cause of death (Brett Diehm, Rod Collins and Tom Schmidt, Barwon Coast, BirdLife Australia and Deakin University)

Fig. 11.3 Examples of dogs on coasts: (**a**), village dog, Kenya; (**b**), dingo, Queensland, Australia; (**c**), domestic pet allowed to roam independently, Victoria, Australia. (Images: M.A. Weston)

for repeat visits to beaches. The project was conducted in accordance with the regulations of Deakin University Animal Ethics permit A45/2006 and Deakin University Human Ethics Exemption 2012–204.

Table 11.1 Classification of certain populations of fully domesticated dogs (after Weston et al. 2011) as invasive species on the Victorian (Australia) coasts within an ecological framework for describing 'invasive' species (after Blackburn et al. 2011 and Colautti and MacIsaac 2004)

Stage of biological invasion	Status of domesticated dogs as found on the Victorian coast
Propagules residing in a donor region	Non-dingo[a] exotic forms outside Australia
Transport	Imported
Introduction	Survival facilitated, deliberately bred, dingos persecuted.
Establishment	Domestication maintained, deliberate regular (mostly accompanied) visits to natural areas such as coasts.
Nature of establishment and spread	Common, widespread, ongoing and increasing.
Impact	Deleterious impacts on wildlife, disease spread, positives for human health and wellbeing, among other impacts.

In the Blackburn et al. (2011) model, domesticated dogs in Australia can be categorised as *"Individuals transported beyond limits of native range, and in cultivation (i.e. individuals provided with conditions suitable for them but explicit measures to prevent dispersal are limited at best)"* (B2)

[a]Dingoes are a specific form of dog isolated from other dogs for c. 5000 years, and are currently extremely uncommon on Victorian coasts (Savolainen et al. 2004)

Fig. 11.4 Selected images of signage associated with different dog regulations on the Victorian coast. (**a**) no dogs; (**b**), seasonal leash access and time restricted leashing. (Images, M.A. Weston)

11.3 Results

11.3.1 Occurrence

Table 11.2 summarises the occurrence of dogs across the six different dog management regulations. Dogs were widespread and common, being sighted on 27.9% (980) of visits and at 86.8% (60) of sites. There were only eight sites where dogs

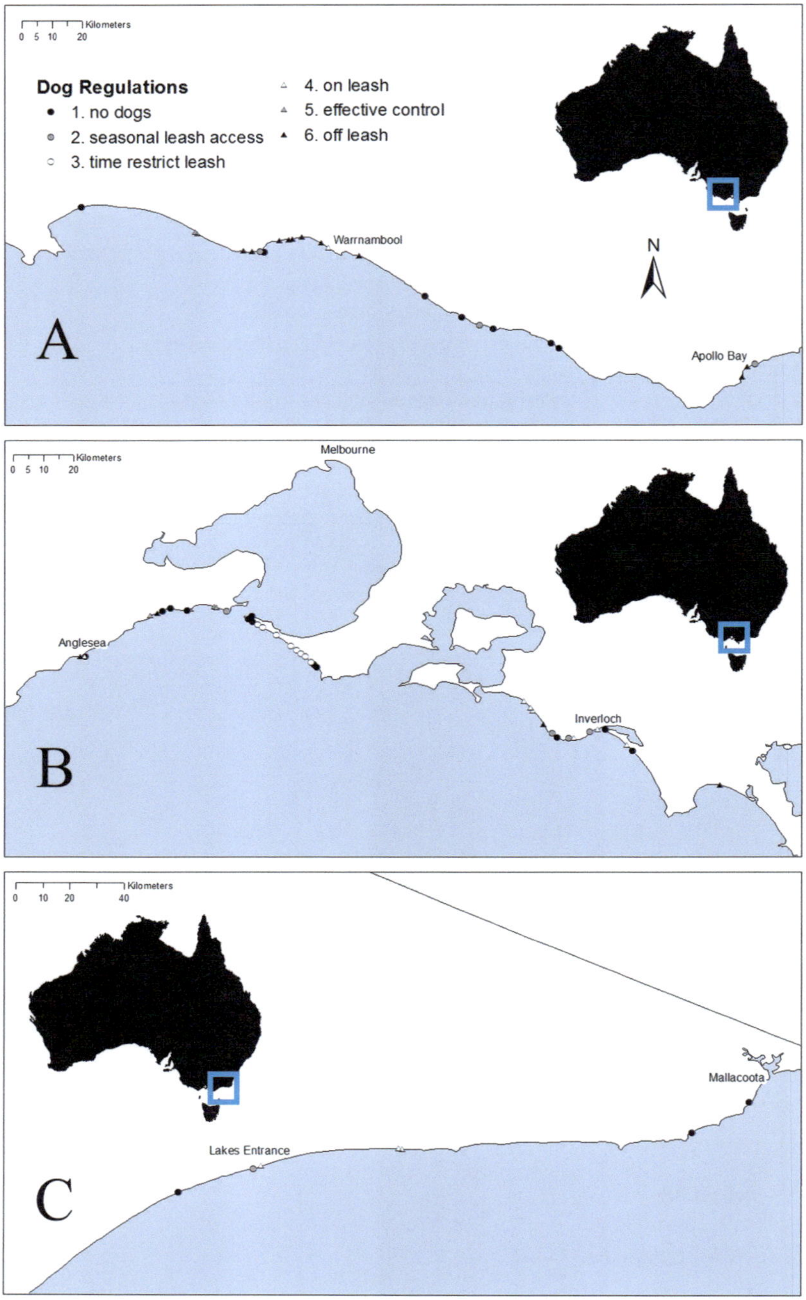

Fig. 11.5 The distribution of sites coded by dog regulations across the: (**a**) West, (**b**) Central, and (**c**) East coasts of Victoria

Table 11.2 The occurrence of dogs, dog prints and leashed versus unleashed dogs across six different dog management regulations

Dog management regulation	Total visits (total number beaches)	Number of visits dogs sighted (number of beaches dogs sighted)	Number of dogs sighted	(a) Model estimates of probability of occupancy by dogs	(b) Model estimates of number of dogs recorded[a]	Percentage of visits with dog prints in the nest zone	Average number of dogs (unleashed; leashed) per visit	Average percentage of dogs leashed
'No dog'	796 (22)	141 (13)	319	0.11	2.27	1.5	2.26 (1.78; 0.48)	24.0
Seasonal leash access	222 (8)	96 (8)	222	0.48	2.33	3.6	2.31 (1.49; 0.82)	35.9
Time restricted leash access	771 (8)	139 (8)	364	0.16	2.63	3.6	2.62 (2.05; 0.57)	23.5
On leash at all times	606 (12)	165 (12)	428	0.22	2.53	2.1	2.60 (1.67; 0.93)	31.3
Effective control	254 (5)	160 (5)	650	0.51	4.03	8.7	4.07 (3.43; 0.64)	17.8
Off leash	867 (14)	279 (14)	715	0.31	2.56	3.5	2.56 (2.03; 0.53)	18.6

GLMM predicted mean estimates are provided for the (a) occupancy of dogs and (b) number of dogs per visit according to management type
[a]Excludes zero counts

Table 11.3 Compliance definitions and rates are provided for four dog management types in coastal Victoria, Australia (n = 1794 visits)

Management type	Definition of non-compliance	Non-compliance; % (n)	Non-compliance (model estimates)[a]	Rank
'No dog'	Any dog observed	17.7 (796)	0.11	4
Seasonal leash access	Dogs outside designated seasonal access or off leash during seasonal access	42.1 (192)	0.55	2
Time restricted leash access	Dogs outside allowed times or off leash during allowed times	20.0 (641)	0.19	3
On leash at all times	Any dog off the leash	79.4 (165)	0.83	1

Off leash and effective control management types are excluded. The frequency of visits in which non-compliance was observed are indicated. GLMM predicted mean estimates are provided for rates of non-compliance and ranked from lowest (1) to highest (4) rates of compliance
[a]0 = compliance, 1 = non-compliance

were never observed and these were all 'No Dogs' areas. The occurrence of dogs (whether or not they were observed on a given visit) was influenced by the prevailing dog management regulation at that beach (GLMM, Wald = 49.47, df = 5, $p < 0.001$; Table 11.2).

On average, 2.11 and 0.64 dogs were off and on leash per visit, respectively. Table 11.2 summarises the sightings of dogs on and off leash (2698 were sighted in total). Excluding zero counts, when dogs occurred, they occurred in different numbers according to the prevailing dog management regulations (GLMM, Wald = 25.66, df = 5, $p < 0.001$; Table 11.2).

11.3.2 Compliance

Compliance was not universal; 26.8% of all observations (n = 1794 visits) involved non-compliance with prevailing dog regulations (Table 11.3). Regardless of dog management regulations, unleashed dogs were more common than leashed dogs (overall, 23.8% were leashed; Table 11.2). A generalised linear mixed model was run using a binomial distribution ('compliance' or 'non-compliance'; dispersion was fixed at 1). Beach site was included as a random factor to account for repeated sampling from sites over time and prevailing dog regulations was included as a fixed factor (with four levels). The analysis revealed significant differences in compliance between the types of prevailing dog regulations (GLMM, Wald = 124.87, df = 1, $p < 0.001$). Model estimates are provided in Table 11.3 where on-leash areas had the highest levels of non-compliance and 'no dog' areas had the highest levels of compliance.

11.4 Discussion

Coastal managers regard dog management as a key management issue (Le Corre et al. 2009), yet few studies examine compliance with dog regulations or factors which might mediate compliance (exceptions include Williams et al. 2009 and Weston et al. 2009). This study has confirmed that (1) dog management regulations alter the degree of dog occurrence and relative abundance, (2) that compliance is at best moderate, and (3) that higher rates of compliance occur under some regulations.

Dogs were widespread across beach sites regardless of the different dog management regulations, and even occurred in 'no dog' areas, albeit at lower frequencies. One explanation for use of 'no dog' areas is the tendency for owners of poorly socialised or aggressive dogs to deliberately visit these areas to avoid other dogs (MA Weston pers. obs.). Use of beaches under different dog management regulations was not equivalent, with some being used more frequently than others. In particular, 'effective control' areas experienced high occupancy by relatively high numbers of dogs, suggesting they were particularly popular, or convenient to access, among dog walkers. Areas with 'effective control' provisions also hosted a relatively high percentage of dog footprints in the protected nesting zones for the threatened species of shorebird we examined, suggesting that in these areas they potentially posed a greater threat compared with the other management regimes. 'Seasonal leash access' areas were also frequently used by substantial numbers of dogs (and were associated with relatively low levels of compliance).

Several processes may explain differences in occurrence and numbers between dog management regimes. Firstly, 'displacement' of dog walkers between beaches may occur which would represent an effective outcome of dog regulations (akin to other recreationists, such as anglers, using regulations to inform their choices of places to fish; Scrogin et al. 2004). However, this study cannot differentiate this from responsive planning which has catered for dog walking demand, such that desired regulations have been designated in areas where local dog walkers seek access (see Maguire et al. 2011). Although confirmation is required, this study suggests that the greatest positive effect of dog regulations may be to divert dog walkers to more appropriate beaches.

As suggested by previous research, compliance with prevailing dog regulations is at best moderate, as it is in many parts of the world (reviewed in Weston and Stankowich 2014). Compliance varied between management areas, and was especially low for leashing in areas where leashing is required at all times. The low compliance evident across the different dog management regulations suggests that these cannot be used as surrogates for dog activity, an approach adopted by some studies of the influence of dogs on wildlife assemblages (Forrest and St Clair 2006).

Dog owners may not leash their dogs because they consider dog exercise important, because leashing is not expected by their peers (social norms), and if they interpret no harm in their dog roaming (Sterl et al. 2008; Williams et al. 2009). In several social studies of beach user attitudes and perceptions, it is apparent that dog walkers significantly differ from non-dog walkers, often underestimating their impacts on threatened species such as the Hooded Plover, and expressing discontent at regulations and exclusions as a whole (Williams et al. 2009; Maguire et al. 2013). The concept of dogs as an invasive species and their off leash activity as a threatening process to wildlife is likely to be novel to most dog owners who perceive their dog as a family member and prioritise their dog's wellbeing ahead of wider animal-wildlife concerns (Cohen 2002; Williams et al. 2009). The higher compliance evident in 'no dog' areas in comparison to areas maintaining on-leash access (albeit in some areas with seasonal or temporal restrictions), poses an interesting conundrum in terms of required levels of protection for sensitive wildlife areas. Long-term conservation programs for beach-nesting birds, for example, focus on achieving coexistence between recreation and wildlife. In line with this, dog owners are requested to leash their dogs when approaching and passing vulnerable beach nesting zones. The observed low compliance with leashing regulations suggests this is an ineffective approach. However, the alternative, prohibiting dog access from these sensitive beaches, is typically met with conflict and division within the community as dog owners are faced with the risk of losing their access (Johnston et al. 2013). One part of the solution is to ensure adequate provision of alternative off-leash areas to divert users away from environmentally sensitive areas.

Processes which may increase compliance with dog regulations, such as education and enforcement (Ormsby and Forys 2010; Young et al. 2011), appear warranted in all dog regulation types. At the time of this study, enforcement data were unavailable, however through liaison with land managers, it was evident that patrols were consistently low to absent across all beaches included in this study (GS Maguire pers. comm.). In a study of beach use and preferences of coastal residents in south east Australia, Maguire et al. (2011) revealed high levels of dissatisfaction at current beach management. The most common suggested improvements were around implementing and enforcing regulations, and improved zoning of activities on beaches, primarily appropriate allocation of zones for activities such as dog walking. There were distinct groups of people wanting dog free access versus those wanting increased off leash access (Maguire et al. 2011). The implementation of education and enforcement strategies is likely to alter the comparative effectiveness of different dog regulations. Delivery of education can occur in many forms from passive information brochures and website information, to targeted community events such as 'Dogs Breakfasts' and one-to-one liaison with a ranger or education officer (Maguire et al. 2013). Surveys of beach users across the Victorian coast have revealed a low opinion of education as a means of behavioural change, but a preference for ranger patrols to provide information but with capacity for enforcement of repeat offenders (Maguire et al. 2013).

A logical extension of this research would be to assess and compare how the various dog regulations are communicated to users across sites. Clarification from land managers of dog regulations was required for several sites in this study, where coastal managers noted that the information was unclear or not readily accessible (GS Maguire pers. comm.). In fact at some sites, the responsible agency for assigning and enforcing dog regulations could not be agreed upon by coastal managers (GS Maguire pers. comm.). Along the coast, signage may or may not be present at access points and this varies greatly in the level of detail provided to dog walkers, from symbols as part of a broader beach regulations 'paddle' to dog walker specific signage with regulations clearly stated and indication of penalties. In some regions, council or shire websites will provide additional information on where dogs can be walked locally, however the quality of this information varies greatly, with some site descriptions being ambiguous and unclear. A review of beach user understanding of regulations would provide land managers with important information on how best to communicate with beach users and which methods most effectively influence human behaviours (McKenzie-Mohr 2011). At present, comprehensive data on education and other on-ground behaviour change programs are unavailable. However, the occurrence of dogs and compliance with regulations is readily measurable. This suggests that management agencies could set compliance triggers and targets to optimise and monitor dog management efforts, and thus, improve their long-term investment and eventual success in dog management.

Beach-breeding shorebird populations face a range of threats, of which dogs are but one (Maguire 2008). Among other threats (e.g., habitat degradation, sea-level rise), other species also collide with or crush them (e.g. humans, horses, unrestrained cattle), prey upon them (e.g. foxes, birds of prey) or disturb them. Resolving problems associated with invasive species is likely to contribute substantially to the conservation prospects of the world's beach-nesting birds. A logical place to start such efforts is with those invasive species under direct human control, such as domesticated animals.

11.5 Conclusions

We argue that dogs on Victorian beaches fit the ecological definition of invasive species, and show that compliance with prevailing rules is, at best, moderate across all available regulatory regimes in the study area. Nuanced regulations optimise coexistence but are associated with lower compliance. The best compliance we observed was in the clear-cut, non-coexistence regime of 'no dogs'. After decades of attempting to improve compliance with dog regulations (Dowling and Weston 1999), managers in Victoria recently banned dogs from an entire coastal National Park (Fig. 11.6). Our results suggest that banning dogs from beaches seems to be one of the only effective ways of managing their occurrence and behaviour. We acknowledge that responsible dog owners who take their dogs to beaches and obey regulations are unduly and unfairly effected by such changes, however, they currently are outweighed by a non-compliant majority. The challenge lies in identifying the triggers that will induce responsible dog owner behaviour so that we can see widespread adherence with regulations, enabling dogs and wildlife to coexist on coasts. A concerted and consistent effort by coastal managers to review current dog regulations and compliance rates, to set measurable targets for compliance in

Fig. 11.6 Signage announcing the banning of dogs from beaches at a major National Park in Victoria, Australia. (Image: M.A. Weston)

dog-wildlife conflict zones, and to invest in incentives for responsible dog walking behaviour need to occur to avoid divided human communities, and ultimately, species extinctions.

Acknowledgements Data were collected by BirdLife Australia and volunteers with the Beach-nesting Birds Program, funded by the Australian Government's Caring for Our Country program. This work was conducted under Deakin University Animal Ethics permit A45/2006 and Deakin University Human Ethics Exemption 2012-204. This work was supported by the Deakin University Centre for Integrative Ecology (CIE) and a School of Life and Environmental Sciences Collaborative Research Grant. Write-up was supported by the Beach Ecology And Conservation Hub (BEACH Venus Bay).

References

Blackburn TM, Pyšek P, Bacher S, Carlton JT, Duncan RP, Jarošík V, Wilson JR, Richardson DM (2011) A proposed unified framework for biological invasions. Trends Ecol Evol 26:333–339

Burrell NS, Colwell MA (2012) Direct and indirect evidence that productivity of Snowy Plovers *Charadrius nivosus* varies with occurrence of a nest predator. Wild 62:204–223

Cohen SP (2002) Can pets function as family members? West J Nurs Res 24:621–638

Colautti RI, MacIsaac HJ (2004) A neutral terminology to define 'invasive' species. Divers Distrib 10:135–141

Dowling B, Weston MA (1999) Managing a breeding population of the Hooded Plover *Thinornis rubricollis* in a high-use recreational environment. Bird Conserv Int 9:255–270

Forrest A, St Clair C (2006) Effects of dog leash laws and habitat type on avian and small mammal communities in urban parks. Urban Ecosyst 9:51–66

Glover HK, Weston MA, Maguire GS, Miller KK, Christie BA (2011) Towards ecologically meaningful and socially acceptable buffers: Response distances of shorebirds in Victoria, Australia, to human disturbance. Landsc Urban Plan 103(3-4):326–334

Gompper M (2014) Free-ranging dogs and wildlife conservation. Oxford University Press, Oxford

Hughes J, Macdonald DW (2013) A review of the interactions between free-roaming domestic dogs and wildlife. Biol Conserv 157:341–351

Iojă CI, Rozylowicz L, Pătroescu M, Niţă MR, Vânau GO (2011) Dog walkers vs. other park visitors perceptions: the importance of planning sustainable urban parks in Bucharest, Romania. Landsc Urban Plan 103:74–82

Johnston C, Briggs J, Wallace P (2013) Dog walking review – Mornington Peninsula National Park. Report to Parks Victoria. Context Consulting, Melbourne

Le Corre N, Gélinaud G, Brigand L (2009) Bird disturbance on conservation sites in Brittany (France): the standpoint of geographers. J Coast Conserv 13:109–118

Lockwood JL, Hoopes MF, Marchetti MP (2013) Invasion ecology. Wiley, Oxford

Maguire GS (2008) A practical guide to managing beach-nesting birds in Australia. Birds Australia report, Melbourne

Maguire GS, Miller KK, Weston MA, Young KA (2011) Being beside the seaside: beach use and preferences among coastal residents of South-Eastern Australia. Ocean Coast Manage 54:781–788

Maguire GS, Rimmer JM, Weston MA (2013) Stakeholder perceptions of threatened species and their management on urban beaches. Animals 3:1002–1020

McKenzie-Mohr D (2011) Fostering sustainable behavior: An introduction to community-based social marketing, 3rd edn. New Society Publishers, Gabriola Island

Miller KK, Ritchie E, Weston MA (2014) The human dimensions of dog-wildlife interactions. In: Gompper ME (ed) Free-ranging dogs and wildlife conservation. Oxford University Press, Oxford, pp 286–301

Morgan R (1999) Preferences and priorities of recreational beach users in Wales, United Kingdom. J Coastal Res 15:653–667

Ormsby AA, Forys EA (2010) The effects of an education campaign on beach user perceptions of beach nesting birds in Pinellas County, Florida. Hum Dimens Wildl 15:119–128

Pereira LCC, JA J'n, Medeiros C, Costa RM (2003) The influence of the environmental status of Casa Caiada and Rio Doce beaches (NE-Brazil) on beaches users. Ocean Coast Manage 46:1011–1030

Savolainen P, Leitner T, Wilton AN, Matisoo-Smith E, Lundeberg J (2004) A detailed picture of the origin of the Australian dingo, obtained from the study of mitochondrial DNA. Proc Nat Acad Sci 101:12387–12390

Schlacher TA, Weston MA, Lynn D, Schoeman DS, Huijbers CM, Olds AD, Masters S, Connolly RM (2015) Conservation gone to the dogs: when canids rule the beach in small coastal reserves. Biodivers Conserv 24(3):493–509

Scrogin D, Boyle K, Parsons G, Plantinga AJ (2004) Effects of regulations on expected catch, expected harvest, and site choice of recreational anglers. Am J Agric Econ 86:963–974

Sterl P, Brandenburg C, Arnberger A (2008) Visitors' awareness and assessment of recreational disturbance of wildlife in the Donau-Auen National Park. J Nat Cons 16:135–145

Teunissen W, Schekkerman H, Willems F, Majoor F (2008) Identifying predators of eggs and chicks of Lapwing *Vanellus vanellus* and Black-tailed Godwit *Limosa limosa* in the Netherlands and the importance of predation on wader reproductive output. Ibis 150:74–85

Valéry L, Fritz H, Lefeuvre JC, Simberloff D (2008) In search of a real definition of the biological invasion phenomenon itself. Biol Invasions 10:1345–1351

VSN International (2005) GenStat for Windows 7th Edition. VSN International, Hemel Hempstead, UK. Web page: GenStat.co.uk

Walsh J (2011) Unleashed fury: the political struggle for dog-friendly parks. Purdue University Press, West Lafayette

Weston MA, Stankowich T (2014) Dogs as agents of disturbance. In: Gompper ME (ed) Free-ranging dogs and wildlife conservation. Oxford University Press, Oxford, pp 94–116

Weston MA, Antos M, Glover HK (2009) Birds, buffers and bicycles: a review and case study of wetland buffers. Vic Nat 126:79–86

Weston MA, Fitzsimons JA, Wescott G, Miller KK, Ekanayake KB, Schneider T (2014) Bark in the park: a review of domestic dogs in parks. Environ Manag 54:373–382

Williams KJH, Weston MA, Henry S, Maguire GS (2009) Birds and beaches, dogs and leashes: dog owners' sense of obligation to leash dogs on beaches in Victoria Australia. Hum Dimens Wildl 14:89–101

Young JK, Olson KA, Reading RP, Amgalanbaatar S, Berger J (2011) Is wildlife going to the dogs? impacts of feral and free-roaming dogs on wildlife populations. Bioscience 61:125–132

Chapter 12
Evaluating How the Group Size of Domestic, Invasive Dogs Affect Coastal Wildlife Responses: The Case of Flight-Initiation Distance (FID) of Birds on Southern Australian Beaches

S. Guinness, W. F. Van Dongen, P.-J. Guay, R. W. Robinson, and M. A. Weston

Abstract Accompanied, domestic dogs frequently disturb birds on coasts, and meet the ecological definition of invasive species. Dogs occur most commonly singly or in 'packs' of two dogs. We examine whether group size (one versus two leashed dogs) influenced Flight-initiation Distance (FID), a measure of wariness towards potential predators, of birds on southern Australian beaches. We report 303 FIDs from 16 species, of which seven species had sufficient data to compare responses between one and two dog approaches. None of the seven focal species varied their FID or escape modality (walk/run versus fly) with one versus two dogs approaching. Birds do not apparently judge risk associated with dogs in relation to 'pack' size. Regulations which reduce the number of dogs walked are therefore unlikely to reduce disturbance of coastal birds. Further studies, using unleashed dogs, and dogs which bark, may evoke greater responsiveness than reported here and may reveal indirect effects of dog group size.

Keywords Beach · Canid · Disturbance · Gulls · Pets · Shorebirds

S. Guinness · M. A. Weston (✉)
Faculty of Science, Engineering and the Built Environment, Centre for Integrative Ecology, Deakin University, Geelong, Australia
e-mail: mweston@deakin.edu.au

W. F. Van Dongen
Faculty of Science, Engineering and the Built Environment, Centre for Integrative Ecology, Deakin University, Geelong, Australia

Institute for Sustainability and Innovation, Victoria University, Melbourne, VIC, Australia

P.-J. Guay · R. W. Robinson
Institute for Sustainability and Innovation, Victoria University, Melbourne, VIC, Australia

C. Makowski, C. W. Finkl (eds.), *Impacts of Invasive Species on Coastal Environments*, Coastal Research Library 29,
https://doi.org/10.1007/978-3-319-91382-7_12

12.1 Introduction

Coastal areas represent habitat for many species of birds, and many are extensively used by humans, thus they represent areas where management of disturbance (disruption of bird behaviour or physiology) may be required (Wallace 2016). Disturbance, under some circumstances, can represent a conservation concern and reduce population viability through processes such as disrupting energy assimilation or expenditure, suppressing reproductive success and/or survival, and even potentially by creating ecological traps (Gill 2007; Sutherland 2007).

Worldwide, coastal areas host many agents of bird disturbance, directly or indirectly associated with humans. While humans are common on coasts, a less appreciated source of disturbance is dogs, and coastal areas host many dogs, on beaches, dunes and throughout small coastal reserves (Lenth et al. 2008; Schlacher et al. 2015; Weston et al. 2014). It is estimated that the population of domestic dogs number over 700 million globally (Hughes and MacDonald 2013). Concern over the impact of dogs on coastal wildlife communities is growing (Hughes and MacDonald 2013; Lenth et al. 2008; Weston et al. 2014; Young et al. 2011), yet few studies exist of wildlife responses to dogs on beaches. Dogs on beaches meet the ecological definition of invasive species and in at least some areas, dogs commonly occur and roam freely, regardless of prevailing regulations (Maguire et al. 2018). Wildlife exhibit a range of responses to external stimuli such as dogs, with the most obvious and measurable responses being behavioral.

12.1.1 Behavioral Responses of Wildlife to Dogs

Wildlife species initiate escape when approached by predators and humans in such a way as to manage the risk of injury or death (Frid and Dill 2002; Samia et al. 2015). Escape responses are sophisticated, and vary among and within species and depending on stimulus-specific attributes (e.g. Glover et al. 2015; Paton et al. 2000; Weston et al. 2012). The response to various stimuli (agents of disturbance) is of interest because it indicates how animals judge risk, and also because those who manage interactions between wildlife and humans (e.g. disturbance), can potentially vary (restrict, modify or encourage) the stimuli which are encountered by wildlife (Weston et al. 2012). Coasts are generally multi-use areas, so management solutions would ideally be those which promote coexistence between wildlife and user groups, especially where spatial restrictions are non-feasible or undesirable. For example, while it may be impossible to exclude people from an area, it may be that their behaviour could be modified to minimise wildlife disturbance (for example, see Lethlean et al. 2016).

A common way to index bird responses to stimuli is to measure the distance at which they initiate an escape response (Flight-initiation Distance, FID; see Blumstein 2016). Another way that birds can adjust their escape is to use different

Fig. 12.1 A dog chasing birds evokes escape, in this case flying, among gulls and migratory shorebirds. (Image: M.A. Weston)

Fig. 12.2 Examples of avian escape among coastal birds. (**a**), flying (Crested Tern *Thalasseus bergii*); (**b**), running (Hooded Plover *Charadrius cucullatus*); (**c**), walking (Australian Pied Oystercatcher, *Haematopus longirostris*); (**d**), swimming (Banded Stilt *Cladorhynchus leucocephalus*). (Images: M.A. Weston)

'modes'; they can run or walk or use the more energetically expensive option of flying (Brackenbury 1984). When they interact with birds, dogs often evoke escape responses (Fig. 12.1). For coastal birds, escape includes walking, running, swimming, diving and flying (Fig. 12.2). Some bird species respond to dogs at greater distances than they do to humans alone (Glover et al. 2011).

12.1.2 The Influence of Pack Size of Dogs on Avian Response

Available studies of avian responses to dogs on coasts use single dogs despite the fact that dogs occur in coastal areas singly and in groups, where (on many coasts)

Fig. 12.3 Images of dogs chasing coastal birds singly (**a**) and occurring in packs (**b**). (Images, M.A. Weston)

they are accompanied by their owners (Weston and Stankowich 2013; Fig. 12.3). Few studies of FID compare the group size of the approaching stimulus (but see Geist et al. 2005; McLeod et al. 2013), yet managing the group size of dogs in coastal areas may represent one practical approach to reducing disturbance to birds which promotes coexistence between recreationists and birds (provided birds respond at shorter distances to single dogs being walked rather than groups of dogs being walked; Weston and Stankowich 2013).

12.1.3 Aims of This Chapter

Here we document the escape distances and type of escape evoked by approaches to birds of a walker with one dog versus a walker with two dogs, to determine if dog pack size influences the response of birds. No previous study of the effect of dog group size on wildlife responses is known to us, yet if an effect exists, limiting the number of dogs walked per walker may promote coexistence between birds and dog walkers on beaches.

12.2 Materials and Methods

Fieldwork was conducted on 30 days during December 2014 and January 2015 between 0800 and 1700 AEST during calm weather, by SG who wore dull clothing, and followed standard protocols for the collection of FIDs (see www.avianbuffer.com). We evoked flight and measured FIDs from any bird species that was associated with beaches and their foreshore areas (only single-species groups were approached when they were roosting or foraging). We collected FIDs from 13 coastal locations on the central Victorian coast west of Melbourne (Australia), from Barwon Heads (38.2814°S, 144.4911°E) to Eastern View (38.4780°S, 144.0288°E). Leashed dogs were permitted at all these locations throughout the year and these areas experience high human traffic. These locations were highly visited by dog walkers. Based on visits to Hooded Plover territories within the study area by citizen scientists, dog walkers were recorded on 34.8% (n = 480 visits) of occasions (BirdLife Australia unpublished data). Most dog walkers had a single dog with them, 97.1% of occasions involved 1–2 dogs (11.7% had two dogs), and no occurrence of a dog walker bringing >3 dogs was recorded (1.29 ± 0.50, mean ± standard deviation, dogs per human when dogs were present).

12.2.1 Design

FIDs were collected in response to three treatments (stimuli), which involved a pedestrian (SG) walking with: (1) a single Labrador ('Dog 1'), (2) a single Roodle (rottweiler x poodle; 'Dog 2'), or (3) both dogs. Both dogs were of similar size weighing approximately 40 kg and 24 kg respectively. Only one treatment was applied per day. After randomly assigning the initial treatment (Dog 1), we systematically alternated between using Dog 1, Dog 2 and both dogs on subsequent field days.

FIDs were collected on non-breeding birds using standard methods (Blumstein 2003; Glover et al. 2011). Briefly, once a focal bird was identified, we recorded its distance from the observer (the starting distance, SD). The observer and either one or two leashed (non-barking) dogs then approached the bird at a constant speed of approximately 1 ms^{-1}. Approach speeds did not differ between the stimuli (means and standard deviations: one dog, 1.0 ± 0.3 ms^{-1}; two dogs, 1.0 ± 0.4 ms^{-1}; $F_{1291} = 0.017$, $P = 0.678$). All distances were measured using a Bushnell Elite 1500 range finder. We also noted the escape method of the focal individual (i.e. walking, running or flying away). We avoided re-sampling the same individuals, and did not sample any individuals of the same species within 50 m of a location at which that species had been sampled. When approaching flocks, we measured the FID of the nearest individual and did not resample any other individuals within that flock. The relatively high abundance of the species sampled made the risk of double sampling negligible. We acknowledge that FIDs do not capture the range of non-escape or indirect responses that might be exhibited, but note that it represents a standard well-regarded index of response.

12.2.2 Statistical Analyses

FIDs were logarithmically-transformed to normality. Although we present FIDs for all species sampled, following McLeod et al. (2013), we only conducted statistical analyses for species for which we had ≥5 FIDs per stimulus (seven "focal" species). We first tested whether FIDs differed in response to Dog 1 and Dog 2 in the single-dog approaches, by conducting GLMs with FID as the dependant variable, dog identity as the fixed variable and SD as the covariate. No species (with $n \geq 5$ approaches per dog) exhibited a difference in FID between dogs (Table 12.S2, in Supporting Information). Using Fisher's exact tests we compared the nature of the flight response (i.e. escape by walking or running vs flying away) within five species with $n \geq 5$ approaches per dog, to test whether focal bird species' mode of escape to a single dog differed between the two dogs (Table 12.S1). We therefore combined the data from both dogs for subsequent analyses. We then tested whether FID varied with the number of approaching dogs, using FID as the dependent variable, number of dogs as the fixed variable and SD as a covariate. To test whether the nature of the flight response (i.e. walking, running or flying) of the focal birds differed between the two stimuli we conducted Fisher's exact tests. All analyses were conducted using SPSS 20.0 (SPSS, Chicago, Illinois, USA).

12.3 Results

We collected 303 FIDs from 16 species (Table 12.1). FIDs varied between the seven focal species (SD, $F_{1,257} = 45.671$, $P < 0.001$; species, $F_{6,257} = 8.712$, $P < 0.001$). Overall, Red-necked Stints displayed the shortest FIDs (14.4 m) and Masked Lapwings the longest FIDs (54.5 m).

For none of the seven focal species did FID vary significantly with stimulus type (Table 12.1, Figs. 12.4 and 12.5). While some differences between means were substantial, large variation often occurred within species in their FIDs. The mean FID for species (where $n \geq 2$ for each treatment) was longer for one dog approaches in eight of ten species. Combined, this suggests that no consistent or substantive increases in FID occurs with two dog approaches.

Some species (e.g. Hooded and Red-capped Plovers) typically responded to the approaches by walking or running away, whilst others (e.g. Australian Magpie and Great Egret) were approximately equally likely to run or fly away (Table 12.2). However, the number of dogs approaching each species did not affect the probability of an aerial versus terrestrial escape mode (Table 12.2).

12.4 Discussion

While it is generally considered that dogs evoke longer FIDs and more extreme responses than pedestrians alone (Weston and Stankowich 2013), we report no difference in FIDs or escape modality for one versus two dogs approaching coastal birds whilst on a leash. McLeod et al. (2013) similarly report no difference in FIDs between single and multiple walkers, although Geist et al. (2005) report that one of their study bird species exhibited longer FIDs when approached by two people rather than one person. A number of studies have demonstrated that birds adjust FIDs in

Table 12.1 Mean flight-initiation distances (± SD) of coastal birds in response to a pedestrian approaching with either one or two leashed dogs

	Flight-initiation distance				Starting distance		
Species	One dog	Two dogs	F	*P*	Mean ± SD	F	*P*
Australian Magpie *Craticus tibicen*[a]	20.7 ± 17.9 (26)	18.6 ± 8.7 (8)	0.046 (1,31)	0.831	66.6 ± 28.2 (34)	2.884 (1,31)	0.100
Australian White Ibis *Threskiornis molucca*[a]	77 (1)				170 (1)		
Caspian Tern *Hydroprogne caspia*[a]	27.6 ± 16.8 (8)	48.5 ± 6.4 (2)			60.1 ± 41.3 (10)		
Common Starling *Sturnus vulgaris*[a]	20 (1)				31 (1)		
Crested Tern *Thalasseus bergii*[a]	29.4 ± 11.3 (5)	25.7 ± 13.3 (3)			70.1 ± 19.4 (8)		
Great Cormorant *Phalacrocorax carbo*[a]	35.0 ± 17.2 (7)	34.0 ± 21.3 (6)	0.229 (1,10)	0.643	81.5 ± 35.4 (13)	9.673 (1,10)	0.011
Great Egret *Ardea modesta*[a]	45.3 ± 19.6 (4)	37 (1)			66.8 ± 34.7 (5)		
Hooded Plover *Charadrius cucullatus*[a]	23.1 ± 10.2 (24)	17.5 ± 7.2 (11)	1.797 (1,32)	0.190	43.2 ± 18.4 (35)	7.178 (1,32)	0.012
Little Black Cormorant *Phalacrocorax sulcirostris* [a]	35 (1)				90 (1)		
Magpie Lark *Grallina cyanoleuca*[a]	16 (1)				43 (1)		
Masked Lapwing *Vanellus miles*	54.5 ± 30.4 (2)				69.0 ± 33.9 (2)		
Pacific Gull *Larus pacificus*[a]	33.3 ± 16.6 (13)	39.8 ± 14.1 (5)	1.303 (1,15)	0.272	89.7 ± 40.7 (18)	7.909 (1,15)	0.013
Red-capped Plover *Charadrius ruficapillus*	21.0 ± 5.9 (30)	19.9 ± 6.4 (13)	1.520 (1,40)	0.225	31.7 ± 9.3 (43)	37.474 (1,40)	< 0.001
Red-necked Stint *Calidris ruficollis*	14.4 ± 5.3 (36)	13.9 ± 5.1 (16)	0.025 (1,49)	0.876	31.9 ± 12.5 (52)	4.255 (1,49)	0.044
Silver Gull *Chroicocephalus novaehollandiae*[a]	22.2 ± 8.3 (48)	21.0 ± 6.7 (22)	0.113 (1,67)	0.738	53.5 ± 27.7 (70)	8.413 (1,67)	0.005
White-faced Heron *Egretta novaehollandiae*[a]	35.9 ± 39.7 (7)	23.5 ± 7.8 (2)			73.3 ± 43.4 (9)		

Mean starting distance (± SD) for each species (with one and two dog approaches grouped) are also shown. Statistical analyses were conducted for seven species for which we had at least five FIDs per stimulus. Values in parentheses after means represent sample sizes, while those for F-values represent degrees of freedom. The tests shown are from a GLM with one versus two dogs as a fixed factor, and SD as a covariate. Species are presented in alphabetical order by common name

[a]indicates those species which have not had FIDs in response to dogs previously recorded (no species has had FIDs evoked by two dogs previously documented)

Fig. 12.4 Seven focal coastal birds in Victoria, Australia, which were included in this study. Only non-breeding birds were approached. Images: Great Cormorant, Daniel Lees (Deakin University); others M.A. Weston

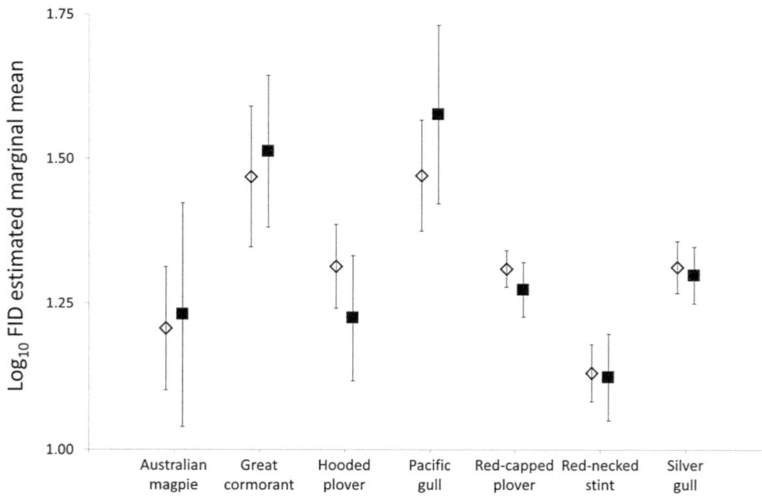

Fig. 12.5 Estimated marginal mean Log_{10} FID evoked by one (open diamonds) or two dogs (black squares) for seven species of coastal birds. Error bars represent 95% confidence intervals

Table 12.2 Escape responses (frequencies) of coastal birds in response to a pedestrian approaching with one or two dogs

	One dog		Two dogs		
Species	Terrestrial	Aerial	Terrestrial	Aerial	*P*
Australian Magpie	13	13	3	5	0.693
Great Cormorant	3	4	1	5	0.559
Hooded Plover	23	1	10	1	0.536
Pacific Gull	5	8	3	2	0.608
Red-capped Plover	28	2	13	0	1.000
Red-necked Stint	28	7	13	2	0.705
Silver Gull	34	13	18	3	0.355

Terrestrial responses refer to cases where focal individuals walked or ran away from the stimulus. Aerial responses refer to flying away from the stimulus. The *P*-values refer to Fisher's Exact tests

relation to the type of stimulus approaching (Glover et al. 2011, 2015; Guay et al. 2014; McLeod et al. 2013). Thus, birds may be adept at recognising types of threats, but may either be unable to count or, regardless of their ability to count, do not judge multiple stimuli to represent a greater risk. While groups of stimuli presumably represent greater visual, auditory and perhaps other cues, the risk between individual stimuli within a given stimulus type is likely to be similar. Thus, while detection distances (after Weston et al. 2012) may be greater for a 'compound' stimulus (i.e., multiple dogs), the risk posed by them may not be amplified above that of a 'single' stimulus.

12.4.1 Future Research

Many dog walkers do not use a leash where mandated to do so, and so their dogs roam free (Williams et al. 2009). Off leash dogs occupy more habitat and observational studies indicate off-leash dogs evoke greater avian responses than on-leash dogs (Weston and Stankowich 2013 but see Miller et al. 2001). Apart from occupying more habitat, and moving more rapidly, birds may perceive walkers with an off-leash dog as two different stimuli approaching from different directions and thus may exhibit longer FIDs or use a different response mode. While we have no data, leashing rates of lone versus groups of dogs may vary, and groups of dogs may be more mobile than lone dogs in beach environments. Future research on the response of birds to off-leash dogs, particularly single versus groups of dogs, would consolidate the current findings.

Given that dog barking alone may cause bird disturbance (Randler 2006), another area of future research would be on barking rates of single versus groups of dogs. While barking has been little studied in general, it clearly functions to communicate with both humans and other dogs (Pongrácz et al. 2010). If groups of dogs bark more frequently or louder (but see Mertens and Unshelm 1996), then they could cause more avian disturbance. Our study involved exclusively non-barking dogs, but an observational study of dog barking behaviour coupled with experiments of the effect of barking on wildlife, may reveal an alternative but plausible mechanism through which dog group size influences wildlife responses.

12.5 Conclusions

Few, if any, invasive species are under such direct control of humans, as are dogs on those coastlines where the dog population is dominated by owned dogs. The number of approaching dogs did not influence the escape distance or mode exhibited by coastal birds, but we used leashed non-barking dogs in our experiment. Codes of conduct, whereby people alter their behavior (e.g. the number of dogs walked at a time) offer promise in reducing wildlife disturbance on coasts (Schlacher et al. 2013). However, given our findings, altering the number of leashed, non-barking dogs walked is unlikely to reduce avian disturbance, at least in the species and sites examined.

Acknowledgements This third year summer research project was funded by a Summer Research Scholarship (Victoria University, College of Engineering and Science), and a Summer Project Prize (Deakin University, School of Life and Environmental Sciences). Deakin Animal Ethics Approval (B32-2012) and DEWLP Permit 10007151 applied to this work. Write-up was supported by the Beach Ecology And Conservation Hub (BEACH, Venus Bay).

Appendix A: Supplementary Tables

Table 12.S1 Mean flight-initiation distances (m) of beach-associated birds in response to a pedestrian approaching with either Dog 1 or Dog 2 (where $n \geq 5$ approaches for each dog)

	Flight-initiation distance				Starting distance	
Species	Dog 1	Dog 2	F	*P*	F	*P*
Australian Magpie	20.1 ± 11.1 (12)	21.1 ± 22.5 (14)	0.259 (1,23)	0.616	4.390 (1,23)	0.047
Hooded Plover	25.3 ± 11.8 (13)	20.5 ± 7.6 (11)	0.506 (1,21)	0.485	8.007 (1,21)	0.010
Red-capped Plover	18.8 ± 3.9 (16)	23.6 ± 6.9 (14)	3.106 (1,27)	0.089	32.087 (1,27)	< 0.001
Red-necked Stint	16.1 ± 5.7 (17)	13.0 ± 4.5 (19)	1.208 (1,33)	0.280	2.267 (1,33)	0.142
Silver Gull	22.0 ± 8.2 (22)	22.3 ± 8.5 (26)	0.146 (1,45)	0.704	8.286 (1,45)	0.006

The results of an ANOVA with logFID as dependent, dog identity as a fixed factor and SD as a covariate are also shown. Values in parentheses for means represent sample sizes, while those for F-values represent degrees of freedom

Table 12.S2 Escape responses of coastal birds (where $n \geq 5$ approaches for each dog) in response to a pedestrian approaching with Dog 1 and Dog 2

	Dog 1		Dog 2		
Species	Terrestrial	Aerial	Terrestrial	Aerial	*P*
Australian Magpie	8	4	5	9	0.238
Hooded Plover	12	1	11	0	1.000
Red-capped Plover	15	1	13	1	1.000
Red-necked Stint	15	1	15	4	0.347
Silver Gull	17	4	18	8	0.505

Terrestrial responses refer to cases where focal individuals walked or ran away from the stimulus. Aerial responses refer to flying away from the stimulus. The p-values refer to Fisher's Exact tests

References

Blumstein DT (2003) Flight-initiation distance in birds is dependent on intruder starting distance. J Wildl Manag 67:852–857

Blumstein DT (2016) Habituation and sensitization: new thoughts about old ideas. Anim Behav 120:255–262

Brackenbury J (1984) Physiological responses of birds to flight and running. Biol Rev 59:559–575

Frid A, Dill LM (2002) Human-caused disturbance stimuli as a form of predation risk. Conserv Ecol 6:11

Geist C, Liao J, Libby S, Blumstein DT (2005) Does intruder group size and orientation affect flight initiation distance in birds? Anim Biodivers Conserv 28:69–73

Gill JA (2007) Approaches to measuring the effects of human disturbance on birds. Ibis 149:9–14

Glover HK, Weston MA, Maguire GS, Miller KK, Christie BA (2011) Towards ecologically meaningful and socially acceptable buffers: response distances of shorebirds in Victoria, Australia, to human disturbance. Landsc Urban Plann 103:326–334
Glover HK, Guay P-J, Weston MA (2015) Up the creek with a paddle; avian flight distances from canoes versus walkers. Wetl Ecol Manage 23:775–778
Guay P-J, McLeod EM, Taysom AJ, Weston MA (2014) Are vehicles 'mobile bird hides'? A test of the 'cars cause less disturbance' hypothesis. Vic Nat 131:150–155
Hughes J, MacDonald DW (2013) A review of the interactions between free-roaming domestic dogs and wildlife. Biol Conserv 157:341–351
Lenth BE, Knight RL, Brennan ME (2008) The effects of dogs on wildlife communities. Nat Areas J 28:218–227
Lethlean H, van Dongen WFD, Kostoglou K, Guay P-J, Weston MA (2016) Joggers cause greater avian disturbance than walkers. Landsc Urban Plan 159:42–47
Maguire GS, Miller KK, Weston MA (2018) Only the strictest rules apply: investigating regulation compliance on beaches to minimize invasive dog impacts on threatened shorebird populations. In: Finkl CW, Makowski C (eds) Impacts of invasive species on coastal environments: coasts in crisis. Springer, Dordrecht
McLeod EM, Guay P-J, Taysom AJ, Robinson RW, Weston MA (2013) Buses, cars, bicycles and walkers: the influence of the type of human transport on the flight responses of waterbirds. PLoS One 8:e82008
Mertens PA, Unshelm J (1996) Effects of group and individual housing on the behavior of kennelled dogs in animal shelters. Anthrozoös 9:40–51
Miller SG, Knight RL, Miller CK (2001) Wildlife responses to pedestrians and dogs. Wildl Soc Bull 29:124–132
Paton DC, Ziembicki M, Owen P, Heddle C (2000) Disturbance distances for water birds and the management of human recreation with special reference to the Coorong Region of South Australia. The University of Adelaide, Adelaide
Pongrácz P, Csaba M, Miklósi A (2010) Barking in family dogs: an ethological approach. Vet J 183:141–147
Randler C (2006) Disturbances by dog barking increase vigilance in coots *Fulica atra*. Eur J Wildl Res 52:265–270
Samia DSM, Nakagawa S, Nomura F, Rangel TF, Blumstein DT (2015) Increased tolerance to humans among disturbed wildlife. Nat Commun 6:8877
Schlacher TA, Weston MA, Lynn D, Connolly RM (2013) Setback distances as a conservation tool in wildlife-human interactions: testing their efficacy for birds affected by vehicles on open-coast sandy beaches. PLoS One 8(9):e71200
Schlacher TA, Weston MA, Lynn D, Schoeman DS, Huijbers CM, Olds AD, Masters S, Connolly RM (2015) Conservation gone to the dogs: when canids rule the beach in small coastal reserves. Biodivers Conserv 24:493–509
Sutherland WJ (2007) Future directions in disturbance research. Ibis 149:120–124
Wallace P (2016) Managing human disturbance of wildlife in coastal areas. N Z Geogr 72:133–143
Weston MA, Stankowich T (2013) Dogs as agents of disturbance. In: Gompper ME (ed) Free-ranging dogs and wildlife conservation. Oxford University Press, Oxford, pp 94–116
Weston MA, McLeod EM, Blumstein DT, Guay P-J (2012) A review of flight-initiation distances and their application to managing disturbance to Australian birds. Emu 112:269–286
Weston MA, Fitzsimons JA, Wescott G, Miller KK, Ekanayake KB, Schneider T (2014) Bark in the park: a review of domestic dogs in parks. Environ Manag 54:373–382
Williams KJH, Weston MA, Henry S, Maguire GS (2009) Birds and beaches, dogs and leashes: dog owners' sense of obligation to leash dogs on beaches in Victoria, Australia. Hum Dimens Wildl 14:89–101
Young JK, Olson KA, Reading RP, Amgalanbaatar S, Berger J (2011) Is wildlife going to the dogs? Impacts of feral and free-roaming dogs on wildlife populations. Bioscience 61:125–132

Chapter 13
Impact of Invasive Nypa Palm (*Nypa Fruticans*) on Mangroves in Coastal Areas of the Niger Delta Region, Nigeria

Aroloye O. Numbere

Abstract Invasive nypa palms (*Nypa fruticans*) are a major threat to mangroves and coastal systems in the Niger Delta region in Nigeria, apart from oil and gas exploration. The palms were first introduced as foreign species to curb coastal erosion over a century ago (i.e. 1906). They later became invasive and started multiplying in the last 30 years. The palms have acclimatized to the coastal environment by developing superior root system, which they use to tap available nutrients. They also have tough and buoyant seeds, which aid in their wide dispersal. These qualities of the palms had made them to have an edge over the mangroves. Oil and gas exploration, which is responsible for numerous oil spillages, is a major cause of mangrove decimation. The establishment of open waste disposal sites in coastal areas have also contributed to the changes in soil and water qualities, leading to further decline in mangroves, with a resultant increase in invasive nypa palms. The palms change the pedology, hydrology and landscape architecture of the coastal environment once they are established. Therefore, a threat to the mangroves is a threat to the entire coastal system, which benefits from the ecosystem services provided by the mangroves. Mangroves may disappear completely from the Niger Delta in the next 50 years if the encroachment of the palms continue unabated. However, this problem can be resolved by the removal of the palms through mechanical, physical or chemical means. Soils on which the palms grow can be excavated to remove the allelopathic properties, after which the palm soil should be replaced with mangrove soil. To ensure smooth re-colonization of the coast, mangroves propagules with good genetic quality should be selected, nurtured and transplanted from the nursery to the coastal areas. The mangrove propagules should be monitored and protected from further invasion by nypa palm after planting.

Keywords Niger Delta · Nypa palm · Invasive species · Hydrocarbon pollution · Mangrove · Exploration · Seismic activities · Oil spillages · Restoration

A. O. Numbere (✉)
Department of Animal and Environmental Biology, University of Port Harcourt, Choba, Nigeria

C. Makowski, C. W. Finkl (eds.), *Impacts of Invasive Species on Coastal Environments*, Coastal Research Library 29,
https://doi.org/10.1007/978-3-319-91382-7_13

13.1 Introduction

Invasive species are plants and animals that are intentionally (aquaculture, agriculture, recreation and ornamental purposes) or unintentionally (humans, airplanes, cars, shirts, ships etc.) introduced into foreign lands (Davis 2009), and become physically and genetically more superior than their host species when they arrive. In addition, invasive species hitch hike to foreign lands on vehicles or vessels (Crosby 1986; Hodkinson and Thompson 1997). Similarly, species are also introduced to foreign land through international trades and travels by humans who carry crops to Africa, Asia, Europe, Australia and America. The foreign species compete with the host species for resource and space, which leads to serious impacts on host species ability to survive. The impact can be direct through predation, competition, parasitism and disease, or indirect through resource competition, trophic cascade, habitat modification (Wootten 1994), ecosystem impact through habitat structure, disturbance regime, nutrient cycling and hybridization. Some of the invasive species are ecosystem engineers, because they change the fundamental aspect of the physical and chemical environment in their new location (Jones et al. 1997). Invasive species also exhibit propagule pressure by producing large pool of source population, which overwhelm the native species. Other impacts of invasive species in coastal areas include: population and community impacts, morphological impacts and genetic and evolutionary impacts. Many studies had come up with some theories concerning the actions of invasive species. Two of these theories are biotic resistance hypothesis and energy release hypothesis. Biotic resistance hypothesis postulates that community diversity affects invasion success (Elton 1958). This means species rich communities are more resistant to invasion. Thus if species are packed into a community they effectively utilize the resources, and prevent the creation of empty niches, which discourage invasive species (Elton 1958). This is because an empty niche will lead to competition for unused and abundant resources. Likewise, consumption by natives reduces invasion success. This is because invading species meet their waterloo in foreign lands when native species fight back. Energy release hypothesis, on the other hand, postulates that invasive species spread rapidly due to the absence of co-evolved natural enemies in the introduced range (Elton 1958; Keane and Crawley 2002). It also means that when there is an abundance of enemies in a native range, they prevent the native species from succeeding whereas the invasive species prospers because by leaving their home base they leave behind most of their natural enemies. However, studies had shown that most introductions fail, for example, starlings, a small to medium sized passerine birds in the family of Sturnidae, failed at its initial introduction.

Nevertheless, nypa palm (*Nypa fruticans*) is one of those species that didn't fail, but rather became very successful at its initial introduction. It is now a major invasive species and a threat to coastal areas in the Niger Delta. This is because the palms have suppressed the growth of mangroves and other coastal species with their explosive population growth. They are able to over crowd the coastal areas because of the high production and effective distribution of their seeds around major creeks and rivers in the region. They grow outwardly from the center of the forest towards the river, constricting the waterways. The palms prevent the growth of other coastal species by occupying every unoccupied inch of land along the coast. However, mangrove growth is different from nypa palm growth because mangrove supports food chains, and food webs and helps in the proliferation of other species in the coastal environment (e.g. epiphytes). For instance, mangrove forests serve as spawning ground for various aquatic organisms, whereas the palms destabilize ecosystem function by pushing away many organisms (e.g. fishes), and making it difficult for local fishermen to have good catch whenever they go for fishing. This results to them sailing further into the Atlantic Ocean to get more catches.

The nypa palms (*Nypa fruticans*) in the Niger Delta region originated from Asia (Jones 1995), but were intentionally introduced into the Niger Delta to fight coastal erosion in 1906 (Keay et al. 1964). The first point of contact of the palms was Calabar, a town in the Niger Delta, from where they spread further south to other coastal towns along the Atlantic Ocean, aided by human activities and tidal currents. One major cause of their spread in the coastal areas is oil and gas exploration activities, which involved forest clearing to create room for seismic work, and deforestation, which changed the landscape of mangrove forest from undisturbed to disturbed state, after the entry of the exploration parties (Fig. 13.1). These actions lead to the proliferation of several alien species such as grasses. Similarly, the arrival of exploration teams led to the construction of boot camps and residential quarters to accommodate field workers, which further attracted a lot of human presence around the forest leading to gradual urbanization of the coastal areas. The creeks and rivers around Okrika, a host town to a major refinery in the Niger Delta, have a lot of oil facilities (e.g. pipelines, well heads, oil jetties etc), which helped to change the configuration of the forest by making it possible for the nypa palms to invade and colonize most areas around the coast (Fig. 13.2). In contrast, Buguma another town has no major oil infrastructure, but few well heads with sub-surface pipelines with limited nypa palm presence (Fig. 13.2). Nothwithstanding, during one of our field trips a nypa palm seedling, about 1 m tall was found growing within the mangrove forest by the sea shore, which most likely was brought in by tidal current. The seedling was uprooted and destroyed, but whether that singular action will prevent further entry and growth of the palms still remains to be seen.

The palms had adapted to the environment and had become a major threat to the native mangroves. The palms are found along the rivers of most of the coastal towns in the Niger Delta region such as New Calabar, Nun, Imo, Saint Barbara, Saint Bartholomew, Orashi, Bonny, Opobo, and Andoni Rivers. They occur in mixed or pure species stands (Wang et al. 2016) at the fringes of the sea where they block water channels and navigational routes. They also clog up drainages, thereby causing the stagnation of fast moving streams. The palms do not only affect coastal environments, they also affect transportation along the river by causing boat accidents, which results in injuries and deaths (Fig. 13.3) because of poor visibility, caused by the blocking of view of boat drivers when they meander through the creeks.

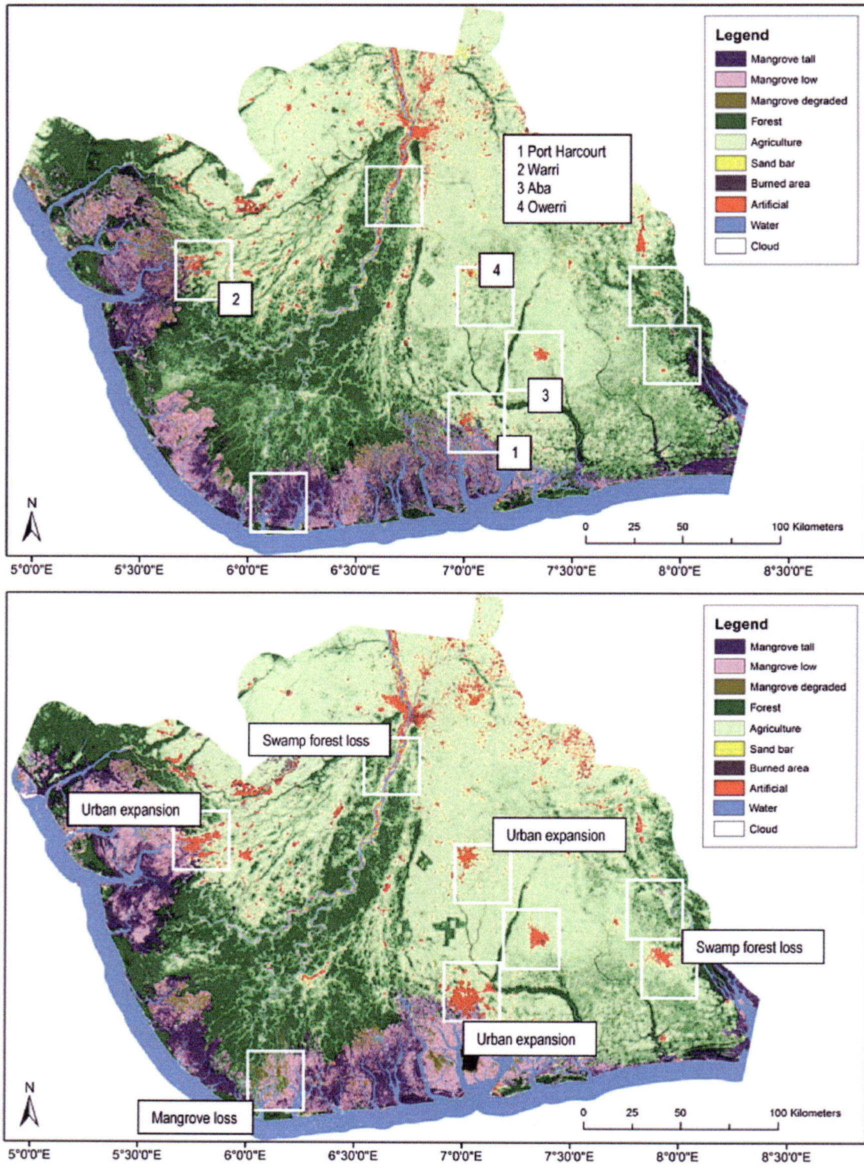

Fig. 13.1 Mangrove forest loss in the Niger River Delta, Nigeria is as a result of urban expansion and other anthropogenic activities (e.g. oil and gas exploration, deforestation, invasive species etc). Mangrove loss is also triggered by agricultural activities leading to the formation of sand bars (Kuenzer et al. 2014), and the reduction of wet mud (Wang et al. 2016)

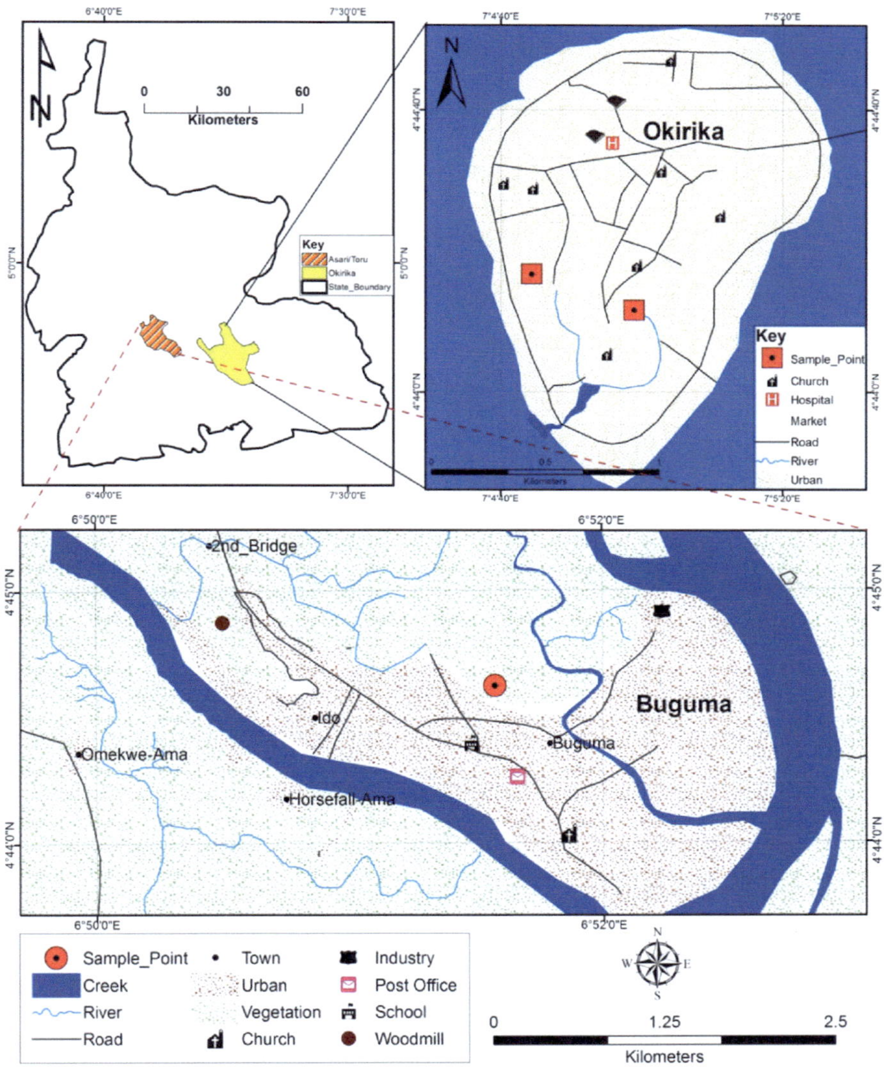

Fig. 13.2 Two study areas in the Niger Delta that had already been invaded by nypa palms (*Nypa fruticans*). The rivers and creeks around Okrika had been over run by the palms, due to years of oil and gas exploration activities which caused crude oil spillages while nypa palm invasion in Buguma had been low because of low oiling activities

Fig. 13.3 Nypa palms invade a small mangrove forests and also constricts a navigational route for sea travelers at Eagle Island, Niger Delta Nigeria. Some sections of the forest had already been dredged on the left side of the picture for the purpose of sand mining and the establishment of an industrial complex

13.2 Factors Influencing Nypa Palm Expansion in the Niger Delta Region

13.2.1 Anthropogenic Factors

Anthropogenic activities contribute greatly to the invasion and spread of nypa palms in the Niger Delta. Oil and gas exploration and exploitation activities which began in 1956 through the striking of the first oil well in Oloibiri town, Niger Delta was particularly a major factor that opened up the mangrove forest to invasion by palms (Figs. 13.4 and 13.5). Exploration for crude oil is one of the first activities that led to the deforestation of vast amount of virgin mangrove forest, aimed at creating a right of way (ROW) passage for seismic lines, laying of pipelines and setting up of base camps (Fig. 13.5). The paths created within the forest do not grow several years after their establishment (Ohimain et al. 2008), but become passages for humans and animals, which use the route created to penetrate deeper into the forest to plunder its resources. Local people take advantage of the cleared area to hunt for animals and set up crop farms while oil workers go into the forest with heavy machinery

Fig. 13.4 Google image of a major crude oil pipeline linking the Port Harcourt refinery and the sea jetty at Okrika, where crude oil is evacuated and transported abroad. Some years ago (i.e. in the year 2010) this location was exclusively a mangrove forest, but of recent most of the mangrove stands had been taken over by the palms. This is because of constant hydrocarbon pollution from broken pipes, and deforestation aimed at clearing the path along pipeline routes, which results in the death of mangrove trees. (Credit: Google Earth)

that disfigures the swamp and introduce foreign species that are picked up along the way.

Hydrocarbon pollution is an outcome of exploration activities, and it occurs during the drilling of oil wells and the transportation of crude oil through pipelines along sea and land routes. Oil spillages occur during pre-exploration, exploration and post exploration stages. The spilled oil changes soil and water qualities and affects the growth of coastal organisms. Reduction in soil quality impedes the growth of native plants, and accelerates the growth of foreign species. Studies had shown that nypa palm seedlings grow better than mangrove seedlings when both are planted in mangrove soil. Similarly, nypa palm seedlings survive longer than mangrove seedlings when both are planted in polluted mangrove soil (Numbere and Camilo 2016a). Contamination of the water body also leads to the death of aquatic organisms and the destruction of the food chain, which depends on the coastal life for survival.

Fig. 13.5 Dredging activity to create pipeline route through the mangrove forest in Bakana town, Niger Delta, Nigeria. (Credit: A.O. Numbere)

Fig. 13.6 Encroachment of nypa palms into mangrove forest (*Rhizophora mangle*) along a waterway leading to Bakana Town, Niger Delta, Nigeria. Invasion of the palms is facilitated by anthropogenic activities such as oil and gas exploration, dredging, sand mining, and urbanization. (Credit: A.O. Numbere)

Fig. 13.7 Building of houses right in the mangrove swamp destroys the native mangrove species and open the way for the invasion of the coastal area by alien species such as nypa palms and a variety of grass species. This location was formerly exclusively occupied by mangrove species, but now only a little patch of mangrove stand still remains at the background of the picture. One of the contributing factors is the use of the location as a refuse dump site, which changes the soil to dirty brown that is mixed with greenish algae (e.g. spirogyra). (Credit: A.O. Numbere)

Dredging activities destroys the native mangrove stands and creates an opening for nypa palm seeds to enter and multiply. In the same vein, the contamination of the soil as a result of oil and gas exploration activities also changes the soil quality leading to the death of the mangroves and the proliferation of the palms as recorded in many sites across the Niger Delta (Fig. 13.6).

Urbanization is a key anthropogenic factor that drives the growth of invasive palms. The palms have more affinity for disturbed than undisturbed soil. Urbanization in this context is the replacment of mangrove forest with urban centres, which promotes the spread of palms by changing the soil quality (Fig. 13.7). This involves the destruction of mangrove forest stands and the building of houses right in the mangrove swamp. This action creates the way for the invasion of other alien species. Conversion of wetlands into place of human habitation leads to the generation of human waste, which further reduces the quality of the environment. A disturbed forest is a fertile ground for the proliferation of the palms. The practice of aquaculture in wetland also degrade native mangrove, and encourage the introduction of foreign species. Mangrove forest had been degraded from high density to low density mangrove for the last 20 years in the Niger Delta (Wang et al. 2016; Kuenzer et al. 2014).

13.2.2 Pedology/Changes in Soil Condition

Mangroves are habitat specialist, and grow only in coastal areas (Nagelkerken et al. 2008). Therefore, a change in soil condition from swampy to sandy soil is detrimental to their survival. Mangroves grow better when they are in their native soil, which is swampier and richer in nutrients than the nypa palm soil. This is in agreement with a previous study which indicates that soil effect had more influence on invasion success than species richness (Stohlgren et al. 1999). The typical mangrove soil is coffee brown in color, slightly muddy and has a pungent ammonia-like smell, which breathes life into the mangrove forest and accelerates the growth of organisms. On the other hand, the palms grow on a variety of soils such as mangrove soil, muddy soil, sandy soil and algae-infested soils. Studies done using soils from different locations show that growth in height of nypa palm seedlings was mainly influenced by soils derived from highly polluted forest than soils derived from lowly polluted forest. Another study also indicates that nypa palms grow better than mangroves in mixed forests (i.e. a combination of mangrove and nypa palm trees growing together) than in pure forest (Wang et al. 2016). This is one of the reasons why the palms out-compete the mangroves when they infiltrate mangrove forest (CEDA 1997). Nypa palm has better growth in mangrove soil than in its own soil. Studies had shown that the growth of nypa palm seedlings in mangroves soils is as a result of the utilization by the palms of the un-used nutrients that are locked within the mangrove soil. One of the conclusions of this study is that nypa palm performed better in mangrove soil and worse in its own soil, in terms of growth in height and production of leaves. Similarly, nypa palm produced more seeds than mangroves in mixed forest. This gives the palms advantage over mangroves because with time the seeds of the palms will germinate and colonize the entire area. This situation is a recurring factor in many locations in the Niger Delta.

During several field trips to many locations it was observed that mangroves growing in core nypa palm forest don't have better growth. This led to the conduction of a pilot study to test the growth performance of mangrove and nypa palm seeds in mangrove and nypa palm soils. We collected soils from both mangrove and nypa palm forest and placed them in a swamp box (Numbere and Camilo 2017b). Twenty five (25) seeds of mangroves (*R. racemosa*) without blemish were planted in each soil, and left under semi-natural conditions for seven months (i.e. March—October). The plants were watered daily with river water collected in-situ, and left in the open, under the elements of the weather. The result was stunning, and provided some answers to why the palms are always performing better than the mangroves, and why the palms had continuously colonized several mangroves forest in the Niger Delta in the last few years. The result showed a robust growth of mangrove seeds planted in mangrove soil and a stunted growth of mangrove seeds planted in nypa palm soil (Fig. 13.8). This experimental growth in a nursery is what has been replicated in the natural environment, which had made the palms a more superior competitor than the mangroves, leading to the domination and eventual elimination

Fig. 13.8 Swamp box experiment showing two partitions of: (**a**), mangrove soil and (**b**) nypa palm soil. Twenty five mangrove seeds were planted in each section. They were watered daily with river water, and monitored for seven months, March—October, 2017. The result indicates that mangrove seeds in mangrove soil had robust growth while mangrove seeds in Nypa palm soil had retarded growth. Result of this study is a classical example of what happens in the natural environment where Nypa palms are quickly colonizing the coastal areas at the detriment of the native mangrove species. (Credit: A.O. Numbere)

of mangroves in several locations. More studies are ongoing with more replications to reinforce the hypothesis that invasion of nypa palms in mangrove forest has a lot to do with changes in soil quality.

13.2.3 Hydrology

Dredging and sand filling activities change the hydrology and affect mangroves and other coastal species (Fig. 13.9) (Ohimain 2004). Sand filling brings in excess sediments along with foreign species into wet land region, which smothers and overwhelm native species. Sandy soil that is pumped from the bottom of the sea unto the shore comes along with foreign species, which invade the coastal areas. Some aquatic organisms such as mangroves do not grow on sandy soil. Moreover, the change in creek size influences tidal flow rate, and affects stream physico-chemistry. Similarly, expansion of the coast through channel canalization (Fig. 13.10) leads to the inadvertent entry of invasive species while reduction in stream size increases tidal pressure with ripple effect on stream chemistry (e.g. salinity) and species population dynamics (e.g. increased competition). Changes in the salinity level of the stream through dredging activities affect species composition and distribution.

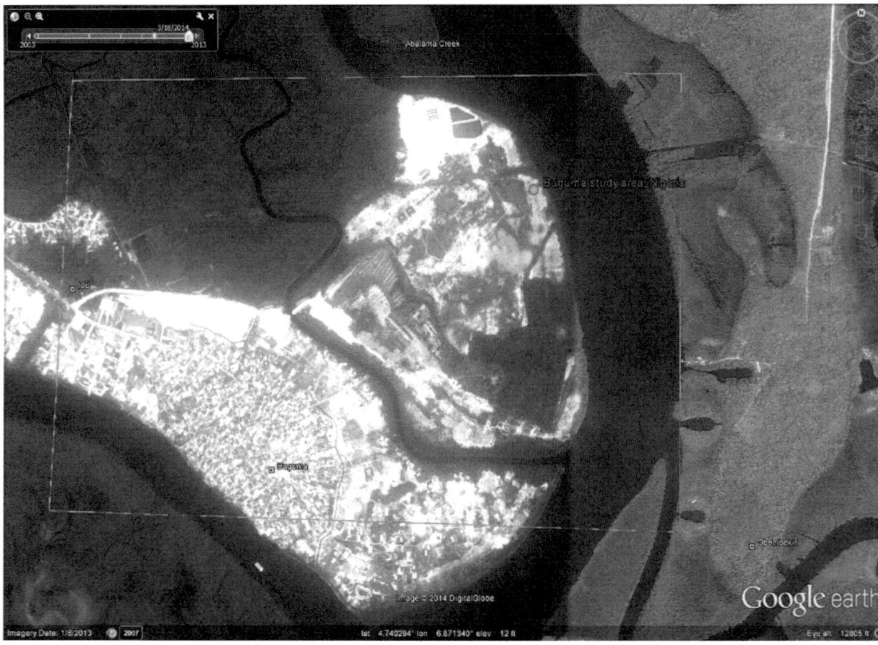

Fig. 13.9 Google Earth image of Buguma, a town in the Niger Delta, Nigeria that was sand filled in 1984 for the purpose of building houses to accommodate increasing population of the local residents. Till date the sand filled area (white patch) has not supported the growth of any mangrove forest, rather different foreign species such grasses had dominated the area covered by the sandy soil. Similarly, some tributaries connected to the river had been blocked by the sand. Some seeds of nypa palms were recently detected on the shores of the river, which indicates that in a matter of some years the remnants of the mangrove forest seen in the picture would be endangered and may go into extinction in the next 20 years if nothing is done. (Credit: Google Earth)

Mangroves are generally halophytic, but the distribution of individual species depends on the salinity level of the location. For instance, *Rhizophora* species are less halophylic whereas *Avicennia germinans* are more halophylic (Lugo 1980), which affects their distribution. During the transition period after dredging, the area is usually invaded by opportunistic foreign species as the native species migrate to a more habitable environment.

Dredging of the coastal areas to lay pipelines disturb the soil structure and composition and destabilize the growth of mangroves and other coastal species. Nypa palms invade areas made bare by deforestation. The dredging activity changes the soil and water quality and prevents the mangroves from growing, leading to their gradual death and loss from the environment (Fig. 13.10). In the same vein, urbanization and improper waste disposal are leading to the disappearance of many mangrove forest along the coastal areas (Fig. 13.11).

Channel reduction changes stream chemistry and size, and lead to the formation of "mangrove islands", which are a preparatory stage for the complete elimination of

Fig. 13.10 Mangrove swamp dredged to lay pipelines that connect an oil tank at the far end of the picture. Years of disturbance of this coastal location at Eagle Island, Niger Delta, Nigeria through sand mining, dredging and improper waste disposal had exposed the mangrove forest to invasion by nypa palms and other alien species as seen on the left side of the picture. (Credit: A.O. Numbere)

small mangrove communities (Fig. 13.12). Stream reduction also constricts space for the free movement of species, and triggers competition amongst aquatic organisms. For instance, some parts of Eagle Island in the Niger Delta, previously had a combination of red (*R. racemosa*) and white (*A. germinans*) mangroves, but now have only white mangroves because of changes in hydro-chemistry as a result of sand dredging and sand filling that took place in the area (Fig. 13.12a). It was observed that the white mangroves performed better in sandy and disturbed soils than the red mangroves. Red mangroves (Rhizophora) are the most dominant coastal species in core undisturbed mangrove soils. Therefore, during dredging and contamination of the coast they become the first victims of long term soil degradation.

The second location is Eagle Island (Fig. 13.12b), which has most of its mangrove stands destroyed. This is because of the toxic effect of sawdust that was dumped into the river by a wood mill industry located nearby. But this physico-chemical change does not affect the palms, which continued to grow as shown in the background in Fig. 13.12b. The palms generally thrive in polluted soil, thus they are not affected by the organic waste materials that are dumped into the river.

Fig. 13.11 A patch of mangrove stand (*Rhizophora mangle*) is boxed into a corner between some urban settlements and invasive species. Already the mangrove stand on the left hand side had withered. Between both stands is a heap of refuse dumped by inhabitants of the adjoining apartments, which had changed the color of the river to green due to algal infestation. The proliferation of urban settlements around the coast had led to an increase in waste generation, which had contributed to the invasion of alien species. (Credit: A.O. Numbere)

13.2.4 Root Micro Structure

Nutrient absorption is carried out by the roots, and determines the rate at which the plant adapt to their environment. Microscopic study of the roots of red mangrove and nypa palm using electron microscope in the laboratory of the Department of Animal and Environmental Biology, University of Port Harcourt, Nigeria, indicates that the roots of palms are structured to carry out better nutrient absorption than roots of mangroves. The root of the palm is light in texture, hollow and straw-like while the root of red mangrove is thick and tightly packed (Fig. 13.13). The root of nypa palm is completely embedded in the soil and grows deep down the soil profile while mangrove roots are mostly adventitious with many parts growing outside the soil in other to carry out aerobic respiration. The palm roots go deep down the soil profile to tap unused nutrients while some roots of mangroves grow at the surface layer. In addition, mangrove roots are woody and less permeable as compared to the roots of nypa palm that are tender and highly permeable.

Fig. 13.12 Changes in coastal hydrology impact (**a**) white mangroves (*Avicennia germinans*) through channel reduction via sand filling and (**b**) red mangroves (*Rhizophora racemosa*) through changes in river physico-chemistry as a result of pollution. Both factors affect the distribution of mangroves leading to the formation of "mangrove islands" and the intrusion of nypa palms as seen in the background in (**b**). Figure 13.12 (**a**) has white mangroves (*Avicennia germinans*) and no red mangroves around because of changes in salinity, while in Fig. 13.12 (**b**) the red mangroves (*Rhizophora racemosa*) are gradually being replaced by nypa palms. Similarly, (**b**) shows one of the few last stands of red mangrove (*R. racemosa*) trees being surrounded by nypa palm in this location in Eagle Island, Niger Delta, Nigeria. (Credit: A.O. Numbere)

Again, in another study mangrove roots were found to have more un-transported nutrient content as compared to nypa palm roots when 20 mangrove and 20 nypa palm root samples were dissected and analyzed for nutrient content in the laboratory. The result indicate that average total nutrient content was more in mangrove roots

Fig. 13.13 Root structure of (**a**) nypa palm and (**b**) mangrove. Nypa palm roots are light and fully embedded in soil while mangrove roots are thick and adventitious. (Credit: A.O. Numbere)

(123.9 ± 121.1 mg/kg) than in nypa palm roots (44.9 ± 40.7 mg/kg). The nutrient content is a combination of Nitrogen, Phosphorous, Potassium and Nitrate. Furthermore, the result of this study indicates that there was no significant difference ($P = 0.55$) in nutrient content between nypa palm and mangrove root. Nonetheless, the low amount of nutrient content in the roots of nypa palm could mean that there is a faster transmission of nutrients from the root to other parts of the plant as compared to the mangrove root. However, out of all the nutrient elements analyzed potassium (K) was the most dominant, and had higher concentration in mangrove root than in nypa palm root.

Fig. 13.14 Nypa palm seeds occur in groups of 20–30 seeds and are tough and have high buoyancy rate, which enables them to float to far distances around coastal areas of the Niger Delta, Nigeria

13.2.5 Seed Buoyancy

The exocarp of nypa palm seed acts like a floater (Fig. 13.14), and enables the seeds to remain afloat for long in the aquatic environment. The buoyancy of the seed makes it to travel thousands of kilometers in and around the coastal regions of the Niger Delta. As a result of this situation, one would hardly find a location around the riverine areas of the Niger Delta without the presence of nypa palm seeds. Another advantage of the seed structure is that it has tough outer covering, which prevents it from being soaked by water or permeated by pollutants such as crude oil. The toughness of the seed also prevents it from being consumed by crabs or other organisms around the mangrove forest, unlike the mangrove propagule that is soft, palatable and edible.

13.2.6 Landscape Changes

Human activities lead to the changes in landscape architecture in coastal areas, which facilitate the invasion of foreign species (Wang et al. 2016). The use of the coastal area as a waste disposal site had led to the complete displacement of mangroves in some locations such as the Eastern Bye Pass section of a prominent Creek in the Niger Delta called the "Ntawogba creek" (Fig. 13.15a). This location was formerly an exclusive mangrove forest. Many other places that were once occupied by mangroves are now completely invaded by the palms in mixed or pure forest stands. As mangroves disappear, the ecosystem services they render also disappear along with them such as fire wood, coastal protection, flood control,

Fig. 13.15 (**a**) A time-line of a former mangrove forest that was completely invaded by nypa palms in 2014; (**b**) The same location after it was sand filled in 2017 (i.e. Eastern-Bye Pass Creek, Niger Delta). Anthropogenic activity such as pollution, improper waste disposal, deforestation etc. contributed to the disappearance of the mangroves from this once vibrant mangrove forest. Since the palms had no economic value it was bulldozed by the local authority and sand filled. It is now a proposed site for constructing industrial and residential quarters. (Credit: A.O. Numbere)

shoreline protection etc. (Polidoro et al. 2010). The palms grow and block drainages (Fig. 13.15a), which are then cleared by dredging causing harm and disfigurement of the landscape. The dredging equipment used in clearing the forest further stresses the coastal environment by destroying benthic organisms. In some cases complete sand filling is done to reclaim the areas lost to nypa palm (Fig. 13.15). Observations had shown that the reclaimed locations are never returned to their original form, but are rather used as platform to construct developmental projects, such as housing complexes, industries, filling stations, offices, apartments, etc.

Invasive nypa palm obstructs waterways and affects navigational activities (Fig. 13.16). The palms constrict the width of the river channels as they grow and expand (Fig. 13.16), which cause maritime accidents of sea crafts. Constriction of the river channel affects the free movement and population structure of aquatic organisms. Canalization and sand dredging lead to sedimentation. This occurs when stands of palms act as barriers to water flow and filter to waste materials carried by river into coastal areas (Fig. 13.17). Accumulation of sand from other locations, especially from dredged soils changes the coastal area to sandy soil, and prevents the growth of mangroves and other aquatic organisms.

Fig. 13.16 A former mangrove forest, over ran by palms in Asarama community in the Niger Delta Nigeria. The pattern of growth of the palm is to stifle other plants around it and to block the water channels thus, preventing the free flow of oxygenated water. Non-flowing water around the forest traps waste, which decomposes to degrade the water quality. This increases the effect of pollutants, which are inimical to the health of the mangroves and the coasts in general. (Credit: A.O. Numbere)

Fig. 13.17 Accumulation of waste materials (log, dirt, branches etc) trapped at the base of nypa palm (*Nypa fruticans*) prevents the free flow of materials in the creek at Okrika, Niger Delta Nigeria. (Credit: A.O. Numbere)

13.3 Transformation of Coastal Areas from an Aquatic to a Terrestrial System

Mangroves were the dominant species along the coastal areas of the Niger Delta prior to the 1990's, but because of anthropogenic activities of oil and gas exploration, urbanization, deforestation and poor waste management, they had been displaced by nypa palms in many locations. The pattern of growth of the palms had made them to obstruct, and change the architecture of the stream channel leading to loss of several coastal species. Local authorities carry out sand filling and dredging of nypa palm invaded areas in other to reclaim the lost mangrove forest as shown in Fig. 13.18a, b. Similarly, private agencies mine sand from coastal areas, which they sell to construction companies for the purpose of building. In addition, the sand-filled locations are set aside for building projects. This situation can be described as a horizontal irreversible coastal change (Fig. 13.18a). This is because the area cannot be reverted to its original form after it has been converted from an aquatic to a terrestrial system. This process thus leads to the total elimination of all aquatic organisms in that coastal system including pelagic and benthic organisms, which is not good for the sustainability of the environment. This is because the coast helps to stabilize the environment by removing atmospheric carbon dioxide, and reducing the impact of global warming. The coast is also a haven for biodiversity such as aquatic and amphibious organisms. Therefore, the loss of the coast is dangerous for the whole environment. However, in this circumstance the best management strategy to adopt is a triangular reversible transformation

Fig. 13.18 Transformation of mangrove forest to (**a**) sand fill and (**b**) pure mangrove forest. Conversion of mangrove forest to sand fill area is a total loss to the coast. This pattern of change (Fig. 13.18) is often observed in the Niger Delta region whenever nypa palm forest is removed from a location. There is often no restorative effort aimed at bringing back the original mangrove forest, rather the area is reclaimed for other developmental projects or urbanization activities. But the best restorative effort is to bring back the area to its original state (Fig. 13.18b). This is to ensure that the mangrove forest does not slip into extinction in some years to come. (Credit: A.O. Numbere)

(Fig. 13.18b). This can be done after the area had been successfully colonized by the nypa palms in mixed or pure forest. To restore the site, the palms should be removed, and in their place mangrove seedlings planted. This can be done when the area is passing through its cycle of change from mangrove to mixed forest or from mixed forest to pure nypa palm forest. Physical, mechanical and chemical means can be used to eliminate the palms. Mangrove soil should then be brought in, and spread all over the area, after which young mangrove seeds are planted to ensure a rejuvenated growth of the mangroves.

13.4 Conditions That Favor Nypa Palm Invasion of Mangrove Forest in the Niger Delta

The entry of nypa palm into the Niger Delta was made possible by an invasion pathway (Carlton and Ruiz 2005) established by the presence of the Atlantic Ocean (Richardson et al. 2000), which facilitated travel from Asia to Nigeria, and made the intentional transfer of the seeds successful (Keay et al. 1964). On arrival to the new habitat the palms got adapted and became physiologically tolerant to the new environment. The life history of the palms matched the new environment. The palms also benefitted from untapped resources in the mangrove forest. These untapped resources are empty niches created by massive deforestation for oil and gas exploration and urbanization. These conditions thus created a disturbed environment, which encouraged a successful invasion, colonization and expansion. Incessant mangrove tree cuttings for fire wood also reduced the native mangrove

Fig. 13.19 A classic example of propagule pressure of nypa palm seeds in a mangrove forest. The palms are waiting to invade and displace the mangroves in a sand filled location in Asarama town, Niger Delta, Nigeria. On the left side of the picture are seeds of the white mangroves (*A. germinans*) hanging on a tree. It is not known if the mangrove seeds can withstand the difficult sandy environment to grow and populate the area, since the muddy mangrove soil on which they thrive is no longer available. (Credit: A.O. Numbere)

species diversity, and led to a geographical isolation that are characteristics of an invaded community. This is the reason why the palms are having better growth in mangrove soil than in nypa palm soil. Field observations also show that the palms produced a lot of seeds as they become adapted to the new environment. In a seed enumeration study in a mixed forest, it was observed that the ratio of the palm seeds to mangrove seeds in a 20 m x 20 m plot was 27:1. This show that the palm seeds in most locations visited out-numbered the mangrove seeds (Fig. 13.19). If this trend continues it won't take too long for the palms to completely overwhelm and colonize the entire region (Wang et al. 2016).

Additionally, continuous anthropogenic disturbance is a major cause of the spread of nypa palms to several other locations after the initial introduction (Kowarik 2003). One of such actions of humans is improper waste disposal within the mangrove forest. This condition makes the environment to be conducive for the palms to grow and proliferate around the coast at the detriment of the mangroves (Fig. 13.20).

Another consequence of urbanization, apart from the physical deforestation of the forest, is the bringing of people closer to the coast so that they will appreciate nature

Fig. 13.20 Waste dump sites in (**a**) nypa palm forest and (**b**) mangrove forest in the Niger Delta, Nigeria. This is one of the consequences of urbanization when people live too close to the coast. The palms grew bigger while the mangroves die gradually with the introduction of waste to the coastal areas. Both areas were exclusively mangrove forest 30 years ago, but continuous dumping of waste converted these areas into a nypa palm (*Nypa fruticans*) forest. (Credit: A.O. Numbere)

better, which is not a bad idea. But the problem is that when people live so close to the coast, and don't have good waste management habit, there is a tendency for them to convert the river to a waste dump site. The coast becomes a victim of increased waste load. In some locations the coast serve as sites for constructing pier latrines especially in the hinter lands where people don't feel the presence of government, in terms of provision of social amenities. The wastes generated when dumped in the river are distributed by tidal currents to other locations. Organic waste when dumped into the river changes the stream chemistry, which leads to eutrophication and increase in growth of foreign species such as water hyacinth, water lilly, macrophytes etc. In Buguma over nine different coastal and non-coastal plant species were found within the mangrove forest. This invasion is as a result of the dredging activity that took place several years ago, which converted the swampy soil to sandy soil (Fig. 13.21).

Fig. 13.21 Different kinds of mangrove and non-mangrove species found in sand filled mangrove forest in Buguma, Niger Delta, Nigeria. (**a**) Nypa palm (*Nypa fruticans*), (**b**) Black mangroves (*Laguncularia racemosa*), (**c**) Conocarpus erectus (**d**) Mangrove associate, (**e**) Red mangrove (*Rhizophora racemosa*) (**f**) Mangrove fern, (**g**) White mangrove (*Avicennia germina*ns), (**h**) *Heriteria littoralis* (**i**) *Acrostichum aureum* spp. (Credit: A.O. Numbere)

13.4.1 Interaction Effect

The combined action of urbanization, improper waste management and pollution (Fig. 13.22) contributes to the proliferation of invasive nypa palms and other alien species in the Niger Delta, Nigeria. These actions are detrimental to the

Fig. 13.22 The interaction effect of (**a**) urbanization and (**b**) hydrocarbon pollution contribute to the proliferation of nypa palms and other alien species in coastal regions of the Niger Delta, Nigeria. Figure 13.22 (**a**) shows houses built after the mangrove forest was cut, dredged and used as site for building residential quarters, Still at the foregound of the picture the remaining mangrove stands had already been invaded by palms and other aliens species while in Fig. 13.22 (**b**) the dark and scorched area is caused by fire from spilled crude oil and on the left of the picture is a nypa palm stand that had already gained a foot hold in the mangrove forest

coastal system because they reduce the aesthetic value and make them vulnerable to total destruction through dredging.

13.4.2 *Impact of Invasive Nypa Palms on Coastal Ecology of the Niger Delta*

The branchless nature of nypa palms excludes a host of coastal organisms that would have benefitted from their presence. For instance, birds can't perch on and build their nest on the palms, which excludes that aspect of the food chain. Similarly, termitarium of termites and nest of ants cannot be found on the palms, since they have no branches and no stem on which to build. Their deep underground roots host little or no shell organisms such as corals, barnacles and mullets. Nypa palms also have low litter fall yearly (Numbere and Camilo 2016), which exclude decomposition activity as experienced in mangrove forest (Numbere and Camilo 2016). In the case where there are a drops of leaves due to wind action, the leaves decompose slowly because of their fibrous nature as compared to mangrove leave litter that is leathery and succulent. One important coastal resource derived from mangroves, apart from fish, is firewood. The palms can't provide it because they lack the stems. The growth of the palms without stem also affects the free movement and distribution of species within the forest. The hard outer covering of the seed (exocarp) makes it difficult for most organisms to feed on it as compared to the tender succulent propagules of mangroves which are consumed by the West African red mangrove crabs (*Goniopsispelii*), and other organism living along the coast.

It was observed from several field trips in this region that the presence of palms in the coastal environment automatically changes the soil chemistry and physical appearance of the soil by converting it to dirty mud. This kind of soil prevents the growth of mangroves and excludes numerous aquatic organisms that usually thrive in mangrove forest swamps. The ability of the palms to trap and accumulate debris within the forest also poses major environmental and ecological problems. This is because the dirty environment makes it less conducive for the breeding and spawning activities of many aquatic organisms especially the fishes.

13.5 Management Strategies

Invasion Control This involves the physical control, which is the destruction of unwanted plants by manual (i.e. digging and pulling) or mechanical (i.e. swamp buggies) methods (Figs. 13.23 and 13.24) or chemical control, which is the use of chemicals such as pesticides and herbicides to destroy alien plants. Biological control involves the enhancement of the genetic quality of native mangrove species through selective production in nurseries to promote robust growth. In addition, the

Fig. 13.23 This area is a creek that was exclusively occupied by mangrove forest some years ago, but have been invaded by the invasive nypa palm (*Nypa fruticans*), which blocked the creek. The palms were destroyed with a swamp buggy to free up the adjoining clogged up canal. (Credit: A.O. Numbere)

Fig. 13.24 This location is the same as Fig. 13.23. It shows a nypa palm resurgence 2 years after the destruction of the forest by swamp buggy. This indicates that the palm stump can proliferate fast if not completely uprooted and destroyed outside the swamp

restored forest is to be filled with core mangrove soils to prevent the negative physicochemical (i.e. allelopathic) effect of nypa palm soil left behind.

Species Based Control This involves experimental research aimed at determining the most effective strategy to embark on based on the particular location. For the nypa palms, studies had shown that, they gain foothold in their new environment by changing the soil chemistry, at the detriment of the host species. Therefore, to eliminate this threat, it is important to remove the nypa palm soils along with the plant completely and to re-introduce mangrove soils to facilitate rapid growth and re-colonization of the area.

Invasion Prevention This involves the identification and regulation of the invasion pathway. For instance dredged spoils, (i.e. waste from dredging activity) that are dumped on fringing mangrove forest can be isolated (Ohimain 2004) and decontaminated of foreign species through heating, chlorination and irradiation etc., to prevent contamination of the host species by invasive nypa palm. This is because "dredged spoil" affects surface topography and hydrology of the coasts (UNEP 2011). Transport trucks and dredging equipments apart from disfiguring the coasts, also carry foreign organisms in soils lodged in their wheels and parts from different locations. Invasion of palms is through dispersal by tidal currents; therefore restored mangrove sites should be fenced off with wire or net gauze to prevent the infiltration of the palm seeds into the newly re-introduced mangrove restoration site.

13.6 Conclusion and Recommendations

Studies had shown that the nypa palms in the Niger Delta had fulfilled the conditions necessary for a successful invasion and colonization. The war therefore, should be taken to them through the adoption of physical, mechanical, chemical and biological method of control. Nypa palm should be removed from all mangrove forests, creeks and coastal areas physically or mechanically using swamp bulldozers. Then their seeds and seedlings should be hand-picked or completely up-rooted and destroyed to prevent resurgence. In 1–3 months time the area should be re-visited and the remaining hidden seedlings that had grown out of the soil pulled out to prevent the re-growth of the palms. The nypa palm soil should be excavated and replaced with mangrove soil before the planting of mangrove seedlings. This is because studies had shown that mangroves don't thrive well on nypa palm soils. The removal of the nypa palm soil will remove the window of opportunity for the re-entry of the palms.

Quick action is therefore, necessary because the palms had already colonized almost a quarter (25%) of the mangrove forest and are in the verge of completely taking over the remaining coastal areas left in the Niger Delta (Fig. 13.25). The palms are advancing rapidly, and if nothing is done, in the next 50 years they might overwhelm and take over the entire coastal areas. But with the timely destruction of the palms and the restoration of mangroves forest, this situation can be turned around for the better.

Fig. 13.25 Nypa palm (*Nypa fruticans*) stands at Asarama River, Niger Delta, Nigeria, overlooking the sea and waiting patiently for an opportunity to take complete control of the mangroves at the adjoining coastal areas. (Credit: A.O. Numbere)

References

Carlton JT, Ruiz GM (2005) Vector science and integrated vector management in bioinvasion ecology: conceptual frameworks. In: Mooney HA, Mack RN, McNeely JA, Neville LE, Schei PJ, Waage JK (eds) Invasive alien species. Island Press, Washington, DC, pp 36–58

CEDA (1997) Coastal profile of Nigeria. Federal Environmental Protection Agency, Abuja

Crosby AW (1986) Ecological imperialism: the biological expansion of Europe, 900–1900. Cambridge University Press, Cambridge

Davis MA (2009) Invasion biology. Oxford University Press, Oxford

Elton CS (1958) The ecology of invasions by animals and plants. Methuen, London

Hodkinson DJ, Thompson K (1997) Plant dispersal: the role of man. J Appl Ecol 34:1484–1496

Jones DL (1995) Palms. Smithsonian Institution Press, Washington, DC

Jones CG, Lawton JH, Shachak M (1997) Positive and negative effects of organisms as physical ecosystem engineers. Ecology 78:1954–1957

Keane RM, Crawley MJ (2002) Exotic plant invasions and the enemy release hypothesis. Trends Ecol Evol 17(4):164–170

Keay RWJ, Onochie CFA, Standfield DP (1964) Nigerian trees. Federal Department of Forestry Research. National Press Limited, Ibadan

Kowarik I (2003) Human agency in biological invasions: secondary releases foster naturalisation and population expansion of alien plant species. Biol Invasions 5:293–312

Kuenzer C, Beijma S, Gessner U, Dech S (2014) Land surface dynamics and environmental challenges of the Niger Delta, Africa: remote sensing-based analyses spanning three decades (1986–2013). Appl Geogr 53:354–368

Lugo AE (1980) Mangrove ecosystems: successional or steady state? Biotropica 12:65–72

Nagelkerken ISJM, Blaber SJM, Bouillon S, Green P, Haywood M, Kirton LG et al (2008) The habitat function of mangroves for terrestrial and marine fauna: a review. Aquat Bot 89 (2):155–185

Numbere AO, Camilo GR (2016) Reciprocal transplant of mangrove (*Rhizophora racemosa*) and Nypa palm (*Nypa fruticans*) seedlings in soils with different levels of pollution in the Niger River Delta, Nigeria. Global J Environ Res 10(1):14–21

Numbere AO, Camilo GR (2017) Mangrove leaf litter decomposition under mangrove forest stands with different levels of pollution in the Niger River Delta, Nigeria. Afr J Ecol 55:162–167

Ohimain EI (2004) Environmental impacts of dredging in the Niger Delta. Options for sediment relocation that will mitigate acidification and enhance natural mangrove restoration. Terra et Aqua 97:9–19

Ohimain EI, Gbolagade J, Abah SO (2008) Variations in heavy metal concentrations following the dredging of an oil well access canal in the Niger Delta. Adv Biol Res 2(5–6):97–103

Polidoro BA, Carpenter KE, Collins L, Duke NC, Ellison EM et al (2010) The loss of species: mangrove extinction risk and geographic areas of global concern. Plos ONE 5:e10095–e10095. https://doi.org/10.1371/journal.pone.0010095

Richardson DM, Pyšek P, Rejmánek M, Barbour MG, Panetta FD, West CJ (2000) Naturalization and invasion of alien plants: concepts and definitions. Divers Distrib 6:93–107

Stohlgren TJ, Binkley D, Chong GW, Kalkhan MA, Schell LD, Bull KA et al (1999) Exotic plant species invade hot spots of native plant diversity. Ecol Monogr 69(1):25–46

UNEP (2011) Environmental assessment of Ogoniland. United Nations Environmental Programme. Available online at www.unep.org/nigeria. Accessed 26 Sept 2016

Wang P, Numbere AO, Camilo GR (2016) Long term changes in mangrove landscape of the Niger River Delta, Nigeria. Am J Environ Sci 12:248–259

Wootton JT (1994) The nature and consequences of indirect effects in ecological communities. Annu Rev Ecol Syst 25(1):443–466

Chapter 14
Acacia spp.: Invasive Trees Along the Brunei Coast, Borneo

Shafi Noor Islam, Siti Mazidah Bin Haji Mohamad, and Abul Kalam Azad

Abstract This chapter reports the results of a study of *Acacia Mangium* and *Acacia Auricaliformis*, which are exotic plantation tree species from Australia that have invaded, and spread within, natural habitats in Brunei Darussalam. There are four types of Acacia trees that are spreading in Brunei, mostly within the coastal areas, the deep forest areas as well as in the urban areas in the country's capital, Bandar Seri Begawan. This study looks into the presently occupied land areas, vegetation cover, land use and landscape changing patterns in the coastal forest areas in Brunei. By clarifying patterns of variation in demographic parameters and hence population growth it is possible to form a connection between qualitative field data, theoretical ideas about invasiveness and rate of spread. The results show that the present growth and expansion rate of the Acacia plant species is alarming for the primary forest as well as for the forest ecology and ecosystems in the coastal forest areas in Brunei. This study makes use of both primary and secondary data sources. The objective of this chapter is to understand the changing patterns of forest vegetation, such as primary forest to secondary forest, and the forest ecosystem in coastal Brunei. Based on this study's findings some applied recommendations have been proposed for the better distribution of the Acacia plant species and maintenance of rainforest vegetation in Brunei.

Keywords Brunei coastal area · Acacia species · Migration · Distribution · Secondary forest · Invasive species · Distribution · Forest Ecology and Ecosystem, Management.

S. N. Islam (✉) · S. M. B. H. Mohamad
Geography and Environmental Studies, Faculty of Arts and Social Sciences (FASS), Universiti Brunei Darussalam, Gadong, Brunei Darussalam
e-mail: shafi.islam@ubd.edu.bn

A. K. Azad
Chemical and Process Engineering, Faculty of Integrated Technologies (FIT), Universiti Brunei Darussalam (UBD), Gadong, Brunei Darussalam

C. Makowski, C. W. Finkl (eds.), *Impacts of Invasive Species on Coastal Environments*, Coastal Research Library 29,
https://doi.org/10.1007/978-3-319-91382-7_14

14.1 Introduction

Brunei Darussalam is a small country located in the north of Borneo, which is an island whose western coastline meets the South China Sea and eastern coastline the Sulu and Celebes Seas (Colonial North Borneo 1953; Brown 1970). Borneo's coastline is formed mainly of alluvial flats, with many creeks and swamps. Its hills are traversed by valleys and occasional plains. Hills and valleys in most cases are covered with dense forest, and there are many rivers (Colonial North Borneo 1953). The main harbor on the west coast is the island of Labuan, which lies to the north of Brunei Bay. Further north, Jesselton, the capital of the Colony, has a well-sheltered harbor for vessels of moderate size, which take away the bulk of the rubber produced on the west coast (Peter 1998; Tyler 2004). At the most northern point of the Colony is Marudu Bay, a former strong-hold of Illanun pirates on its western shore, and eleven miles from the entrance is Kudat Harbor (Colonial North Borneo 1953).

Brunei has one of the largest tropical rainforest land areas in the world. Over 65% of the land is covered with tropical rainforest, i.e. land with green vegetation, which become part of the hotspots of tropical biodiversity and green spaces in the Southeast Asian region. There are central mountain ranges, from four to six thousand feet in height, rising somewhat sharply from the low-ranging hills nearer to the coast. Brunei has a flat coastal plain along the South China Sea in the western part of the country with some hills further inland (Tyler 2004; Brunei Museum 2005). Mountains rise in the eastern segment of Brunei, which is also the most undeveloped and inaccessible part of the sultanate. An equatorial climate gives the area abundant rainfall, and most of the country remains heavily forested (Figs. 14.1, 14.2, and 14.3), with mangrove swamps within the coastal region.

Fig. 14.1 Topographic view within Champion Deltaic Morphology of Kampong Katimahar, Tutong District, in Brunei Darussalam. (Source: Picture collected by Author 2017)

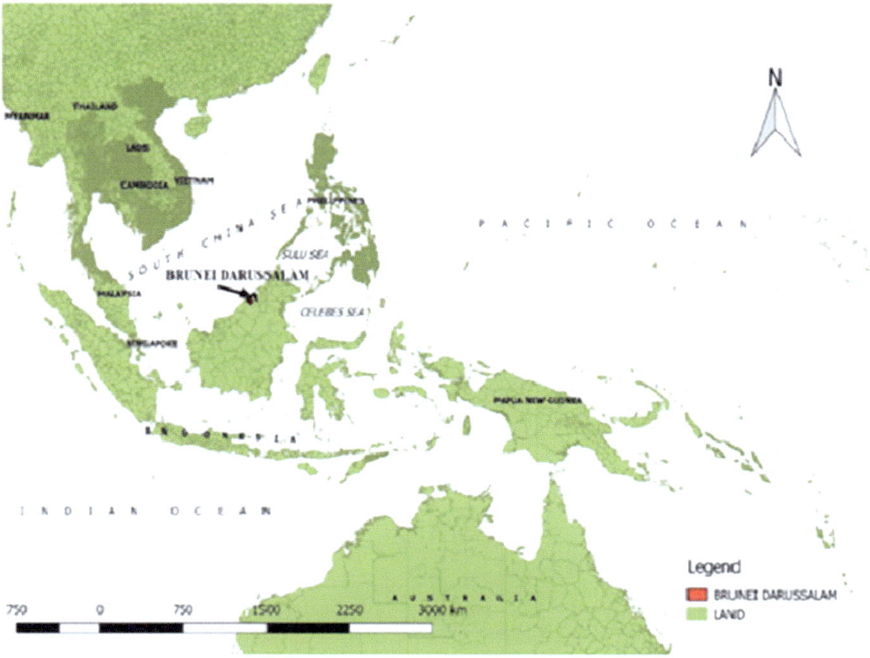

Fig. 14.2 The Geographical location of Brunei Darussalam and Borneo Island in South East Asian region

Brunei consists of four districts: (1) the Belait District in the southwest, which is the largest, (2) the Tutong District in the middle, (3) the Brunei-Muara District that surrounds the capital, Bandar Seri Begawan, and (4) the separate Temburong District in the east (Figs. 14.1, 14.2, and 14.3) (CDD 1993, 2008; Ibrahim 2005; Khatib and Sirat 2005). The population of the country is 423, 200 (as of 2017) and most of the people live in the capital, which is also where its chief port of Muara is located, 25 mi (41 km) to the northeast. Bandar Seri Begawan consists of great contrasts, such as that between the Sultan's palace, the glittering Sultan Omar Ali Saifuddien Mosque and the world's largest stilt village, Kampong Ayer. This village, in existence for the past 400 years and provides housing for about 30,000 inhabitants, was declared a national monument in 1987 and is the most popular tourist attraction in the country (Brown 1970:132; Hassan 1988; CDD 1993, 2008; Ibrahim 2005).

The official name of the country is Negara Brunei Darussalam, whereby "Negara Brunei" means the state of Brunei and "Darussalam" means "abode of peace" in Arabic (Colonial North Borneo 1953; CDD 1993). Its full name, Brunei Darussalam, is a compound of a Sanskrit name (possibly meaning "sea people" or is derived from the name of a local tree that also gave its name to the entire island of Borneo) and was added by Muslim sultans in the fifteenth century (Hassan 1988; CDD 1993, 2008; BBYB 2015). Brunei's 600-year-old monarchy and centuries-spanning history and cultural heritage have been steadfastly upheld to this day (Mahmud 2017).

Fig. 14.3 The Geographical Location and Characteristics of Brunei Darussalam. (Map Source: http://www.maps-of-the-world.net/maps/maps-of-asia/maps-of-brunei/detailed-tourist-map-of-brunei.jpg)

The total land use pattern shows that only 2.08% land are being used as arable land and 0.87% land are being used for permanent crop cultivation, whereas the remaining 97.05% land are still forest land or are being used for other purposes (2005). The topography of Brunei consists of hilly lowlands in the western region, rugged mountains in the eastern region and swampy tidal plains in the coastal area (Figs. 14.1 and 14.2) (Hassan 1988). Brunei has an equatorial type of climate characterized by high temperatures, high humidity and heavy rainfall (CDD 1993, 2008) (Fig. 14.3) (Hassan 1988). The natural resources are mainly petroleum, natural gases and timber, which are recognized as high potential resources in Brunei.

Mountains of moderately to severely deformed Eocene to Miocene sediments occupy most of the southern part of the Temburong District (Fig. 14.2), including the Forest Reserve (Hutan Simpanan) of Batu Apoi (Earl of Cranbrook and Edwards 1994). The summit levels range from 700–900 m, but reach more than 1000 m at Bukit Tudal (1181 m) and Bukit Retak (1618 m), with the highest elevation in Brunei being Bukit Pagon near the Sarawak border (1850 m) (Earl of Cranbrook and Edwards 1994). Most of the coastal forest region which was totally occupied by rainforest is now degrading in forest density and native plant species are also shrinking. This scenario is observed all over the Brunei coastal forest region.

14.2 Aim and Objectives

The Acacia trees were initially planted at the farms' periphery for hedge/shade purposes, as a source of nitrogen supplement and for coastal erosion control. The trees have now encroached on these farms. The aim of this study is to:

(i) Find the use and size structure population, distribution pattern and rates of growth of the Acacia species in the coastal areas in Brunei.
(ii) Investigate the distributional classification of Acacia plants species in Brunei.
(iii) Highlight the Acacia plant species and its ecological impact in Brunei.
(iv) Make recommendations for better management of tropical rainforest vegetation in Brunei, particularly its protection from invasive plant species.

14.3 Geographical Location and Physiographical Characteristics of Brunei Darussalam

Brunei is located between 4° N and 5.8° N latitudes and, 114.6° E and 115.4° E longitudes on the northwest coast of Borneo (maps shown in Figs. 14.2 and 14.3). The area of the country is 2055 square mi (5770 square km) (land: 5270 sq. km; water: 500 sq. km) (Hassan 1988; CDD 1993, 2008; Ibrahim 2005; Khatib and Sirat 2005; Omar 2005). The country shares a 266 km (165 mi) border along the southern and eastern sides with the Malaysian state of Sarawak, which has an enclave along the Limbang River that splits Brunei into two separate parts (Figs. 14.2 and 14.3) (Brown 1970:132; Hassan 1988; CDD 1993, 2008; Ibrahim 2005; Khatib and Sirat 2005). The country's 161-km (100 mi) coastline faces the South China Sea from near the mouth of the Baram River in the southwest to the headland of Muara in the northeast (Fig. 14.3).

To the east of Muara is Brunei Bay, a large and protected shallow embayment. Brunei's tropical climate is seasonal with high temperatures, rainfall is year-round (Turnbull 2017) and climate appears to be highly favorable for the growth of Acacia species (Peter 1998). Figs. 14.2 and 14.3 show the geographical and geomorphological overview of Brunei. Figure 14.3 also displays the four districts' territorial boundaries and geographical features and landmarks. There are five major rivers that dominate the floodplain morphology in Brunei, namely Brunei River, Belait River, Temburong River, Tutong River, and Pandaruan River. The Brunei River is situated within the three major deltaic regions, which are Baram River Delta, Champion River Delta and Meligan River Delta (UNESCO 1972, 1999, 2004, 2008).

The Baram River Delta has covered most of the southwest corner of the floodplain in the Baram River catchment area and the Champion River Delta has covered the middle part as well as Tutong, Brunei-Muara and the northern part of the Belait District. The Temburong River has covered, and connects most of the Temburong District area (Fig. 14.3) (Hassan 1988; CDD 1993, 2008).

The country as a whole appears as a flat coastal plain that rises to mountainous regions in the east and hilly lowlands in the west (Figs. 14.1, 14.2, and 14.3) (Hassan 1988). As for the elevation of the country, the lowest point is at the South China Sea level, which sits at zero (0) m and the highest point of the country is Bukit Pagon at 1850 m.

14.4 Importance of and Threats to Forest Resources in Brunei Darussalam

Like many other ecosystems, mangroves provide goods and services for the people and so are an important source especially to the communities living in the coastal areas. Participants from the studied villages understand the importance of mangrove functions and the benefits to their livelihoods, hence, mangrove benefits are listed based on common topics that the participants discussed during the interviews (importance of mangrove benefits; relaxation and recreation; tourist attraction; water filtration and coastal protection; shrimps; crabs; shellfish; fish; honey; hunting; medicinal purposes; leaves; woods and timber; and fruits). Some benefits, however, are not included, such as spiritual value, due to its vagueness and consequent difficulty in rating its importance as found during the pilot study. Nevertheless, it will not be overlooked in this study as it will be discussed in the following subtopic (Figs. 14.2 and 14.3). The data also suggests that the community is aware of the ecological function of mangroves and the cultural services that they provide, even though the categories are limited and specific compared to more scientific categories. Regardless, it shows that the people have a significant understanding of the importance of the mangrove ecosystem and the benefits it provides (Islam 2014; Siti NorAqilah 2017).

It is clear that people are less dependent on mangrove forest resources compared to fishery resources. This could be the reason people collect mangrove forest resources less frequently than they do fishery resources. Development and modernization could also be important factors causing the lower dependency on forest resources due to the increasing number of different construction material, adequate healthcare in Brunei, better access to different sources of nutrition as well as changing preferences (Siti NorAqilah 2017). Additionally, activities such as hunting and honey harvesting are less significant since the fauna in Brunei is limited compared to that in other mangrove forests such as Sundarban Mangrove in India and Bangladesh. Also, honey harvesting is less-practiced due to the value of the mangrove forest (Siti NorAqilah 2017). People believe that honey harvesting is damaging to the forest and leads to the degradation of its ecosystem. It was also found that the residents understand the ecological functions of the mangrove as a water filtration system, as coastal protection, as well as its aesthetic value for recreation and tourism (Siti NorAqilah 2017). This could also be a contributing factor for showing certain appreciation for the mangrove forest or rainforest as they have an understanding of, and relationship with, the forest. The native plant species in Brunei play significant roles in maintaining the balance in the local ecology as well as supporting the community's livelihoods, many of which depend on the native medicinal plant species (Peter 1998; The Brunei museums 2005). However, these species are now under threat as coastal forests in Brunei are degrading due to the spread of the invasive Acacia plant species.

14.5 Climate, Weather and River System in Brunei Darussalam

The climate in Brunei is tropical equatorial and humid subtropical at higher altitudes with heavy rainfall. Brunei has two seasons (Yogerst 1994; Islam 2014), whereby the dry season is extremely hot (24–36 °C or 75.2–96.8 °F) and the rainy season is generally warm and wet (20–28 °C or 68.0–82.4 °F). The following shows the regional climate characteristics of the four districts in Brunei (Brown 1970:132; Hassan 1988; CDD 1993, 2008; Ibrahim 2005):

- The Brunei-Muara District, where Bandar Seri Begawan is located, is humid tropical on the coast and in the lower-altitude north, whereas the central area of the district is humid subtropical (20–36 °C or 68–97 °F).
- The Tutong District is tropical – hot in the north and warm in the south. (22–32 °C or 71.6–89.6 °F).
- The Belait District is tropical – hot in the north and slightly warm in the south. (25–37 °C or 77.0–98.6 °F).
- The Temburong District is humid subtropical on the higher-altitude south and humid tropical on the coastal and lower-altitude north (18–29 °C or 64–84 °F).

14.5.1 Climate in Brunei Darussalam

Climatic variations in Brunei follow the influence of the monsoon winds. The northeast monsoon blows from December to March, while the southeast monsoon occurs around June to October (BBY 2014). The total average annual precipitation is an estimated 2722 mm. There are two rainy seasons: from September to January and from May to July (Omar 1981). The temperature is relatively uniform throughout the year, with an annual average of 27.9 °C, ranging from 23.8 to 32.1 °C. The drought months of March and April are the warmest. Owing to high temperatures and rainfall, humidity is high throughout the year with an average of 82% Peter 1998).

Fig. 14.4 Acacia plant species are continuously growing at the coast as well as in the other land areas (Acacia Mangium species in the UBD campus in 2017)

14.5.2 River System in Brunei Darussalam

The Belait, Tutong and Temburong Rivers are the main rivers in Brunei. These rivers and their tributaries are part of two major drainage basins: the Baram drainage basin in the west and the Brunei Bay drainage basin in the east (Fig. 14.3). The Belait River catchment, with an area of approximately 2300 km^2, is the largest in Brunei, and is part of the 21,500 km^2 Baram basin (Kershaw 2001). Its headwaters are separated from the Baram River by the topographical relief of the Belait anticline (the Belait hills, Fig. 14.3) (Yunos 2011, 2017). The Temburong River drains a basin of some 1100 km^2 (nearly the whole of the Temburong District). The Tutong River drains an area of around 1300 km^2 in the central part of the country directly into the South China Sea (Mahmud 2017). The Baram River drainage basin and Baram Delta is dominated by the Baram River and its tributaries (Edson 2004). The Baram is one of the larger rivers in Borneo, feeding a delta outbuilding into the South China Sea. The Limbang, Temburong, Trusan, Padas and Klias Rivers are the most important rivers in the Brunei Bay drainage basin. These rivers are smaller, with relatively immature catchment areas. They discharge their sediment load into Brunei Bay.

Figures 14.4 and 14.5 demonstrate the two university campuses which are located in the coastal region of Brunei. The university campuses were built in the dense, native forest areas, and while there are new, on-going developments to the campus, the campus vegetation, as well as the immediate coastal areas, are currently being dominated by invasive Acacia plant species (Figs. 14.2 and 14.3).

Fig. 14.5 Acacia plant species are now available in everywhere in the coastal forest areas and other parts of land areas (*Acacia mangium*). (Photo by author 2017: Acacia in the UBD Campus)

14.6 History of Acacia Species in Brunei Darussalam

The Acacia species are shrubs and trees belonging to the Acacia genus which are native to Australia, Africa, Madagascar, land throughout the Asia-Pacific region and the Americas. The Acacia species can adapt to a wide range of tropical, subtropical and temperate environments, and this adaptability has made them most popular for planting on degraded lands in the Asia-Pacific region and other parts of the world (Turnbull et al. 1997; Turnbull 2017; Figs. 14.6 and 14.7).

Acacia is a large genus of 1350 different species, which makes it one of the largest plant taxa in the world that can grow in warm, tropical environments and even in deserts. They are most diverse in Australia, with close to 1000 species recorded and

Fig. 14.6 The coastal area of Brunei Darussalam and Acacia species are standing as protector of coastal erosion (*Acacia mangium*). (Photo by author 2017)

Fig. 14.7 The protected coast in South China Sea with huge Acacia Plant Species in Brunei coast, students of Geography and Environmental Studies of UBD, Brunei are investigating biodiversity and environmental coastal wetland characteristics. (Photo by author 2017)

are the largest genus of vascular plants in Australia (ANBG 2017). The most popular types of Acacia species based on their physical appearance and special characteristics are flat-topped Acacia, Swollen-thom Acacia, Koa Acacia and flowering Acacia (ANBG 2017; TAS 2017). Figure 14.8 shows the mature Acacia species, which

Fig. 14.8 Showing the protected coast in South China Sea where the Acacia Plant Species started to grow and students of Geography and Environmental Studies of UBD, Brunei in investigation mode. (Photo by author 2017)

grows scattered around the UBD campus (Figs. 14.4 and 14.5) as well as in the coastal region in Brunei (Figs. 14.6 and 14.7).

Acacia trees, or in the native language, *Pohon jati,* are an invasive plant which was introduced as plantation trees to the Borneo island, initially to the Malaysian states of Sabah and Sarawak in the early 1980s and then in Brunei in the early 1990s (Turnbull et al. 1997).

14.7 Acacia Plant Species in the Coastal Region in Brunei Darussalam

Brunei is one of the regions the world with high forest cover where tropical forests covering 75% of the country's total land area (Forest Department 2009). The Acacia species was first introduced to the country when *Acacia Mangium* was first planted to produce high-quality wood for the timber industry (Turnbull and Crompton 1998; Turnbull 2017).

Acacia Mangium was introduced in Brunei for production of timber and furniture. It was also planted for restoration of forest lost during the construction of the 40-km Tutong to Muara coastal highway. The road connects the capital city of Brunei, Bandar Seri Begawan, to regional towns in the Tutong and Belait Districts. Figures 14.9, 14.10, 14.11, 14.12, 14.13, 14.14, and 14.15 display the plantation, distribution and growth pattern of the Acacia species in the coastal region in Brunei. It was determined that the chosen Acacia trees, *Acacia Mangium*, *Acacia Cincinnata*

Fig. 14.9 The Api-Api wetlands area in the coastal region in Brunei with wetland forest where acacia plant species are gradually establishing acacia forest in Api-Api wetland area. (Photo by author 2017)

Fig. 14.10 The Api-Api wetland area in the coast in South China Sea with Acacia Plant Species in Brunei coast (*Acacia cincinnata*). (Photo by author 2017)

Fig. 14.11 The protected coast in South China Sea with huge Acacia Plant Species in Brunei coast and students of Geography and Environmental Studies of UBD, Brunei investigating biodiversity and environmental coastal wetland characteristics (*Acacia auriculiformis*). (Photo by author 2017)

Fig. 14.12 The newly protected coast in South China Sea with Acacia Plant Species in Brunei coast. (Photo by author 2017)

Fig. 14.13 Acacia Plant Species densely growing in the UBD Campus (*Acacia Auriculiformis*). (Picture Source by author 2017)

and *Acacia Auriculiformis*, were planted along the highway as they are fast-growing trees that can prevent soil eroding on the highway barrier, which was becoming an environmental problem (Osunkoya et al. 2005; Forest Department 2009).

Figures 14.9, 14.10, 14.11, 14.12, and 14.15 show the coastal wetlands and forest areas where the students investigated the wetland, forest ecosystem and ecosystem services. Under the module wetland ecology and management a field trip was organized at the Api- Api Forest wetlands areas in Brunei Muara District in Brunei Darussalam. The coastal forest biodiversity issue was also one of the focal points of investigation for the Bachelor students in Geography and Environmental Studies at the University of Brunei Darussalam. The Acacia, other invasive plants and animals were also investigated in the Api-Api coastal forest wetland areas (Figs. 14.9 and 14.10) in the Brunei-Muara District. Initially, the Api-Api wetland and forest areas (Figs. 14.9, 14.10, and 14.12) were highly dense rainforest areas and were of primary

Fig. 14.14 Acacia Plant Species densely fast growing trees in the every corner of UBD campus (*Acacia holosericea* in Brunei Darussalam). (Source: Photo by author 2017)

forest status in the Brunei-Muara District. However, their characteristics are rapidly changing, and as a result, at present the status of the forest areas in the coastal region is becoming secondary. The students were also investigated the invasive plant species and biodiversity status in forest wetland areas in the coastal wetland areas (Figs. 14.8, 14.9, 14.10, and 14.11).

In the mid-1990s, two other fast-growing species, *Acacia auriculiformis* and *Acacia cincinnata,* were cultivated along with *Acacia mangium* to rehabilitate vegetation along the Tutong-Muara highway to prevent soil erosion (Osunkoya et al. 2005). The plantation of the Acacia species in the Brunei coastal region went very well as these species can grow at very fast rates by protective nitrogen and thus can grow and establish themselves even in nutrient-poor soils. These planted species managed to reduce the soil erosion problem as well as closed up the open canopy along the coastal highway in Brunei. But the Acacia species have spread to other habitats in Brunei, particularly in degraded forests and lands, and they are commonly seen as roadside vegetation throughout Brunei (Osunkoya et al. 2005). There are four species of the Acacia currently found in Brunei, namely (i) *Acacia Mangium* (Figs. 14.5 and 14.15), (ii) *Acacia auriculiformis* (Figs. 14.11 and 14.13), (iii) *Acacia cincinnata* (Fig. 14.10), and (iv) *Acacia holosericea* (Fig. 14.14). Of these, *Acacia mangium* appears to be abundant, *Acacia auriculiformis* and *Acacia cincinnata* are fairly abundant, whereas *Acacia holosericea* is less commonly found in Brunei.

Fig. 14.15 The Api-Api Wetland Forest area in Muara District of Brunei Darussalam an area dominated by Acacia invasive plant species and students of Geography and Environmental Studies of UBD, Brunei investigating biodiversity and forest characteristics. (Photo by author 2017)

14.8 Uses and Potentiality of Acacia Species in Bioenergy Generation

Acacia wood is suitable for making paper pulp or even woodchips. The wood can also be useful for making furniture. As it is a very fast-growing tree, more bark can be extracted, hence increasing the production of timber (Turnbull et al. 1998). Additionally, the bark of the Acacia trees can be made into firewood as it has a very high affinity to fire. The leaves of the tree can be used as forage and the fallen leaves can be collected to produce fuel. Another use for the bark is for the collection of *Acacia mangium* honey (Lemmens et al. 1995).

Furthermore, the Acacia species meet the criteria to be seen as an ideal energy crop, criteria which were set by McKendry in 2002. Benefits include (i) high yield of biomass per hectare, (ii) less energy impact to grow, (iii) minimum cost requirement for cultivation, and (iv) production of biomass with minimum contaminants and minimum external nutrient requirements to grow (Mckendry 2002). As an alternative management solution to this Acacia invasion, these Acacia can potentially be used to provide biomass supply for bioenergy production through thermochemical conversion processes. The standing biomass of the species per area could be estimated by the harvest method or by using allomatic equations, which are an effective way of estimating the biomass of trees (MacDicken 1997). The Acacia species would also need to show a satisfactory quantity of biomass production per hectare to ensure the sustainability of the feedstock for biofuels production processes. As the Acacia species are evergreen trees and can generate large quantities of biomass per annum, the potential of this management solution is noteworthy. *Acacia mangium* can grow up to 30 m in height with a straight trunk and the colour of the bark varies from pale grey-brown to brown. The mature *Acacia mangium* tree can reach up to 60 m. Mature *Acacia mangium* trees drop their leaves and instead the stalks are modified into leaf-like structures called phyllodes which are found to be 25 cm in length and 10 cm in breadth approximately, depending on the growing soil conditions (Turnbull et al. 1997; Turnbull 2017). On the whole, the Acacia plant species could help to generate bioenergy, wood supply, honey collection and, considering the rapid fast growing tendency of the Acacia forest in Brunei, could even help to protect the country's energy sector. This bioenergy generation could be a considerable benefit from the increasingly spreading Acacia plant species in Brunei.

14.9 Results and Discussion

The long-term national development plan by the government of Brunei – also known as "Brunei Vision 2035" – aims to reduce energy intensity by 45% (with 2005 as the base year) and to generate 10% of energy requirements from renewable resources (APER 2012). As the Acacia species have been able to successfully establish themselves in degraded habitats in Brunei, and the *Acacia mangium* and *Acacia auriculiformis* in particular can exhibit high biomass production within a few years of their growth, they could potentially help Brunei in achieving its aims to reduce CO_2 emissions whilst also ensuring that the country's biodiversity is not adversely impacted by their invasiveness (Ahmed et al. 2018).

The Brunei-Muara District in particular has been impacted negatively by the Acacia plant species' rapid growth and high density. This has been overserved in the Api -Api coastal forest wetland area in Brunei. After the investigation it has been stated by the students and interviewees in the case area where they stated that the primary forest status are reducing and Acacia is spreading very fast in the case area.

Under this Acacia plant species project some interviews were arranged with academic experts and professional peoples those are involved with plants and forest resources as well with Acacia species. Within this initiative Mrs. Rosle was selected as one of the interviewees who is a traditional medicinal plants practitioner in Brunei and living in the coastal area in Brunei. This has been observed in the areas surrounding Ms. Rosle's (a medicinal garden owner in Brunei Muara District) home garden. She explained that the Acacia species is a danger to all other plant species that are in her home garden. Ms. Rosle is a traditional medicinal plant practitioner in the Brunei-Muara District. Her family has more than 200 years' worth of indigenous knowledge and experience in the traditional practices of medicinal plant species. She had a medicinal plant garden with more than 100 medicinal plant species available. According to Ms. Rosle, she noticed 5 years ago that some *Acacia mangium* plants were starting to grow in her garden. She observed that within 2 years, the plant species gradually formed a huge shadow that eventually engulfed the whole garden. Within 2 years she lost 16 different medicinal and native plant species that were very rare and valuable such as special of banana trees. It was a great loss to her business and she was still, at the time of interview, trying to recover and restore her garden to its former condition prior to the invasion of the *Acacia mangium* plants.

One of the ways in which the negative impact of the Acacia species invasion can be curtailed is to make use of the trees, for example, in energy generation. Developing renewable energy will also effect economic growth, which is a positive impact that has been demonstrated in many countries around the world. This is a worthy possibility in the case of Brunei, especially if energy generation as well as entrepreneurships related to the maintenance, management and use of the Acacia species were to be developed. This way, the available biomass resources can be optimally utilized and, at the same time, positively contribute to the growth of the local economy in a number of ways such as creating new employment opportunities in various sectors including engineering, agriculture, transportation and research and development.

Other benefits of Acacia trees are that they can reduce soil erosion and that they can be used as a boundary or as a windbreak. The trees can also fix atmospheric nitrogen and produce a rich harvest of litter, which increases soil biological activity and rehabilitate the physical and chemical properties of the soil (Otsamo et al. 1995). Figure 14.15 shows the Acacia plant species' secondary forest development process in the Api-Api wetland and forest area in the Brunei-Maura District.

Despite these benefits, it should also be noted that the Acacia plant species can play an influential role in the spread of forest fires (Turnbull et al. 1998). Forest fires (also known as "wildfires" and "wild land fires" in North America and "bushfire" in Australia) are a widespread phenomenon and not unfamiliar to Brunei. A combination of natural and human factors including dry weather conditions, El-Nino, open burning and recreational fires influence the triggering of forest fires. Uncontrollable fires have damaging social and economic effects, such as curtailing valuable or necessary activities relating to forestry, significantly reduce natural resources such as timber, and property loss (Turnbull et al. 1998; Turnbull 2017). A forest fire starts when an oxidizing agent, which is usually oxygen (O_2), is resent into the air, which then fuels any substance that can burn the forest areas.

Forest fires also produce haze and this can affect both humans and animals that live nearby the affected forest areas. Haze can induce harmful conditions in humans such as eye inflammation, nasal irritation, throat irritation and inflammation of lung tissue. Forest fires will also cause disturbances to animals living in, or near to, the affected areas, which may cause them to decrease in number due to migration or death. This may ultimately result in some animal species becoming endangered. The biodiversity of the heath forests in Brunei will also be reduced due to haze – invasive plants such as the Acacia species will take advantage of the reduction in heath forest areas in order to thrive since they are capable of growing in various soil nutrient levels.

14.10 Recommendations and Conclusion

Water and forest are the dominant elements in the geography of Brunei as well as in the human ecology of both Brunei and the whole of coastal northwest Borneo. In Brunei, the settlement patterns, farming, communications and economic activities have been shaped by the geographical patterns of coast and interior jungle. The latter has generally repelled dense settlement, cultural interpretation and patterns of secondary farming whilst the former has been a magnet for trade, settlement, some semblance of urban life and cultural exchange. Since Brunei is centrally-positioned within the Southeast Asian region, its favorable location enables the country to serve as an entrepôt. This facilitates mobility for traders active in other parts of Borneo such as in the eastern and southern regions of the island as well.

The government agencies are only able to partially manage the geographic and coastal environment for the time being. But the overall recent economic, engineering, geomorphological as well as urban developments, and urban morphological changing patterns indicate the future trend of Brunei's progressing development.

Acacia trees now bring unexpected consequences to Brunei's biodiversity as it has become as alien invasive species in the country, which was discovered by a research project in University of Brunei Darussalam in 2014 (UBD News 2014). Due to its ability to survive in harsh environmental conditions and its regenerative quality, Acacia was first introduced to Brunei to restore and rehabilitate degraded habitats. But in reality the Acacia's invasiveness has resulted in significant loss in native plant diversity in coastal forests like Kerangas (make up about 1% of Brunei's forests).

The impact of the rapid regeneration of the Acacia species is likely to have huge negative impacts on Brunei forests and biodiversity in general, which threatens primary forests in Brunei. Acacia increases the flammability of an invaded area, which is one of the main causes of forest fires in the coastal forest areas in Brunei. The densest Acacia forest can be found in the coastal areas of Brunei from the Brunei-Muara District to the Belait District; and these coastal forest areas are thus at high risk of forest fires and are declared as fire-prone zones. The areas of local settlement in the coastal areas are also threatened due to the rapid expansion of the Acacia species.

On the whole, this study found that the expanding growth of the Acacia species is a threat to forest biodiversity, forest status, agricultural crops and forest fire safety. From investigations by the International Union for Conservation of Nature (IUCN), it was found that there are six invasive plant species that have already entered

Brunei, yet IUCN still did not include the Acacia as an invasive plant species. It is important to address Acacia as the invasive plants in the coastal areas of Brunei that have high growth rates. They also affect other habitats and are prone to fire which causes pollution as well as loss of habitat. There is a need for rehabilitation of the sustainable green forest in Brunei. The present status and the development of four different types of the Acacia plant species in Brunei is gradually creating threats for the coastal native plant species as well as the coastal forest ecology of Brunei, so it must control or use as a biomass feedstock for biofuel or electricity production.

References

ANBG (Australian National Botanic Gardens) (2017) Information about the Acacia species n.d. http://www.anbg.gov.au/acacia/. Accessed 19 Jan 2018

APERC (Asia Pacific Energy Research Centre) (2012) Asia Pacific energy research Centre, Energy demand and supply outlook of Brunei Darussalam 5th edition, 2012

Ahmed A, Bakar MSA, Azad AK, Sukri RS, Mahlia TMI (2018) Potential thermochemical conversion of bioenergy from Acacia species in Brunei Darussalam: a review. Renew Sust Energ Rev 82:3060–3076

BBY – (Borneo Bulletin Yearbook) (2014) Brunei 2035 development, education and health- seeds of success, Brunei Press Sdn Bhd, Bandar Seri Begawan, Brunei Darussalam, pp 1–192

BBYB – (Borneo Bulletin Year Book) (2015) The guide to Brunei Darussalam 2015, Brunei Press Sdn Bhd, BSB, Bandar Seri Begawan, Brunei Darussalam, pp 1–232

Brown DE (1970) Brunei the structure and history of a Borneoan Malay Sultanate. Brunei Museum, The Government Printing Department, Bandar Seri Begawan, Brunei Darussalam

CDD – (Curriculum Development Department) (1993) History for Brunei Darussalam (Secondary 3). EPB Pan Pacific, Singapore, pp 1–144

CDD – (Curriculum Development Department) (2008) History for Brunei Darussalam: sharing our past. (Secondary 1 & 2). EPB Pan Pacific, Singapore, pp 1–108

Cranbrook E, Edwards DS (1994) Belalong: a tropical reinforest, the royal geographical society. J Trop Ecol, London, & Sun Tree Publishing, Singapore 11(2):314

Edson G (2004) Heritage, pride or passion, product or service? Int J Herit Stud 10(4):333–348

Hassan ABH (1988) Elementary geography for Brunei Darussalam. Federal Publications (s). Ltd, Singapore, pp 1–51

Ibrahim HALH (2005) Culture and counter cultural forces in contemporary Brunei Darussalam. In: Thumboo E (ed) Cultures in ASEAN and the 21st century. Uni Press, Singapore, p 22

Islam SN (2014) Biosphere reserve, heritage identity and ecosystem services of Sundarbans transnational site between Bangladesh and India, In: Conference proceeding of world heritage alumni conference from 22–25 October, 2014. pp 224–244. http://www.IAWHA.com/

Kahar SR (2002) Invasion by the exotic Acacia mangium willd in highly degraded health vegetation within Berakas forest Reserve, Brunei, BSc Thesis, Universiti Brunei Darussalam, Brunei Darussalam

Kershaw R (2001) Brunei, Malay, Monarchical, Micro-state. In: Funston J (ed) Government and politics in southeast Asia. Seng Lee Press Ptc Lad, Singapore, pp 1–35

Khatib MABP, Sirat DPHA (2005) Some aspect of the cultural trends for tomorrow's Brunei-A strategy. In: Thumboo E (ed) Cultures in ASEAN and the 21st century. Uni Press, Singapore, pp 35–49

Lemmens RHMJ, Soerianegara I, Wong WC (1995) Plant resources of South-east Asia no. 5(2). Timber trees: minor commercial timbers. Backhuys Publishers, Leiden, pp 1–655

MacDicken KG (1997) A guide to monitoring carbon storage in forestry and agroforesty projects, 1997. Winrock International Institute for Agricultural Development, Arlington

Mahmud R (2017) The Borneo Bulletin, 16 October 2017 http://borneobulletin.com.bn/new-bridge-a-boon-to-business-commuters/

McKendry P (2002) Energy production from bioenergy (part 1): overview of biomass. Bioresour Technol 2002(83):37–46

Omar M (1981) Archaeological excavations in Protohistoric Brunei. Meseum Brunei, The Government Printing Department, Brunei, pp 1–91

Omar DPHM (2005) The making of a national culture: Brunei's Experience. In: Thumboo E (ed) Cultures in Asean and the 21st century. Uni Press, Singapore, pp 5–21

Osunkoya OO, Othman FE, Kahar RS (2005) Growth and competition between seedling of an invasive plantation tree, Acacia mangium, and those of a native Borneo heath-forest species, Melastoma beccarianum. Ecol Res 2005(20):205–214

Otsamo A, Adjers G, Hadi TS, Kuusipalo J, Tuomela K, Vuokko R (1995) Effect of site preparation and initial fertilization on the establishment and growth of four plantation tree species used in reforestation of imperata cylindrica (L.) Beauv. dominated grasslands. For Ecol Manag 73:271–277

Peter C (1998) Nature, western attitudes since ancient times. University of California Press Ltd, London, pp 1–17

Siti NorAqilah BHA (2017) Mangrove dependency in Brunei Darussalam: a comparative study of three villages on the Brunei Bay. In: Unpublished Bachelor thesis in the Department of Geography and Environmental Studies, Faculty of Arts and Social Sciences, University of Brunei Darussalam, pp 1–84

The Brunei Museums (2005) The Brunei museums – 40th anniversary of the Brunei museums department (1965–2005). Perletakan Oscar Sdn Bhd, Malaysia, pp 1–115

TFD (The Forestry Department) (2009) The 5th national report to the convention on biological diversity of Brunei Darussalam. Vol. 2009.n.d.

Turnbull JW (2017) Australian Acacias in developing countries. In: Proceedings of an international workshop at foresrty training centre, Gympie, Old, 4–7 August 1986, organized by Australian Centre for International Agricultural Research, printed by Brown Prior anderson Pry Ltd, Victoria, Australia, Proc. no.16, pp 1–196

Turnbull JW, Crompton HR, Pinyopusarerk K (1997) ACIAR proceedings, recent developments in Acacia planting. In: Proceedings of an international workshop held Hanoi, Vietnam, 27–30 october 1997. Hanoi, Vietnam, 1998

Turnbull JW, Midgley SJ, Cossalter C (1998) Tropical acacia planted in Asia: an overview. In: Turnbull JW, Crompton HR, Pinyopusarerk K (eds) Recent developments in Acacia planting, vol 82. ACIAR proceedings, ACIAR, Canberra, pp 14–28

Tyler R (2004) Brunei Darussalam- the marking of a modern nation. Bandar Seri Begawan, pp 1–58

Tylor EB (1991) Primitive culture, vol 1. John Murray, Albemarle Str. W, London

TAS (Types of Acacia Species) (2017) Types of Acacia species based on their physical appearance and special characteristics n.d. http://www.2020site.org/trees/acacia.html. Accessed 12 Dec, 2017

UNESCO- (United Nations Educational Scientific and Cultural Organization) (1972) Convention concerning the protection of the world heritage cultural and natural heritage, Paris

UNESCO- (United Nations Educational Scientific and Cultural Organization) (1999) World Heritage Centre, Report 1999/3. Paris, pp 1–12

UNESCO- (United Nations Educational Scientific and Cultural Organization) (2008) Policy document on the impacts of climate change on world heritage properties. World Heritage Centre, Paris

UNESCO– (United Nations Educational Scientific and Cultural Organization) (2004) Brunei Darussalam UNESCO country programming document 2013–2016. Organized by UNESCO Jakarta in Bandar Seri Begawan, pp 1–31

UBD News- (University of Brunei Darussalam News) (2014) UBD Discovers Acacia's Threats on Native Kerangas forests
Yogerst J (1994) The Golden legacy-Brunei Darussalam. Syabas, Bandar Seri Begawan, pp 1–205
Yunos R (2011) Our Brunei heritage- a collection of Brunei historical accounts. Perpustakaan Dewan Bahasa dan Pustaka Brunei, Bandar Seri Begawan, pp 1–405
Yunos R (2017) The Golden Islamic heritage of Brunei Darussalam- a collection of selected historical articles. UNISSA Press, Gadong, pp 1–93

Index

A

Acacia auriculiformis, 42, 468, 469, 471
Acacia mangium, 465, 469–472
Acacia spp, vii, 456–474
Adriatic coast, vi, 130
African grass, vi, 316–362
Agricultural expansion, 129
Agricultural revolution, 36
Algae, 21, 31, 152, 155, 180, 186–188, 196, 208, 209, 223, 248, 374, 376, 379, 382, 383, 385–390, 392, 393, 433
Alien, v, 4, 69, 90, 114, 172, 250, 427, 473
Alien species, v–vii, 4, 6, 15, 21, 24, 29–31, 40, 42, 44, 90, 91, 100, 102, 110, 114–115, 117, 119–122, 125, 130–132, 134–135, 138, 139, 172, 173, 175–179, 182, 200, 204, 205, 220, 224, 225, 228–231, 250, 427, 433, 437, 438, 449
Amphibalanus amphitrite, 250, 262
Angiosperm tree, 6, 7
ANOVA, 120, 127, 130, 333–335, 343, 351–355, 357, 423
Anthropogenically disturbed, 30, 32
Anthropogenic dispersal mechanisms, 29
Anthropogenic influences, 4, 69, 78, 80
Anthropogenic pressure, 68, 190
Apex invasive predators, 29, 32
Aquacides, 33
Aquaculture, 174, 178–180, 183, 185, 193, 201, 206–208, 212, 217, 229, 326, 328, 329, 359, 374, 426, 433
Arborescent bryozoans, 248, 261, 262
Argentina, 375, 378, 379, 383–384, 387
Atlantic Ocean, 38, 211, 270, 427, 445
Australia, vii, 13, 15, 30, 32, 34, 157, 159, 187, 251, 252, 318, 319, 374, 399, 401, 403, 406, 408, 414–422, 426, 463, 472
Australian pine, 5–9, 30, 33, 34, 41

B

Bahamas, 6, 270, 274, 278, 283–285, 287, 289
Ballast water, 146, 159, 162, 178, 180, 195, 220, 221, 248, 251, 374, 384–385, 392
Bangladesh, vi, vii, 64–81, 184, 192, 202–204, 211, 460
Barnacles, 154, 155, 180, 195, 196, 248, 249, 254, 256, 261, 262, 450
Beach, vii, 7, 88, 114, 147, 182, 374, 399, 414
Beach erosion, vii, 374, 393
Beach vegetation, 107
Bioaccumulation, 152, 218, 223, 249
Biochemistry, 31
Biodiversity, v, vi, 11, 31, 64, 67, 68, 75, 77–80, 89, 90, 94, 109, 110, 114, 130, 147, 152, 160, 172–175, 177, 179, 180, 182–184, 186–198, 204–206, 208–210, 212, 221, 223, 224, 227–231, 269–275, 277, 281–289, 316, 327, 331–334, 337, 338, 340, 341, 349, 358–362, 382, 393, 444, 456, 464, 467–471, 473
Biodiversity loss, 173, 174, 229
Biofilm, 248, 249, 255, 257
Biofouling, vi, 159, 196, 199, 247, 248, 251–255, 260–262
Biogeography, 4

C. Makowski, C. W. Finkl (eds.), *Impacts of Invasive Species on Coastal Environments*, Coastal Research Library 29,
https://doi.org/10.1007/978-3-319-91382-7

Bioinvasion, v–vii, 8, 31–35, 39, 41, 172–180, 182, 184, 186–188, 190–193, 195, 197–202, 204–206, 208–214, 218, 220, 221, 224, 227–231, 270, 382
Bioinvasion countermeasures, 33–35
Biological control agents, 12, 24, 32, 34
Biomagnification, 218, 219, 223
Biomass, 17–20, 31, 178, 179, 193, 197, 231, 252, 270, 282, 284, 326, 330, 333–336, 341, 348, 357, 359, 360, 383, 386, 388, 390, 392, 393, 470, 471, 474
Biome, 29
Biotic resistance hypothesis, 426
Biotransformation, 223
Bird disturbance, 414, 422
Borneo, vii, 456–474
Brazilian pepper, 5, 10–12, 30, 33–35
Broadleaf paperbark tree, 5, 13–15
Brunei Darussalam, vii, 456, 457, 459–462, 464, 465, 468–470, 473
Bufo marinus, 22–24, 32, 34
Burmese python, 6, 27–29, 32, 33, 35, 40, 41

C

Calcareous algae, 248
Calcareous tubeworms, 248, 261, 262
Canid, 413
Canopy growth, 11
Caribbean Sea, vi, vii, 88, 270
Carrying capacity, 37, 39
Casuarina equisetifolia, 5–9, 30, 33, 42
Category I exotic species, 6
Cayman island, 274, 278, 287, 289
Chemical countermeasures, 33
China, 30, 32, 34, 36, 151, 221, 316, 456, 459, 460, 462, 464–467
Chromolaena ordorata, 95, 97, 109
Climate change, v, 64–81, 110, 114, 209, 210, 229, 361
Cluster analysis, 273, 287, 289
CO_2 output, 37
Coastal management, 392, 393, 398
Coastal reserves, 414
Coastal zone, v, 64–81, 88, 182, 200, 205, 209, 216, 225, 229
Coast Conservation Act, 109
Coatings, 188, 248, 249, 254–262
Colombian Caribbean, 270, 273–275, 288, 289
Colonization, 159, 173, 175, 177, 194, 210, 221, 229, 248, 261, 318, 341, 385, 392, 452
Control strategies, vi, 316–362
Convention of Biological Diversity (CBD), 90, 172, 175, 179, 182, 206, 228
Coral reef, vi, 21, 22, 89, 147, 197, 269–275, 277, 281–289, 361
Corophid amphipods, 250
C4 perennial species, 317–319
C4 photosynthesis, 357
Crabs, 51, 180, 186, 187, 192–195, 198, 215, 248, 441, 450, 460
Crassostrea, 152–155, 158, 159, 162–164
Cuscuta campestris, 91, 95, 97, 99

D

Delta, vi, vii, 26, 27, 64–66, 68, 71, 72, 74, 76, 78, 79, 147, 182, 184, 191, 201, 205, 212, 215, 274, 426–428, 430–453, 459, 462
Direct competition, 9, 146
Displacement, 9, 173, 178, 195, 205, 318, 319, 407, 441
Distribution, 15, 16, 75, 88, 90, 100, 119, 160, 180, 187, 190, 193, 195, 198, 199, 204–206, 209, 210, 213, 225, 262, 270, 286, 322, 360, 374, 383, 399, 406, 427, 435, 439, 450, 459, 465
Disturbance, 16, 26, 41, 109, 114, 130, 132, 134, 135, 137, 138, 160, 175, 197, 202, 209, 251, 317, 327, 333, 341–343, 349, 374, 383, 398, 414, 416, 422, 426, 437, 446, 472
Domestic animal management, 397
Domestic dogs, 398, 414
Drupes, 10–12
Dune, vi, 8, 9, 25, 30, 31, 88, 89, 108, 129, 183, 184, 190, 214, 328, 330, 331, 350, 401, 414

E

Echinochloa pyramidalis, vi, 316–362
Ecological balance, 29, 31, 41, 185
Ecological change, 65, 208, 382, 386
Ecological footprint, 37, 40
Ecological framework, 29, 403
Ecological restoration, 164
Ecological traps, 414
Eco-management strategies, 230
Ecosystem, v, 4, 64, 89, 139, 146, 172, 248, 269, 316, 374, 426, 460
Ecosystem services, 64, 65, 67, 68, 79, 81, 224, 228, 316, 327, 341, 349, 359, 360, 362, 442, 468

Ecotonal succession, 31
Ecotourism, 183, 199, 212, 214, 216, 229, 358
Eichhornia crassipes, 16–18, 30, 42, 69, 90, 91, 192, 202, 209
Encrusting bryozoans, 248, 257, 261
Endangered species, 9, 215
Endemic, 4, 79, 89, 91, 93, 97, 100, 283
Energy release hypothesis, 426
Environmental impacts, vi, 176, 249, 316–362, 380, 398
Environmental management systems, 110
Environmental perturbation, vi, 172, 173, 175–180, 182–185, 187, 190–196, 198–201, 204–206, 208–213, 215, 218–221, 224, 227–231
Environmental resource base, v, 4, 6, 8–12, 14–16, 18, 19, 21, 22, 24, 26–34, 36, 37, 39–42, 44, 46
Epiphytes, 427
Escape modality, 418
Estuary, 72, 146, 151, 179, 183, 184, 190, 213, 216, 219, 220, 224, 384
Europe, 31, 37, 99, 114, 129, 139, 159, 163, 252, 316, 374, 426
Eutrophication, 31, 176, 208–210, 214, 229, 344, 376, 392, 448
Exotic, vii, 4, 6, 8, 9, 17, 19, 25, 26, 28, 30–35, 40, 41, 44, 46, 89, 97, 107, 172, 173, 176–179, 195, 197, 201, 206, 207, 209, 224, 228, 317, 326, 332, 346, 349, 358, 359, 374, 383, 384, 386, 392, 399
Exotic fauna, 34, 46
Exotic flora, 33, 40, 42

F
Fecundity, 8, 22
Fisheries, 67, 68, 72, 78, 102, 109, 147, 148, 160, 164, 179, 180, 185, 193, 212, 316, 326, 374
Flight-Initiation Distance (FID), vii, 414–423
Floodplain, 64–66, 78, 79, 319, 325, 459
Florida Everglades, 5, 9, 15, 18, 20, 27, 29, 30, 32, 39–41
Florida Keys, 5, 6
Forest ecology, vii, 474
Fouling associates, 248
Functional diversity, 115, 120, 127, 137, 139
Fungal agents, 12

G
GIS, 117
Grooming, vi, 251
Gulf of Mexico, 107, 274, 279, 288, 289, 327, 328, 330–334, 336–339, 341–349, 359, 362
Gulls, 415, 419, 421, 423

H
Habitat destruction, 172, 173, 382
The Habitats Directive, 114, 116
Herbaceous neophytes, 115
Herbicides, 33, 333–335, 346, 347, 349, 358, 450
High biomass production, 334
High vegetative propagation, 357
Homeostasis, 3
Homo sapiens, 29, 37
Human interference, 32, 41
Human pressure, 182, 317
Hurricanes, 8, 71, 270, 316, 328, 361
Hydrilla, 18–20, 30, 34, 35, 40–42
Hydrilla verticillata, 18, 19, 30, 34, 42
Hydrocarbon pollution, 431, 449
Hydrodynamically vulnerable, 392
Hydroids, 248, 250

I
Imbalance of Abundances with higher cClades Index (IAC), 120, 122, 129, 135, 137
Indigenous, vi, 4, 6–8, 11, 12, 15, 17, 19, 24, 26, 28–32, 35, 39–42, 89, 99, 146, 153, 154, 158–163, 166, 205, 208, 224, 228, 230, 322, 327, 348, 374, 472
Indigenous flora, 9, 12, 31
Indigenous vegetation, 7, 12, 30
Indo-Pacific, 31, 50, 153, 159, 270
Industrial Revolution, 40
Infestation, 15, 18, 19, 32, 230
Introducing disruptor, 29
Invasive grasses, vi, 317, 327, 333, 334, 351
Invasive species, v, 4, 64, 90, 152, 172, 248, 270, 320, 381, 399, 414, 426, 473
Invasive wildlife, v, vii
Ipomea pes-capre, 94, 95, 102–104

K

Kelp, vi, 387
K-T boundary, 4
K-T Mass Extinction Event, 4

L

Lagoon, 16, 20, 64, 68, 88, 330, 346
Land reclamation, 317
Land use, 70, 118, 119, 121, 129, 130, 174, 208, 209, 329, 330, 346, 458
Lantana camara, 43, 90–92, 97–99, 102, 109
Legislative control measures, 33
Lionfish, vi, 5, 19–22, 31, 41, 52, 270–275, 277, 281–289
Low reproductive potential, 231

M

Macroalgae, 374, 386
Macrofouling, 248, 249, 255, 257, 260–262
Malacofauna, 145
Management, 8, 64, 92, 139, 149, 173, 271, 325, 375, 398, 414, 444, 459
Mangrove, vi, 9, 64, 88, 147, 172, 316, 426, 456
Manila Bay, vi, 150
Mar del Plata, 383
Marine bioinvasions, 34, 193
Marine cane toad, 22–24, 32, 35, 41
Marine non-indigenous species (MNIS), 146, 147, 152, 157–161, 163, 165
Marine shipping industry, 247
Marine-related infectious diseases, 316
Marshes, 15, 88, 89, 91, 187, 209, 322, 341, 342, 346, 347
Mean Pairwise Phylogenetic Distance Index (MPD), 120, 129, 137
Mechanical control measures, 33
Mediterranean, 50, 115, 116, 158, 160, 374
Melaleuca quinquenervia, 5, 13–15, 30, 43
Mexico, vi, 36, 88, 97, 107, 274, 279, 288, 289, 316, 318, 320, 322, 326–328, 330, 346, 348, 349, 358, 359, 362
Migration, 39, 221, 342, 472
Mikania micrantha, 45, 90, 91, 94, 95, 98, 102, 109
Montenegro, vi, 115, 117, 118
Mussels, 148, 154, 157, 158, 160–163, 165, 180, 210, 248, 250, 252, 257
Mytella charruana, 153, 157, 159–164

N

Native, v, 4, 64, 89, 114, 159, 172, 250, 270, 317, 374, 403, 422, 426, 458
Native fauna, 250, 386
Native flora, 4, 70, 71, 80, 173, 374
Natural system, 36, 37, 176, 316, 327, 347–349
Negative cascade effect, 11
New Zealand, 187, 251, 316, 374, 382, 383
Niger Delta, vii, 426–428, 430–453
Nigeria, vii, 36, 326, 426–428, 430–453
Nile monitor, 5, 26, 27, 32–34, 40, 41
Nitrogen fixation, 119, 127, 132
Non-indigenous species transport (NIS), 250, 260, 262, 334
Non-native, v, 4, 7, 10, 14, 16, 27, 29, 31–33, 35, 40, 41, 46, 92, 94, 102–105, 107–109, 173, 195, 227, 229, 250, 374, 382
Nutrient levels, 151, 160, 248, 473
Nypa fruticans, vii, 191, 211, 426, 427, 430, 431, 433–438, 441–453
Nypa palm, vii, 426, 427, 430, 431, 433–438, 441–453

O

Office of Naval Research (ONR), 262
Oil and gas exploration, 427–430, 432, 433, 444, 445
Oil spillages, 429, 431, 449
Open waste disposal, 425
Opuntia stricta, 91, 92, 95, 97, 98, 102, 109
Out-compete, 6
Oysters, 46, 148, 154, 155, 158, 164, 196, 206, 210, 248

P

Patagonian coast, vi, 374
Pathogens, 12, 174, 206, 221, 230, 385
PCB levels, 164
Pedology, vii, 434
Pennisetum purpureum, 43, 317, 360
Perna perna, 162, 250
Pets, 27, 28, 40, 41
Phalaris arundinacea, 317
Philippines, vi, vii, 32, 47, 146, 147, 149, 152, 154, 158, 160, 162–165
Phylogenetic diversity, 120, 124, 129, 137, 139
Phylogenetic Evenness-Abundance Index (PAE), 120, 122, 129, 135

Piscivorous species, 289
Plant communities, vi, 6, 17, 34, 42, 44, 91, 114, 115, 118, 127, 135, 137–139, 178, 328, 334
Pollutants, 151, 218, 344, 346, 441, 443
Predators, 19–21, 25–29, 31, 32, 34, 41, 174, 193, 194, 196, 230, 261, 270, 272, 275, 282, 288, 341, 358, 398, 414
Prey items, 21, 274, 283, 289
Primary vectors, 250, 262
Pterois volitans, vi, 5, 19–22, 31, 52, 269–275, 277, 281–289
Puerto Deseado, 383, 384
Puerto Madryn, 374–379, 383, 388–391, 393
Python molurus bivittatus, 6, 27, 28, 32

R
Recruitment, vi, 154, 155, 158–160, 162, 163, 174, 191, 193, 198, 252, 270, 283, 288, 289, 379, 381
Red imported fire ant, 24, 32, 41
Regulation compliance, vii, 398, 399, 401, 403, 405–409
Resident species, 146, 178, 179
Restoration, 15, 75, 164, 228, 318, 327, 329, 331, 332, 349–352, 354–357, 359, 452, 465

S
Salinity, 16, 67, 71, 72, 74–79, 114, 115, 119, 122, 146, 158–162, 176, 183, 187, 191, 199, 210–213, 248, 250, 322, 382, 384, 435, 439
Sand dunes, vi, 88, 89, 108, 114, 183, 214
Schinus terebinthifolius, 10–13, 30, 33, 43
Scorpaenid, 272, 289
Seagrass, 147, 281, 361, 393
Sea-level rise (SLR), 70–72, 89, 409
Sea stars, 248
Seaweed, 180, 206, 248, 374, 386, 389, 390, 392, 393
Secondary forest, 472
Sedimentation, 147, 152, 160, 180, 182, 191, 212, 261, 317, 328, 443
Sediment transport, 208, 388, 392
Seed, 6, 8, 10, 12–16, 34, 75, 89, 109, 119, 207, 323, 326, 336, 350, 359, 385, 427, 433–436, 441, 445, 446, 450, 452
Seismic activities, 427
Shade treatments, 349
Shannon diversity index, 94, 107, 338
Ship ballast, 29, 31, 146
Ship hulls, vi, 256
Shorebirds, vii, 25, 398, 399, 401, 403, 405–409, 415
Simpson Diversity Index, 105, 107
Snails, 89, 196, 248, 386
Soil physicochemical changes, vi, 343, 351
Solenopsis invicta, 24–26, 32
South Florida, v, 4–41
Spartina alterniflora, 317
Sphagneticola trilobata, 46, 91, 95, 97, 99, 102
Spinifex littoreus, 94, 96, 102, 103
Sri Lanka, vi, vii, 19, 88–90, 92–97, 99, 100, 102, 103, 105–110
Stable isotopes, 200, 272, 274, 285, 286, 289
Sundarbans coastal zone, v, 64–81
Swarm, 19, 22, 26, 38
Synergistic impact, vi, 172–231

T
Temperature, 116, 119, 127, 152, 187, 193, 198, 209–211, 214, 248, 328, 344, 379, 381, 384, 458, 459, 461
Threatened species, 26, 288, 327, 398, 401, 407, 408
Tidal flat, 64, 68, 158
Tributyltin (TBT), 248, 249
Tripterpenes, 12
Tropical wetlands, vi, 316–362
Tubastrea coccinea, 250
Tunicates, 155, 157, 248, 250, 259

U
Undaria pinnatifida, vi, 389
United Kingdom, 37, 40
United States, 16, 17, 19, 24, 28, 30–32, 34, 35, 38, 187, 206, 221, 251, 287, 289, 316, 318, 326
Urban settlements, 36, 149, 438
Urban sprawl, 26, 36, 37, 149

V
Varanus niloticus, 6, 26, 32
Vertical accretion, vi

W
Water hyacinth, 16, 18, 30, 34, 35, 40, 41, 91, 448
Water quality, 18, 74, 78, 147, 152, 161, 163, 164, 194, 208, 316, 327, 343–347, 362, 436, 443

West Bengal, vi, 77, 172–231
Wetland hydrology, vi, 349
Woody neophytes, 115
World Heritage Site (WHS), vi, 14, 39, 64, 65, 67, 68, 71, 77–79, 183

Y
Yellowtail snapper, 21

Z
Zauclophora pelodes, 34
Zone, v, 64–81, 88, 89, 91, 105, 106, 109, 120, 135, 138, 150, 182, 184, 187, 191, 193, 195, 199, 200, 205, 206, 209, 212, 213, 216, 221, 223, 225, 229, 325, 330, 333, 343, 346, 359, 379, 381, 389, 405, 407, 408, 410
Zoo, 187, 192, 196, 218, 223, 252, 382

MIX
Papier aus verantwortungsvollen Quellen
Paper from responsible sources
FSC® C105338

If you have any concerns about our products,
you can contact us on
ProductSafety@springernature.com

In case Publisher is established outside the EU,
the EU authorized representative is:
Springer Nature Customer Service Center GmbH
Europaplatz 3, 69115 Heidelberg, Germany

Printed by Libri Plureos GmbH
in Hamburg, Germany